No. 2673
$28.95

HOME ELECTRICAL WIRING AND MAINTENANCE

MADE EASY

AMERICAN ASSOCIATION FOR VOCATIONAL INSTRUCTIONAL MATERIALS (AAVIM)

TAB BOOKS Inc.
Blue Ridge Summit, PA 17214

AMERICAN ASSOCIATION FOR VOCATIONAL INSTRUCTIONAL MATERIALS
The National Institute for Instructional Materials
120 Driftmier Engineering Center
Athens, Georgia 30602

The American Association for Vocational Instructional Materials (AAVIM) is a non-profit national institute.

The institute is a cooperative effort of universities, colleges and divisions of vocational and technical education in the United States and Canada to provide for excellence in instructional materials.

Direction is given by a representative from each of the states, provinces and territories. AAVIM also works closely with teacher organizations, government agencies and industry.

Part 1: Electrical Wiring

AUTHOR
Thomas S. Colvin, Executive Director, Georgia Electrification Council

EDITORIAL AND DESIGN STAFF

Third Edition
Editor
Richard M. Hylton
Art and Graphics
James E. Wren
Donna D. Pritchett

First and Second Edition
Editor
J. Howard Turner
Art and Graphics

James E. Wren
George W. Smith, Jr.
Thomas B. Brown

CONSULTANT
James H. Allison, Associate Professor, Agricultural Enginnering, The University of Georgia

Part 2: Maintaining the Lighting and Wiring System

AUTHOR
Thomas S. Colvin, Research and Development Specialist

GRAPHIC DESIGN

George W. Smith, Jr., Art Director, AAVIM.

CONSULTANT
James M. Allison, Assistant Professor, Agricultural Engineering Department, University of Georgia.

FIRST PRINTING

Printed in the United States of America

Library of Congress Cataloging in Publication Data

Home electrical wiring and maintenance made easy.

Includes index.
1. Electric wiring, Interior. I. American Association for Vocational Instructional Materials.
TK3285.H66 1986 621.319'24 86-3756
ISBN 0-8306-0473-1
ISBN 0-8306-2673-5 (pbk.)

Contents

Part 1

Electrical Wiring

Preface

This manual will be useful for those who are preparing to become electricians and for do-it-yourselfers who may be installing electrical wiring. All procedures recommended in this manual are based on provisions of the National Electrical Code® . This edition has been revised to include all pertinent changes in the 1984 National Electrical Code® which remains in effect through 1987. Wiring installed according to procedures given should provide for an installation that meets Code requirements. Consult the National Electrical Code® for any needed additional information. Procedures are given for complete installation of the wiring systems for residences, utility buildings and service areas. In some localities, wiring by unlicensed persons may be prohibited. Check your local and state laws.

All procedures are fully explained and illustrated. If proceeding on your own, you should pay close attention to all safety procedures as outlined in this book.

After reading this book, you will be able to perform the following:

—Read and interpret wiring plans.
—Locate and mark routes for small appliance, general purpose and individual circuits.
—Install device (switch) boxes and outlet boxes.
—Install 120-volt, 120/240-volt and 240-volt circuits.
—Ground the electrical system and equipment.
—Connect receptacles, switches and fixtures for each circuit.
—Determine type and size of service entrance equipment to install.
—Install service entrance equipment using cable or conduit with overhead or underground conductors.
—Install ground fault circuit interrupter.
—Install conduit.
—Estimate wiring costs.
—Install wiring for agricultural and other utility buildings.

NOTE: SI (metric) conversions are given in approximate values. SI terms are given first and U.S. Customary terms in parentheses.

Canadian Key

A Canadian Supplement is available for this book which gives the Canadian Code requirements wherever it differs from the U. S. Code. The small block shown in the left margin above is the key that appears in every section where there is a difference between the two codes.

To find applicable Canadian Code requirements where the key (block) appears in Part 1, Electrical Wiring, Canadian readers should:

(1) Turn to Table of Contents in the Canadian Supplement.

(2) Find section heading and page numbers corresponding to those in Part 1, Electrical Wiring.

(3)Turn to section and page indicated to find Canadian Code requirements.

Introduction

A wiring system that is well planned and installed correctly will supply power for all of the owner's electrical needs (Figure 1). It will also provide for additions that might be required in the future. To be assured that wiring is installed correctly, most communities have adopted a *wiring code* that governs the installation of all types of electric wiring. Local codes may have requirements that are more strict than the National Electrical Code®.

There are two organizations with which you should be familiar:

- **National Fire Protection Association (NFPA).**
- **Underwriters Laboratories (UL).**

The **National Fire Protection Association,** under the auspices of the American National Standards Institute (ANSI), sponsors the **National Electrical Code**® which provides minimum standards for the safe elec-

FIGURE 1. A well planned housewiring system supplies power for all electrical needs.

FIGURE 2. The National Electrical Code gives requirements for safe installation of electrical wiring.

trical wiring installations (Figure 2). The National Electrical Code® (NFP 70) is abbreviated "NEC" but is usually referred to as the "Code." When used in this manual, the Code means the National Electrical Code®. A condensed version of the Code (NFP 70A) which covers wiring procedures for one or two family dwellings is also available from NFPA.

Code tables included in the manual are copyrighted by the NFPA and are reprinted by special permission of the NFPA.

A revised version of the Code is issued every three years. Copies may be purchased by writing the "National Fire Protection Association," 470 Atlantic Avenue, Boston, Massachusetts 02210.

The purpose of the Code is to assure safe installation of electric wiring for the protection of life and property. It does not have the force of law unless it is adopted by the governing bodies of the state, county, or local body where it will be applied. Local authorities may include additional requirements.

The National Electrical Code® is referred to or quoted in the manual in many places and interpretation may sometimes be implied. The reader should understand that any interpretation of the Code, implied or otherwise, is the interpretation of AAVIM and is not official. AAVIM cannot accept liability from the use of the book. However, the book has been reviewed by several dozen individuals from throughout the United States and Canada in an effort to make the information correct and understandable. Among those who reviewed the book are members of the National Electrical Code Committee, electrical contractors, electrical engineers and engineering professors, electrical inspectors, and vocational instructors from high schools, vocational-technical schools and junior colleges.

You should have copies of both the Code and the condensed version and know how to use them. Keep up-to-date copies on hand.

The **Underwriters Laboratory** (UL) tests electrical wiring materials. The purpose of the tests is to determine if the products meet minimum standards for safety and quality which are established either by the laboratory, or by ANSI. You should look for the UL tag or stamp when buying wiring materials (Figure 3). This is your assurance that the products meet **minimum** safety standards. It does not mean that all items that are UL listed are the same quality. Materials made by several manufacturers may all carry the UL stamp. But one product may last much longer in service than the other. Any material that does not carry the UL tag or stamp may not be safe. It will not be approved for use by most electrical inspectors.

FIGURE 3. Underwriters Laboratory stamp or tag assures only that the product meets minimum safety standards.

Most localities require the electrician or owner to obtain a **permit** before starting electrical work. The permit is authorization to begin the wiring installation. One or more inspections are usually made by the local inspector. One may be made at the "rough-in" stage and another upon completion of the installation. The **local inspector** is the final authority in interpreting local codes and the National Electrical Code®.

Procedures for electrical wiring are given under the following headings:

I. Reading and Interpreting Wiring Plans.
II. Installing Wiring.
III. Connecting Circuits.
IV. Installing Service Entrance Equipment.
V. Installing Metallic Conduit.
VI. Estimating Wiring Costs.
VII. Wiring Applications for Rural Areas.

I. Reading and Interpreting Wiring Plans

The purpose of this section is to provide basic understanding of wiring plans. Beginning with simple circuits, you will follow step-by-step the development of a house wiring plan.

On most house wiring jobs, the electrician works from the wiring plan shown on the floor plan of the house (Figure 4). **Symbols** are used to represent ceiling outlets, switches, convenience outlets (receptacles) and electrical equipment (Figure 5). Three of the most common symbols are the ceiling outlet, the switch and the duplex receptacle (Figure 6).

Upon successful completion of this section, you will be able to **identify electrical symbols** and **trace the route of each type of circuit** in a **house plan.**

Reading and interpreting wiring plans are discussed under the following headings:

A. Understanding Circuits.
B. Identifying Small Appliance Circuits.
C. Identifying General Purpose Circuits.
D. Identifying Individual Circuits.
E. Identifying Structural Parts of a House.

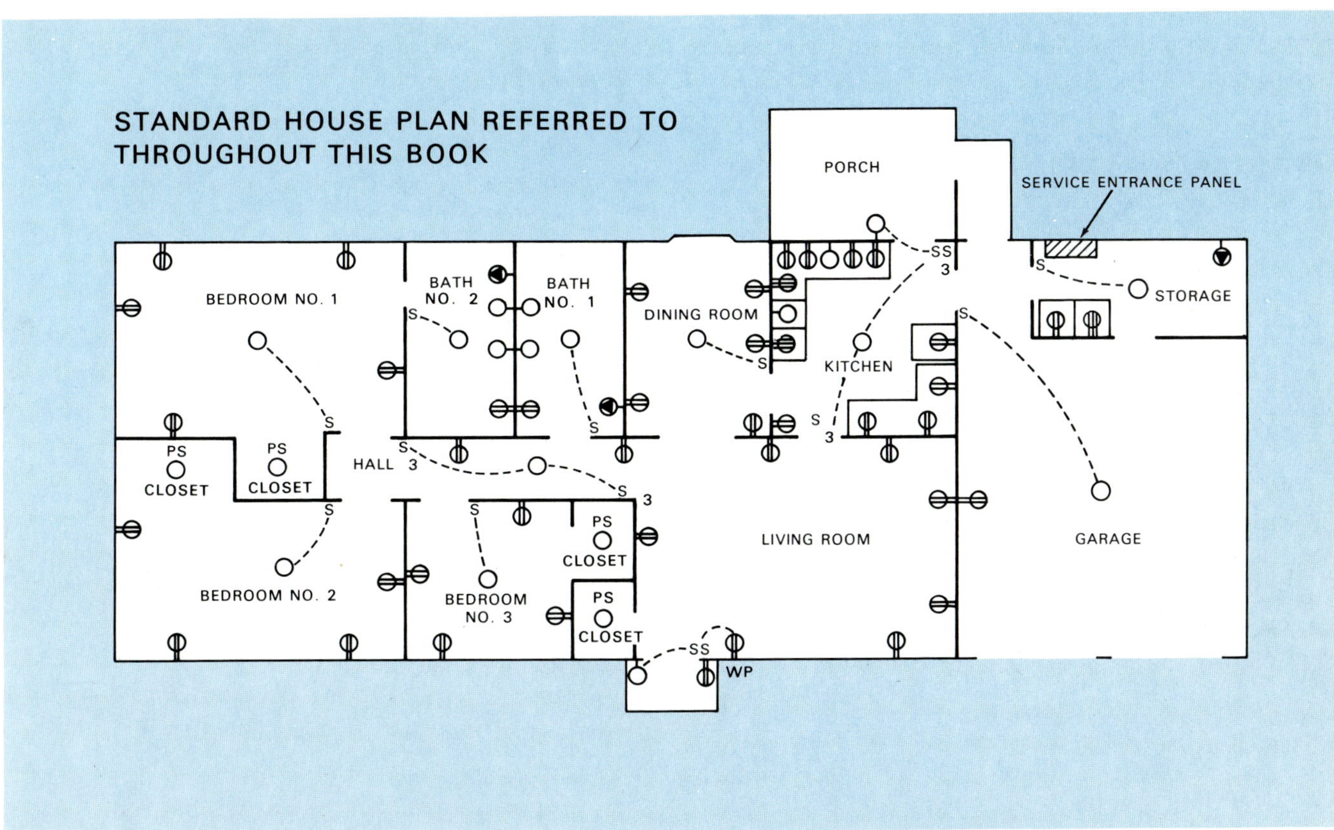

FIGURE 4. The electrician usually works from a wiring plan shown on the house floor plan.

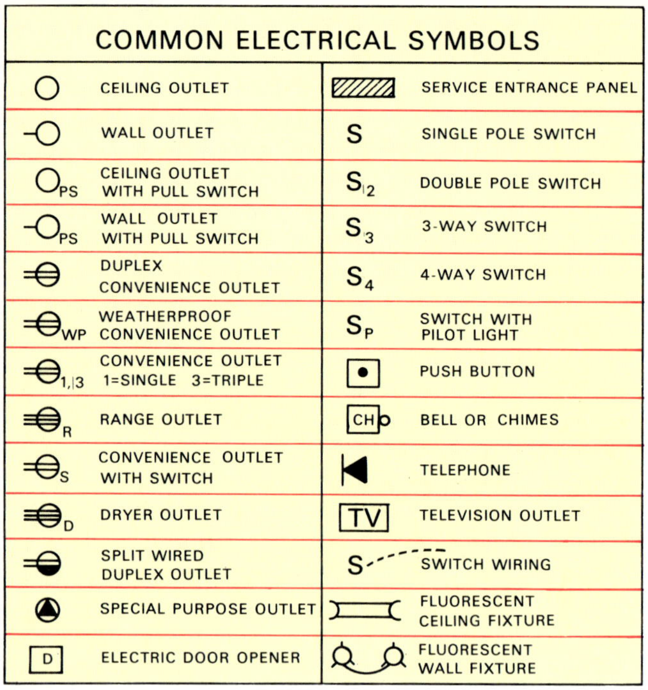

FIGURE 5. Examples of symbols used to represent electrical outlets, switches and equipment.

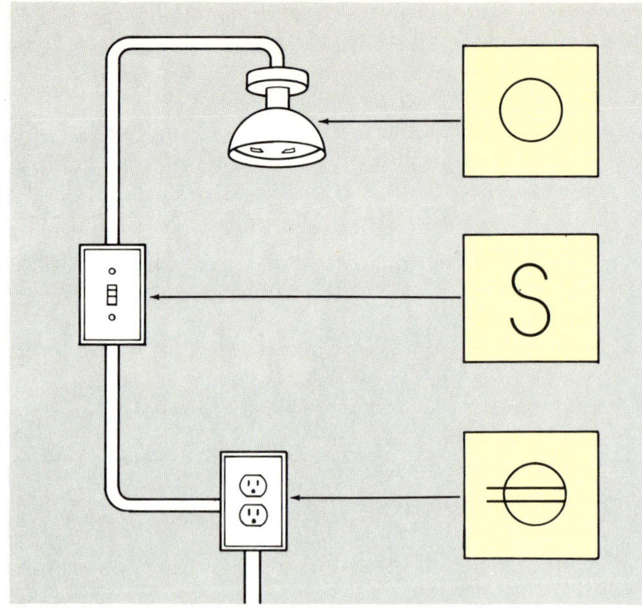

FIGURE 6. The most common symbols are the ceiling outlet, the switch and the duplex receptacle.

A. Understanding Circuits

Understanding circuits is discussed as follows:

1. Types of Electrical Circuits.
2. Current Carrying Capacity (Ampacity) of Electrical Circuits.

1. TYPES OF ELECTRICAL CIRCUITS

An electrical circuit may be defined as two or more conductors (wires) through which current flows from a source to one or more outlets.

To help you understand what is meant by the **source,** the sketch in Figure 7 shows the wiring from the transformer of the power supplier to the service equipment. From the transformer, the current flows through the meter to the **service entrance panel** (SEP) inside the house. This panel is considered the source for every circuit in the system. From the service entrance panel, the circuits branch out to all parts of the house. In fact, in residential wiring, most circuits that begin at the service entrance panel are known as **branch circuits.**

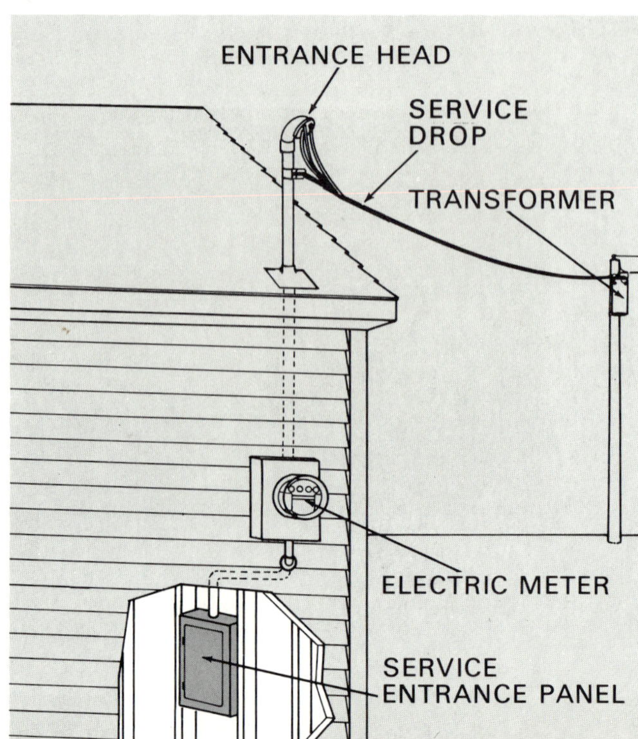

FIGURE 7. The source of power is the service entrance panel (SEP) which receives power from the supplier lines.

To have a circuit of any kind, you must have two or more conductors starting at the SEP and extending to one or more outlets. A **conductor** is any type of material that is used to carry electrical current. In most instances in this book, they are called "wires," but the terms "wire" and "conductor" are used interchangeably at times. Most circuits are made up of two or more insulated wires bound together in a "cable."

Types of electrical circuits are discussed as follows:

a. Branch Circuits.

b. Feeder Circuits.

a. Branch Circuits

A **branch circuit** is the circuit between the last fuse or circuit breaker and the outlet(s). Most circuits installed in homes are branch circuits. They run from the fuse or circuit breaker in the SEP to one or more outlets, such as a light or receptacle (Figure 8).

Having several circuits in a system makes it possible to use smaller and less expensive wire. One reason for this is to **keep costs down**. If only one circuit is used, larger and more expensive wire and materials are required.

There are three common branch circuits. They are:

—Small appliance.

—General purpose.

—Individual equipment.

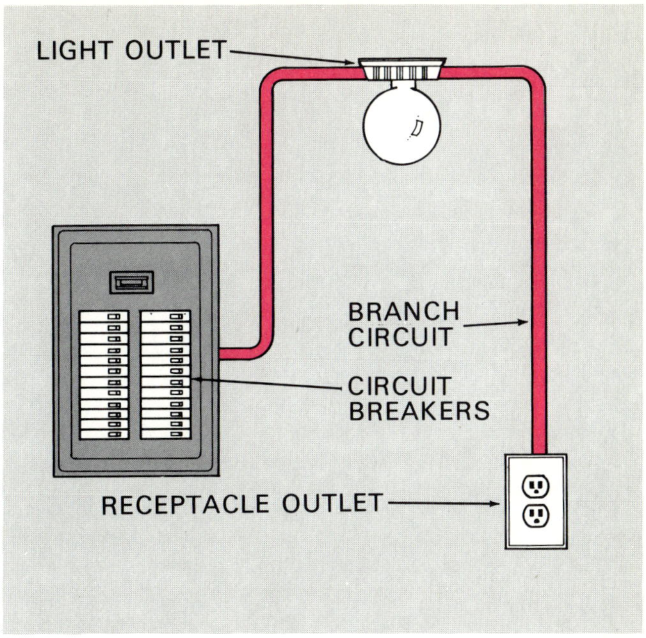

FIGURE 8. A branch circuit starts at the fuse or circuit breaker in the SEP and runs to one or more outlets.

b. Feeder Circuits

A feeder circuit is the circuit between the SEP and the fuse or circuit breaker protecting a branch circuit. In larger homes or structures, a feeder circuit is sometimes required (Figure 9).

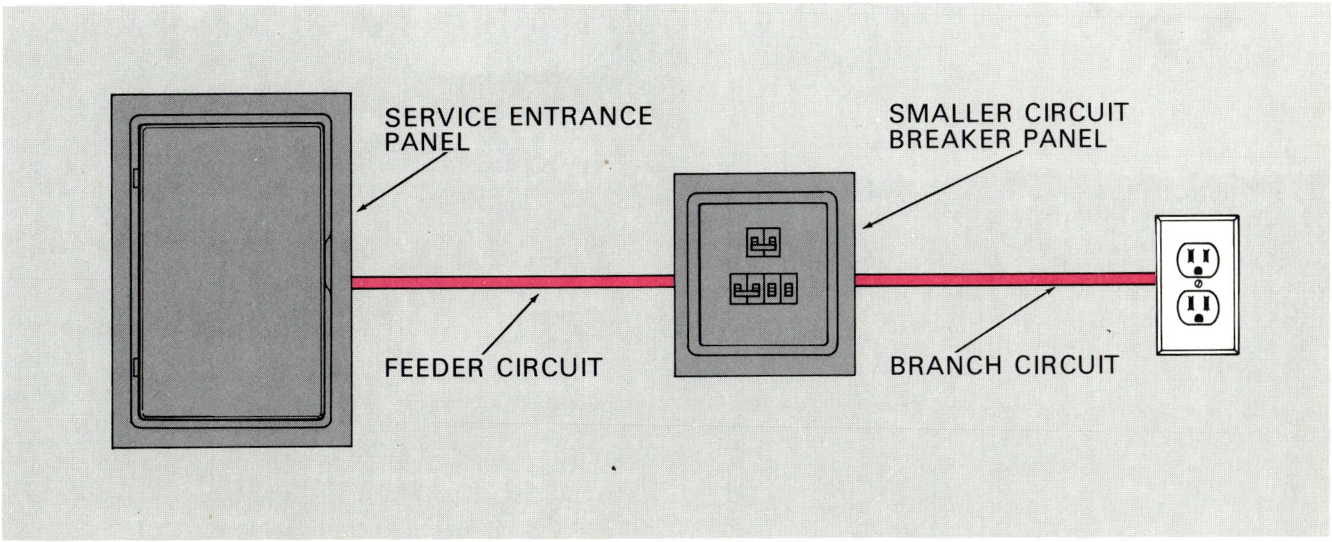

FIGURE 9. A feeder circuit extends between the SEP and a fuse or circuit breaker protecting a branch circuit.

2. CURRENT CARRYING CAPACITY (AMPACITY) OF ELECTRICAL CIRCUITS

The **total amount of current required** in all parts of the system is known as the **electrical load**. The term is usually shortened and referred to as the "**load**." Different kinds of **circuits** are needed to carry the load in each room or combination of rooms.

The wiring in each circuit must be protected by an overcurrent device, either a **fuse** or a **circuit breaker**, sized to protect the circuit wiring (Figure 10). If the circuit becomes overloaded, or a short circuit develops, the fuse or circuit breaker will open to stop the flow of current. The circuit breaker is normally used in newer installations.

In addition to keeping costs down, another reason for providing several branch circuits is for **safe operation** of the system. Dividing the current flow among several circuits helps prevent overloading of any single circuit. By dividing load requirements into several circuits, a fuse or circuit breaker can be provided for each circuit (Figure 11). If trouble develops on one circuit, only that circuit will be out of operation when the fuse blows or circuit breaker trips. The load should be evenly divided among the branch circuits within the SEP.

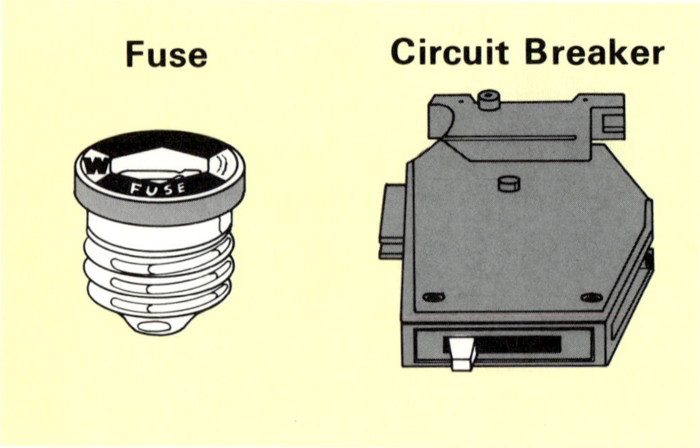

FIGURE 10. Each circuit must be protected by either a fuse or a circuit breaker.

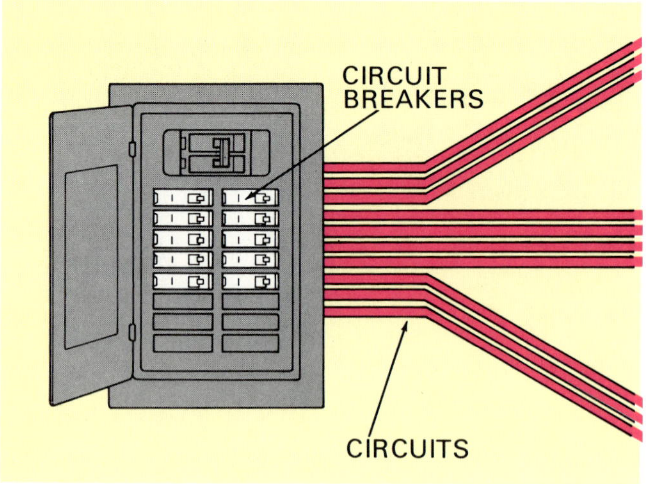

FIGURE 11. By dividing load requirements into several circuits, a fuse or circuit breaker can be provided for each circuit.

A large diameter water pipe can carry more water than a small pipe. In a similar way, large diameter wire will allow more electrical current to flow than a small wire (Figure 12). The load-carrying capacity of wire is measured in **amperes**. The load-carrying capacity of the wire stated in amperes is **ampacity**. The code defines ampacity as, "the current in amperes a conductor can carry continuously under conditions of use without exceeding its temperature rating." As you would expect, the large diameter wire costs more than small diameter wire. Therefore, to keep costs down, the electrician chooses wire of the correct ampacity for each circuit. For example, the circuit for an electic range in the kitchen must carry one of the largest loads used in the house. Therefore, the circuit for the range

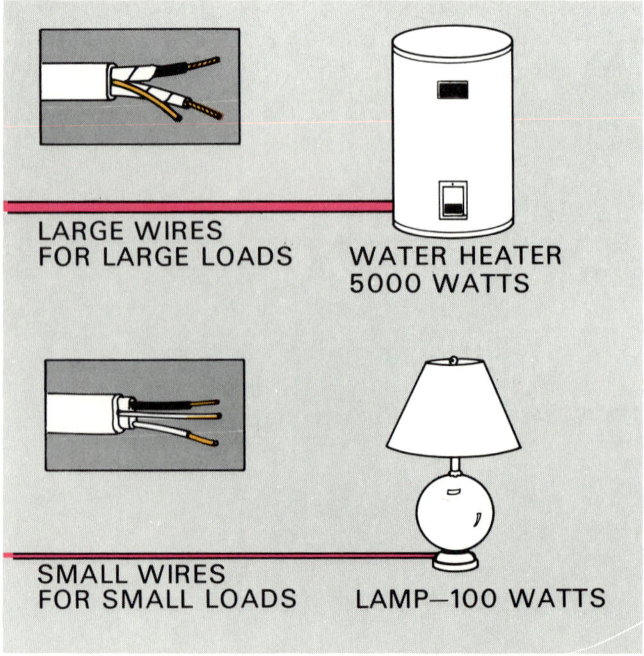

FIGURE 12. A large diameter wire will carry a larger current load than a small wire.

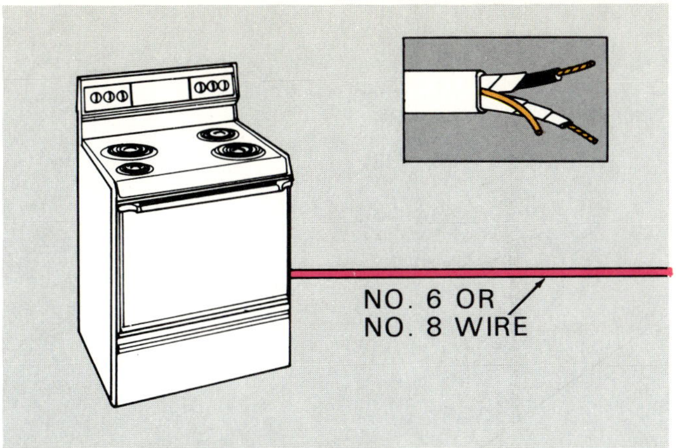

FIGURE 13. The wiring for a range must be large gage to provide the ampacity needed.

must be wired with large gage wire to provide the ampacity required (Figure 13). When using American Wire Gage (AWG) ratings, the smaller the number, the larger the wire. Thus, No. 8 AWG has a larger diameter than No. 10 AWG but smaller than No. 6 AWG (Figure 14). All numbers referring to wire size indicate the American Wire Gage size.

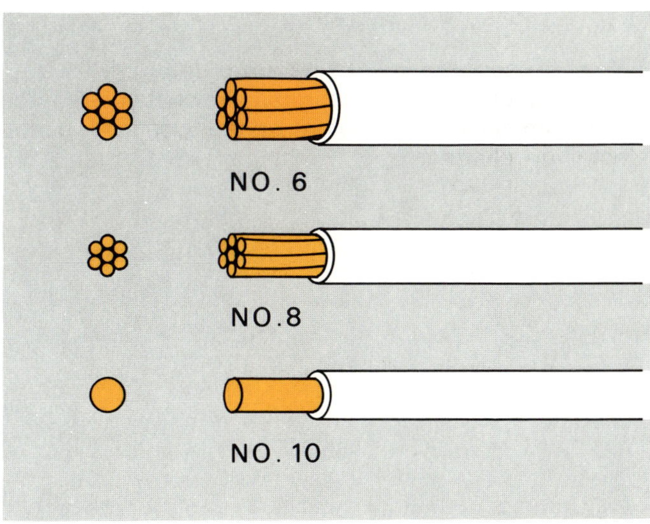

FIGURE 14. The smaller the AWG number, the larger the wire.

General purpose circuits for lighting carry the smallest loads in the house. The Code calls them lighting circuits but to call them general purpose circuits describes them better. They do supply lighting, but vacuum cleaners, radios, stereos, televisions and many other items are also connected to them. Number 12- or 14-AWG is used for general purpose circuits (Figure 15).

The **small appliance** circuit carries a larger load than the general purpose circuit but a much smaller load than the range circuit.

Fuses and circuit breakers are rated in amperes. The rating used depends on the ampacity of the circuit it protects. Wire size and overcurrent protection for residential circuits are given in Table I.

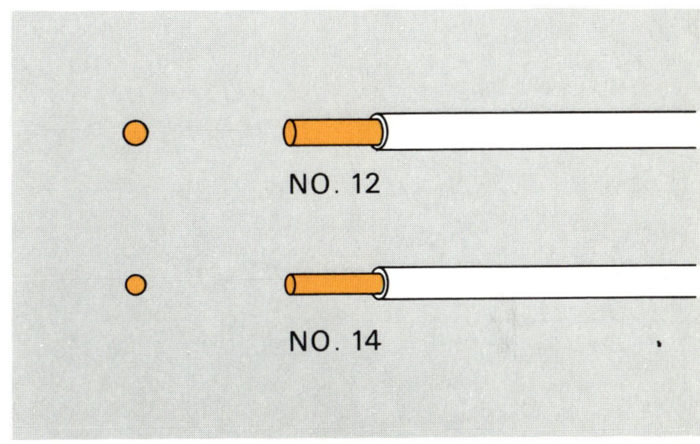

FIGURE 15. The Code allows the use of either No. 12 or No. 14 AWG for general purpose circuits.

TABLE I. WIRE SIZE AND OVERCURRENT PROTECTION FOR RESIDENTIAL CIRCUITS

Kind of Circuit	AWG Wire Size* (copper)	Maximum Overcurrent Protection (amperes)
General Purpose	14 12	15 20
Small Appliance	12	20
Individual Equipment	10 or larger	30 or larger

*The larger the wire, the smaller the AWG number.

♣B. Identifying Small Appliance Circuits

Small appliances include **mixers, blenders, toasters, coffee makers, waffle irons, fry pans, refrigerators** and many others.

The Code requires that two or more **20-ampere small appliance circuits** be provided for receptacle outlets, in the kitchen, pantry, breakfast room and dining room. The wire size for the small appliance circuit must be at least No. 12 (Figure 16). It is protected by a 20-ampere fuse or circuit breaker.

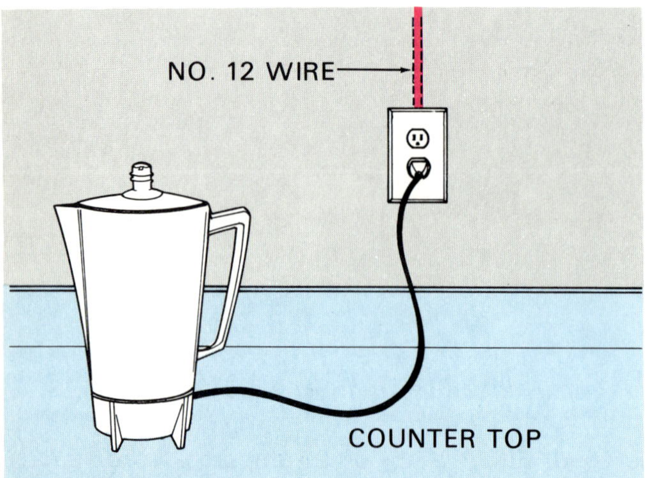

FIGURE 16. Wire size for small appliance circuits must be at least No. 12 AWG.

The Code also requires that a separate 20-ampere circuit be provided to supply the laundry receptacles. No other outlets can be connected to it.

At least two small appliance circuits must serve the **countertop kitchen receptacles** but they may also extend into the other rooms named above to serve receptacle outlets only. If more than two small appliance circuits are installed, the additional circuits may serve receptacle outlets in the other rooms specified above, if desired.

No lighting or other outlets are permitted on small appliance circuits with two exceptions. A clock outlet that supports and supplies power to a clock may be connected to a small appliance circuit and to outdoor receptacle outlets (Figure 17).

It may seem strange to include the refrigerator on a small appliance circuit, because of its size. But it often requires less current than many of the small appliances, such as a roaster or fry pan.

Small appliance outlets are easily identified on the plan. No lighting outlets are included. All convenience outlets (receptacles) indicated for the kitchen, pantry, breakfast room and dining room make up the small appliance circuits (Figure 18).

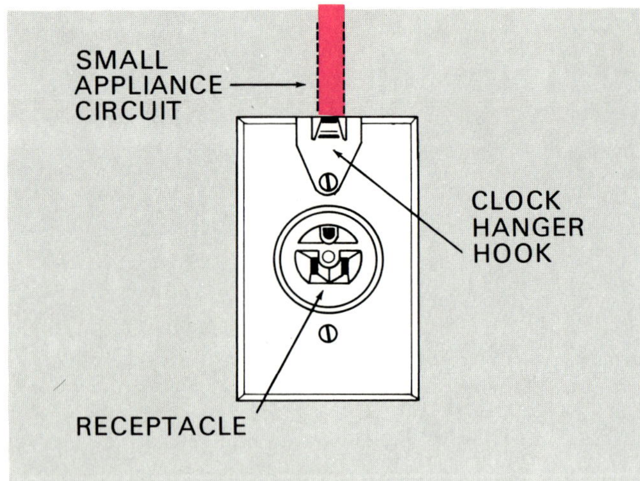

FIGURE 17. A clock outlet may be connected to a small appliance circuit.

The parts of a small appliance circuit are shown in Figure 19. The circuit begins at a circuit breaker in the service entrance panel and continues through a stud wall to the kitchen counter where receptacles are installed.

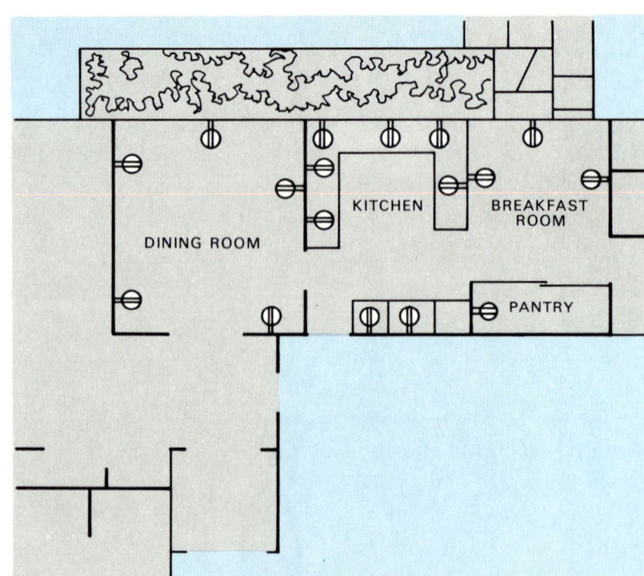

FIGURE 18. Small appliance circuits consist of all convenience outlets (receptacles) in the kitchen, pantry, breakfast room and dining room.

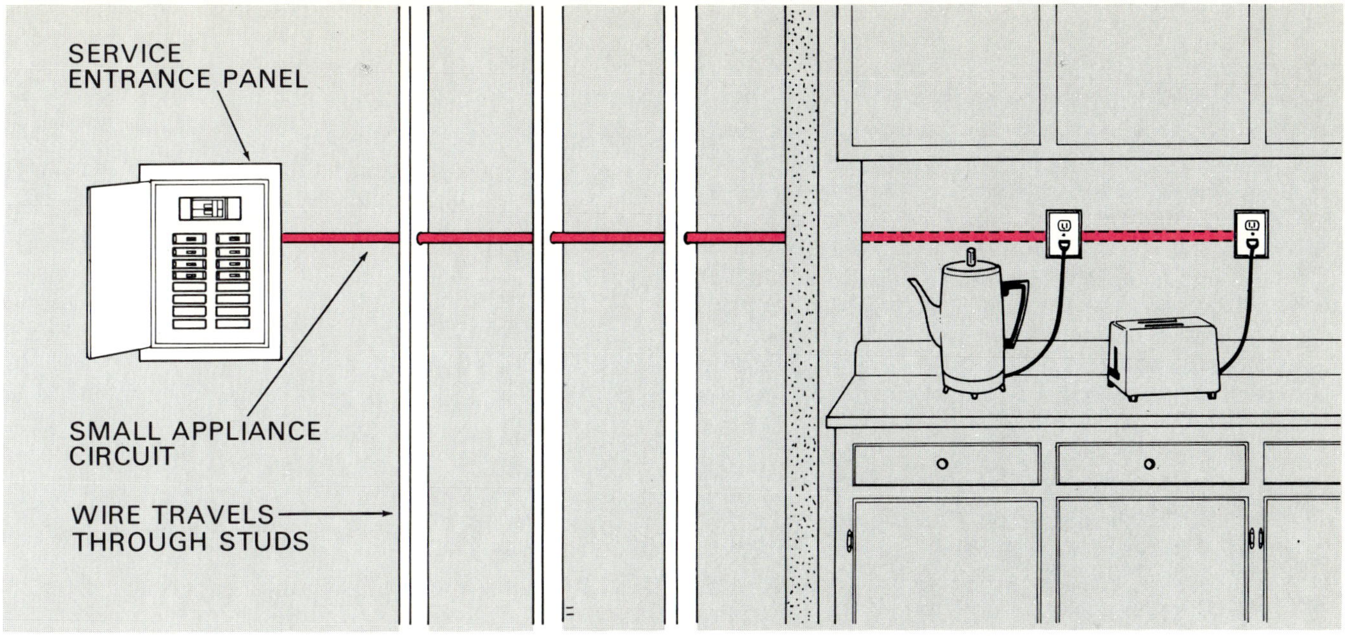

FIGURE 19. A small appliance circuit begins at the SEP and runs through studs to kitchen receptacles.

C. Identifying General Purpose Circuits

General purpose circuits make up the largest part of the wiring system for a home. Included are all ceiling and wall mounted lighting outlets and most receptacles (Figure 20). General purpose circuits are protected by either a 15- or 20-ampere fuse or circuit breaker, depending on whether No. 12 or No. 14 wire is used. Either a 15- or 20-ampere fuse or circuit breaker may be used with No. 12 wire. But No. 14 wire must be protected by a 15-ampere or smaller fuse or circuit breaker.

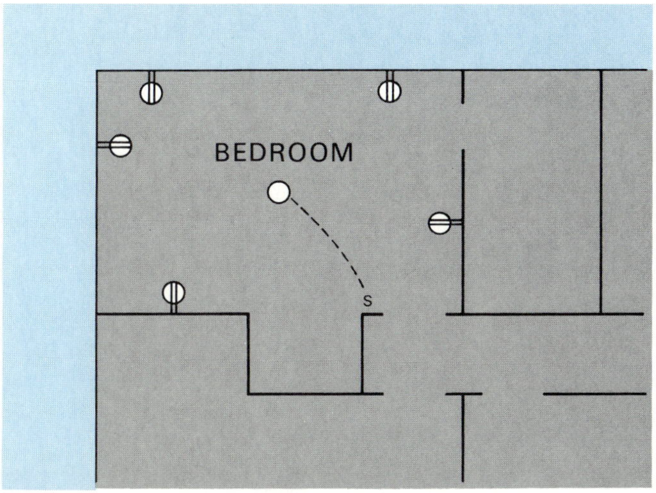

FIGURE 20. General purpose circuits such as ceiling lights and receptacles in a bedroom make up the largest part of the wiring system for a home.

Identifying general purpose circuits is discussed as follows:

1. Determining Number of General Purpose Circuits.
2. Dividing Wiring into General Purpose Circuits.

1. DETERMINING NUMBER OF GENERAL PURPOSE CIRCUITS

How do you decide the number of general purpose circuits you will need for a house? The wiring plan shows the location of each outlet in the house. However, most plans do not indicate which outlets will be combined to make up a circuit. The electrician must decide how the circuits will be grouped to include all outlets called for on the wiring plan.

Code Section 220-2 provides detailed procedures for computing the required number of circuits. Instructions in the use of those procedures are given in Section IV, "Installing Service Entrance Equipment." The following procedures provide a less complicated method that is fairly accurate:

Most authorities agree that there should be a **minimum** of **one** general purpose circuit for every 46.5 square meters (500 square feet) of floor space. On this basis, a house containing 185.8 square meters (2,000 square feet) would require four general purpose circuits (2000 ÷ 500 = 4) (Figure 4). Most house plans will show the number of square meters (or square feet) in the house. If not, it can easily be found by multiplying the outside dimensions, leaving out open porches and garages. However, you should include any area which might be developed for future use, such as an unfinished attic or basement. If you have such an area, you should run one or more separate circuits to it. Provide at least one outlet if the area is used for storage or contains equipment requiring servicing, for example, a furnace. When the area is finished at a future date, the wiring can be extended to run additional outlets without danger of overloading.

An example of the above procedures is as follows:

1. *Find the number of square meters (or square feet) in the house to be wired by multiplying the length × width.*

   A house 12.1m x 15.2 m (40′ x 50′) contains 183.9 square meters (2000 square feet.)

   A two-story house 8.5 m × 12.2 m = 103.7 square meters (28′ × 40′ = 1120 square feet), multiplied by 2 for both floors = 207.4 square meters (2240 square feet).

2. *Divide floor area by 46.5 square meters (500 square feet) per circuit.*

   185.8 ÷ 46.5 (2000 ÷ 500) = 4 general purpose circuits (Figure 4), or 207.4 ÷ 46.5 (2240 ÷ 500) = 5 circuits (Always go to next highest number of circuits if division shows a fraction.)

2. DIVIDING WIRING INTO GENERAL PURPOSE CIRCUITS

After you have calculated the number of general purpose circuits required, your next step is to divide the outlets into the correct number of circuits. The total load, as nearly as practicable, must be evenly proportioned (balanced) among the number of circuits calculated.

When laying out general purpose circuits, you might not have the same number of outlets on each circuit. Some will have more than the average number of outlets computed from Table II. Some will have a smaller number. To avoid overloading, do not exceed 10 outlets per 15-ampere circuit or 13 per 20-ampere general purpose circuit. When grouping outlets to make up a circuit, keep in mind the combinations that will require the least amount of wire to connect. Remember, every circuit begins at the SEP.

Some rooms may have part of the outlets on one circuit and part on another. This is necessary to balance the load properly, and is desirable. With practice, you will soon be able to lay out the lighting circuits in a way that is both economical and serviceable.

NOTE: Some local codes do not allow lighting outlets and convenience outlets to be wired on the same circuit.

Architects and electrical contractors sometimes use a system of lines drawn on house plans to indicate the exact route for each circuit. Some plans also indicate the number and size of wires for each circuit.

One procedure that may be used to balance the load for general purpose circuits is as follows:

1. *Count the number of outlets shown on the wiring plan and divide by the number of circuits required.*

   For this problem, assume 15-ampere circuits. Include in your count all ceiling and wall lights and all convenience outlets except **small appliance** and **individual** outlets. Individual outlets include those for the range, water heater, clothes dryer, and similar appliances.

   A table may be helpful that shows the following:

TABLE II. NUMBER OF OUTLETS IN GENERAL PURPOSE CIRCUITS (Figures shown are taken from the house in Figure 4)

Location		Ceiling and Wall Lighting Outlets	Convenience Outlets*
Bedrooms & Closets	No. 1	2	5
	No. 2	2	4
	No. 3	2	4
Hall & Entry		3	3
Baths	No. 1	3	1
	No. 2	3	1
Living Room		1	6
Dining Room		1	–
Kitchen		3	–
Laundry or Utility Room		1	–
Outdoor Entry Lights and Garage		3	1
Total Outlets in Each Column		24	25
Total Outlets in Both Columns		49	

*Note that convenience outlets on circuits for small appliances and for individual circuits are not included in dining room, kitchen and laundry room.

2. *Total the two columns and add them together.*
3. *Divide by the number of circuits called for to find the average number of outlets per circuit.*

   If the average is more than 10 outlets per circuit, it is a good practice to add one or more circuits.

   In Table II, the total number of general purpose outlets is 49. Divide by 4 (number of circuits computed on a square meter or square foot basis), you will find the average number of outlets per circuit is 12.2. As suggested above, if the average number of outlets per circuit is more than 10, another circuit should be added. By doing so, the number of general purpose circuits is now five for the house in Table II. Divide 49 outlets by 5 circuits (49 ÷ 5 = 9.8) and the average number of outlets per circuit is under ten. This number decreases the possibility of some circuits being overloaded when placed in service. It is less expensive in the long run to allow some extra capacity in new homes than to add on later.

To meet Code requirements, the wiring plan must show receptacle outlets installed so that no point along the wall line is more than 1.83 m (6 feet) from a receptacle including any wall space 610 mm (2 feet) or more in width. Kitchen countertop receptacles must be installed at each counter space wider than 305 mm (12 inches).

D. Identifying Individual Circuits

An **individual** branch circuit is one that is wired directly from the entrance panel to one outlet or directly connected to only one appliance or item of electrical equipment (Figure 21). It is installed for items that require larger amounts of power than small appliance and lighting loads. For some individual circuits, larger gage wire and larger ampacity circuit breakers or fuses are required. Also, some appliances and equipment on individual circuits must be wired for 240 volts, or a combination of 120/240 volts.

As a general rule, the following are wired as separate or individual circuits:

—Range, self contained.
—Range, built-in oven; separate range top.
—Range, built-in double oven; separate range top.
—Central electric heat.
—Clothes dryer.
—Water heater.
—Air conditioner.
—Dishwasher.
—Food freezer.
—Garbage disposer.
—Trash compactor.
—Fixed bathroom heater.
—Furnace motor and controls.
—Motors ½ hp and over.

Although not required by the Code, some power companies now recommend individual 20-ampere circuits for bathrooms. One reason is that many handheld hair dryers are rated at 1500 watts, which requires 13 amperes. A small additional load would overload a 15-ampere circuit.

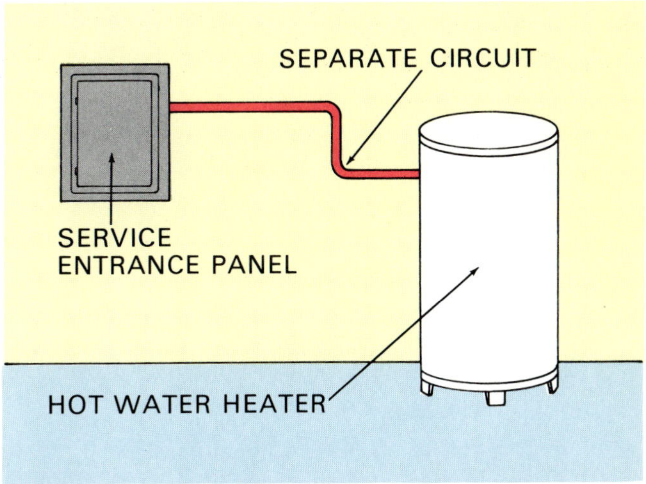

FIGURE 21. An individual circuit has separate wiring from the service entrance panel to the outlet for only one appliance or item of electrical equipment, such as the water heater.

E. Identifying Structural Parts of a House

House plans contain many terms that refer to parts of the house. You will become familiar with most of these as you gain experience. However, there are a few basic structural parts you should be able to identify from the start. Knowing the names of the parts will help you understand the location of electrical circuits more clearly.

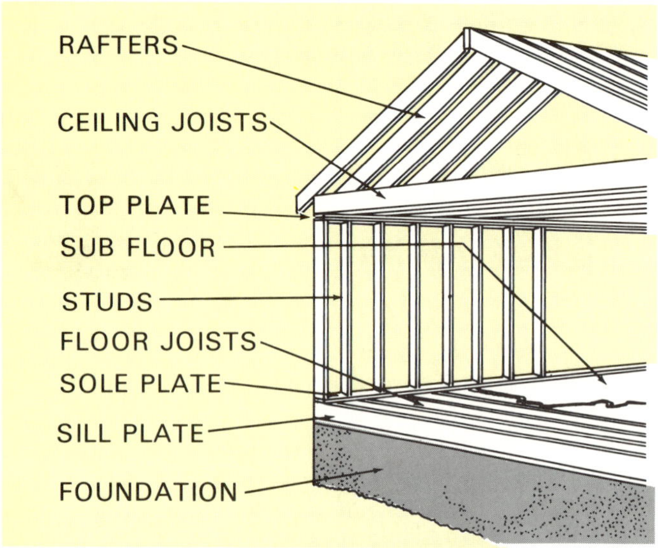

FIGURE 22A. Much of the wiring system is attached to studs, floor and ceiling joists and top plates and sole plates.

On new work, you will run wiring and attach devices to the open framing (Figure 22). The horizontal members are **ceiling joists** and **floor joists.** These are usually 5.1 cm × 15.2 cm (2″ × 6"), 5.1 cm × 20.3 cm (2″ × 8") or 5.1 cm × 25.4 cm (2″ × 10") lumber.

The same structure has a **top plate** usually made up of two 5.1 cm × 10.2 cm (2″ × 4″'s) and a **sole plate** made of one 5.1 cm × 10.2 cm (2″ × 4″) (Figure 22). The lower plate is often referred to as the **sole.**

The uprights are the wall studs, usually constructed of 5.1 cm × 10.2 cm (2″ × 4″'s). Studs and plates in outside walls may be 5.1 cm × 15.2 cm (2″ × 6″'s) to allow for more insulation. Firestops, 5.1 cm × 10.2 cm (2″ × 4″) cross members, may or may not be included. Much of the wiring and many electrical devices are attached to the studs.

The **subfloor** is the first layer of flooring material over the floor joists (Figure 22). Several kinds of material may be used for the subfloor. They include 1.2 m × 2.4 m (4′ × 8′) sheets of plywood and 1.2 m × 2.4 m (4′ × 8′) sheets of particle board (pressed wood chips). Lumber in 2.5 cm × 15.2 cm (1″ × 6″) dimensions may also be used.

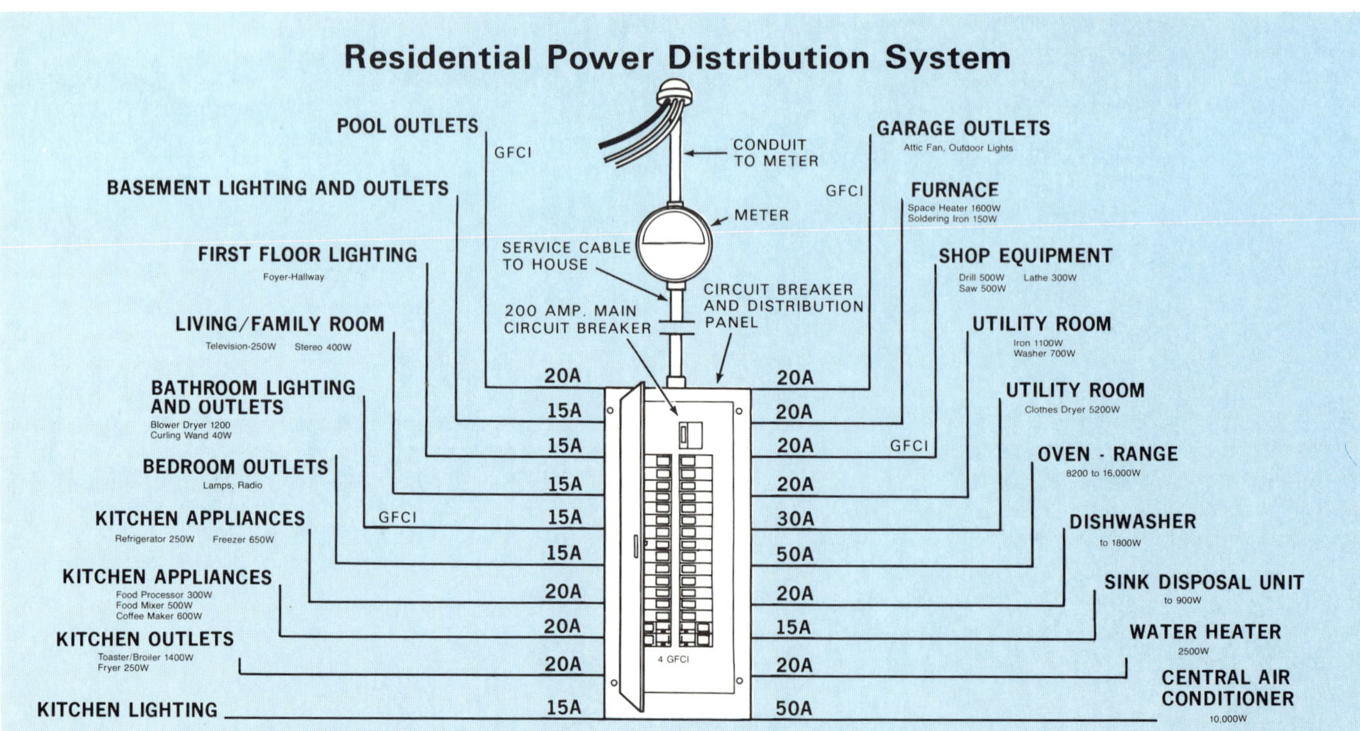

FIGURE 22B. Circuits in the completed residential wiring system are sized to meet the requirements of family living and leisure.

II. Installing Wiring

Wiring procedures discussed in this section are those used to install **nonmetallic sheathed cable** (Figure 23). Local codes in some areas may require the use of other kinds of material such as conduit (metal or plastic) (Figure 24). Flexible armored cable may also be used (Figure 25).

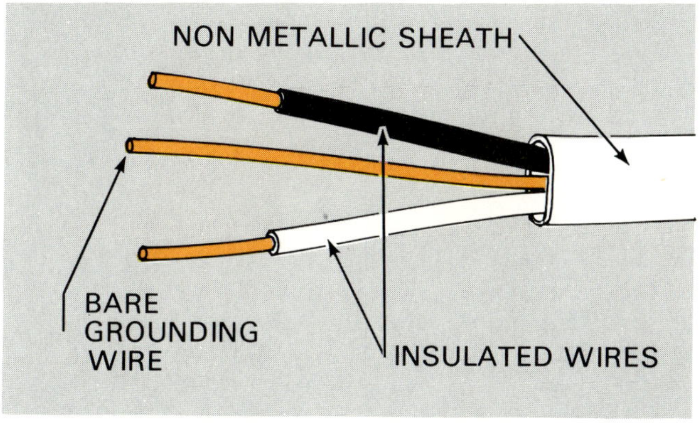

FIGURE 23. Nonmetallic sheathed cable is most often used for house wiring.

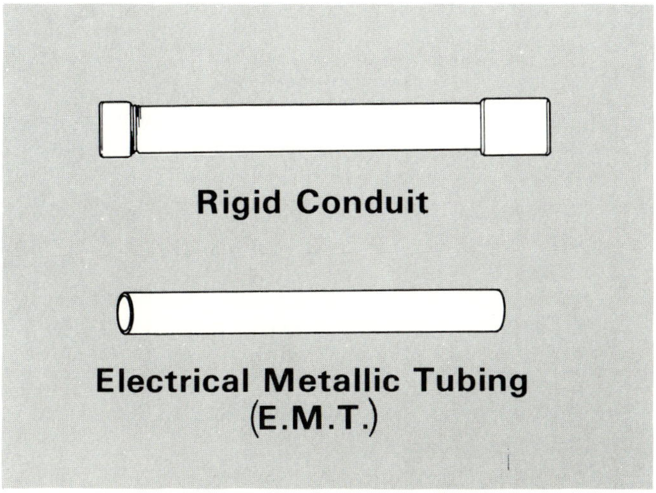

FIGURE 24. Some local codes require the use of conduit.

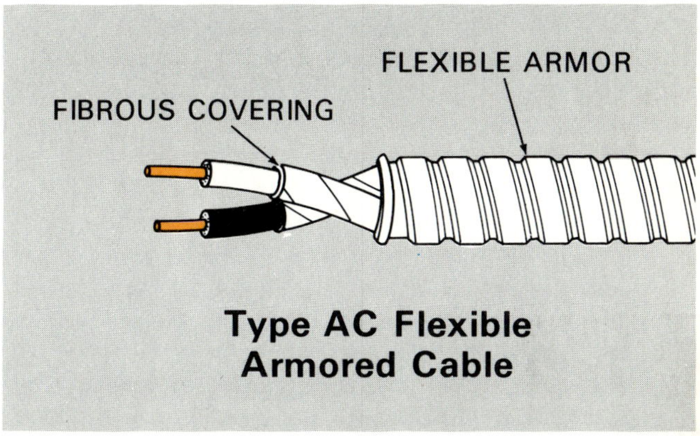

FIGURE 25. Flexible armored cable is required by some local codes.

Nonmetallic sheathed cable is most often used for house wiring (Figure 23). If you learn the correct procedures for installing nonmetallic sheathed cable, you will be able to install other kinds of wiring with a minimum of additional instruction.

Upon successful completion of this section, you will be able to **select** the **correct type** and **size** of **cable** and **other material required, locate** and **mark** the **location** for the **service entrance panel** and **lo-**

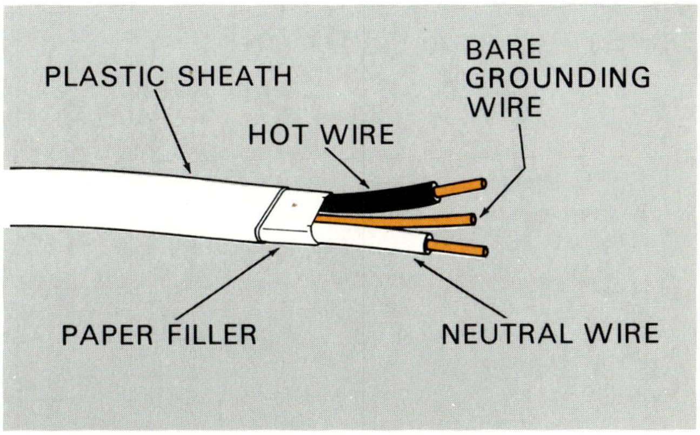

FIGURE 26. Nonmetallic sheathed cable is made up of plastic sheathing, paper filler, insulated wires and a bare grounding wire.

cate and **mark routes** for **all types** of **circuits.** You will be able to **install outlet boxes, switch boxes** and **junction boxes, provide access** for cable, and **pull** and **install cable** to **all types** of **outlets.**

Installing wiring is discussed under the following headings:

A. Observing Safety Practices.
B. Selecting Tools and Equipment.
C. Locating Circuits.
D. Installing Device Boxes and Outlet Boxes.
E. Providing Access for Nonmetallic Sheathed Cable.
F. Pulling Cable for Circuits.
G. Anchoring Cable.

A. Observing Safety Practices

In addition to following safety requirements of the Code, there are some common sense safety rules that should be followed at all times. The electrician should develop safe working habits that will protect against accidents common to many workers. Some of these practices are:

1. *Wear appropriate clothing, shoes and headwear for the job to be done (Figure 27).*

   A **hard hat** and **steel-toed shoes** will help protect the worker from falling objects. **Rubber heels** and **soles** without nails help to insulate against shock. **Clothing** should **fit well** and be free of flapping pockets that might be caught on projections. Do not wear metal rings and watchbands that might get caught on something. Wear **safety glasses** when doing work where flying particles or fluids might cause problems.

FIGURE 27. Wear appropriate clothing, shoes and hat for personal safety.

2. *Use tools that protect you.*

   Use only **UL listed power tools. Double insulated power tools** offer valuable protection against electric shock (Figure 28). Hand tools such as screwdrivers and pliers should have **insulated handles** for further protection against shock.

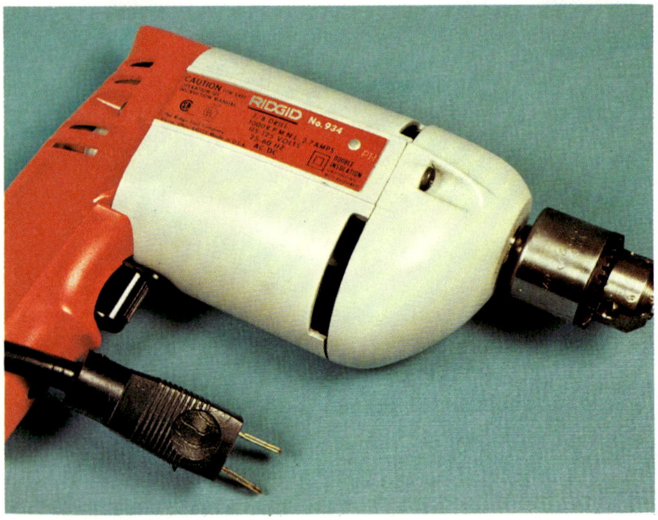

FIGURE 28. Use only UL listed tools, preferably double insulated.

3. *Keep tools in good condition.*

   Cutting tools should be kept sharp. Any damaged or broken tool should be repaired or replaced. Keep screwdriver blades correctly ground and use the correct size. Be sure all cords and plugs on non-double insulated power tools are grounding type and in good condition. Keep them in good condition by removing plugs correctly and by protecting cords. Grasp the plug and pull directly on the plug to remove it. Never jerk on the cord. If extension cords must cross a traffic area, protect them with planks (Figure 29) or other means.

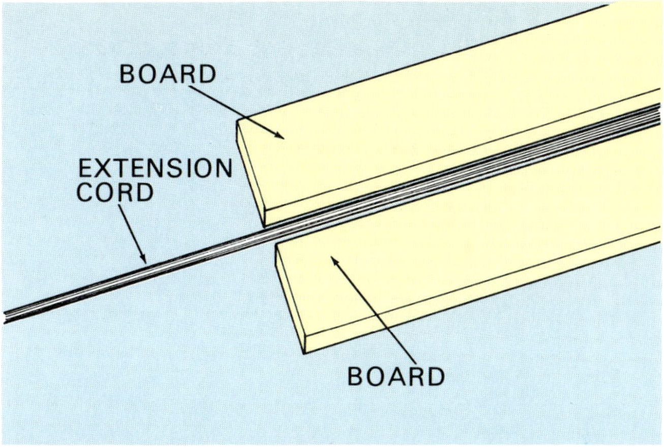

FIGURE 29. Protect extension cords if used in a traffic area.

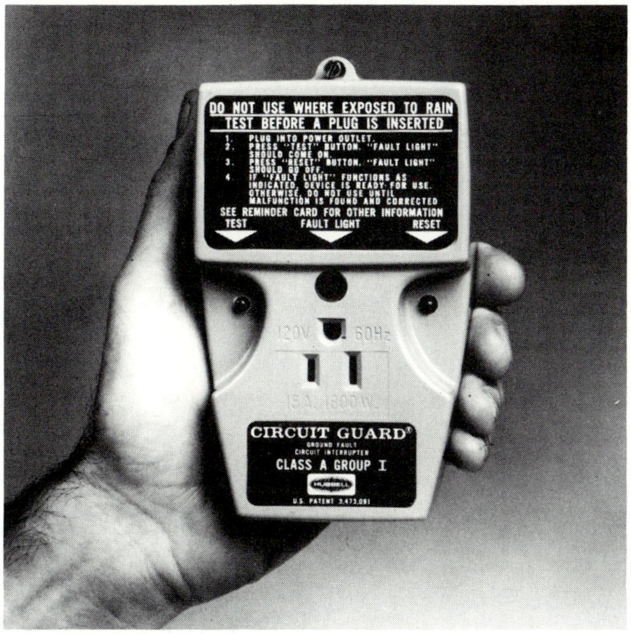

FIGURE 30. Use portable Ground Fault Circuit Interrupters if the power source is not protected permanently at the service entrance.

FIGURE 31. Keep your work area clear to prevent falls.

FIGURE 32. Use ladders safely.

4. *Use Ground Fault Circuit Interrupter for power tools.*

   Portable Ground Fault Circuit Interrupters (GFCI) are available that may be moved to any outlet for protection (Figure 30). If the power source is not protected by GFCI, the portable type is required for construction sites.

5. *Keep the work area clear (Figure 31).*

   Developing good housekeeping habits can prevent injuries caused by tripping and falling.

6. *Use ladders safely (Figure 32).*

   Use only wooden or fiberglass ladders for electrical work.

   Check ladders regularly to see that they are in good condition. Repair loose or broken rungs immediately. Use ladders with rubber or non-slip feet. Have someone hold the ladder at the base if necessary. Be sure no tools are left on step ladders and do not use the top two steps to stand. Be sure all braces are locked in place.

7. *When working in a damp location, take precautions against shock (Figure 33).*

   Don't stand on wet ground or a damp floor when using power tools without protecting yourself. Stand on a rubber mat, wooden platform or other non-conducting material for protection.

8. *Never work on a "hot" electrical circuit (Figure 34).*

   Disconnect power to the branch circuit before you work on wiring or equipment. To do this, turn the main service-entrance switch off. Then lock the switch cover in place to prevent someone from turning on the switch while you are working on the electrical system. If for some reason this cannot be done, disconnect power to the branch circuit where you will be working. Remove the branch-circuit fuse(s) or turn the circuit breaker to the "off" position. Then test the circuit to be sure it is not live by plugging in a lamp or circuit tester (Figure 35).

FIGURE 33. A rubber mat or wooden platform helps protect you on wet ground or damp basement when using power tools.

FIGURE 34. Disconnect the circuit to be worked on and lock the cover whenever possible.

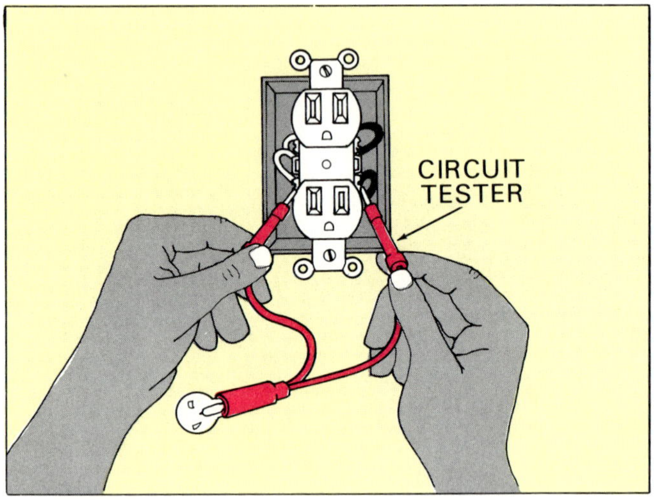

FIGURE 35. Test the circuit to see if it is energized.

B. Selecting Tools and Equipment

Basic tools needed by the electrician are (Figure 36):

—Screwdrivers.

Flat blade Phillips and Robertson in several sizes.

—Pliers.

Pliers of various types probably rank first among tools frequently used by electricians. **Lineman's** pliers are available in several sizes with 18 cm and 20 cm (7″ and 8″) sizes being most popular. The flat jaws will bend, grip, pull and twist light or heavy cable and wires. The side cutting jaws will cut cable up to large sizes and the shoulders will crush insulation to make cutting easier. **Locking** pliers are used to grip and hold cable, tighten connectors and pull cable. **Diagonal cutters** have exposed tapered jaws that are convenient for cutting cable ends in boxes and other places where close fitting connections are made. **Long nose pliers** (often called needle nose) are useful for bending loops in wire to fit under screws. **Round nose pliers** make a desirable rounded loop in wire to fit under screws. **Water pump (channel locking) pliers** are used to tighten locknuts and cable connectors.

All pliers should have insulated handles. Slip-on insulated covers are available for most sizes.

—Wire stripper.

This tool quickly removes insulation without damage to the wire. Some types include wire cutting jaws.

—Wrenches.

Adjustable jaw should have several sizes—20 cm and 25 cm (8″ and 10″) are preferred by many. **Pipe wrench** is used to grip and turn conduit.

—Drilling tools.

Portable electric drills and **carpenter's brace and bits** are used to drill holes. Several sizes and types of bits may be used. Hole-saw bits may be used to make larger holes. Various types of auger bits are used for drilling holes in framing. Carbide-tipped bits are required to drill into masonry. Also, star masonry hand bits are often used.

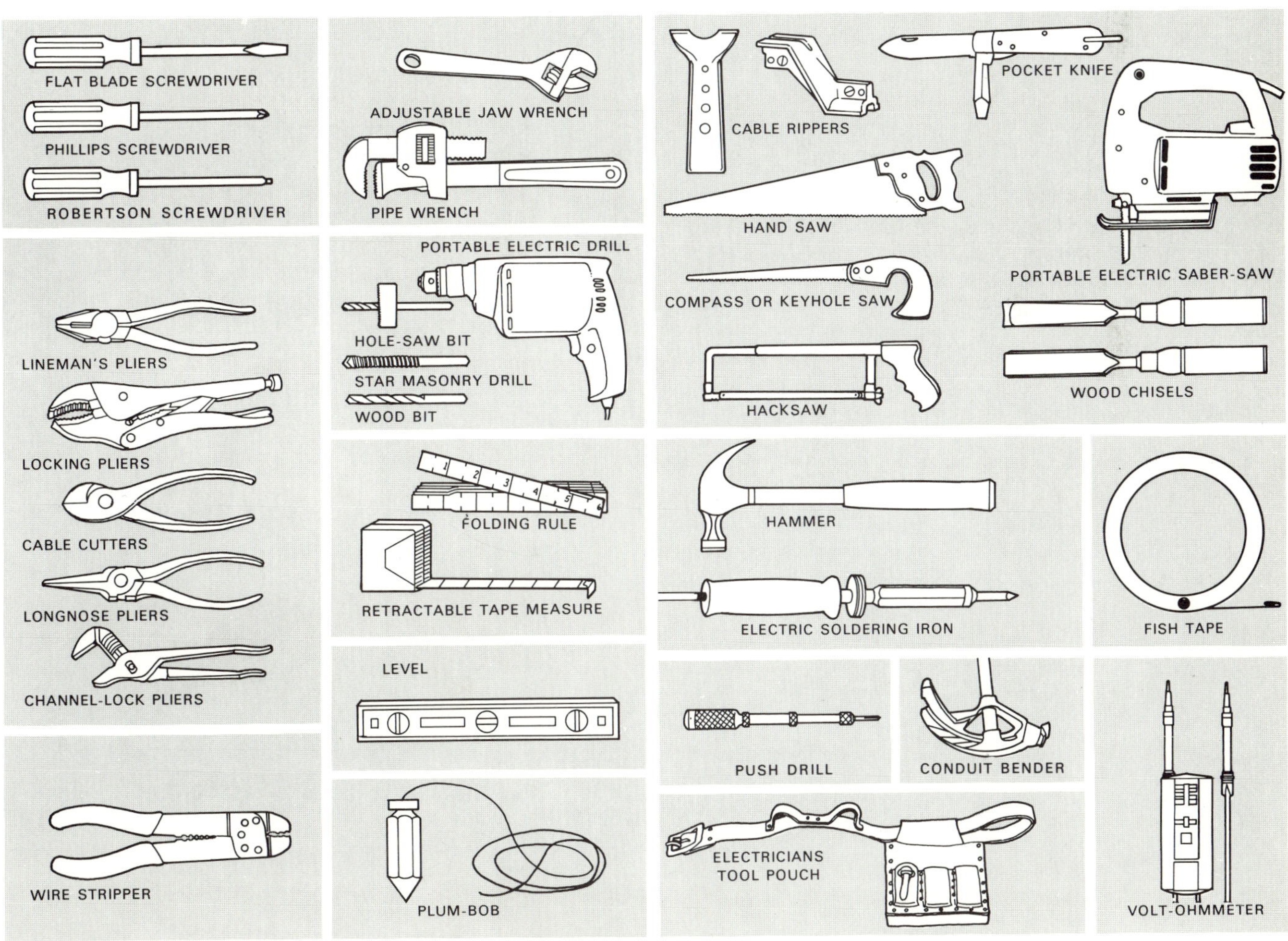

FIGURE 36. Hand tools commonly used by electricians.

—**Measuring tools.**

Folding extension rules made of wood, metal or plastic are usually six or eight feet in length. The six-inch extension slides out to measure box set-out. Metal tapes with spring return are 1.8 to 6.1m (6 to 20 feet) in length. Fifteen-meter or 30-meter (50-foot or 100-foot) **tapes** may be metal or fabric and usually have a crank return. The combination square is useful for simple squaring.

—**Level.**

These come in many sizes. They are useful for horizontal leveling or for vertical plumb position.

—**Plumb bob.**

Line with weight attached to find true perpendicular line.

—**Cutting tools.**

The cable ripper will quickly slit the sheath (outer jacket) on cable. A heavy duty **pocket knife** may also be used to slit sheath and may be used to taper insulation on wire ends. The standard **carpenter's crosscut saw** is for sawing wood. The **compass saw** or keyhole saw is a small tapered saw for rough cutting holes in flooring, paneling and plasterboard. The **hacksaw** is used for metal cutting, including large cable or conductors. **Wood chisels** are used to cut or gouge wood. The **electric saber saw** is very versatile. It may be used for cutting wood, metal and plasterboard.

— **Hammer** is used for nailing boxes and other devices in place and for driving staples.

—**Soldering iron.**

Soldered connections are not now required in most localities but may be called for under some conditions. Soldered connections are not allowed for certain connections.

—**Tool pouch.**

Use a belt-mounted leather pouch with space to carry "most-used" tools.

—The **fish tape** is used to pull or push wire through conduit.

—The **volt-ohmmeter** is used to test circuits for voltage and continuity.

—**Crimping tool.**

Available in several sizes, they are used for crimping connectors on wires.

C. Locating Circuits

On most wiring jobs, you will have a plan showing the location of electrical outlets in the house. You must decide how to route the branch circuits from the service entrance panel to each outlet shown on the plan (Figure 37). All circuits discussed in this section are branch circuits.

Locating circuits is discussed under the following headings:

1. Marking Location for Service Entrance Panel.
2. Locating and Marking Routes for Small Appliance Circuits.
3. Locating and Marking Routes for General Purpose Circuits.
4. Locating and Marking Routes for Individual Circuits.

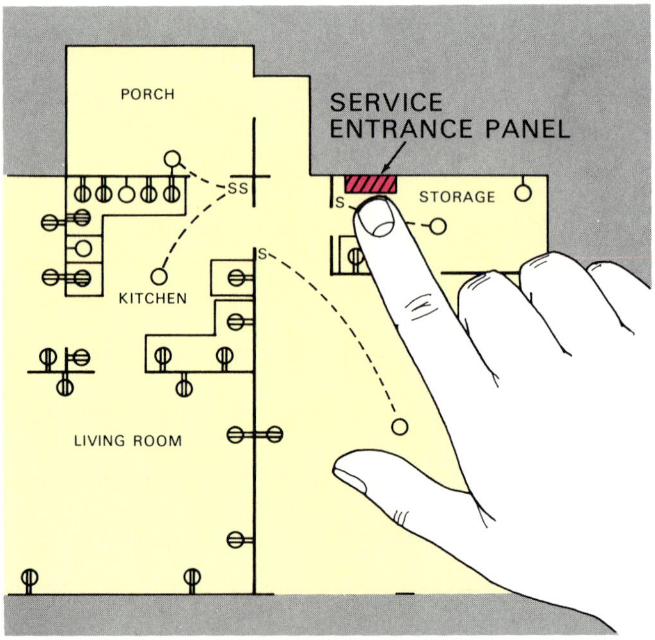

FIGURE 37. Circuits must be routed from the SEP to outlets shown on the plan.

1. MARKING LOCATION FOR SERVICE ENTRANCE PANEL

The procedures described in this section are part of the "roughing-in" process in building a house. Proceed as follows:

1. *Locate position of SEP (Figure 38).*

   The SEP should be located as near the service entrance as possible. Check with your power supplier to determine where the electric service will enter the house. It is also desirable to locate the SEP near the appliances that use the most power. The reason is that these appliances require large conductors which are expensive. They include kitchen appliances, electric water heater, clothes dryer and electric furnace.

   The SEP must be easily accessible at all times. Keep in mind the future use of the area when locating the panel. The Code prohibits locating the SEP in a clothes closet.

2. *Mark the location selected on a stud or wall using pencil or crayon (Figure 39).*

   Mark the location of the SEP cabinet for either surface mount or flush mount as required. The bottom edge of the cabinet should be about 102 – 122 cm (40 – 48 inches) above floor level or as required.

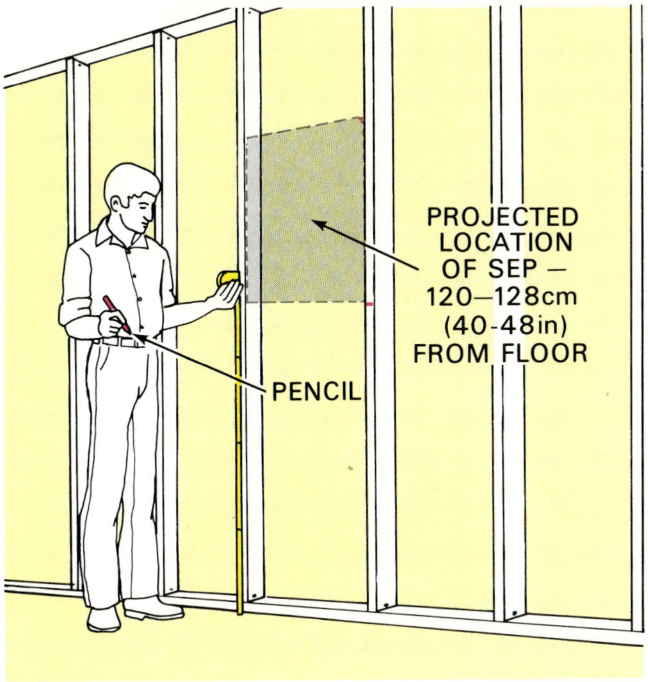

FIGURE 39. Using crayon or pencil, mark the location of the SEP.

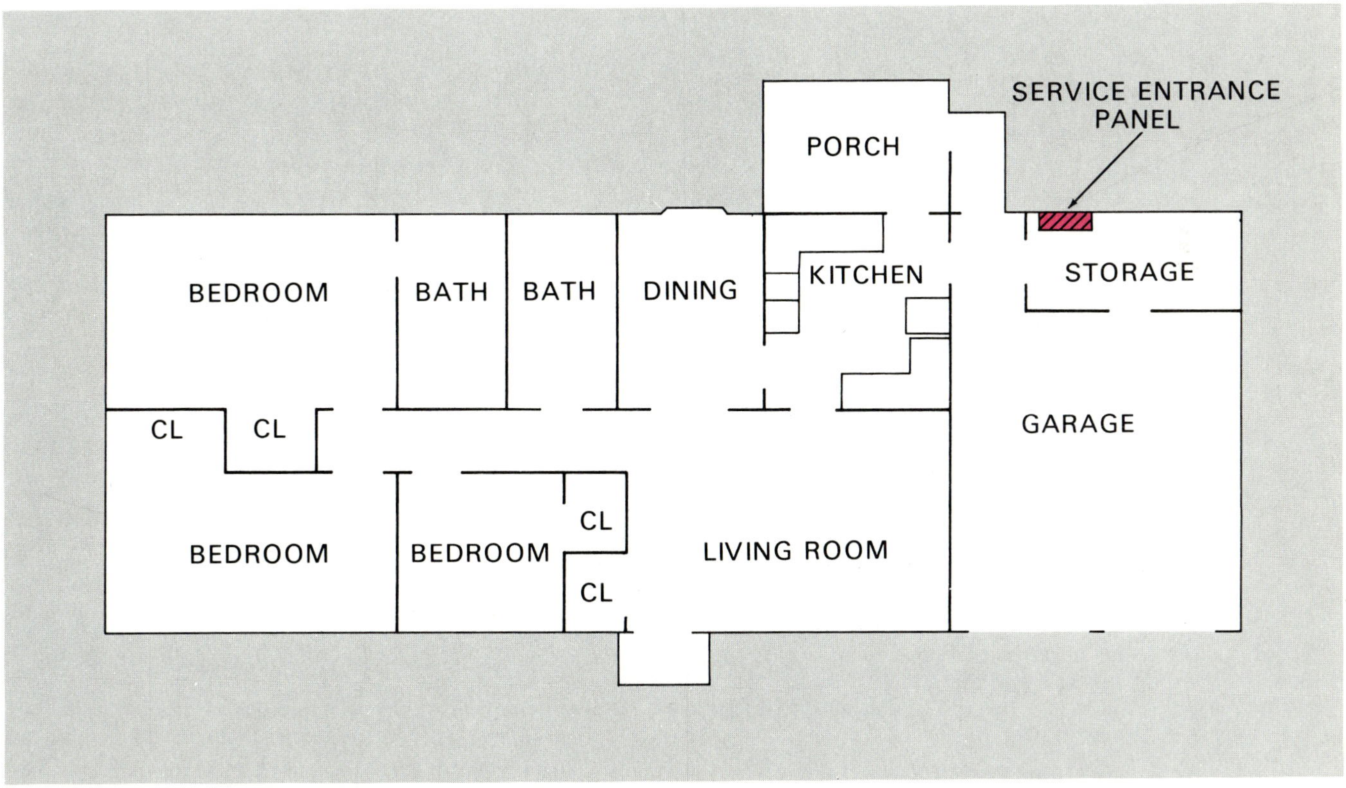

FIGURE 38. Locate the SEP as near as possible to the point where the service entrance conductors enter the house.

2. LOCATING AND MARKING ROUTES FOR SMALL APPLIANCE CIRCUITS

When you first begin locating and marking circuits, it may be helpful to indicate the location of each one on your floor plan similar to those used in this manual. As you gain experience you will be able to mark and follow circuit routes on the framing members.

There are several different ways to route (run) circuits in most wiring jobs and each one may be satisfactory. The route for **small appliance** circuits should be fairly easy to locate if Code requirements are followed.

Locating and marking routes for small appliance branch circuits are discussed under the following headings:

a. Locating and Marking Receptacles.
b. Locating and Routing Small Appliance Circuits.

a. Locating and Marking Receptacles

Your house plan shows where each receptacle outlet is to be located. Mark each of the outlets on the stud where the box is to be mounted (Figure 40). Each receptacle on the kitchen counter should be located so the bottom of the box will be approximately 20 cm (8 inches) above the counter top. All other receptacles should be 30 to 36 cm (12 to 14 inches) from the floor. Some people use the length of the hammer handle as a quick measure of correct height.

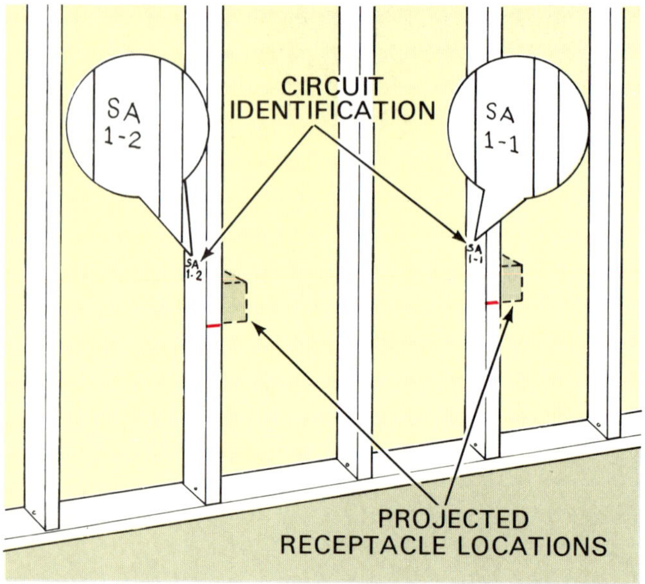

FIGURE 40. Mark the location of each receptacle using your house plan as a guide and identify the circuits. Mark the stud small appliance 1 (SA-1-1) for the first outlet, (SA-1-2) for the second outlet etc.

b. Locating and Routing Small Appliance Circuits

The following example is based on the plan shown in Figure 41. Proceed as follows:

1. *Determine number of small appliance circuits needed.*

   On the plan in Figure 41, only the kitchen and dining room have small appliance outlets, so two circuits should be enough. The small appliance circuit shown in red is divided between the kitchen outlets and the dining room outlets. Small appliance circuit No. 2 contains only kitchen receptacles. The laundry circuit is also shown in black in Figure 41.

2. *Locate and mark small appliance circuit No. 1.*

   For example, in Figure 41, circuit No. 1 is shown in red. A circuit or part of a circuit between outlets is often referred to as the **circuit run.** In some instances, it will be called a cable run or just a run.

   (1) *Mark circuit run beginning at kitchen receptacle outlet nearest SEP at the end of the counter.*

   Using a crayon, identify circuit No. 1 on the studs near the locations of the receptacles (Figure 40). For example, starting with receptacle No. 1 (SA1–1), receptacle No. 2 (SA1–2) and continue until all outlets on the circuit have been identified.

   Route the cable one of two ways to reach the first outlet. Run either overhead through the attic or under the floor and up.

   (2) *Mark circuit run to outlet No. 2.*

   From the first outlet, run the cable through the studs above the counter top to the second outlet (Figure 42).

   (3) *Mark circuit run to first and second dining room outlets.*

   Run the cable through the stud to the first dining room outlet and overhead to the second dining room outlet.

   (4) *Mark circuit run to dining room outlet No. 3.*

   To reach the third outlet across the room, again run either overhead through the attic or under the floor. Figure 43 shows the cable run from the attic through the plate down to

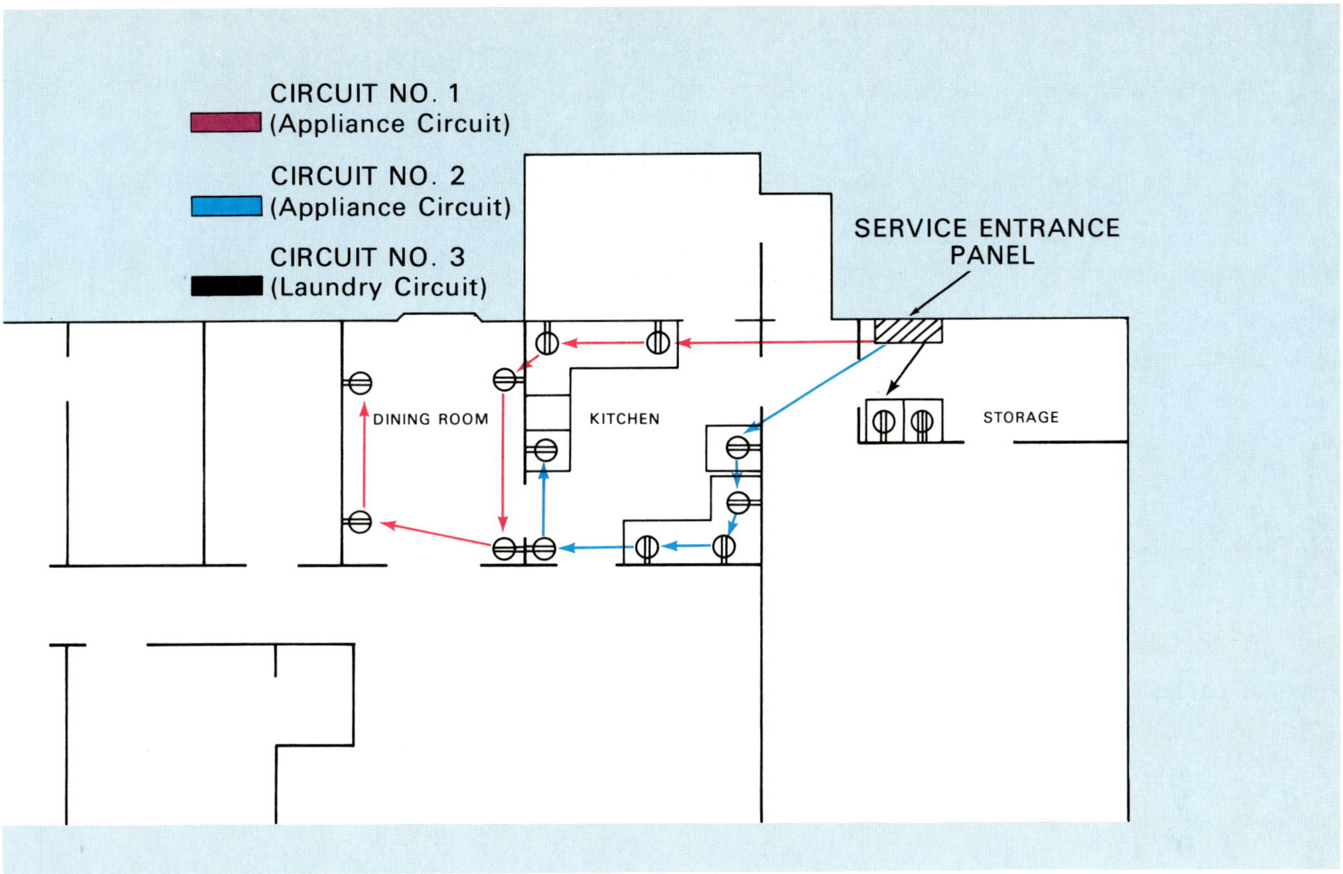

FIGURE 41. Small appliance circuits Nos. 1 and 2 and the laundry circuit.

the third dining room outlet, and through the studs to dining room outlet No. 4. This completes small appliance circuit No. 1.

3. *Locate and mark small appliance circuit No. 2.*

   Small appliance circuit No. 2 is shown in blue. (Figure 41). Identify the circuit on the stud as described for circuit No. 1.

   (1) *Run small appliance circuit No. 2 from the service entrance panel overhead to the refrigerator outlet.* (Some prefer to place the refrigerator on a separate circuit).

   (2) *Run through studs to the other three outlets above the counter top.*

   (3) *From the last counter-top outlet, run the cable overhead and down to the wall outlet beside the door.*

   To reach the outlet above the counter top across the second doorway, run the cable overhead and down to the outlet to complete circuit No. 2.

4. *Locate and mark the laundry circuit.*

   Run the laundry circuit overhead to the outlet for the clothes washer (Figure 41). The clothes dryer is often connected to a 120/240-volt circuit so it must be wired separately. It is discussed under Individual Circuits on page 34.

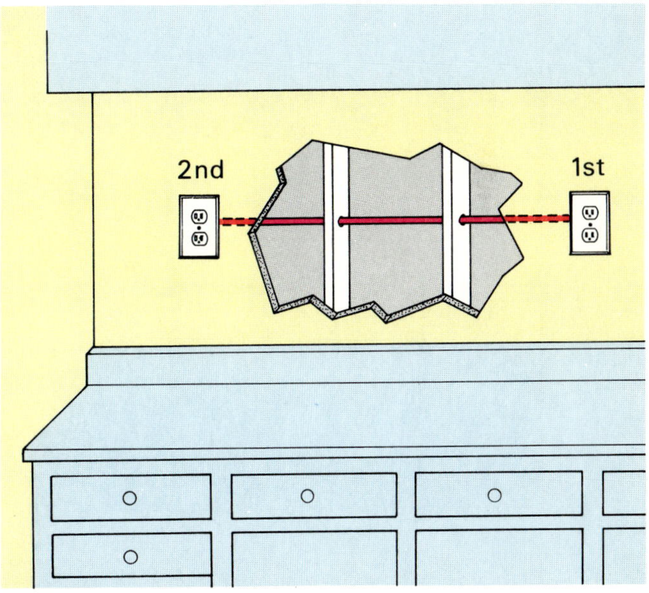

FIGURE 42. Small appliance circuit No. 1 runs through studs above the kitchen counter to the second outlet.

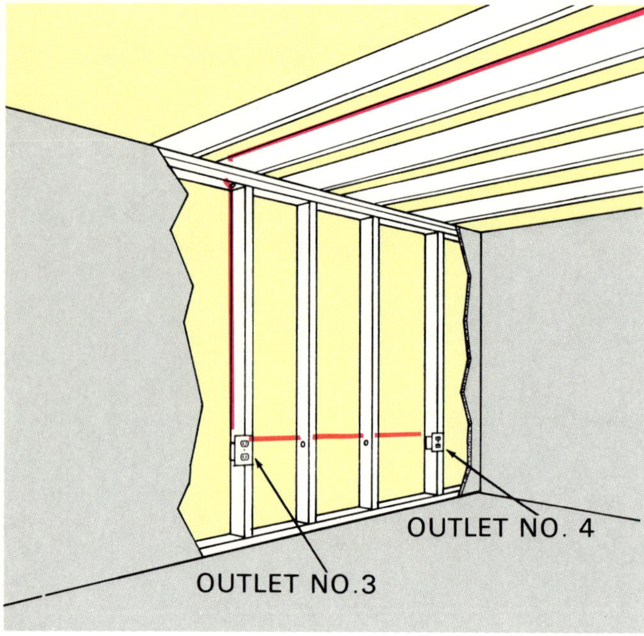

FIGURE 43. Cable routed overhead from the attic to dining room outlets three and four.

3. LOCATING AND MARKING ROUTES FOR GENERAL PURPOSE CIRCUITS

General purpose branch circuits include lighting outlets, and most convenience outlets, such as those used for TV, radio, stereo, vacuum cleaners, and hair dryers.

Figure 44 shows a suggested layout for the five general purpose circuits. Several other combinations are possible that would work as well. Each circuit begins at the outlet most convenient to the SEP as indicated by the arrow.

No. 12 wire is often required as the minimum wire size for general purpose circuits by local codes. However, the Code permits the use of No. 14 wire for general purpose circuits.

In this section, you will lay out and mark a practical and economical route for each of the five general purpose circuits in the wiring plan (Figure 44). Keep in mind that the target number is 10 outlets or less per circuit for No. 14 wire and 13 outlets or less for No. 12 wire. Mark

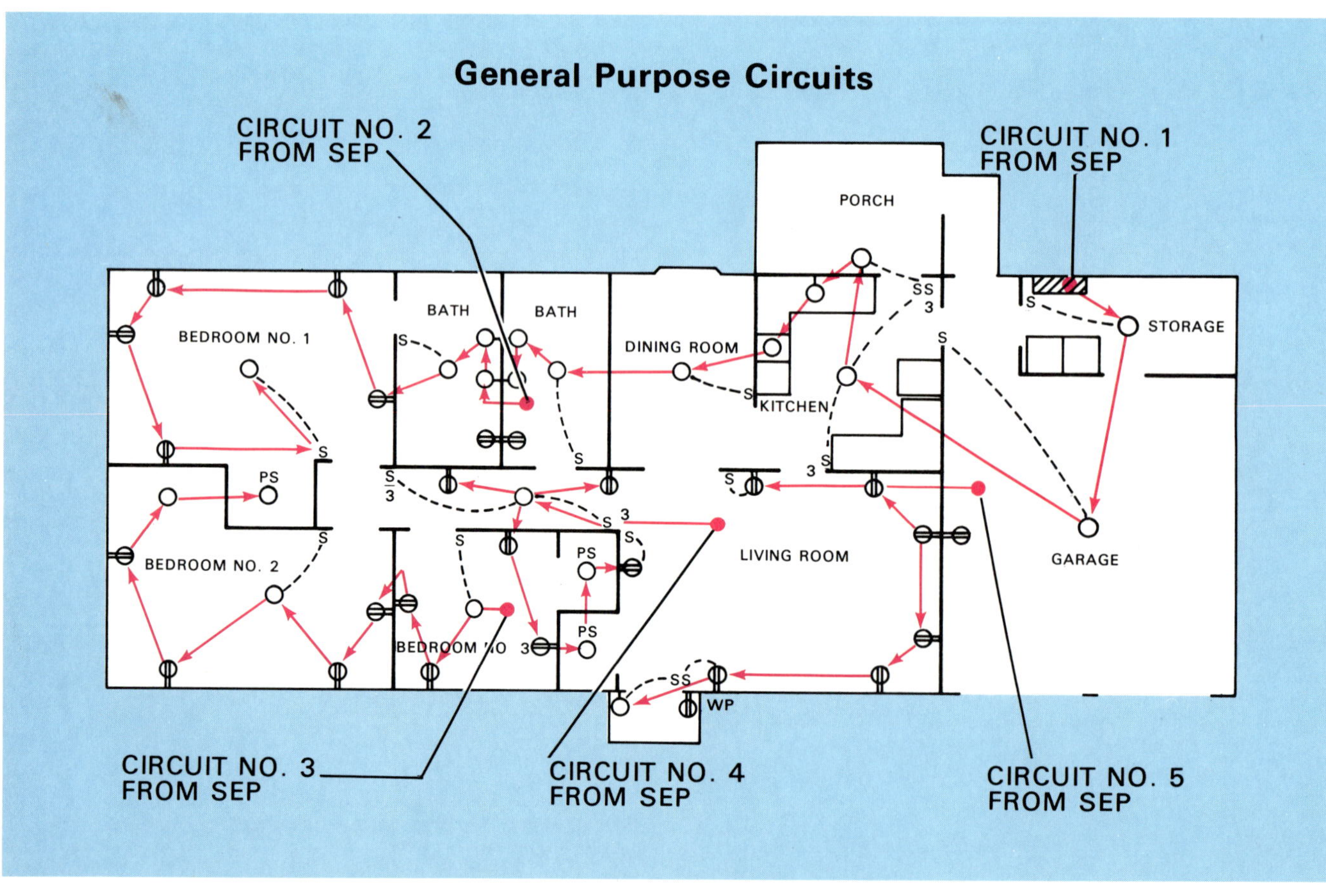

FIGURE 44. The standard house plan contains five general purpose circuits.

the switch and receptacle locations as each circuit is completed. Switch height should be 122 cm (48 inches) from floor to center of box and on the latch side of the door. Some outlets may have more than one switch.

Locating and marking routes for general purpose circuits are discussed as follows:

a. Marking Location for Overhead Lighting Outlets.
b. Locating Circuit No. 1.
c. Locating Circuit No. 2.
d. Locating Circuit No. 3.
e. Locating Circuit No. 4.
f. Locating Circuit No. 5.

a. Marking Location for Overhead Lighting Outlets

Overhead lighting outlets are usually located in the center of the room. To mark the location for the overhead lighting outlet, proceed as follows:

1. *Using a chalk line or establish a diagonal line from one corner to another on floor.*
2. *Repeat for the other two corners.*
3. *Mark the spot where the lines cross (Figure 45).*
4. *Locate the ceiling outlet above this point by means of a plumb bob, or extend a ruler upward to mark the spot on a ceiling joist.*

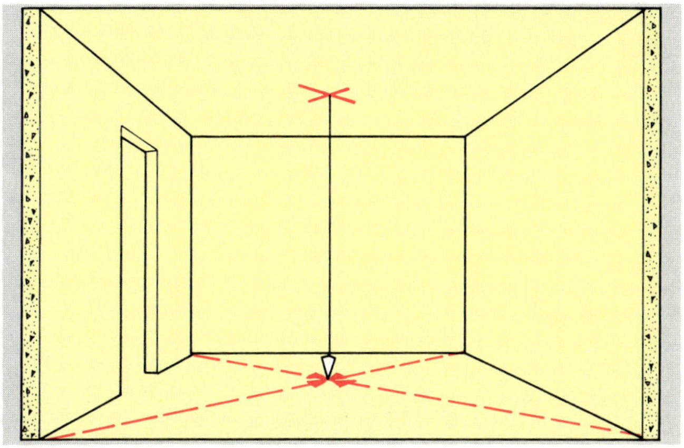

FIGURE 45. Draw diagonal lines from each corner to locate lighting outlet in center of room.

b. Locating Circuit No. 1

The lighting outlets for the storage room, garage, bathroom, kitchen and dining room are part of general purpose circuit No. 1. Locate and mark each outlet and switch on the circuit.

Some of the circuit runs could be under the floor if you have a crawl-space or basement. Other parts of the circuit may be overhead or through the studs. Proceed as follows (Figure 46):

1. *From the SEP, run the circuit to the ceiling light outlet in the storage room.*

   Identify the circuit by marking the outlets and receptacle locations in sequence. For example, (GP1-1, GP1-2, GP1-3, etc.)

2. *Extend to the garage ceiling light.*
3. *From the garage, run the circuit to the ceiling light in the kitchen.*
4. *From the kitchen ceiling light, run the circuit to the wall light on the porch.*
5. *Return to the other two kitchen lights.*
   (1) *Run the circuit first to the light above the sink.*
   (2) *Extend to the light above the range.*
6. *Continue the circuit to the dining room ceiling light.*
7. *From the dining room, run to the ceiling light in bathroom No. 1.*
8. *From the bathroom ceiling light, extend the run to the two wall lights, one on each side of the bathroom mirror.*

   The circuit ends at the light on the left. This completes a circuit containing 10 outlets.

9. *Mark location of switches.*

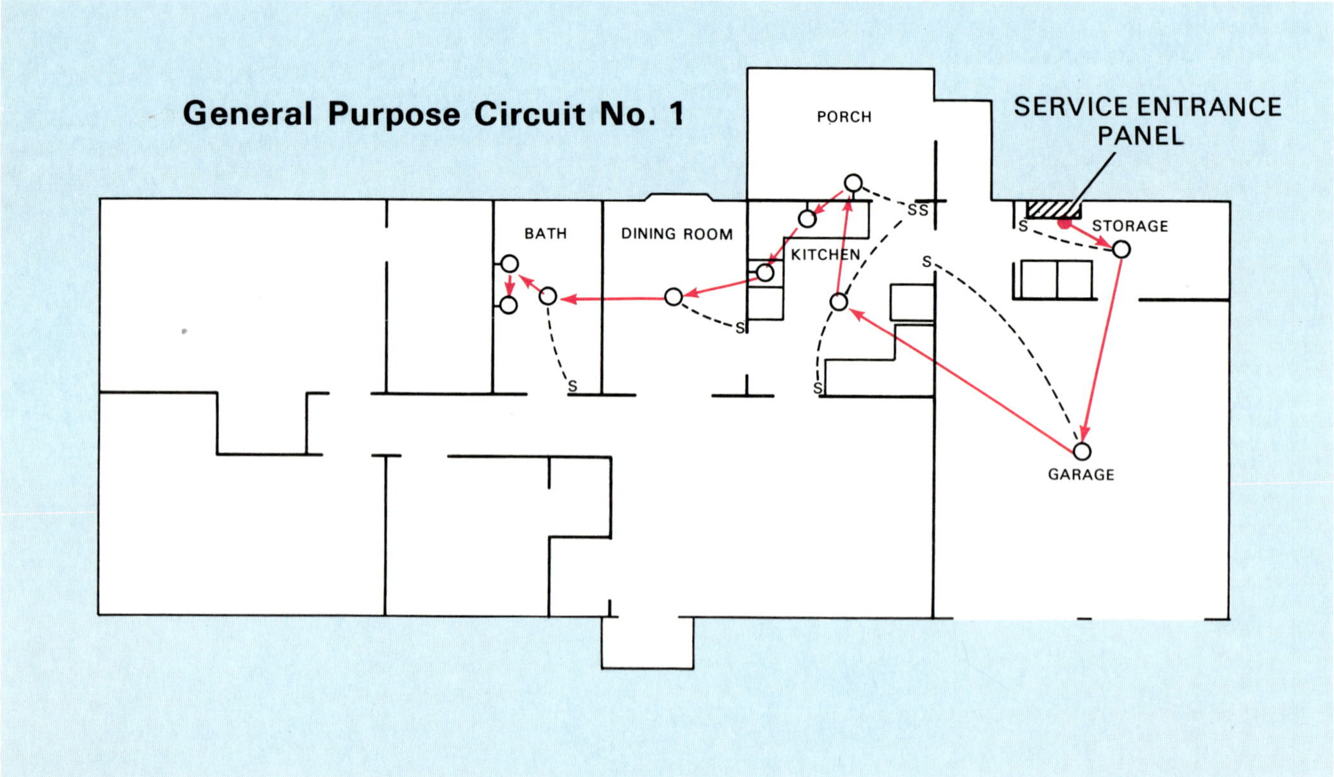

FIGURE 46. Beginning at the SEP, general purpose circuit No. 1 extends from a lighting outlet in the storage room to others in garage, kitchen, porch, dining room and bathroom No. 1.

c. Locating Circuit No. 2

Circuit No. 2 (Figure 47) shows light outlets and convenience outlets mixed on the same circuit. Check your local code to see if both types of outlets are permitted on the same branch circuit.

In circuit No. 2, some of the circuit runs could be under the floor, overhead or through the studs (Figure 48).

The arrow extending from the wall lighting outlet in bathroom No. 2 indicates the starting point for circuit No. 2 (Figure 47). To show the entire circuit would require a line drawn from the SEP to the bathroom outlet.

Proceed as follows to locate circuit No. 2:

1. *Beginning at the wall light in the bathroom, extend the circuit to the wall light on the opposite side of the mirror.*

   Identify circuit by marking the location of each outlet and receptacle as follows: GP2-1, GP2-2, GP2-3, etc.

2. *Run next to the combination ceiling light and fan (Figure 47).*

   Note: A combination ceiling light, fan and heater may require a separate circuit, depending on heater wattage.

3. *From the bathroom ceiling light, run to the nearest bedroom receptacle.*
4. *The remaining receptacles are then connected in order.*

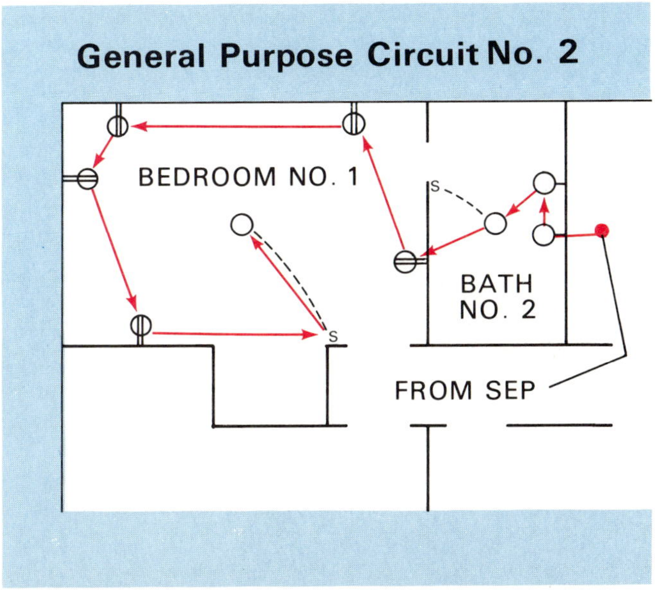

FIGURE 47. Circuit No. 2 includes lighting outlets and receptacles in bathroom No. 2 and bedroom No. 1. The starting point of circuit No. 2 is indicated by the arrow at the bathroom wall lighting outlet.

   (1) *Make the run between the two outlets on the same wall through the studs (Figure 48).*
   (2) *Run the other two through the studs or overhead (Figure 48).*

5. *From the last receptacle, run to the switch and from there to the ceiling light.*
6. *Mark location of switches.*

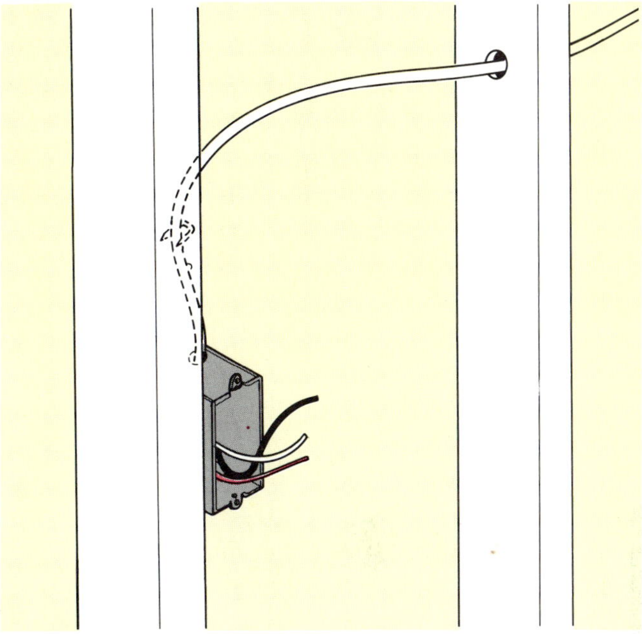

FIGURE 48. Cable run through holes in studs.

d. Locating Circuit No. 3

Circuit No. 3 also includes both ceiling lights and receptacles (Figure 49). Note that it includes all out-

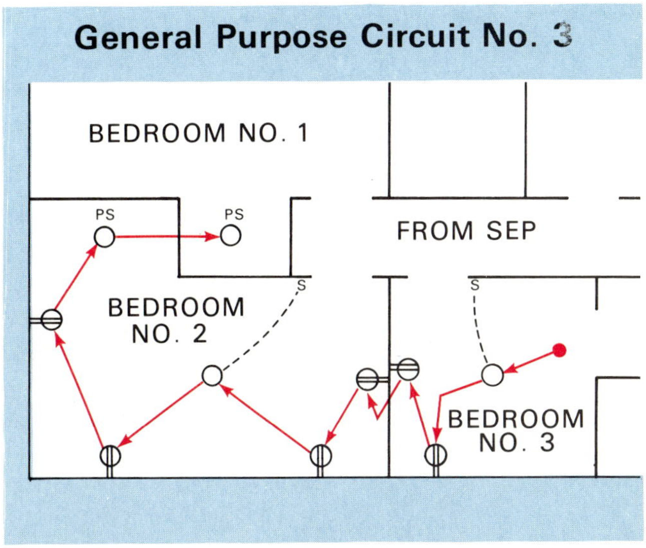

FIGURE 49. Circuit No. 3 includes lighting outlets and receptacles in closets and in bedroom No. 2 and part of bedroom No. 3.

lets in one bedroom and part of another. Proceed as follows:

1. *From the SEP, start with the ceiling light outlet indicated by the arrow.*

   Identify circuit No. 3 at each outlet or receptacle location.

2. *Run next to a bedroom receptacle.*
3. *From there extend to a second outlet that is back-to-back with another in the adjoining bedroom (Figure 50).*

   This is a common arrangement that saves cable.

4. *Run to a second receptacle in the bedroom (Figure 49).*
5. *Run to the ceiling light outlet.*
6. *Extend to the other two receptacles.*
7. *Add the ceiling lights in the two closets to complete the circuit containing 10 outlets.*
8. *Mark location of switches.*

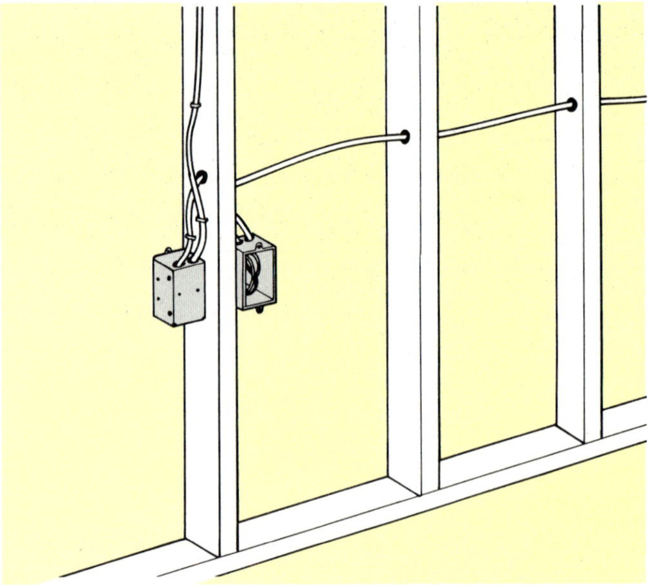

FIGURE 50. Back to back receptacles on opposite sides of the stud serving different rooms.

e. Locating Circuit No. 4

Circuit No. 4 is the most compact of all. It covers the hall light and outlets, the remainder of bedroom No. 3 and one living room receptacle (Figure 51). Proceed as follows:

1. *Begin with the hall ceiling light switch at the arrow.*

   Identify circuit No. 4 at each outlet and receptacle location.

2. *Run to the hall ceiling outlet.*

   Here you find a situation not previously covered. From the hall ceiling outlet the circuit runs in three directions. This is a common arrangement that you will use often.

3. *Extend one run to the first receptacle in the hall. Extend another to the second hall receptacle.*
4. *Run next to the nearest receptacle in bedroom No. 3.*
5. *From the receptacle, run to the second bedroom receptacle.*
6. *Extend to the two closet ceiling lights.*
7. *Run to the receptacle on the living room wall. This complete circuit contains 8 outlets.*
8. *Mark location of switches.*

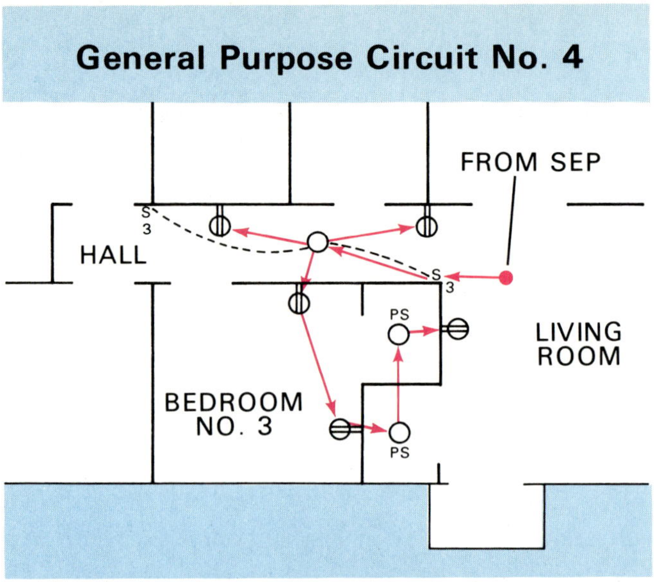

FIGURE 51. Circuit No. 4 serves the hall light and outlets, part of the bedroom and one living room receptacle.

f. Locating Circuit No. 5

Circuit No. 5 is made up of the receptacles in the living room plus one wall light outlet outside the front door (Figure 52). Proceed as follows:

1. *Begin the circuit at the receptacle indicated by the arrow.*

   Identify circuit No. 5 at each outlet and receptacle location.

   The circuit runs two ways from the starting point.

2. *Run next to the two outlets near the kitchen and dining room doors.*
3. *On the other side, run to the remaining receptacles in living room and garage.*
4. *Extend to the outside receptacle and wall light.*
5. *Mark location of switches.*

This completes a layout of the 5 general purpose circuits. As you have seen, none of the circuits has more than 10 outlets. Circuit Nos. 1 and 3 have exactly 10. The others contain 9, 8 and 8, leaving room to add outlets if needed at some future date.

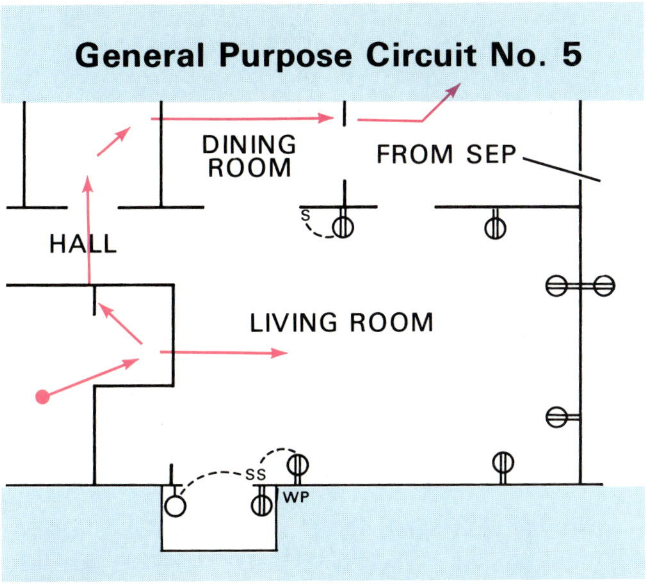

FIGURE 52. Circuit No. 5 includes the receptacles in the living room plus the outside light and receptacle near the front door.

4. LOCATING AND MARKING ROUTES FOR INDIVIDUAL CIRCUITS AND FOR GROUND FAULT CIRCUIT INTERRUPTERS

Locating the Marking Routes for Individual Circuits and for Ground Fault Circuit Interrupters are discussed under the following headings:

a. Locating and Marking Routes for Individual Circuits.

b. Locating and Marking Routes for Ground Fault Circuit Interrupters.

a. Locating and Marking Routes for Individual Circuits

Individual branch circuits are installed for *heavy duty* appliances and equipment. According to the Code, each must have its own circuit (Figure 53). This means that a separate cable is run from the SEP to the outlet for the heavy duty item. Some are connected direct without use of plug or receptacle. Another difference in some individual circuits is the higher voltage. All small appliance and general purpose circuits are wired for 120 volts. Several items on individual circuits require 240 volts or a combination of 120/240 volts. The 240-volt circuits or 120/240-volt circuits may call for larger wire sizes and higher capacity fuses or circuit breakers. Some common individual circuits are indicated by arrows.

Individual circuits for the following items are usually wired at 120 volts with **No. 12 wire** and protected by a **20-ampere** fuse or circuit breaker.

—Garbage disposer.*
—Dishwasher.*

*Many local codes allow the garbage disposer and dishwasher to be wired on the same circuit.

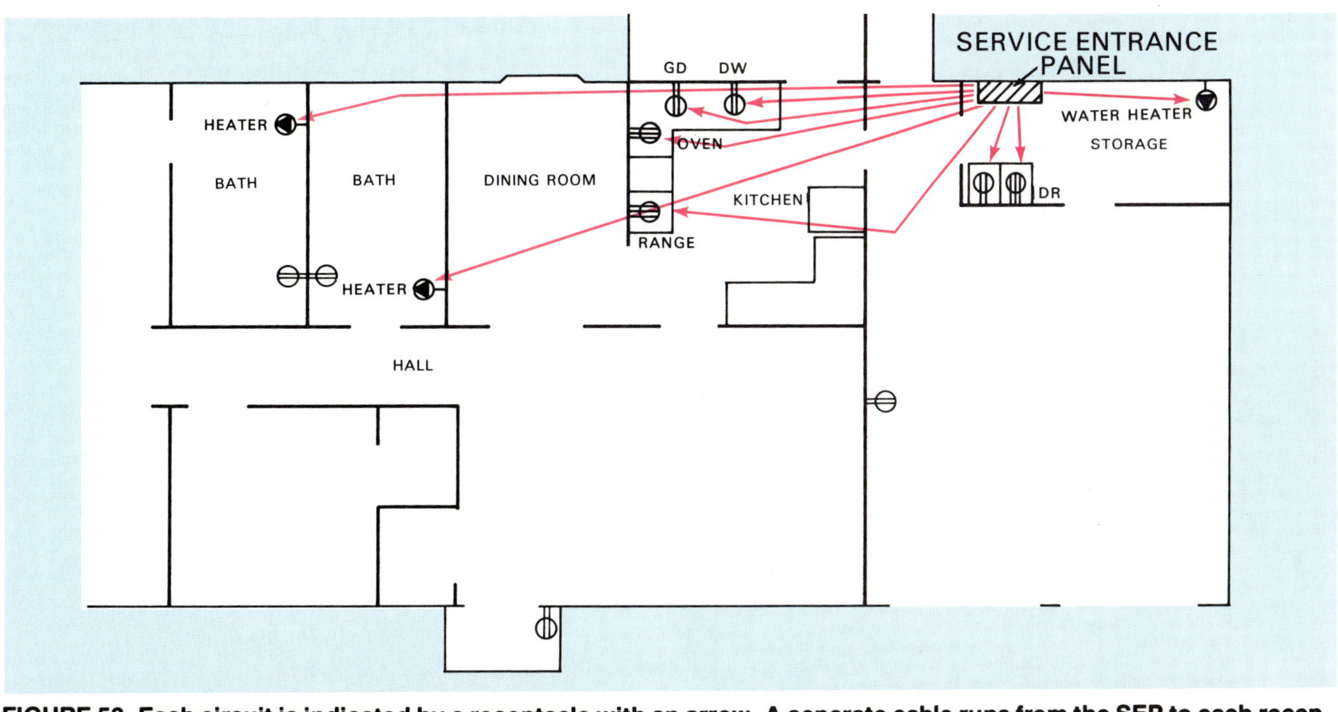

FIGURE 53. Each circuit is indicated by a receptacle with an arrow. A separate cable runs from the SEP to each receptacle so marked. Receptacles not marked with an arrow are protected by a GFCI, which may be installed at each outlet or in the SEP.

TABLE III. EXAMPLES OF VOLTAGES, WIRE SIZES AND OVERCURRENT PROTECTION FOR INDIVIDUAL CIRCUITS

Usual Voltage	Circuit	Copper Wire Size (AWG)	Fuse or Circuit Breaker (Amps)
240	Air conditioner (large)	12	20
240	Water heater	12/10	20/30
120/240	Separate oven	10	30
120/240	Counter top cooking unit	10	30
120/240	Clothes dryer	10	30
120/240	Self contained range	6	50
240	Central air conditioning or heat pump	8 or 6	40–60
240	Water pump	12	20
240	Central electric furnace	6	50

—Trash compactor.
—Motor on fuel-fired furnace.
—Room air conditioner.
—Bathroom heaters.

The following items are usually wired on individual circuits at 240 volts or 120/240 volts. Wire sizes and fuse or circuit breaker sizes as shown in the chart are common examples (Table III).

Food freezers may also be on a separate circuit to eliminate the possibility of a power loss due to other items on the circuit. However, it is a good idea to include **another outlet** on the freezer circuit to indicate power loss, such as a clock or a light.

When marking the locations for each of the special purpose outlets, you should identify the 120/240-volt or 240-volt circuits. A different size and type of outlet is required for each.

The individual circuits in Figure 53 are as follows:

1. *Beginning in the storage room, locate the water heater on an individual 240-volt circuit.*
   Identify individual circuits by marking the outlet location, such as (WH).
2. *Locate the dryer on an individual 120/240-volt circuit.*
   The **washer**, previously mentioned, is on a 120-volt circuit.
3. *In the kitchen, place the separate oven and countertop cooking unit on 120/240-volt circuits.*
4. *Locate both the dishwasher and garbage disposal on 120-volt circuits, (or combine if local code permits).*
5. *Locate the bathroom heaters normally on 120 volts. They may be connected to 240 volts if so designed.*

b. Locating and Marking Routes for Ground Fault Circuit Interrupters

In addition to the individual circuits already mentioned, many people prefer to run individual circuits to outside receptacles. An example is one near the front door and one in the garage. **All outdoor receptacles must be protected by Ground Fault Circuit Interrupters (GFCI).** At least one receptacle in a garage must be protected by a GFCI. Receptacles for plug-connected garage doors and receptacles intended for food freezers and laundry equipment in a garage are not required to have GFCI protection. The purpose of the GFCI is to provide protection for people who may be using a faulty tool or equipment or may accidentally misuse a tool that could cause a dangerous shock. A shock received when standing on wet ground can be severe enough to cause death. The GFCI device may be installed at each outlet (Figure 54), or in the SEP (Figure 55).

Each convenience outlet in the bathroom must also be protected by a GFCI. In the house in Figure 53, there is one in each bath. One circuit to the two outlets should be satisfactory. Both may be protected from a single GFCI device installed in the service entry panel. Procedures for GFCI installation are discussed in "**Section IV. Installing Service Entrance Equipment.**"

FIGURE 54. Type of GFCI installed at outlet.

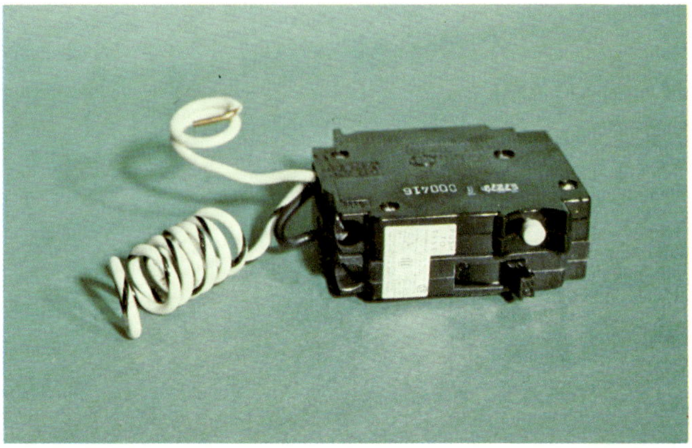

FIGURE 55. Type of GFCI used in service entrance panel.

D. Installing Device Boxes and Outlet Boxes

Installing device boxes and outlet boxes is often among the first tasks given the beginning electrician. **A device is a unit of the electrical system which is intended to carry, but not use, electricity.** Examples of devices are switches and receptacles (Figure 56).

FIGURE 56. Switches and receptacles are examples of the devices installed in device boxes.

Some electricians prefer to drill the holes for the cable before mounting the boxes. This is a matter of personal preference and can be done either way. **All connections and splices must be made in a box.** Your house wiring plan shows you where each box must be mounted.

Installing device boxes and outlet boxes is discussed under the following headings:

1. Types and Sizes of Device Boxes and Outlet Boxes.
2. Number of Conductors Allowed in Boxes.
3. Installing Switch Boxes and Handy Boxes.
4. Installing Outlet Boxes.

1. TYPES AND SIZES OF DEVICE BOXES AND OUTLET BOXES

Most boxes used in house wiring are made of galvanized sheet steel (Figure 57). Another type of box made of durable plastic material is growing in use for residential wiring. Plastic boxes (nonmetallic) may only be used with nonmetallic sheathed cable or nonmetallic (plastic) conduit (Figure 58). Boxes of three basic shapes and many sizes are available for house wiring.

Types and sizes of device boxes and outlet boxes are discussed as follows:

a. Types and Sizes of Device Boxes.
b. Types and Sizes of Outlet Boxes.

FIGURE 57. Metal boxes made of galvanized sheet steel.

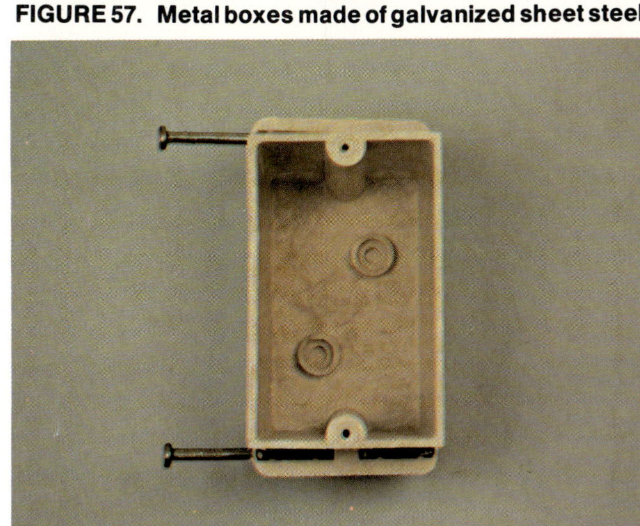

FIGURE 58. Nonmetallic boxes made of durable plastic.

a. Types and Sizes of Device Boxes

Types and sizes of device boxes are discussed as follows:

(1) Types of Device Boxes. The box in Figure 59 is a **device** box. It is usually called a **switch box,** but it is more than that. It is used to install switches and for several other purposes. For example, you will use more of them to install duplex receptacles than switches. They are referred to as switch boxes in the following discussion.

Switch boxes are available with square or beveled corners and with mounting brackets of various types (Figure 60). Other types are those with brackets on top and bottom with 16d nails attached (Figure 61) and those with holes through the sides for nails (Figure 62).

FIGURE 59. A common type of device box, usually called a switch box.

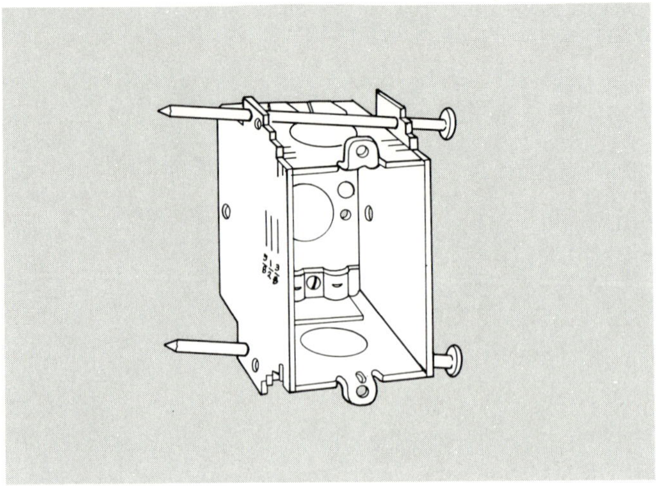

FIGURE 61. A switch box with square corners and nails through brackets at bottom and top for mounting.

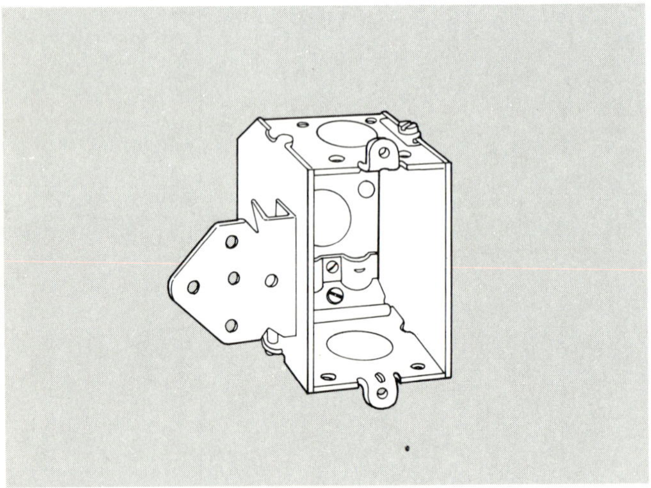

FIGURE 60. A device (switch) box with beveled corners and mounting bracket.

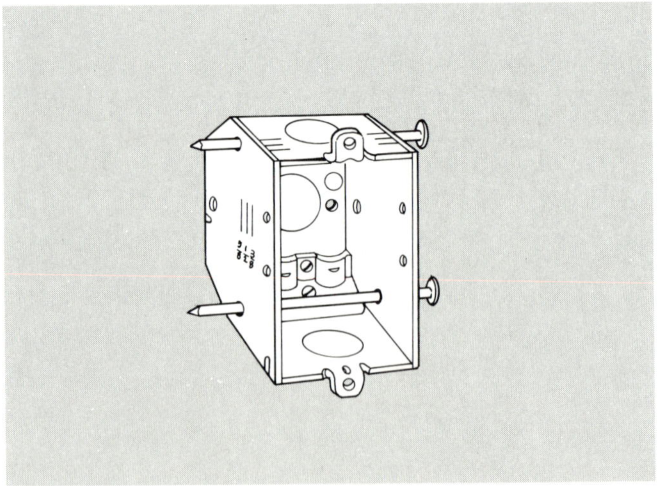

FIGURE 62. A switch box with beveled corners and nail holes through the sides of the box for mounting.

Detailed features of the switch box are seen in Figure 63. **Knockouts** are the coin-sized plugs shown on the sides, bottom and rear of the box. A sharp blow knocks out the plug which has been cut out except for a narrow strip (Figure 64). Removal of the plug makes an opening where needed to bring in cable or conduit. **Knockout openings that are not used must be plugged with knockout seals** (Figure 64 inset).

Pryouts (rear, top and bottom) are smaller knockouts that are also present on some boxes. They are for nonmetallic cable installation and are always in pairs. Remove those needed by inserting a screwdriver in the slot, then pry up and twist (Figure 65).

Mounting ears make mounting and adjusting easier. They are especially useful when installing a switch box in an existing (older) home (Figure 66).

FIGURE 63. Knockouts are the coin-size circles on the sides, bottom and rear of the box.

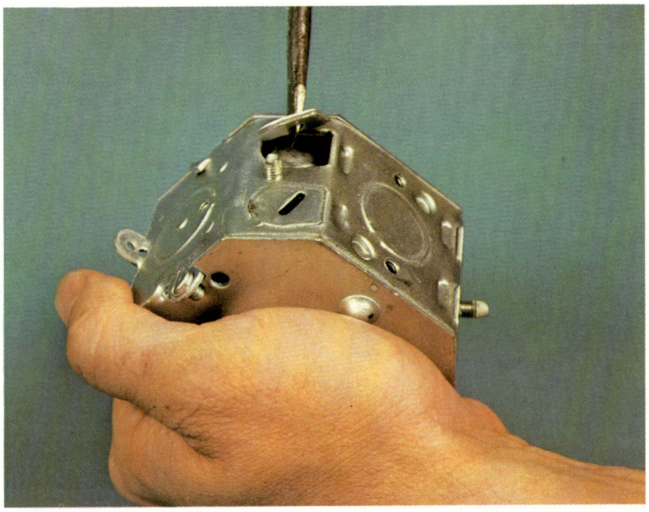

FIGURE 65. Pryouts are smaller openings present on some boxes that are removed by prying and twisting with a screwdriver.

FIGURE 64. A sharp blow knocks out the plug to make an opening. The knockout seal is used to plug unused openings.

FIGURE 66. Mounting ears are used to mount and adjust the box.

The Code now requires that a way be provided for connecting a grounding wire in each metal box. In most boxes, this will be a tapped (threaded) hole that will accept a screw for the grounding terminal (Figure 67).

The pre-installed **pigtail grounding wire** saves time when connecting the circuit grounding wire to the box, as explained in the next section on installing boxes (Figure 67).

The **grounding clip,** seen in Figure 68, is another item that may be used to connect the circuit grounding wire to the box.

The **cable clamp** is tightened against the cable near where it enters the box. Several other types are available (Figure 69). All are held in place by screws except the "**quick clamp**" which is already under tension (Figure 70).

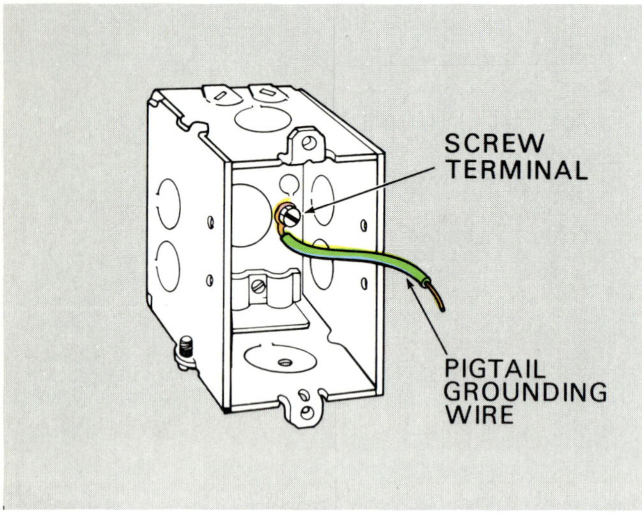

FIGURE 67. Grounding wire screw terminal installed in the tapped hole.

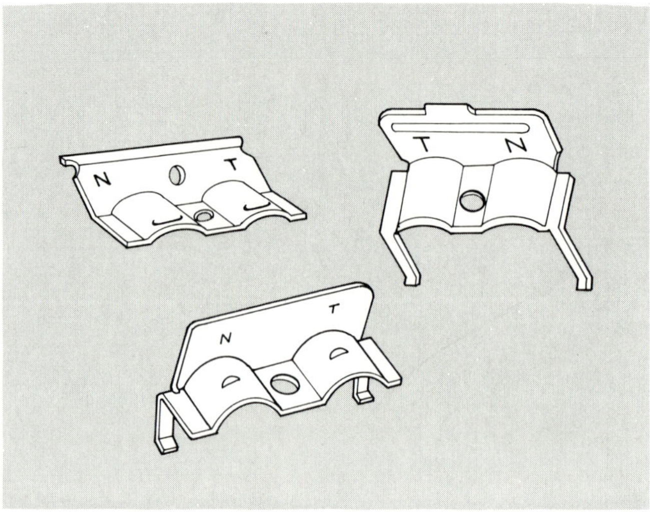

FIGURE 69. The cable clamp grips and holds the cable in the box when the screw is tightened.

FIGURE 68. A grounding clip may be used to connect the grounding wire to the box.

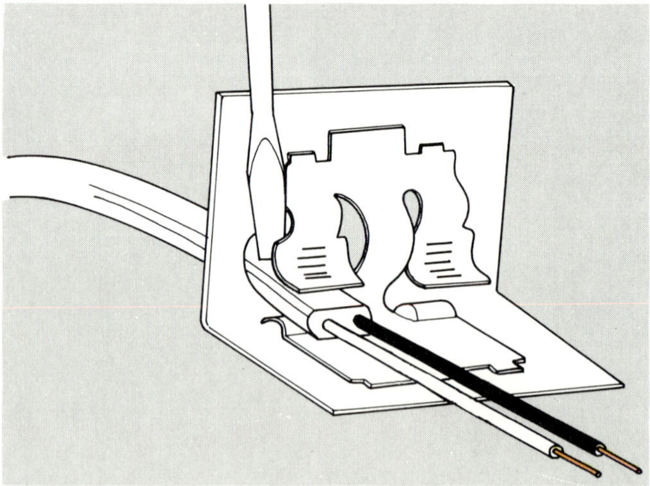

FIGURE 70. A quick-clamp type cable holder does not require a screw.

A **cable connector** may be installed where the knockout opening is used to bring in cable (Figure 71). Cable clamps inside the box are not required where connectors of this type are installed.

(2) Sizes of Device Boxes. The dimensions for device (switch) boxes are as follows: The sides are 7.6 cm (3 inches) high and the width 5.1 cm (2 inches). They are available with depths from front to rear of 3.8 to 8.9 cm (1½ to 3½ inches). More of the larger sizes such as 7.6 × 5.1 × 5.1 cm (3 × 2 × 2 inches) and 7.6 × 5.1 × 8.9 cm (3 × 2 × 3½ inches) are used (Figure 72).

Switch boxes may be purchased in sizes to accommodate two or more switches or receptacles side by side. Or you may make your own "gang boxes" as they are called. Remove one side on each box and put them together to provide space for two devices. You may gang as many as you need using the same procedure (Figure 73). To mount gang boxes, use metal or wood strips between studs to hold them in place (Figure 74).

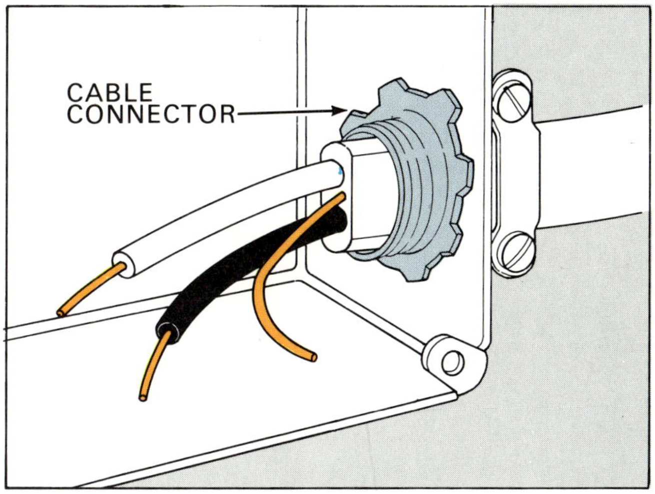

FIGURE 71. A cable connector installed in a knockout opening to hold the cable in place.

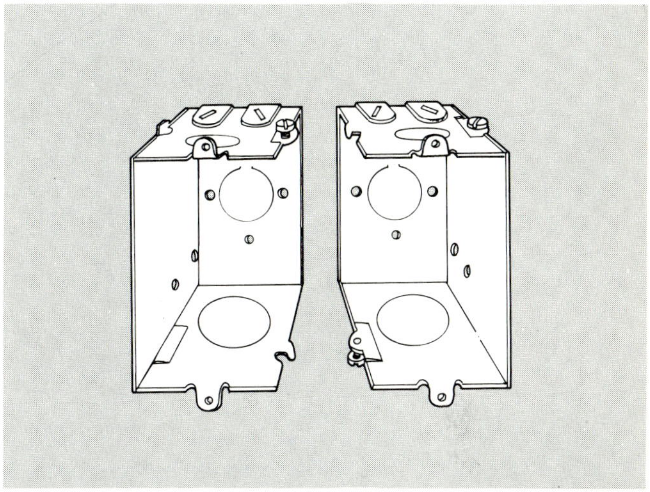

FIGURE 73. Two or more metal boxes may be fastened (ganged) together by removing one side on each box.

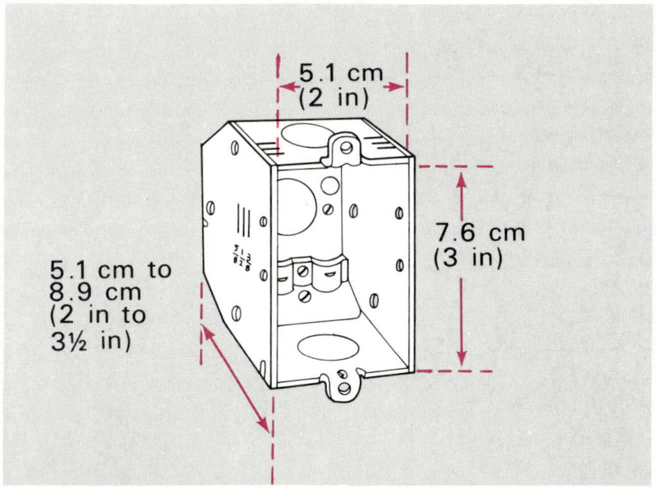

FIGURE 72. A large size switch box used in house wiring.

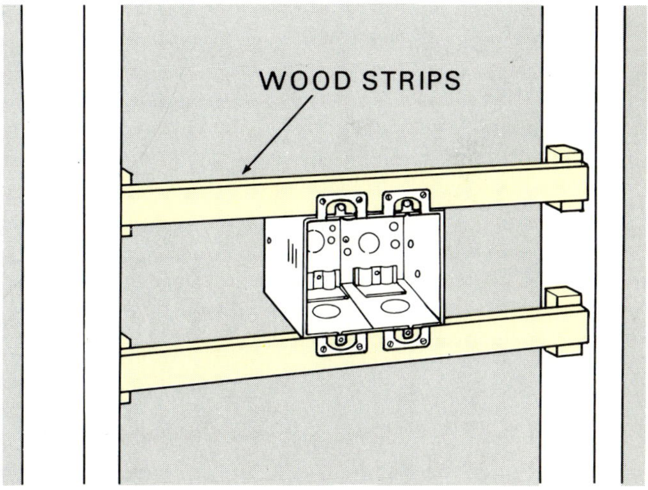

FIGURE 74. Ganged boxes may be mounted on metal or wood strips between studs to hold them in place.

In garages, basements and utility buildings, you may need a switch or receptacle that can be **surface mounted.** It is best to use a box known as a **handy box** for such locations (Figure 75). The handy box is a little larger than the standard switch box. Sizes are from 10.2 × 5.4 × 3.8 cm (4 × 2⅛ × 1½ inches) to 10.2 × 5.4 × 5.4 cm (4 × 2⅛ × 2⅛ inches). Device and handy box **covers** are available for use with either switches or receptacles (Figure 76).

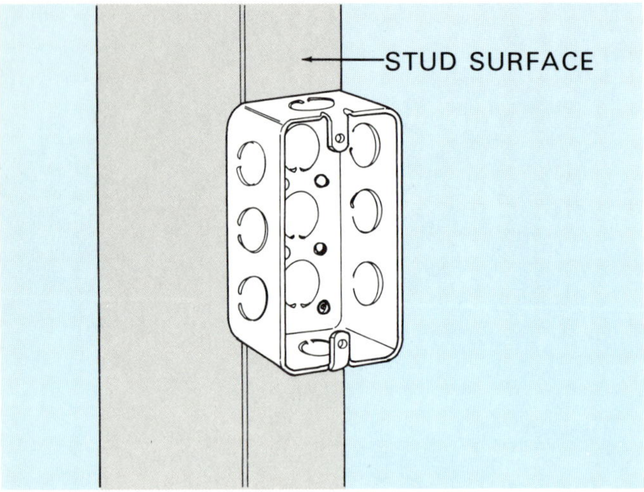

FIGURE 75. Handy boxes may be used for surface mounting switches or receptacles in such places as garages, basements and utility buildings.

FIGURE 76. Covers available for use on handy boxes to mount switches or receptacles.

b. Types and Sizes of Outlet Boxes

Types and sizes of outlet boxes are discussed as follows:

(1) TYPES OF OUTLET BOXES. Outlet boxes may be octagonal or square (Figure 77). The octagon-shaped box is the common choice for overhead lighting fixtures and other locations requiring more room than a switch box provides. They are available with mounting brackets similar to those used on switch boxes. They may also be designed with the same type of knockouts, clamps and other fittings (Figure 78). Switches and receptacles may be mounted in the octagon and square boxes if they are fitted with the correct type of cover (Figure 79). These larger boxes provide room for more wiring inside.

FIGURE 77. Outlet boxes may be octagonal or square.

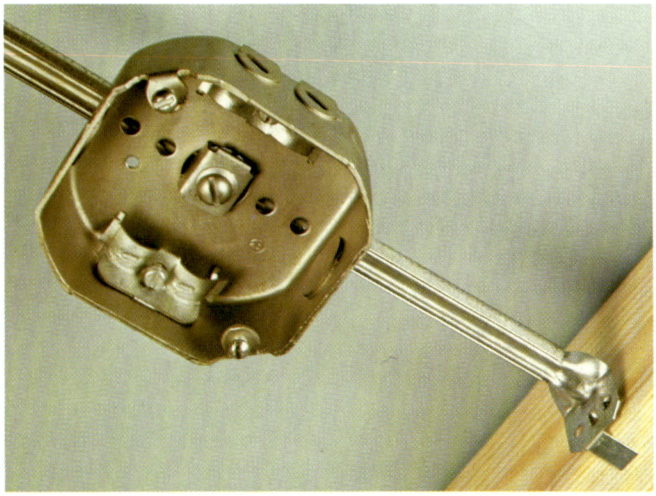

FIGURE 78. Octagonal boxes are often used for ceiling light fixtures.They may have brackets and fittings similar to those on switch boxes.

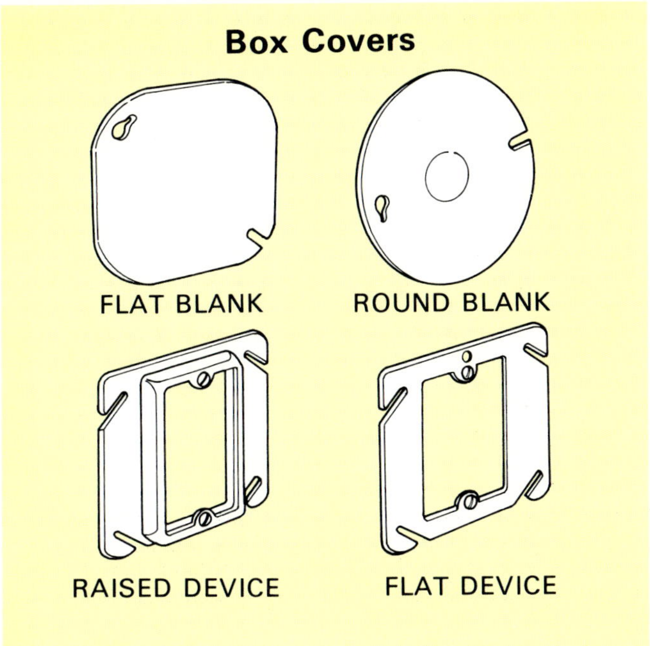

FIGURE 79. Types of outlet box covers.

(2) SIZES OF OUTLET BOXES. Sizes of outlet boxes and device boxes are shown in Table IV. If you find during installation that you need more space than is available in a standard size box, you can add an extension ring to increase the capacity (Figure 80). Extension rings are also used to bring the edge of the box out to the surface when plaster or tile edges are beyond the box opening.

Table IV gives the number of wires the Code allows to be installed in each box where there are no fittings or devices such as fixture studs, cable clamps, switches and receptacles. The number of wires allowed depends on the size of the box and the size of the wires going into the box. However, most boxes have at least some number of cable clamps, studs and other items, in addition to the wires. The number of wires allowed in a box must be reduced to make room for these extra items.

The reductions required for items other than conductors can make a big difference in a size of the box needed. Detailed rules are given in the next section.

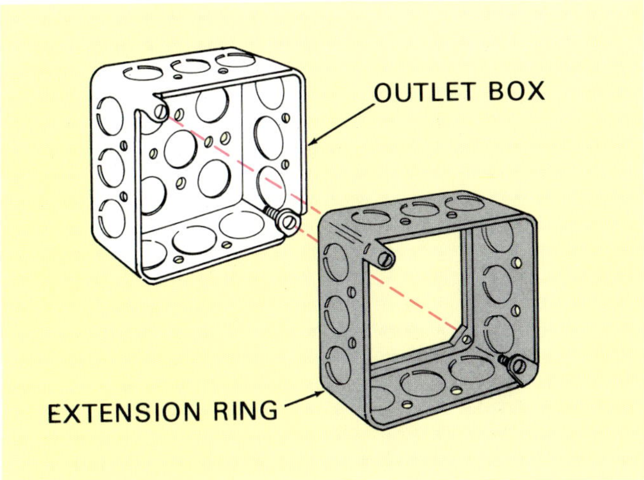

FIGURE 80. An extension ring added to a box to increase capacity or to extend box to flush surface.

TABLE IV. MAXIMUM NUMBER OF CONDUCTORS TO BE INSTALLED IN METAL BOXES
[Customary Dimensions excerpted from Code Table 370-6(a)]

Reprinted by permission from NFPA 70, National Electrical Code®, Copyright 1984, National Fire Protection Association, Boston, MA

Box Dimenson Trade Size or Type		Capacity		Maximum No. of Conductors (AWG Nos.)			
Centimeters	Inches	CC	Cu In	14	12	10	8
10.2 × 3.2	4 × 1¼ Round or Octagonal	205	12.5	6	5	5	4
10.2 × 3.8	4 × 1½ Round or Octagonal	254	15.5	7	6	6	5
10.2 × 5.4	4 × 2⅛ Round or Octagonal	352	21.5	10	9	8	7
10.2 × 3.2	4 × 1¼ Square	295	18.0	9	8	7	6
10.2 × 3.8	4 × 1½ Square	344	21.0	10	9	8	7
10.2 × 5.4	4 × 2⅛ Square	497	30.3	15	13	12	10
11.9 × 3.2	4 11/16 × 1¼ Square	418	25.5	12	11	10	8
11.9 × 3.8	4 1/16 × 1½ Square	484	29.5	14	13	11	9
11.9 × 5.4	4 11/16 × 2⅛ Square	688	42.0	21	18	16	14
7.6 × 5.1 × 3.8	3 × 2 × 1½ Device	123	7.5	3	3	3	2
7.6 × 5.1 × 5.1	3 × 2 × 2 Device	164	10.0	5	4	4	3
7.6 × 5.1 × 5.7	3 × 2 × 2¼ Device	172	10.5	5	4	4	3
7.6 × 5.1 × 6.4	3 × 2 × 2½ Device	205	12.5	6	5	5	4
7.6 × 5.1 × 7.0	3 × 2 × 2¾ Device	229	14.0	7	6	5	4
7.6 × 5.1 × 8.9	3 × 2 × 3½ Device	295	18.0	9	8	7	6
10.2 × 5.4 × 3.8	4 × 2⅛ × 1½ Device	169	10.3	5	4	4	3
10.2 × 5.4 × 4.8	4 × 2⅛ × 1⅞ Device	213	13.0	6	5	5	4
10.2 × 5.4 × 5.4	4 × 2⅛ × 2⅛ Device	238	14.5	7	6	5	4

2. NUMBER OF CONDUCTORS ALLOWED IN BOXES

The Code requires boxes to be large enough to provide free space for all wires brought into the box. Table IV shows the maximum number of wires by sizes that can be brought into metal boxes of various dimensions, according to the Code.

The number of wires allowed per box according to Table IV must be reduced by one for the following contained in a box:

—**For one or more studs or cable clamps or hickeys, reduce by one for each type of device. If one of each is included, reduce by three (Figure 81).**

—**For each strap containing one or more devices such as a switch or receptacle, reduce by one (Figure 81).**

—**For one or more grounding wires entering the box, reduce by one.**

—**For each conductor originating outside the box and ending inside the box, reduce by one.**

If no part of a conductor leaves the box, no reduction is required for that conductor, for example, a pigtail ground wire (Figure 82).

The importance of the Code restrictions may be seen in the following example. Suppose you want to install a wall-mounted light in a bathroom as shown in the house plan, using No. 12 wire. Following Code requirements in the table, you must reduce the number of conductors allowed in the box as follows (Figure 83):

—For one stud, reduce number allowed by 1
—For two cable clamps, reduce number allowed by 1
(a reduction of one, whether one or two cable clamps)
—For one strap, reduce number allowed by 1
—For one or several grounding wires, reduce number allowed by 1
—Total reduction 4

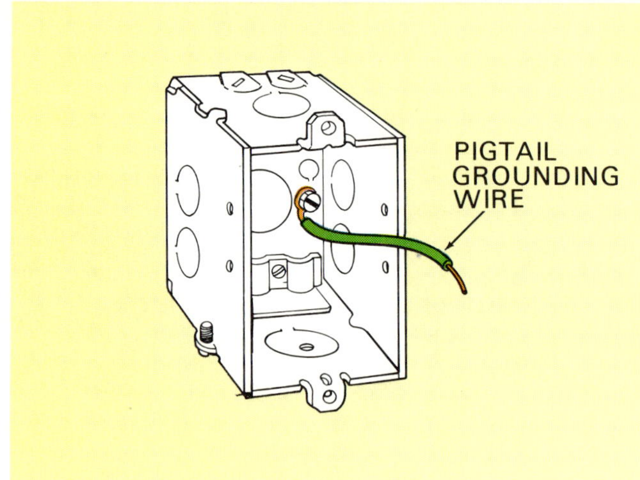

FIGURE 82. A pigtail ground wire is confined to the box, therefore, it is not counted against the maximum number of conductors to be allowed in a box.

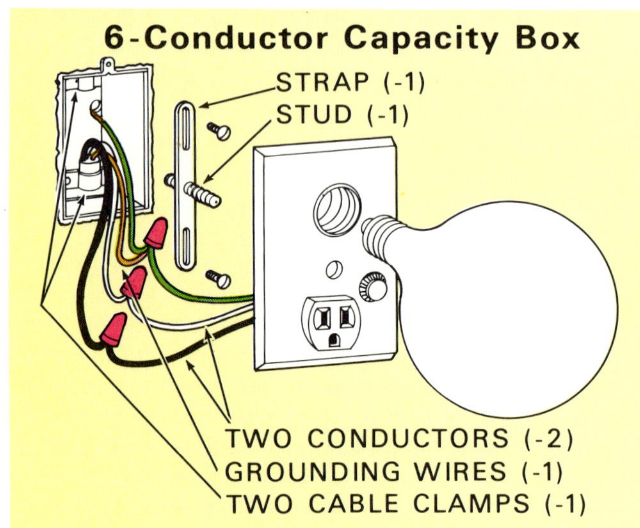

FIGURE 83. Box must be large enough for 6 conductors for this 2-conductor light outlet.

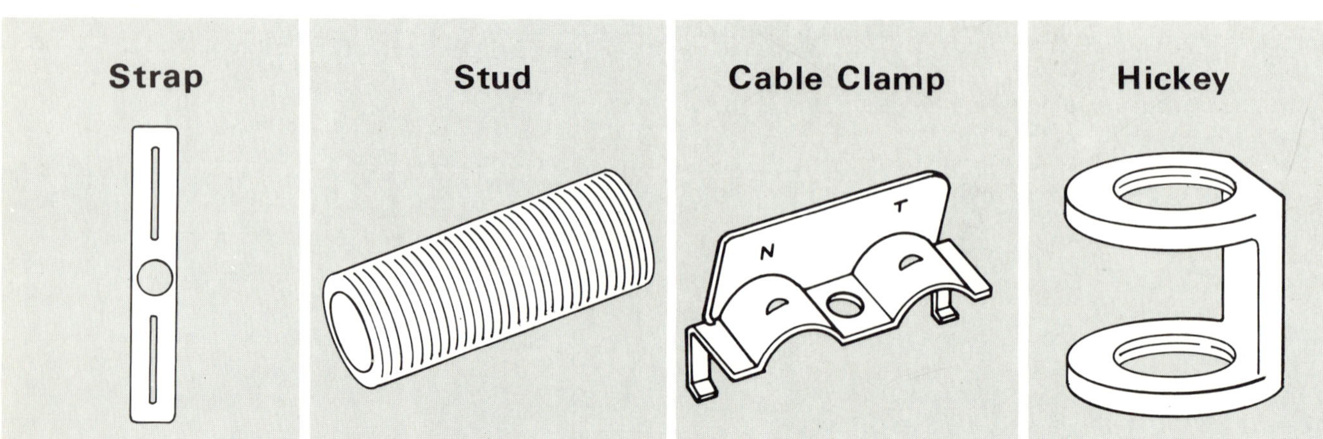

FIGURE 81. Strap, stud, cable clamp and hickey. Reduce the number of conductors allowed in a box by one for one or more of each.

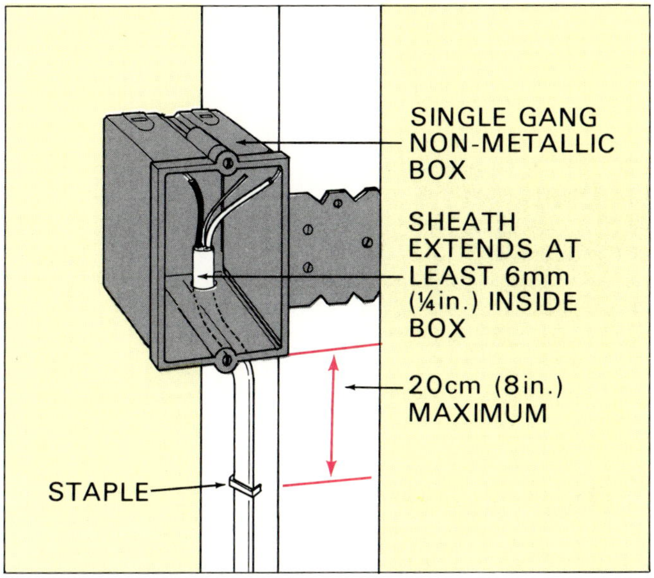

FIGURE 84. Single gang nonmetallic box, without cable clamp.

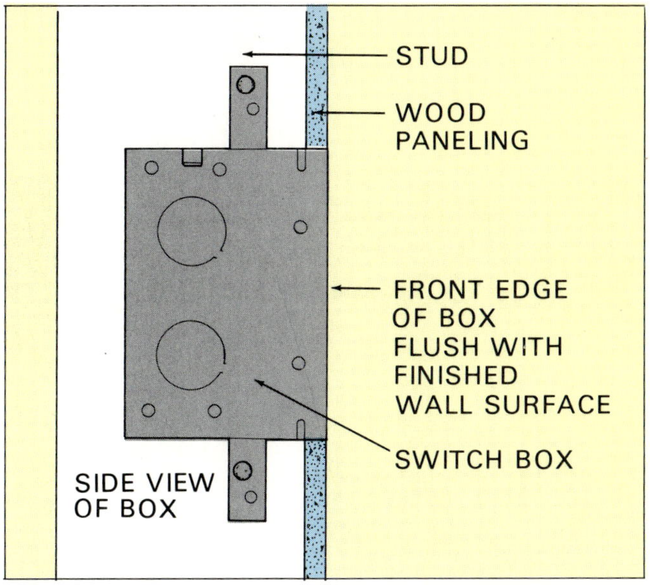

FIGURE 85. Boxes must be mounted flush with the edge of finished wall surface on combustible materials.

To allow space for the two conductors required for the light outlet, you must have a device box large enough for six conductors, according to the table.

The minimum size of the device box for No. 12 wire is 7.6 × 5.1 × 6.4 cm (3 × 2 × 2 ¾ inches) (Table IV). For No. 14 wire, the minimum size device box is 7.6 × 5.1 × 5.7 cm (3 × 2 × 2 ½ inches).

Using nonmetallic boxes will allow more wires to be installed than in a metal box of the same size because the clamps may be omitted.

Where nonmetallic sheathed cable is installed in **single-gang nonmetallic boxes,** the Code does not require the cable to be clamped to the box, provided the cable is supported within 20 cm (8 inches) of the box and where the sheath extends into the box at least 6mm (¼ inch) (Figure 84). In all other instances, cable must be fastened to nonmetallic boxes.

3. INSTALLING SWITCH BOXES AND HANDY BOXES

The Code requires that switch boxes and other outlet boxes be mounted flush with the edge of the wall surface on combustible materials such as wood paneling (Figure 85). On plaster, concrete or gypsum wallboard, the box may be recessed not more than 6mm (¼ inch) in back of the finished surface. Many boxes are manufactured with depth gage marks on the side. The marks save time by eliminating the need for measuring at each mounting. The lines on some boxes show the depth of each mark, for example, 1, 1.3, and 1.6 cm (⅜, ½, and ⅝ inches) (Figure 86). Others have the lines without the figures.

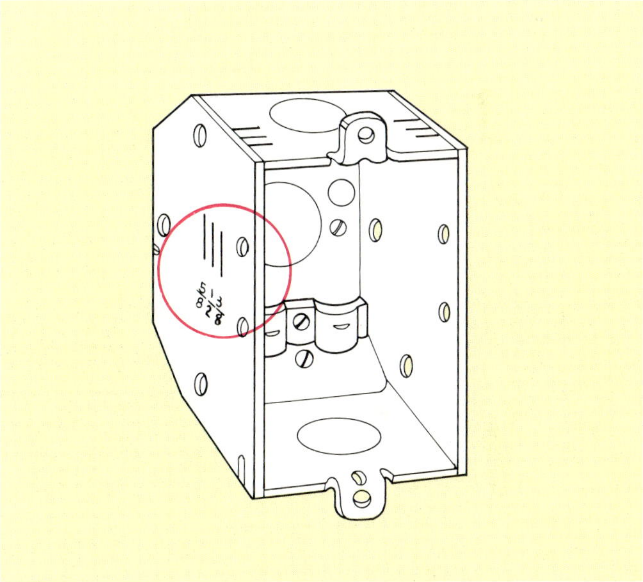

FIGURE 86. Box with depth gage marks on the side, eliminating the need for measuring at each mounting.

Installing switch boxes and handy boxes is discussed under the following headings:

a. Mounting Switch Boxes on Studs.
b. Mounting Switch Boxes in Old Buildings.
c. Installing Switch Boxes Between Studs.
d. Installing Handy Boxes.

a. Mounting Switch Boxes on Studs

The most common method of installing a switch box is to nail it to the stud. Keep in mind that switches and receptacles are both commonly mounted in switch boxes.

Proceed as follows to mount switch boxes on studs:

1. *Remove knockouts or pryouts.*

   It is best to remove **knockouts** and **pryouts** before mounting the box. Remove only those needed. Any knockout and pryout openings not used must be covered with knockout seals. Determine where the openings are needed and loosen the knockout plug with a sharp blow. Remove with pliers. To remove pryouts, insert a screwdriver in the slot, twist and pry.

2. *Determine mounting depth.*

   (1) *Check depth gage markings on side of box and select line that corresponds to needed depth.*

   For example, if wallboard or plaster is 1.3 cm (½ inch) thick, align the box on that mark (Figure 87).

   (2) *If the box has a side-mounted bracket, be sure to buy boxes with brackets that are backset to the correct depth for flush mounting with finished wall surface (Figure 88).*

   (3) *If no marks are present, measure the depth with a ruler or gage.*
   Mark the line (Figure 89).

3. *Nail box to stud.*

   (1) *Drive two nails through side holes in the box (Figure 90) or through bracket top and bottom (Figure 91).*

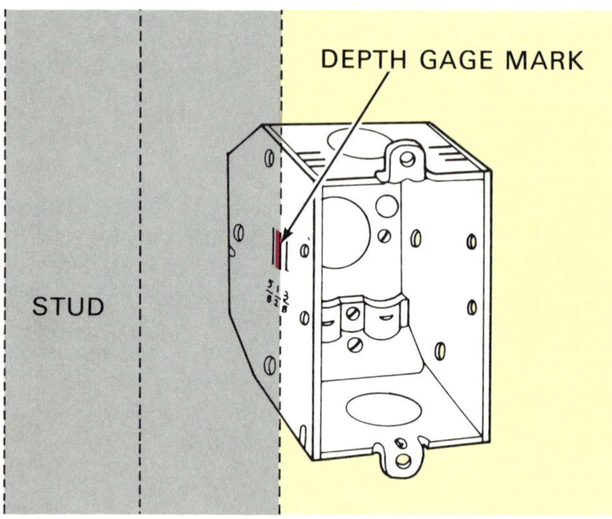

FIGURE 87. Marker on the box aligned with front edge of the stud.

FIGURE 89. Marking mounting depth for wallboard thickness.

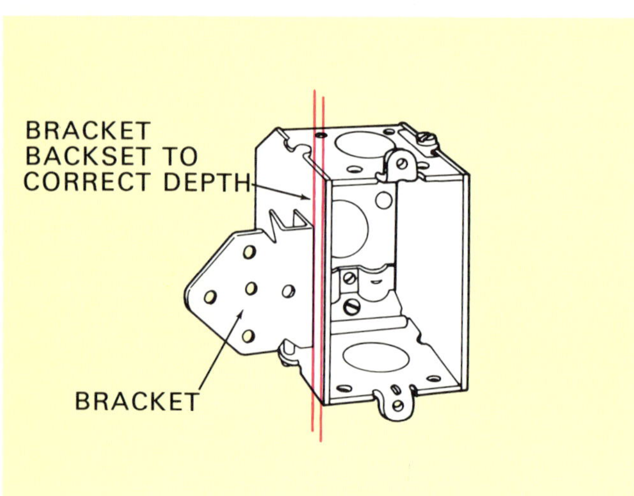

FIGURE 88. Box with side-mounted brackets backset for wallboard thickness.

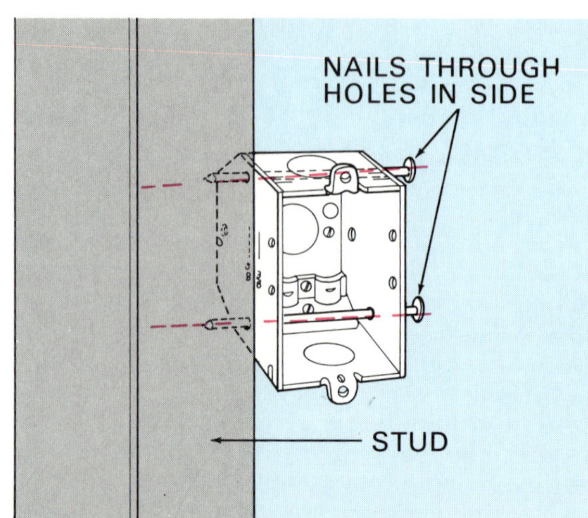

FIGURE 90. Nails driven through holes on sides of box.

No. 10d galvanized nails are preferred by many. The galvanized coating gives them added holding strength.

(2) *To mount side-bracket box, hold bracket against stud and tap bracket points into place.*

Drive nails into holes in bracket (Figure 92).

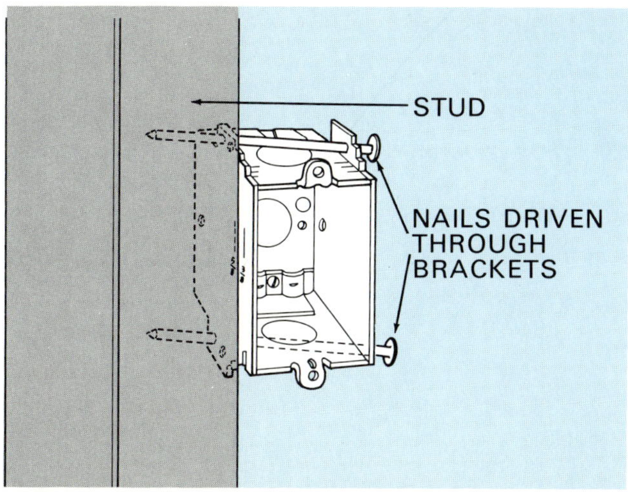

FIGURE 91. Nails driven through brackets on top and bottom of box.

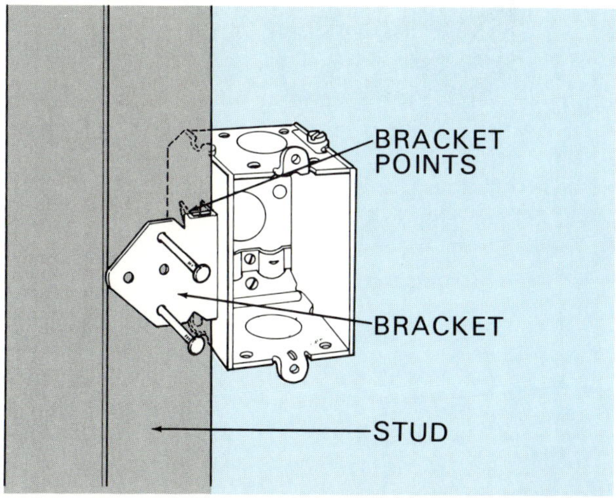

FIGURE 92. Side-bracket box mounted by tapping bracket points into stud, then driving nail through the bracket holes.

b. Installing Boxes in Old Buildings

To mount a new box in an old building, a special type of box or supports may be required. Proceed as follows:

1. *Use a template or outline of your box to mark lines to be cut for opening in the wall.*
2. *Drill holes at opposite corners to start the cut.*
3. *Finish cut with a keyhole saw or other type of saw (Figure 93A).*
4. *Insert a special box with cable installed into the opening.*

For a box with side-bracket supports, tighten the screws to bring the brackets up snug against wall (Figure 93B). Several other types of supports are available. Follow directions included with your box to install.

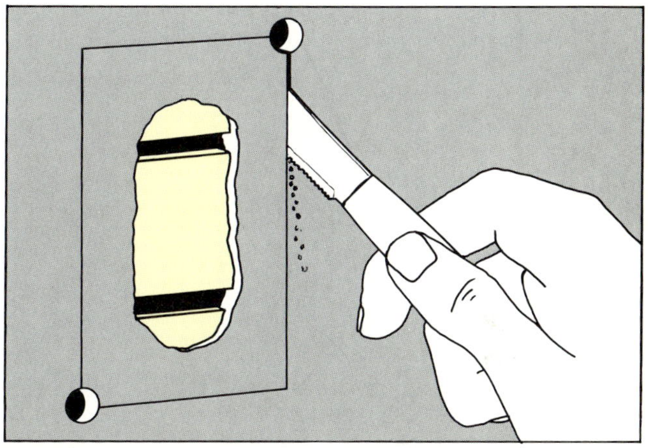

FIGURE 93A. Keyhole saw or hacksaw blade will cut hole for box.

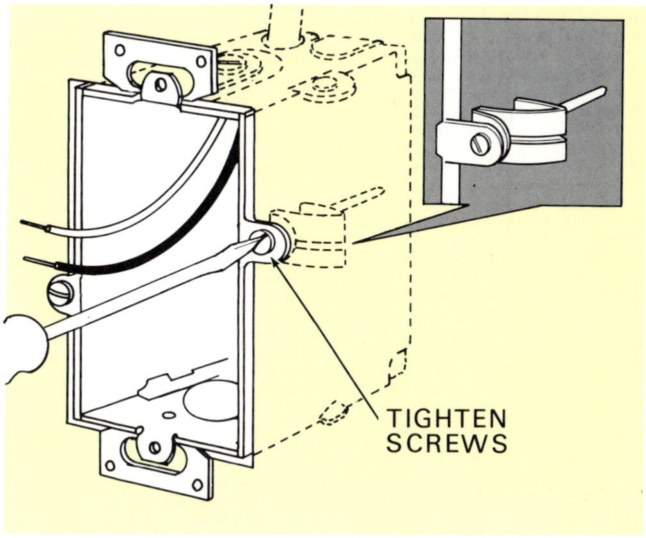

FIGURE 93B. Tighten screw on each side of box to pull the brackets up tight against the inside wall.

c. Installing Switch Boxes Between Studs

You will sometimes find that you need to mount a switch box or gang boxes at a point that falls between two studs. When that happens, you must mount a support between the studs (Figure 94). It may be a steel support strip with mounting holes punched or two wooden strips at least 2.5 cm (1 inch thick) (Figure 95). Proceed as follows:

1. *Mark location for each strip.*
   Use the switch box as a spacer.
2. *Nail strips in position between studs.*
3. *Locate one or more boxes at the desired point on the support.*
4. *Fasten box to strip with metal screws or wood screws.*

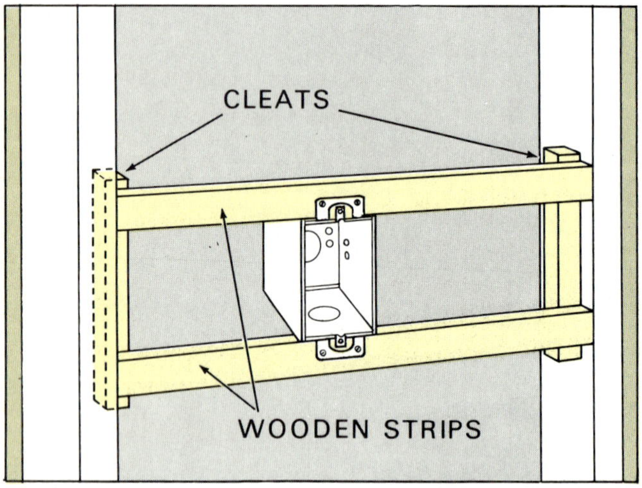

FIGURE 94. Box mounted between studs.

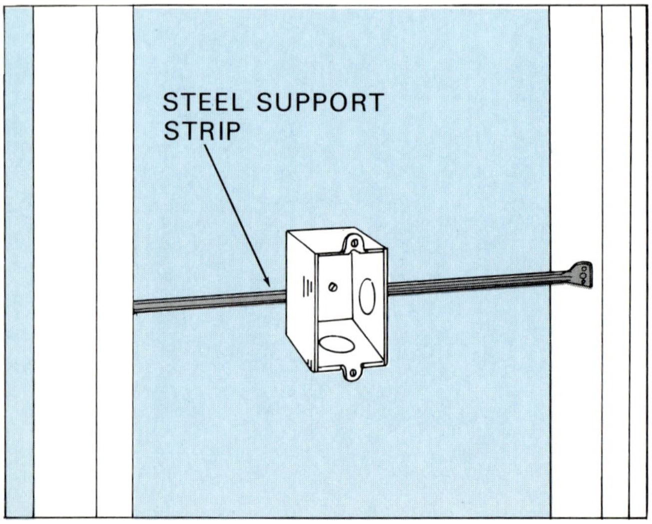

FIGURE 95. Box mounted between studs on steel support strip.

d. Installing Handy Boxes

Handy boxes are surface mounted in such places as garages, basements, utility buildings, and agricultural buildings. Proceed as follows:

1. *Remove knockouts needed.*
2. *Place handy box in position as marked.*
3. *Nail two or more 2.5 cm (1 inch) nails through holes located in bottom of box or use screws for mounting (Figure 96).*
4. *For concrete blocks, drill holes where required and place inserts for screws (Figure 97).*
   Mount box and fasten with screws.

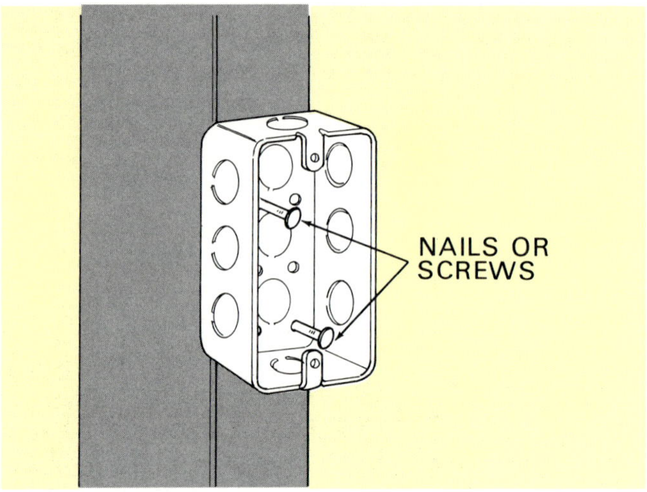

FIGURE 96. Handy box nailed directly to face of stud.

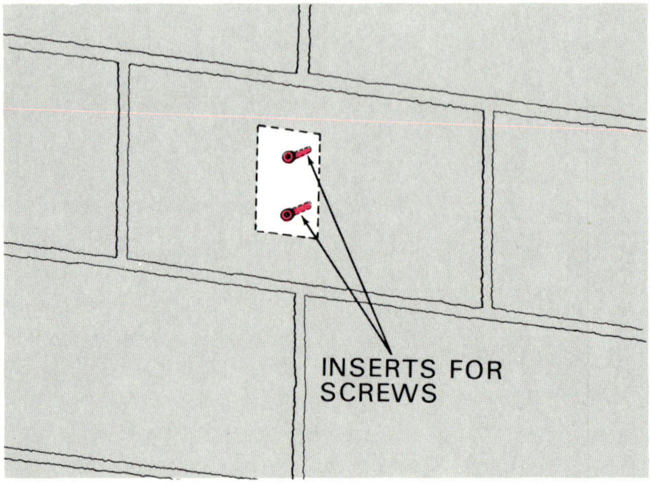

FIGURE 97. Holes drilled in concrete blocks and inserts placed for screws.

4. INSTALLING OUTLET BOXES

You now have procedures for installing a switch box or handy box at every point in the house where a switch or receptacle is to be installed. However, one 120-volt receptacle in the standard house plan must have a larger box to accommodate more wiring. Refer to Figure 52. Note the outlet indicated by the arrow as the starting point for circuit No. 5. The cable enters the box from the SEP. Wiring must then be installed for a run in two directions. One run is to the receptacle near the dining room door. Another run is to the rest of the circuit. A switch box would not provide enough space for the required wiring. A square outlet box, size 10.2 × 10.2 × 5.4 cm (4 × 4 × 2⅛ inches) or similar size should be installed to allow room for the wiring required and for the duplex receptacle. Attach special-made cover for mounting the receptacle (Figure 98).

Outlet boxes may be **octagonal** in shape or square. Procedures for installation are the same for both types. The octagonal box is the usual choice for overhead lighting outlets. The square boxes are used where more space is needed. Both types are available with brackets similar to those used for switch boxes. Others are mounted on bar hangers.

Installing outlet boxes is discussed under the following headings:

a. Installing Outlet Boxes with Mounting Brackets.
b. Installing Outlet Boxes with Bar Hangers.

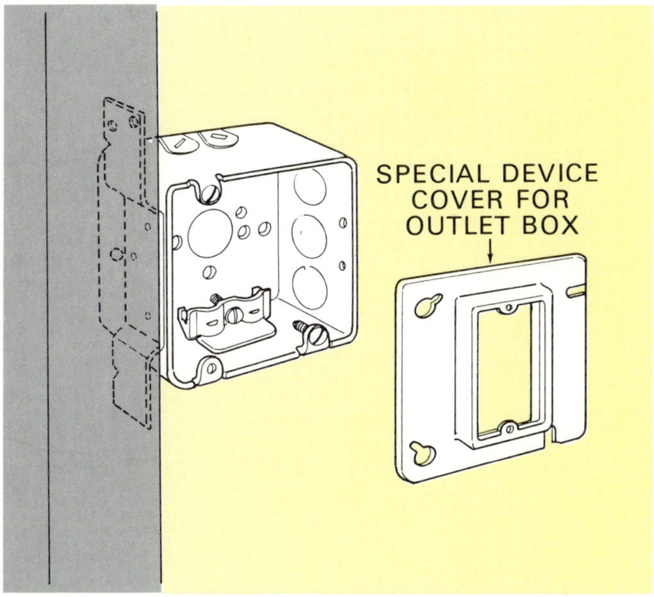

FIGURE 98. Special cover for receptacle mounting in large box.

a. Installing Outlet Boxes with Mounting Brackets

Proceed as follows:

1. *Remove knockouts as needed.*
2. *Line up bracket with stud.*

   Use brackets that are set at the correct depth for the wall covering.
3. *Tap in the bracket point on the bracket to hold the box in position.*
4. *Nail the bracket to the stud using two or more nails (Figure 99).*

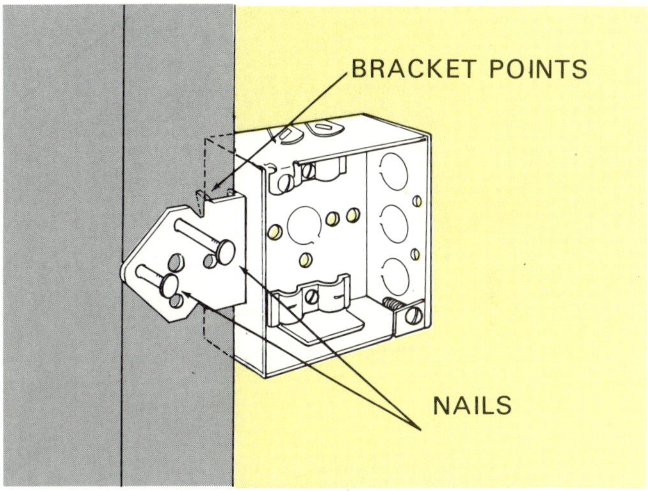

FIGURE 99. Bracket of outlet box nailed to the stud.

b. Installing Outlet Boxes with Bar Hangers

When used for mounting overhead fixtures, the octagonal box may be installed several different ways. Most often it will be mounted with a metal bar hanger attached between the joists. The bar and hanger may be **bought separately** and assembled on the job or you may buy it assembled and ready to hang. The bar hanger may be (1) **adjustable,** or (2) **solid.**

(1) ADJUSTABLE BAR HANGERS. Proceed as follows to attach and install the **adjustable bar hangers,** if not assembled.

1. *Remove center knockout from box (Figure 100).*
2. *Remove clamping screw and fitting from bar (Figure 101).*
3. *Place box on clamp, insert screw and fitting (Figure 102).*
4. *Tighten screw loosely against fitting (Figure 103).*

   The box must move freely to adjust the position after hanging.
5. *Mount hanger bar between joists.*

   Adjust to desired depth of recess. Adjust bar to fit between joists and nail at each end (Figure 104).

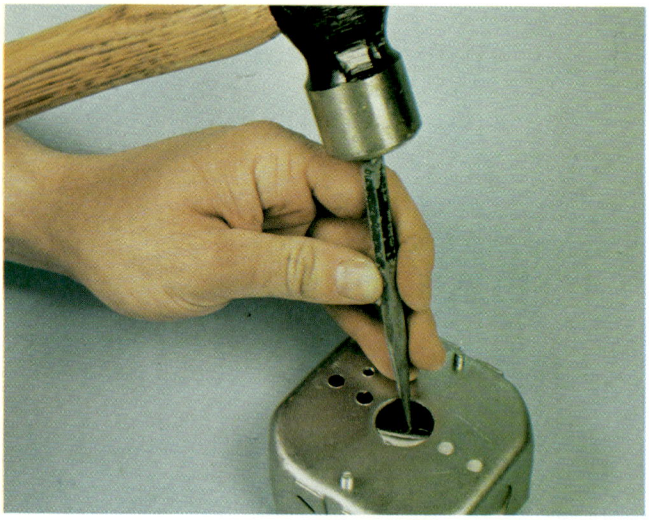

FIGURE 100. Removing center knockout from box.

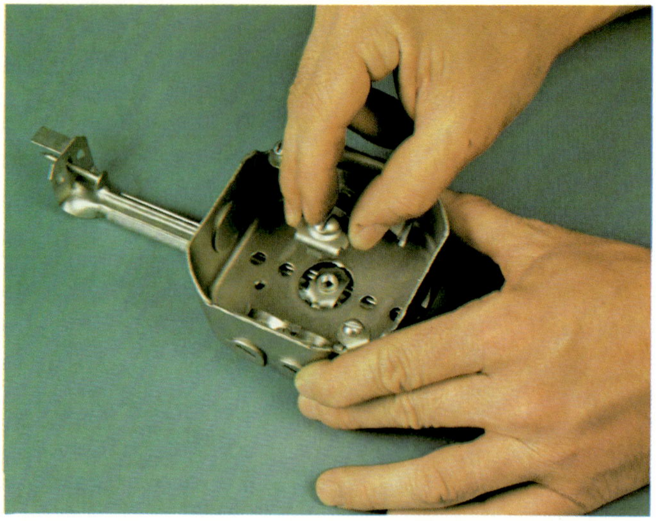

FIGURE 102. Box placed on the clamp and screw and fitting inserted.

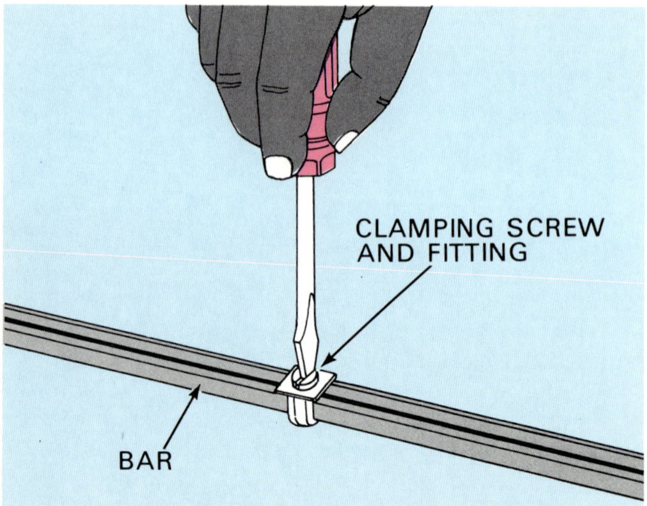

FIGURE 101. Removing clamping screw and fitting from bar.

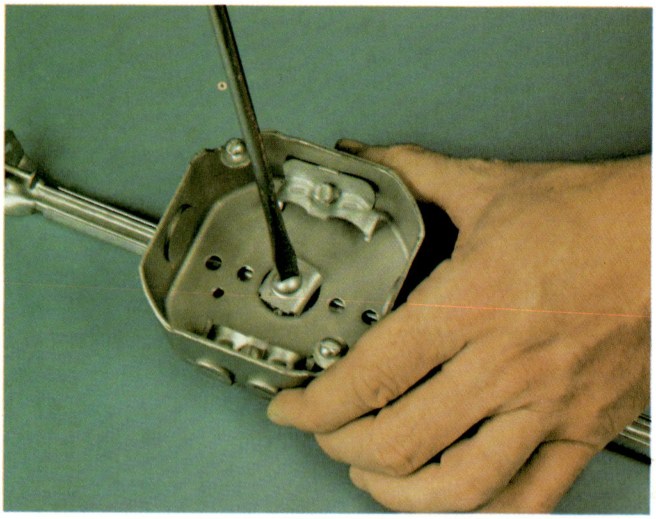

FIGURE 103. Screw tightened loosely to allow adjustment after hanging.

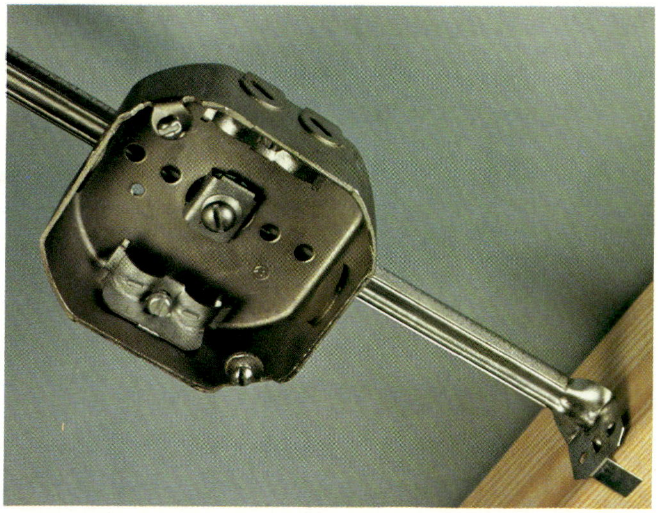

FIGURE 104. Hanger bar placed in position and nailed at each end between joists.

6. *Move box to desired position on bar and tighten screw firmly against fitting.*

   If adjustable bar and hanger are **already assembled,** follow only steps 5 and 6 above.

   Several types of pre-assembled and pre-wired **recessed ceiling fixtures** are available. The mounting procedure for adjustable hanger bars on these units is the same as for octagon box hangers except that there are two hanger bars to adjust (Figure 105). Procedures for steps 5 and 6 will vary depending on how the fixture is fastened to the hanger bar. Tighten screws at side of each bar.

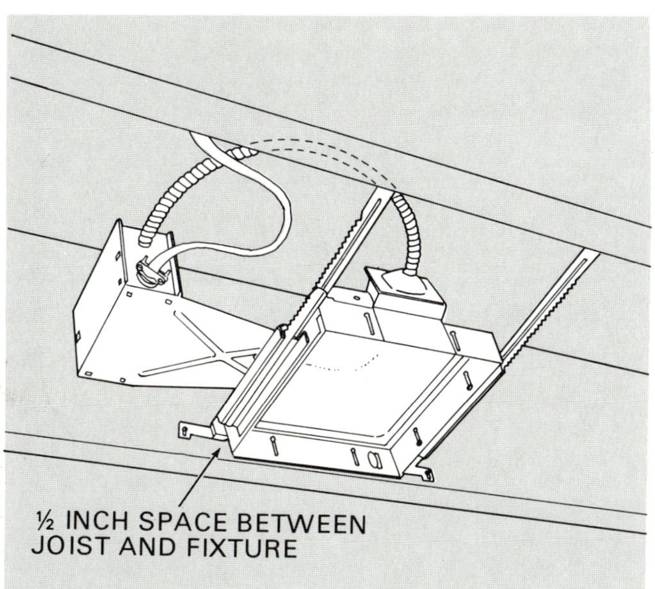

FIGURE 105. Hanger bar adjusted on each side of preassembled, recessed ceiling fixture.

(2) SOLID BAR HANGERS may be straight or offset (Figure 106). They may be ordered in overall lengths of 46, 61 and 76 cm (18, 24 and 30 inches). Assembly procedures are the same as for adjustable bars if not assembled. Proceed as follows:

1. *Follow steps (1) through (4) as for unassembled adjustable bar hangers.*
2. *To hang the bar, either straight or offset, place the bar in position across the joist edges and nail each end.*

   Cut notches if needed.
3. *Move the box to the desired position and tighten the screw against the fitting (Figure 107).*
4. *To hang pre-assembled fixture, place fixture in position between joists and nail both bars at each end (Figure 108).*

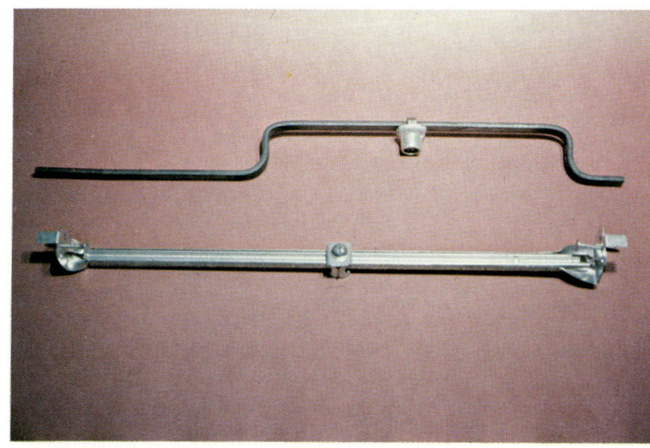

FIGURE 106. Straight and offset solid bar hangers.

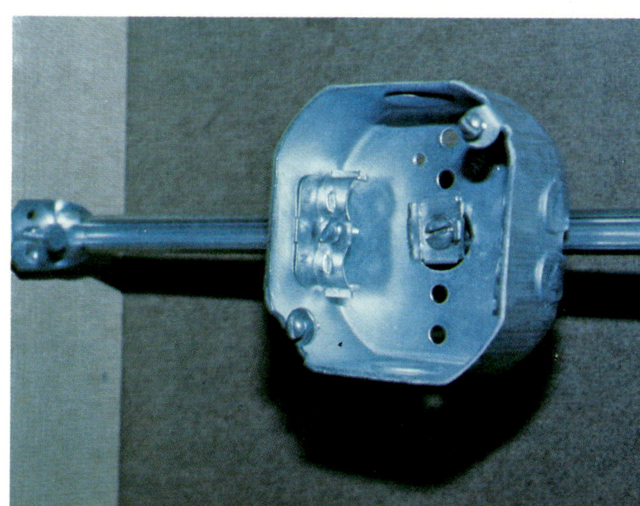

FIGURE 107. Box placed at desired position on the bar and screw tightened.

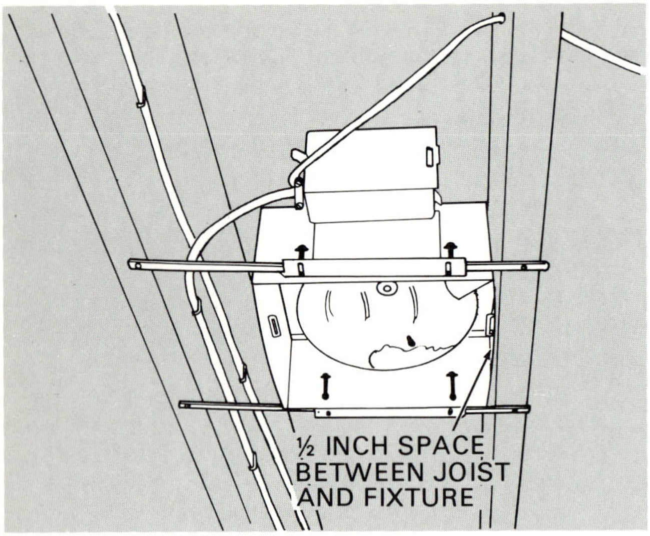

FIGURE 108. Preassembled fixture in position.

E. Providing Access for Nonmetallic Sheathed Cable

Most residences are wired with nonmetallic sheathed cable. It is one or more wires wrapped in insulation.

Providing access is another way of saying that you must provide a path for a circuit of nonmetallic sheathed cable to follow. From the SEP to the outlets, cable must be protected in some way from possible damage after it is installed.

Cable protection may be provided by drilling holes or cutting notches in studs, joists and rafters. In attics and unfinished basements, cable may be placed on running boards or between guard strips.

Providing access for nonmetallic sheathed cable is discussed as follows:

1. Drilling Holes in Framework.
2. Cutting Notches in Framework.
3. Installing Guard Strips.
4. Installing Running Boards.

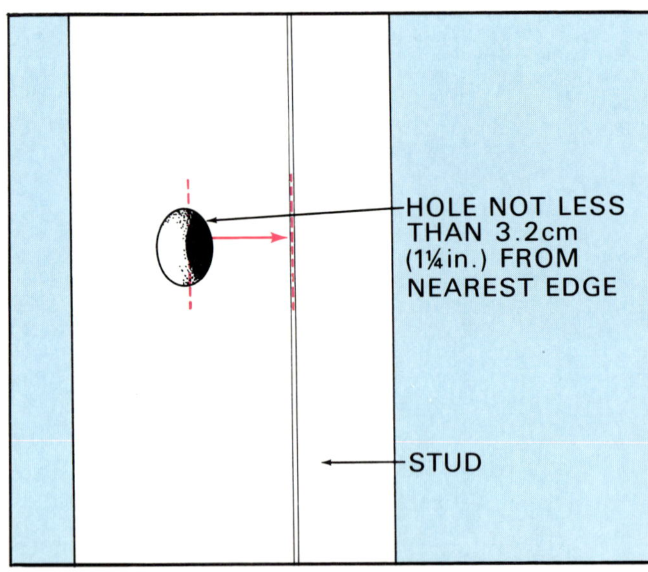

FIGURE 109. Hole in stud for supporting cable.

1. DRILLING HOLES IN FRAMEWORK

Holes drilled for cable installation must be placed at the approximate center of the framing member. The edge of holes in studs must be not less than 3.2 cm (1 ¼ inches) from the nearest edge of the stud (Figure 109). If less than 3.2 cm (1¼ inches), the cable must be protected from nails by either a bushing or a steel plate at least 1.6 mm (1/16-inch) thick (Figure 110). Holes in the center are at neutral axis and do not weaken the structure in horizontal framing members.

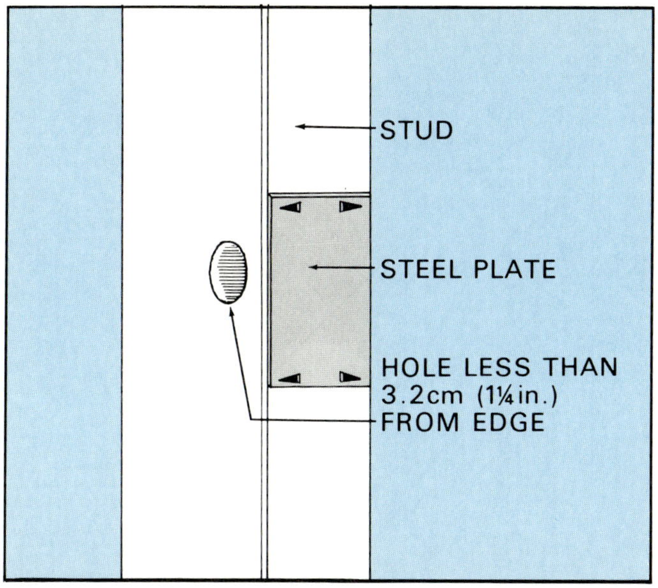

FIGURE 110. Steel plate for protecting cable from nails if hole less than 3.2 cm (1¼ inches) from nearest edge of stud.

Having the right equipment for the job makes the drilling go faster. The hand operated brace and bit has largely been replaced by power drills for drilling holes in framing. A power drill equipped with a boring bit cuts holes fast. A 1.6 cm (⅝-inch)-bit will fit most requirements for cable openings in houses. The flat bits are available in sizes from 6 mm (¼ inch) up (Figure 111).

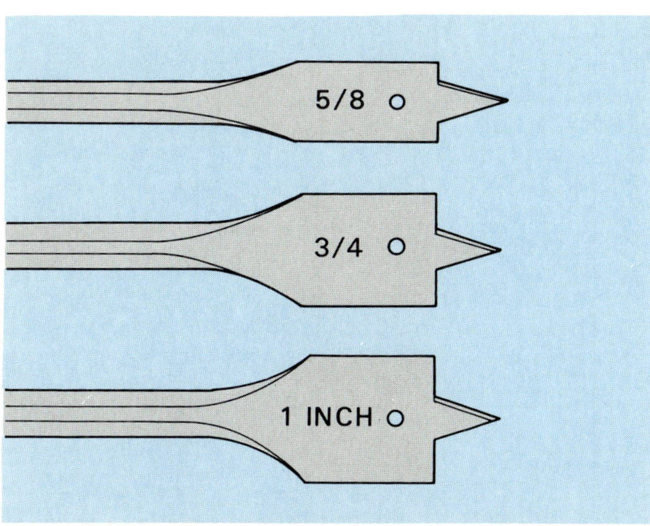

FIGURE 111. Bits for cutting holes in framing.

With switch boxes, receptacle boxes and outlet boxes in place, you may drill holes between them to provide access for the cable. Proceed as follows:

1. *Check circuit routes to see where drilling is required.*

   Refer to Figure 41 for small appliance circuit No. 1. This circuit has 6 outlets. Figure 42 shows how the circuit is routed and, therefore, where holes are needed in studs above the kitchen counter. Figure 43 shows a continuation of the same circuit with holes drilled in a dining room wall.

2. *Drill holes for small appliance circuits.*

   Locate as shown in Figure 42. Line up the circuit route by sight. Drill a hole in the center of each stud, or at least 3.2 cm (1¼ inches) from the nearest edge. Keep all holes in line between the boxes. Holes that are out of line are harder to pull cable through. They also require more wire. Where it is necessary to change elevation of holes, make gradual changes without sharp turns (Figure 112).

FIGURE 112. Holes drilled for changes in elevation to avoid sharp turns.

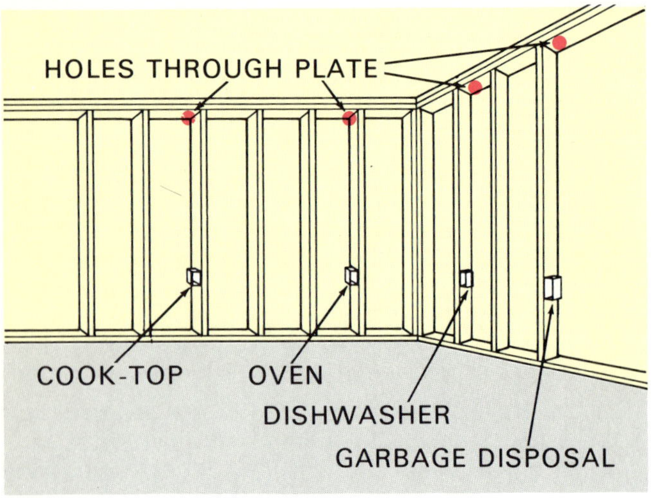

FIGURE 113. Holes drilled in top plate for individual overhead circuits to reach outlets on studs.

3. *Drill a hole in the plate for each individual kitchen circuit directly above the outlet box (Figure 113).*

   Note the location of the four individual kitchen circuits, the garbage disposer, dishwasher, oven and cooking top. The oven and cooking top will each have an individual circuit that runs overhead from the SEP. The dishwasher and garbage disposer may be combined on one circuit if desired. If so, drill holes between the outlets for the connecting cable.

4. *Drill required holes for switches.*

   Check the location of each switch outlet in the kitchen and dining room and drill a hole in the plate directly above each one for overhead wiring (Figure 114).

5. *Drill holes for general purpose circuits (Figure 44).*

   Follow procedures as for small appliance circuits.

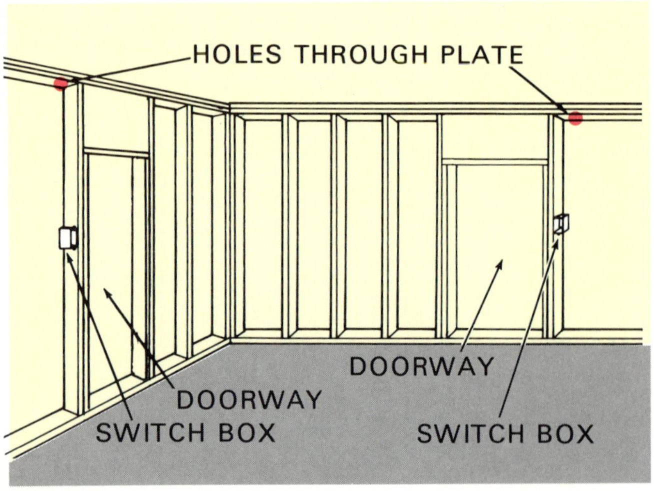

FIGURE 114. Holes drilled in top plate for switches.

2. CUTTING NOTCHES IN FRAMEWORK

The Code permits the installation of cable in notches cut into the framing if it can be done without weakening the building structure. However, you must protect the cable from screws and nails with a steel plate at least 1.6 mm (1/16-inch) thick at those points (Figure 115).

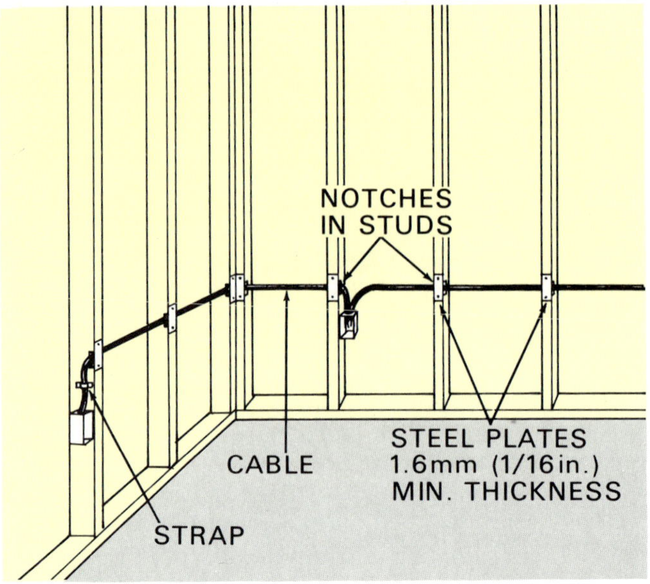

FIGURE 115. Notches in studs for cable route covered by steel plates.

Recent developments in housing design may bring about greater use of notching or drilling as a means of improving the efficiency of insulation. Many house plans designed for maximum energy conservation now call for 5.1 × 15.2 cm (2″ × 6″) studs in the outside walls. The wider studs permit installation of six inches of insulation. Wiring installed in holes through the center of studs would interfere with installing batt or roll-type insulation and with its efficiency after installation. **Notching or drilling such studs for cable at the bottom** would eliminate much of the problem (Figure 116). A similar problem is present with 5.1 × 10.2 cm (2″ × 4″) studs when insulation is installed. A portable saw or drill should be used for speedier notching or drilling.

Two methods for notching or drilling are given as follows:

a. Before Stud Installation.
b. After Stud Installation.

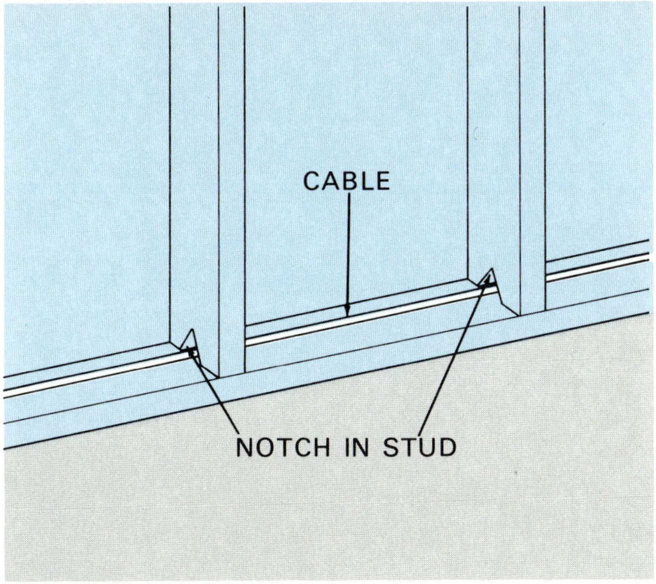

FIGURE 116. Studs notched at the bottom for cable installation permit more efficient use of insulation between studs.

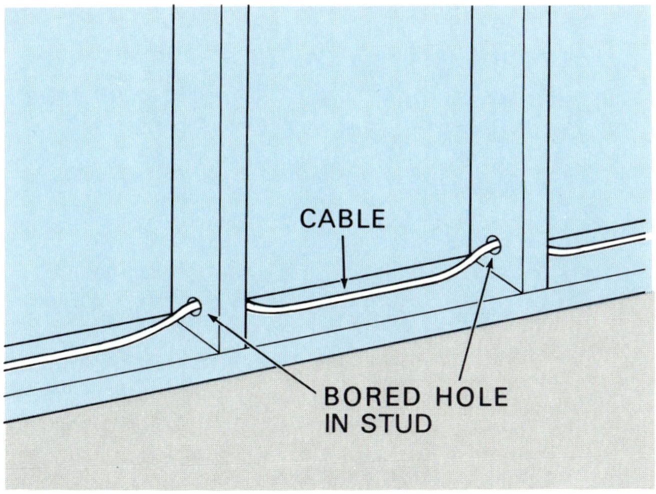

FIGURE 117. Holes drilled at bottom of studs.

a. Before Stud Installation

Proceed as follows:

1. *Before erecting each stud, cut a V-shaped notch or drill a hole in one end.*

   The mouth of the cut can be 2.5 cm (1 inch) wide on 5.1 × 10.2 cm (2″ × 4″) studs and much wider on 5.1 × 15.2 cm (2″ × 6″) studs. Taper the cut to the center of the stud (Figure 116).

   For drilling, a 5⁄8-inch hole is satisfactory (Figure 117).

2. *After studs are erected, insert the cable through the opening.*

   The cable rests on the sole plate out of the way of insulation.

b. After Stud Installation

Proceed as follows to notch or drill installed studs:

1. *Line up the notches by sight.*
2. *Cut notch as wide and deep as required for a loose fit of cable.*
3. *Use a wood chisel to remove the chip (Figure 118).*
4. *Repeat for all remaining cuts.*
5. *Install an individual steel plate 1⁄16-inch thick at each cut (Figure 119).*
6. *If using a drill, make the hole as near the bottom of the stud as possible.*

FIGURE 118. Removing the chip with a wood chisel.

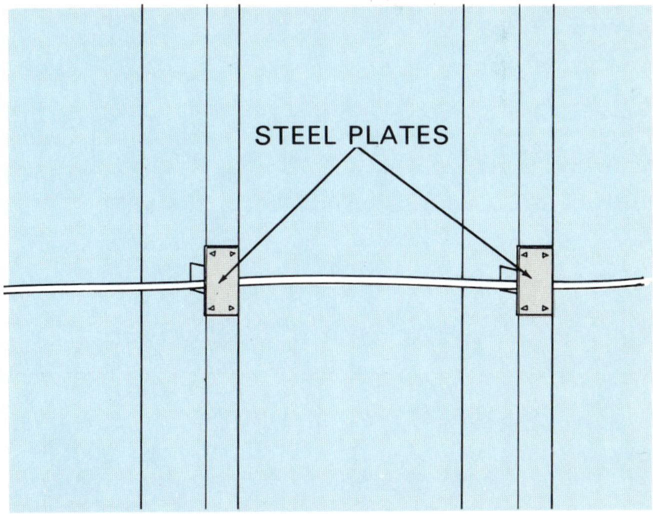

FIGURE 119. Steel plates installed over the notches.

3. INSTALLING GUARD STRIPS

If an attic is not easily accessible and not floored, cable must be protected only within 1.8 m (six feet) of the attic entrance. In other parts of the attic, cable may be run across the top of the ceiling joists. If the attic is easily accessible by stairs or permanent ladder, a cable run across the face of floor joists, rafters or studding must be protected either by running it through holes drilled in ceiling joists or by providing guard strips that are at least as high as the cable (Figure 120).

Proceed as follows to install guard strips:

1. *Determine route to be followed by cable run.*
2. *Nail strips approximately 1.3 cm (½-inch) apart for the length of the run.*
3. *The cable is installed between strips and fastened to framing.*

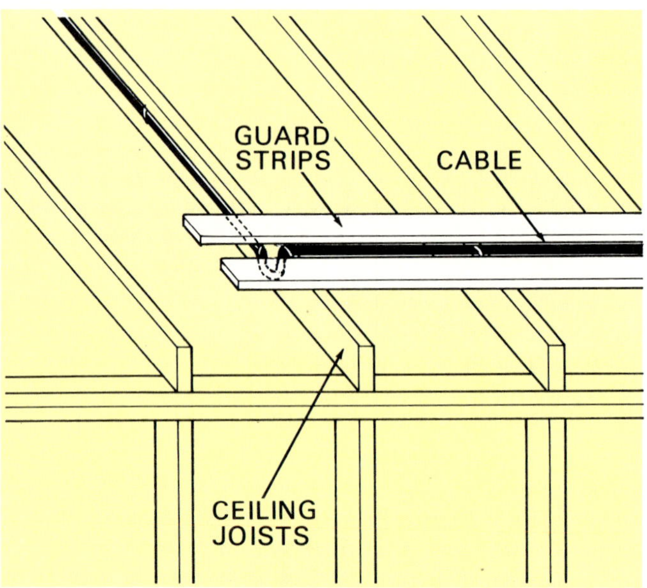

FIGURE 120. Guard strips installed in the attic.

♣4. INSTALLING RUNNING BOARDS

When cable is run across joists in unfinished basements, you are permitted to fasten cables of two No. 6 or three No. 8 conductors or larger, directly to the lower edges of the joists (Figure 121). Smaller cables must be run either through holes drilled in joists or on running boards (Figure 122). A running board may be 2.5 × 10.2 cm (1″ × 4″) or other size board, placed end to end over the cable run. Cable is stapled to the board.

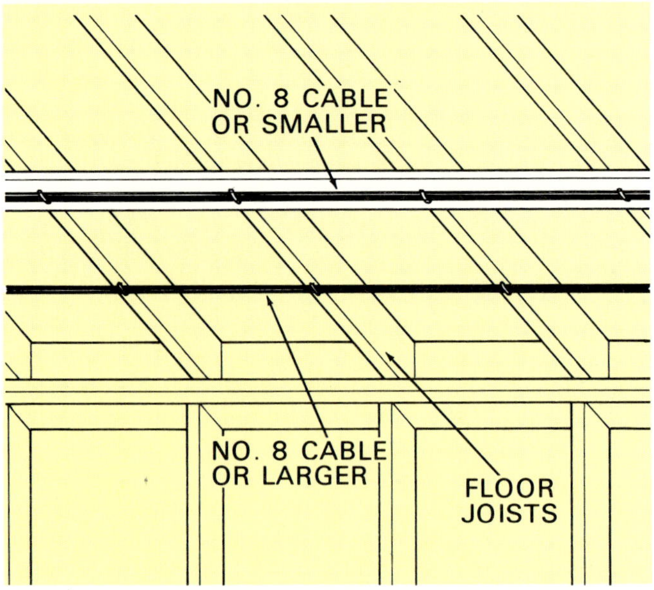

♣ **FIGURE 121. In unfinished basements, cable size No. 8 or larger may be fastened directly to lower edge of the joists. Smaller cable must be fastened to running boards.**

Proceed as follows to install running boards.

1. *Determine route to be followed by the cable.*
2. *Nail running board in place (Figure 121).*

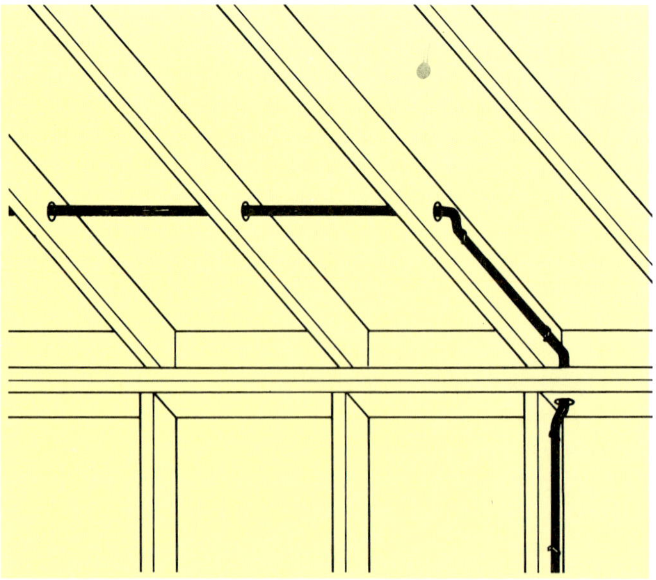

FIGURE 122. Cable run through holes in joists.

F. Pulling Cable for Circuits

Your next step is to pull the cable for each circuit. Pulling cable for circuits is discussed under the following headings:

1. Types and Characteristics of Cables.
2. Removing Cable from the Roll.
3. Pulling the Cable.

1. TYPES AND CHARACTERISTICS OF CABLES

Types and characteristics of conductors and cable are discussed as follows:

a. Types of Cable.
b. Characteristics of Cable.

a. Types of Cable

Types of cable are discussed under the following headings:

(1) Types of Wire.
(2) Characteristics of Wire and Insulation.
(3) Characteristics of Cable.
(4) Nonmetallic Sheathed Cable.
(5) Cable Identification.
(6) Cable Ampacity.

(1) TYPES OF WIRE are discussed as follows:

—Copper Wire.
—Aluminum Wire.

More **copper wire** is used in house wiring than any other type (Figure 123). It is a better conductor of electrical current than **aluminum wire** and gives fewer problems after it is installed. But in larger wire sizes, copper is considerably more expensive than aluminum or copper-clad aluminum (aluminum wire coated with copper). On installations requiring wire sizes **larger than No. 8,** such as feeder lines and service entrances, aluminum is gaining in popularity because it is less expensive.

Number 14 is the smallest size copper wire the Code permits for house wiring. However, many local codes require that nothing smaller than No. 12 copper be used.

Because aluminum is a poorer conductor of electrical current than copper, aluminum wire must be larger than copper to carry the same current load. Generally, aluminum wire must be one trade size larger than copper to be equivalent. In an installation where No. 8 copper is satisfactory, a No. 6 aluminum wire would be required to carry the current load (Figure 124). In the same manner, No. 10 aluminum could be substituted for No. 12 copper wire.

NOTE: Size increments come from ampacity tables in the Code. For example, the difference between AWG sizes 6 and 8 or between 8 and 10 is generally considered to be one size. The Code ratings are listed only for commonly manufactured sizes.

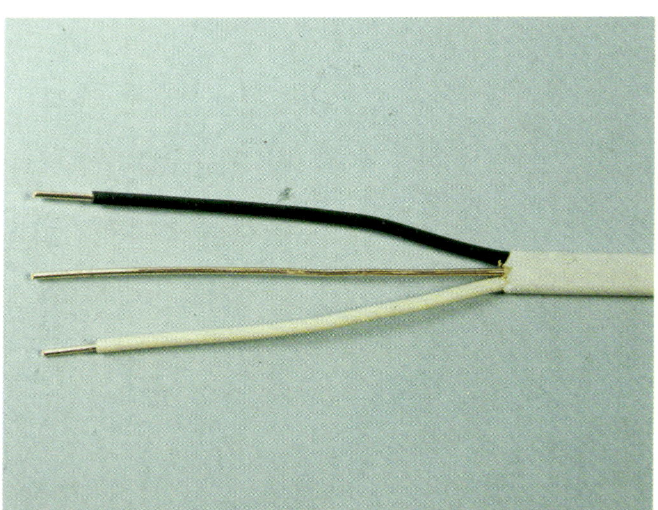

FIGURE 123. Copper wire is most often used in house wiring.

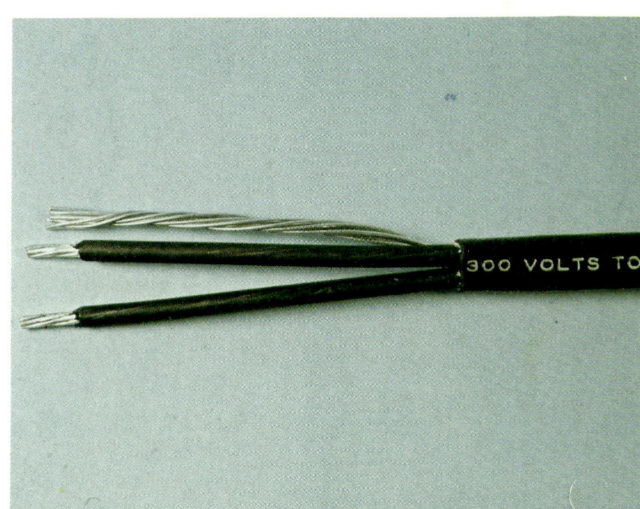

FIGURE 124. In an installation where No. 8 AWG copper wire is required, aluminum wire must be No. 6 AWG to conduct the current load.

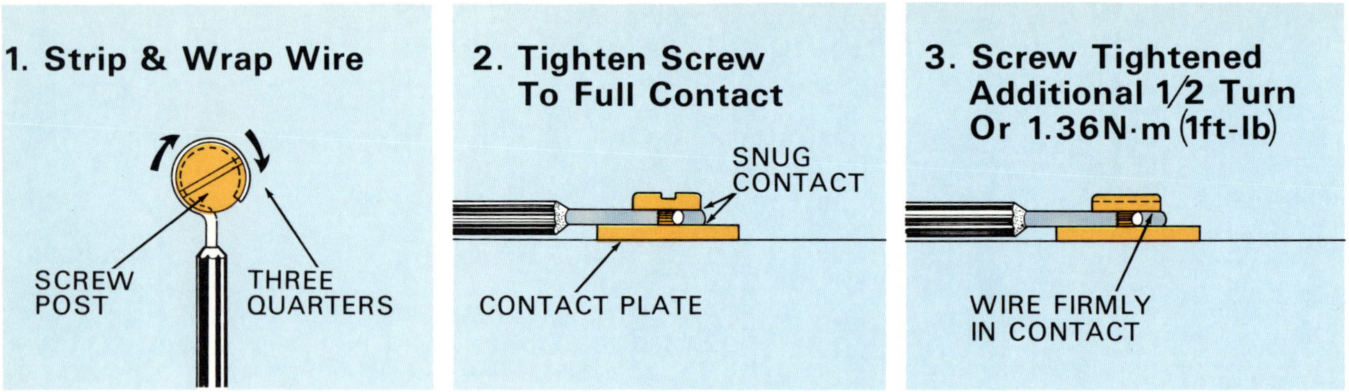

FIGURE 125. Steps required to connect wire to a terminal screw correctly.

All references to wire size refer to copper wire unless aluminum is indicated. If aluminum wire is to be used instead, remember that it must be one size larger.

Terminations of smaller sizes of **aluminum wire**, such as those used for interior wiring of homes, have caused problems in many installations in the past. One reason is that older types of switches, receptacles and other devices not made especially for use with aluminum proved unsatisfactory. Expansion and contraction at terminals caused connections to become loose and begin arcing.

Another cause of problems with copper or aluminum wiring is poor workmanship. Installers must be especially careful when making connections. A terminal screw that is not tight, or a wire loop overlapped under a terminal screw, is likely to cause problems.

Steps for properly connecting wire to a terminal screw are shown in Figure 125.

Newer **receptacles** and **switches** made for use with either copper or aluminum and rated 15- or 20-amp are marked CO/ALR. Always use devices carrying only this mark with aluminum cable (Figure 126). **Equipment** rated 30 amperes and over with terminals listed for either copper or aluminum wire is labeled AL-CU.

Copper-clad aluminum wire is aluminum wire that has a copper coating (Figure 127). It may be used with all wiring devices currently listed by UL for use with copper, including the push-in backwired type. Ampacity ratings for copper-clad aluminum are the same as for uncoated aluminum.

(2) CHARACTERISTICS OF WIRE AND INSULATION. To understand the types of cable, you must become familiar with the types of conductors (wires) and insulation used to manufacture cable. Table 310–13 of the Code contains several pages of descriptive information on conductors used for general wiring. The information in Table V is an excerpt from the Code table which describes conductors used in house wiring. It lists trade names, types of wire, maximum operating temperature, types of locations where each may be used, the insulating material of each wire type and its characteristics, and the type of outer cover (sheath).

FIGURE 126. Connect aluminum wire to devices marked CO/ALR only.

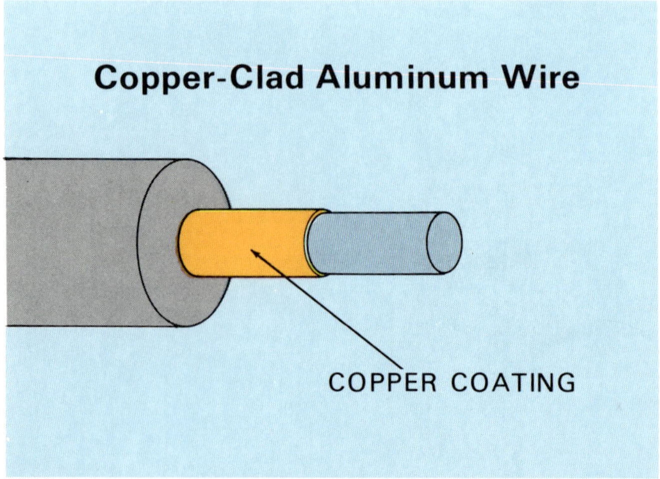

FIGURE 127. Copper-clad aluminum wire is aluminum wire that has a copper coating.

TABLE V. CONDUCTOR APPLICATION AND INSULATIONS
(Excerpts from Code Table 310–13)

Trade Name	Type Letter	Max. Operating Temp.	Application Provisions	Insulation	AWG or MCM	Thickness of Insulation Mils	Outer Covering
Moisture-Resistant Latex Rubber	RUW	60°C 140°F	Dry and wet locations.	90% Unmilled, Grainless Rubber	14-10 8-2	18 25	Moisture-resistant, flame-retardant, non-metallic covering
Thermoplastic	T	60°C 140°F	Dry locations.	Flame-Retardant, Thermoplastic Compound	14-10 8 6-2 1-4/0 213-500 501-1000 1001-2000	30 45 60 80 95 110 125	None
Moisture-Resistant Thermoplastic	TW	60°C 140°F	Dry and wet locations.	Flame-Retardant Moisture-Resistant Thermoplastic	14-10 8 6-2 1-4/0 213-500 501-1000 1001-2000	30 45 60 80 95 110 125	None
Heat-Resistant Thermoplastic	THHN	90°C 194°F	Dry locations	Flame-Retardant, Heat-Resistant Thermoplastic	14-12 10 8-6 4-2 1-4/0 250-500 501-1000	15 20 30 40 50 60 70	Nylon jacket or equivalent
Moisture- and Heat-Resistant Thermoplastic	THWN	75°C 167°F	Dry and wet locations.	Flame-Retardant, Moisture- and Heat-Resistant Thermoplastic	14-12 10 8-6 4-2 1-4/0 250-500 501-1000	15 20 30 40 50 60 70	Nylon jacket or equivalent
Moisture- and Heat-Resistant Cross-Linked Synthetic Polymer	XHHW	90°C 194°F 75°C 167°F	Dry locations. Wet locations.	Flame-Retardant Cross-Linked Synthetic Polymer	14-10 8-2 1-4/0 213-500 501-1000 1001-2000	30 45 55 65 80 95	None
Underground Feeder and Branch-Circuit Cable-Single Conductor. (For Type UF cable employing more than one conductor see Article 339.)	UF	60°C 140°F 75°C** 167°F	See Article 339	Moisture-Resistant Moisture- and Heat-Resistant	14-10 8-2 1-4/0	*60 *80 *95	Integral with insulation
Underground Service-Entrance Cable-Single Conductor. (For Type USE cable employing more than one conductor see Article 338.)	USE	75°C 167°F	See Article 338	Heat- and Moisture-Resistant	12-10 8-2 1-4/0 213-500 501-1000 1001-2000	45 60 80 95 110 125	Moisture-resistant non-metallic covering [See 338-1 (2).]

*Includes integral jacket.

**For ampacity limitation, see 339–1(a).

***Insulation thickness shall be permitted to be 80 mils for listed Type USE conductors that have been subjected to special investigations. The nonmetallic covering over individual rubber-covered conductors of aluminum-sheathed cable and of lead-sheathed or multiconductor cable shall not be required to be flame retardant. For Type MC cable, see Section 334–20. For nonmetallic-sheathed cable, see Section 336–2. For Type UF cable, see Section 339–1.

In the manufacture of nonmetallic-sheathed cable, two or more insulated conductors such as those in Table V are encased in an outer sheath to make a cable.

Some characteristics of insulation on conductors included in Table V are:

(a) Moisture Resistant Latex Rubber is for use in dry and wet locations.

(b) Thermoplastic* Type T is for use in dry locations, including most applications in house wiring. A wet basement area would be an exception.

(c) Moisture Resistant Thermoplastic* Type TW is suitable for both dry and wet locations. This is the type most commonly used for conductors in nonmetallic cable.

(d) Heat Resistant Thermoplastic* Type THHN can withstand higher temperatures but is suited only for dry locations. It is often used in conduit because the insulation is thinner. Thus more conductors can be installed in a given conduit.

(e) Thermoplastic* Types THW and THWN may both be used in dry and wet locations. Both are moisture and heat resistant thermoplastic* material.

(f) Type XHHW is a new type of cross-linked synthetic polymer material that is moisture and heat resistant.

The thickness of the insulation on each type depends on the wire size.

(g) Type UF is used for direct burial of underground feeder circuits, such as between the residence and a separate utility building or garage. It cannot be buried in concrete.

(h) Type USE is for direct burial of underground service entrance, feeder circuits, and for indoor circuits.

(3) CHARACTERISTICS OF CABLE. Table V describes characteristics, insulations and conditions for use of individual wires. Most house wiring is done with **cable** made of the individual wires. Each type of cable also has its own characteristics and conditions for use. **The type of outer covering, or sheath, and the type and number of wires contained in the cable determine where and how the cable may be used.**

A cable containing two No. 12 wires is called a 12-2 cable. A cable having three No. 12 wires is a 12-3 cable. If cable also contains a grounding wire, it is called a 12-2w/g (with ground) (Figure 128) or a 12-3w/g (with ground).

The Code requires color coding of insulated wires in a cable. They are usually colored as follows:

—**Two wires**—one black, one white** (Figure 129).

—**Three wires**—one black, one white**, one red (Figure 130).

*Thermoplastic insulation may stiffen at temperatures colder than minus 10°C (plus 14°F), requiring care be exercised during installation at such temperatures. Thermoplastic insulation may also be deformed at normal temperatures where subjected to pressure, requiring care be exercised during installation and at points of support.

**A natural gray color may be used in place of white, but white is much more common.

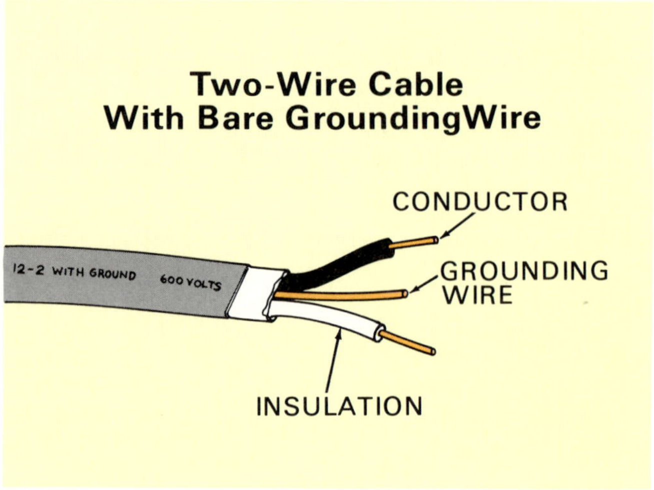

FIGURE 128. A 12-2 cable with ground wire.

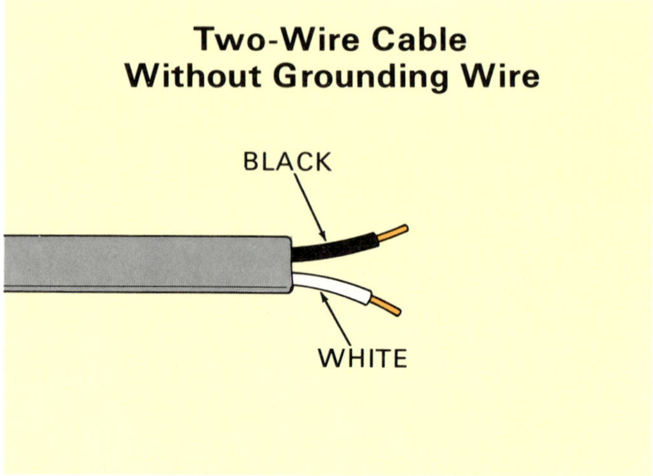

FIGURE 129. In a two-wire cable, the colors are usually one black, and one white.

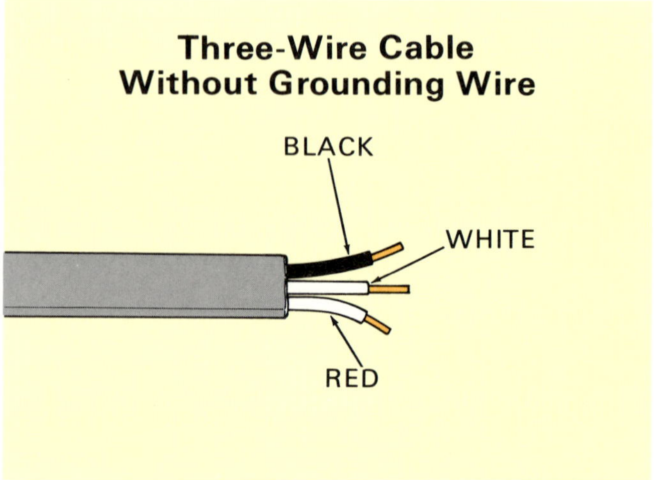

FIGURE 130. Wires in a three-wire cable are usually colored black, white and red.

If the cable contains an insulated grounding wire, it is green or green with yellow stripes (Figure 131). More often the grounding wire is bare (Figure 132).

(4) NONMETALLIC SHEATHED CABLE is often called Romex, which is the trade name of one manufacturer. The sheath on nonmetallic-sheathed cable is moisture-resistant, flame retardant, nonmetallic material. The sheath is usually made of plastic on cable used for house wiring.

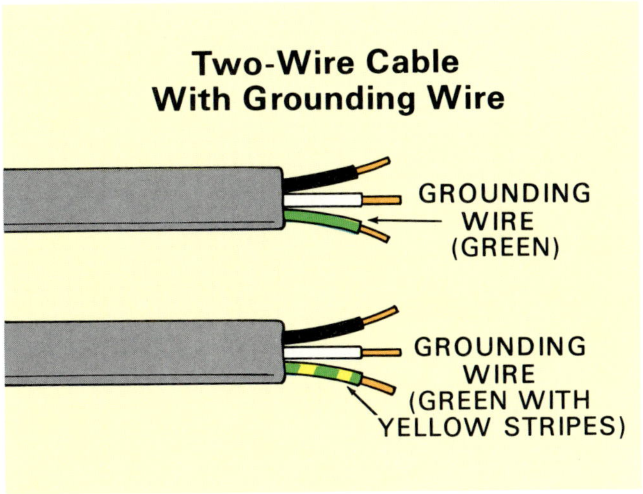

FIGURE 131. An insulated grounding wire in a cable is green or green with yellow stripes.

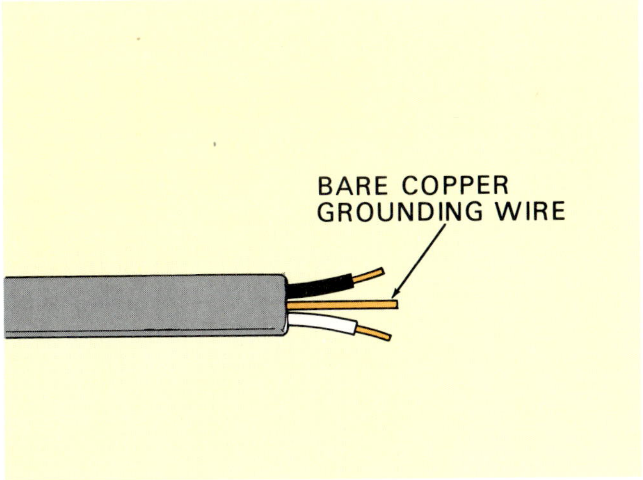

FIGURE 132. The grounding wire in non-metallic sheathed cable is usually bare.

Nonmetallic-sheathed cable is available with either two or three insulated conductors. Also, it is available with an additional grounding wire which may or may not be insulated (Figure 133). Most cable installed in a house must have a grounding wire to meet local code requirements.

Two types of nonmetallic-sheathed cable are used in house wiring. Both types are manufactured in copper in sizes 14 through 2 (Figure 134). If the conductors are aluminum, or copper-clad aluminum, the available sizes are 12 through 2. They are described as follows:

(a) Type NM.

(b) Type NMC.

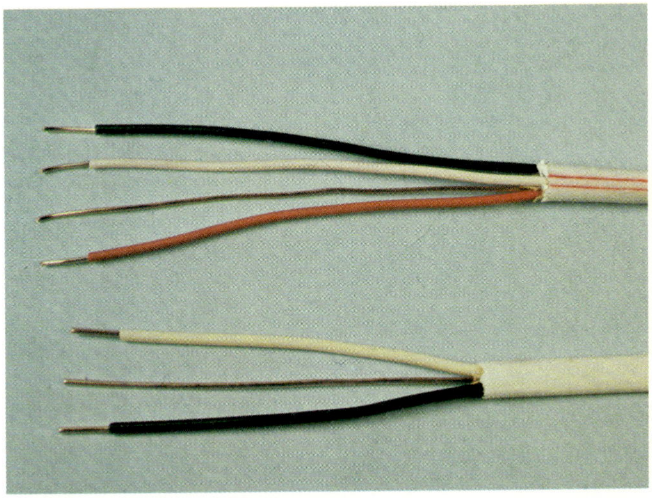

FIGURE 133. Nonmetallic sheathed cable is available with either two or three insulated conductors and with or without a bare grounding wire.

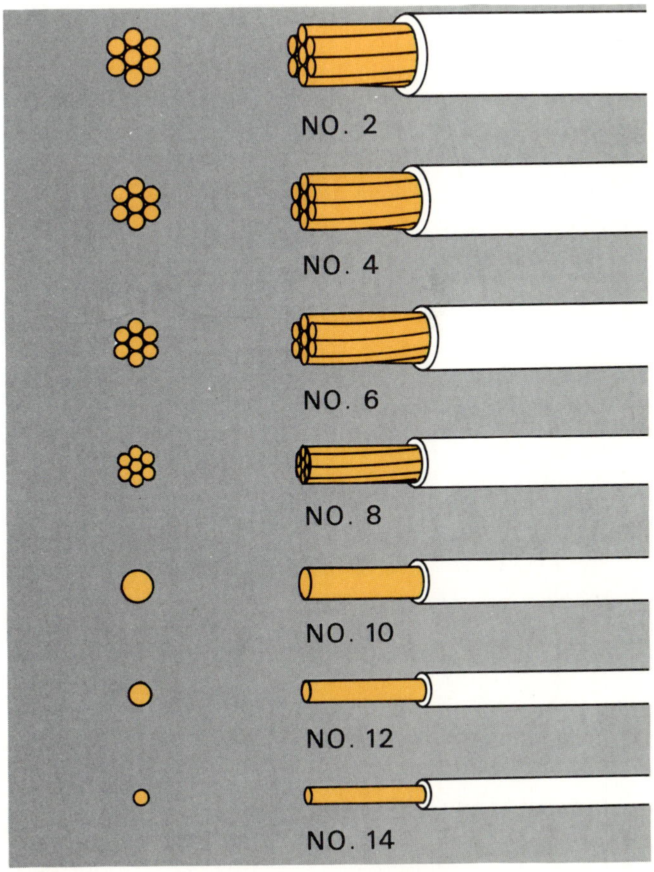

FIGURE 134. Available sizes of conductors in nonmetallic sheathed cable are 2 through 14 if copper and 2 through 12 if aluminum or copperclad aluminum.

(a) **Type NM nonmetallic-sheathed cable** is most often used for house wiring. It may be installed for both exposed and concealed work in normally dry locations. This includes air voids in masonry block or tile walls if excessive moisture is not present. Type NM may **not** be installed where it is exposed to corrosive fumes or vapors. Neither can it be imbedded in masonry, concrete or plaster.

(b) **Type NMC nonmetallic-sheathed cable,** in addition to meeting the requirements for NM, must also be fungus resistant and corrosion resistant. It may be used in both exposed and concealed work in dry, moist, damp or corrosive locations, and in outside and inside walls of masonry and block or tile. Type NMC is not readily available in many localities.

(5) CABLE IDENTIFICATION. Every cable must be marked to show the following (Figure 135):

—Letters indicating the type of wire.

—Maximum working voltage.

—Manufacturer's name, trademark or other identifying marking.

—The AWG or circular mil size. Surface markings shall appear at least every 61 cm (24 inches).

(6) CABLE AMPACITY. Table VI gives allowable ampacities of different types of cable, with not more than three insulated conductors in a cable or conduit, AWG sizes 18-2000 MCM. The figures apply to conductors in a cable or to those directly buried in the earth.

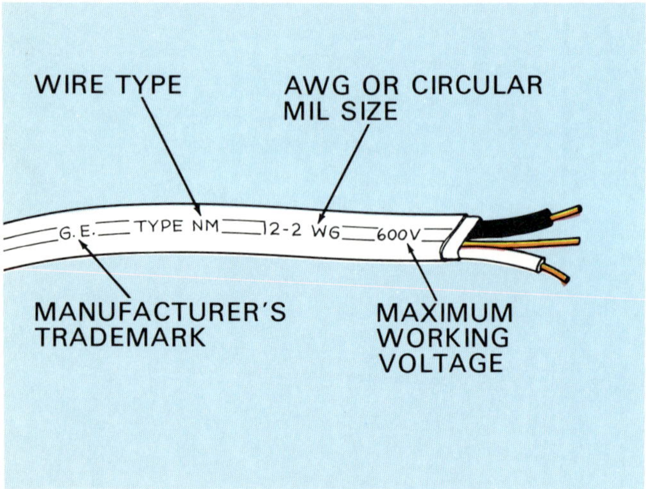

FIGURE 135. Cable identification.

2. REMOVING CABLE FROM THE ROLL

Nonmetallic-sheathed cable is brought to the job as a roll or coil inside a box or on a reel. The way it is removed from the roll can make a difference in the ease of handling.

Pulling a length of cable for a long run is usually easier if you unwind the cable from the inside of the roll. Most cartons have a circle marked to be cut out for the opening. Proceed as follows:

1. *Rotate the end of the cable against the direction of the coil as you remove it from the inside of the roll (Figure 136).*
2. *Cut the length required to reach from the SEP to the first box in the circuit run.*

   If using a reel, pull from outside of reel to the outlet and cut needed length.

   Cable may be spliced only inside a box. Be sure you allow enough wire to reach from the SEP to the first box and from one box to another in each circuit. Also, at each box allow 15 to 20 cm (6 to 8 inches) of cable for making splices and connections inside the box. Much longer lengths are needed for connections inside the SEP cabinet, 61 to 91 cm (24 to 36 inches).

NOTE: No device connections will be made at either the SEP or the boxes until all plastering, wallboard or paneling is in place. Damage to switches and receptacles is less likely if installed after plaster and/or wallboard are in place.

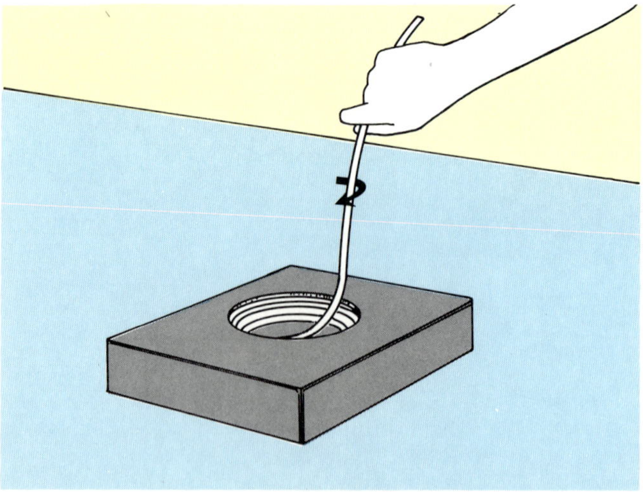

FIGURE 136. Rotate the cable against the direction of the coil as you remove it from inside the coil.

TABLE VI. ALLOWABLE AMPACITIES OF INSULATED CONDUCTORS RATED 0–2000 VOLTS, 60° TO 90°C (From Code Table 310–16)

Not More Than Three Conductors in Raceway or Cable or Earth (Directly Buried), Based on Ambient Temperature of 30°C (86°F)

Size	Temperature Rating of Conductor See Table 310–13								Size
	60°C (140°F)	75°C (167°F)	85°C (185°F)	90°C (194°F)	60°C (140°F)	75°C (167°F)	85°C (185°F)	90°C (194°F)	
AWG MCM	TYPES †RUW, †T, †TW, †UF	TYPES †FEPW, †RH, †RHW, †RUH, †THW, †THWN, †XHHW, †USE, †ZW	TYPES V, MI	TYPES TA, TBS, SA, AVB, SIS, †FEP, †FEPB, †RHH, †THHN, †XHHW*	TYPES †RUW, †T, †TW, †UF	TYPES †RH, †RHW, †RUH, †THW, †THWN, †XHHW, †USE	TYPES V, MI	TYPES TA, TBS, SA, AVB, SIS, †RHH, †THHN, †XHHW*	AWG MCM
	COPPER				ALUMINUM OR COPPER-CLAD ALUMINUM				
18	....	....	....	14	....	....	....	....	....
16	....	....	18	18	....	....	....	....	....
14	20†	20†	25	25†	....	....	....	....	....
12	25†	25†	30	30†	20†	20†	25	25†	12
10	30	35†	40	40†	25	30†	30	35†	10
8	40	50	55	55	30	40	40	45	8
6	55	65	70	75	40	50	55	60	6
4	70	85	95	95	55	65	75	75	4
3	85	100	110	110	65	75	85	85	3
2	95	115	125	130	75	90	100	100	2
1	110	130	145	150	85	100	110	115	1
0	125	150	165	170	100	120	130	135	0
00	145	175	190	195	115	135	145	150	00
000	165	200	215	225	130	155	170	175	000
0000	195	230	250	260	150	180	195	205	0000
250	215	255	275	290	170	205	220	230	250
300	240	285	310	320	190	230	250	255	300
350	260	310	340	350	210	250	270	280	350
400	280	335	365	380	225	270	295	305	400
500	320	380	415	430	260	310	335	350	500
600	355	420	460	475	285	340	370	385	600
700	385	460	500	520	310	375	405	420	700
750	400	475	515	535	320	385	420	435	750
800	410	490	535	555	330	395	430	450	800
900	435	520	565	585	355	425	465	480	900
1000	455	545	590	615	375	445	485	500	1000
1250	495	590	640	665	405	485	525	545	1250
1500	520	625	680	705	435	520	565	585	1500
1750	545	650	705	735	455	545	595	615	1750
2000	560	665	725	750	470	560	610	630	2000
	AMPACITY CORRECTION FACTORS								
Ambient Temp. °C	For ambient temperatures other than 30°C, multiply the ampacities shown above by the appropriate factor shown below.								Ambient Temp. °F
31–40	.82	.88	.90	.91	.82	.88	.90	.91	87–104
41–45	.71	.82	.85	.87	.71	.82	.85	.87	105–113
46–50	.58	.75	.80	.82	.58	.75	.80	.82	114–122
51–60	....	.58	.67	.71	....	.58	.67	.71	123–141
61–70	....	.35	.52	.58	....	.35	.52	.58	142–158
71–80	....	....	.30	.41	....	....	.30	.41	159–176

†The overcurrent protection for conductor types marked with an obelisk (†) shall not exceed 15 amperes for 14 AWG, 20 amperes for 12 AWG, and 30 amperes for 10 AWG copper; or 15 amperes for 12 AWG and 25 amperes for 10 AWG aluminum and copper-clad alumninum after any correction factors for ambient temperature and number of conductors have been applied.

*For dry locations only. See 75°C column for wet locations.

3. PULLING THE CABLE

When pulling cable, be careful not to damage insulation. This is where holes drilled in line, or with gradual changes in elevation make the job easier. When turning corners, the loop must not be pulled too tight. The insulation will break if turns are too short (Figure 137). The Code states that no bend shall have a radius less than 5 times the diameter of the cable.

Pulling cable is discussed under the following heading.

a. General Procedures for Pulling Cable.

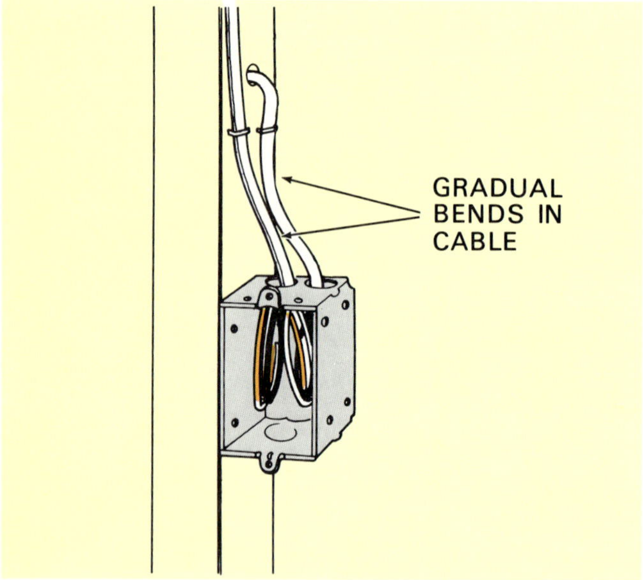

FIGURE 137. Make gradual bends in cable to prevent damage to insulation.

a. General Procedures for Pulling Cable

Proceed as follows:

1. *You may start at the SEP or at a box to be wired.*

   In this example, assume you will start at the SEP pulling No. 12-2w/g.

   Allow enough wire at the SEP to go around the inside of the box to reach the terminals when installed (Figure 138).

2. *Pull the cable from the coil through holes, notches, alongside guard strips or over running boards.*

   Pull the cable through each opening to the outlet or switch box.

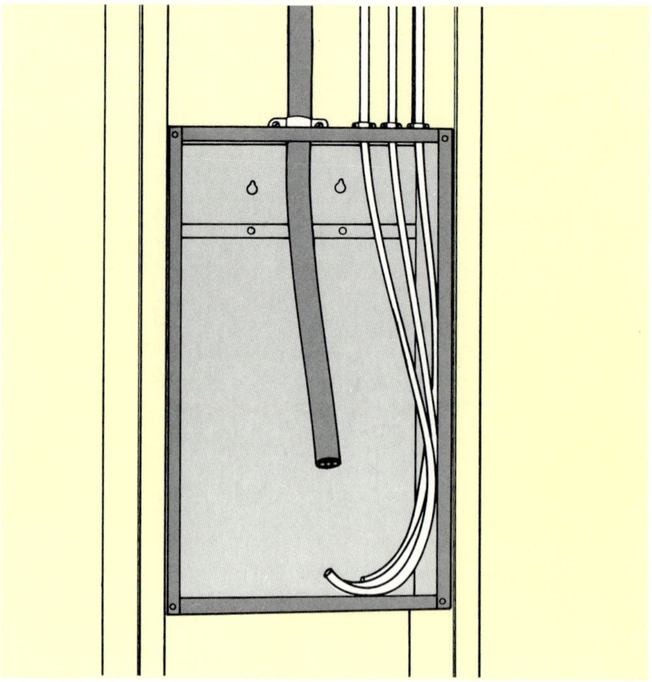

FIGURE 138. Allow enough cable for connecting inside the SEP.

3. *Pull the cable through the knockout hole in the box.*

   Allow an extra 15 to 20 cm (6 to 8 inches) for splicing and connections.

4. *Cut cable for correct length at the roll.*

5. *Remove cable sheathing.*

   Many prefer to remove the **cable sheathing** at this time and fold the individual wires back into the box. Insulation will be removed from the ends of wire later. Figure 139 shows sheathing removed from cables in the box. To remove the sheathing leave 0.6 to 2.5 cm (¼ to 1 inch) of the cable sheathing exposed past the clamp on the inside of the box (Figure 139).

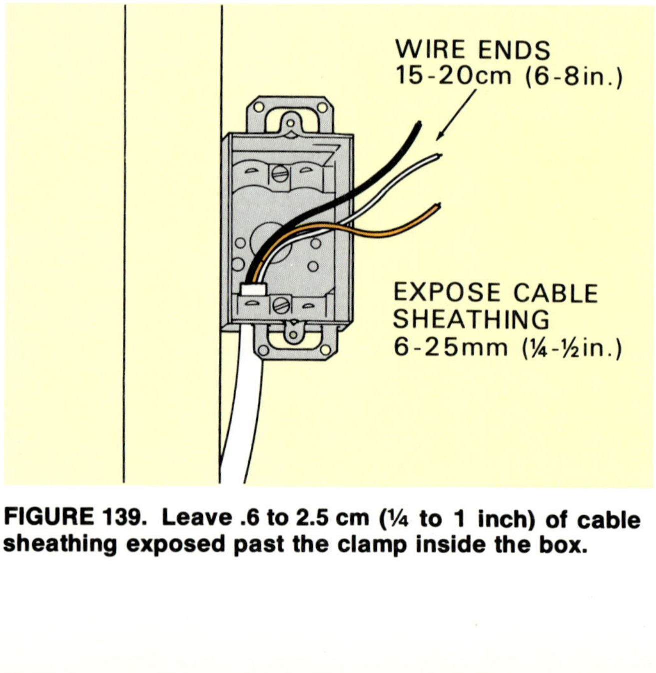

FIGURE 139. Leave .6 to 2.5 cm (¼ to 1 inch) of cable sheathing exposed past the clamp inside the box.

The sheathing can be cut with a knife. Some prefer to use a cable ripper. Instructions for use are given in the following section, "Connecting Circuits."

6. *Fasten cable in boxes.*

   Cable must be securely fastened to every **metal** box. For **nonmetallic** boxes, the Code requires that NM cable sheath extend through the knockout opening at least .6 cm (1/4 inch) inside the box. The cable is not required to be fastened to the nonmetallic box when it is secured with a strap or staple within 20 cm (8 inches) of a single-gang box (one box mounted alone). For nonmetallic boxes **other than single-gang**, the cable must be fastened to the box.

   Two types of anchors (fasteners) are used for metal boxes. They are:

   (a) Connectors.

   (b) Clamps.

   (a) To install cable using connectors, proceed as follows:

   (1) *Slide the connector over the cable 15 to 20 cm (6 to 8 inches) from the end and tighten locking screws (Figure 140).*

   This allows 15 to 20 cm (6 to 8 inches) of cable inside the box for connections.

   (2) *Insert cable and connector through the knockout and install locknut (Figure 140).*

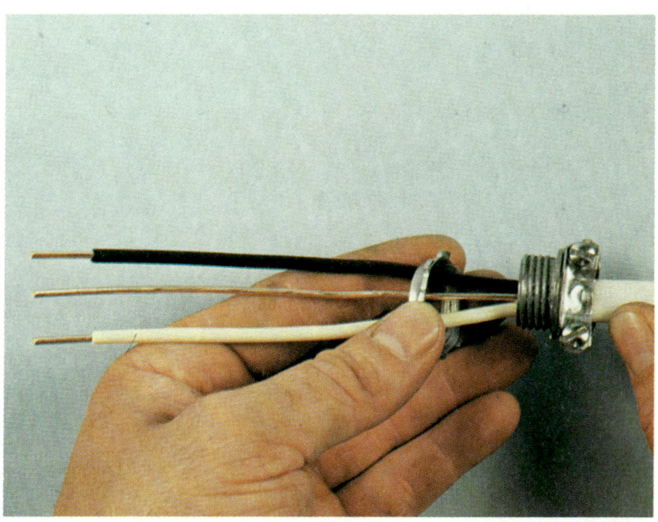

FIGURE 140. Cable and connector to be installed through knockout.

   (3) *Tighten locknut with screwdriver (Figure 141).*

   Some prefer to install the connector through the knockout first, then install cable and tighten connector screws.

   (4) Several UL-listed **plastic** connectors are now marketed that snap into a knockout. For example, one type of plastic connector does not require any tools for installation. It snaps into the knockout (Figure 142). When the cable is inserted, the connector clamps it tightly. Another type snaps into place and a

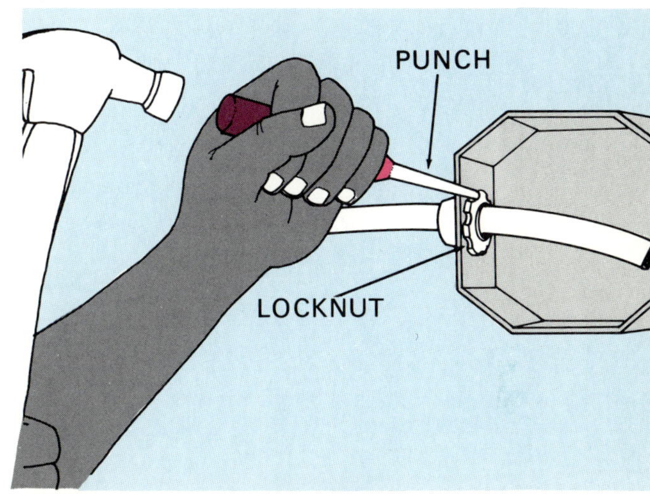

FIGURE 141. Tightening locknut.

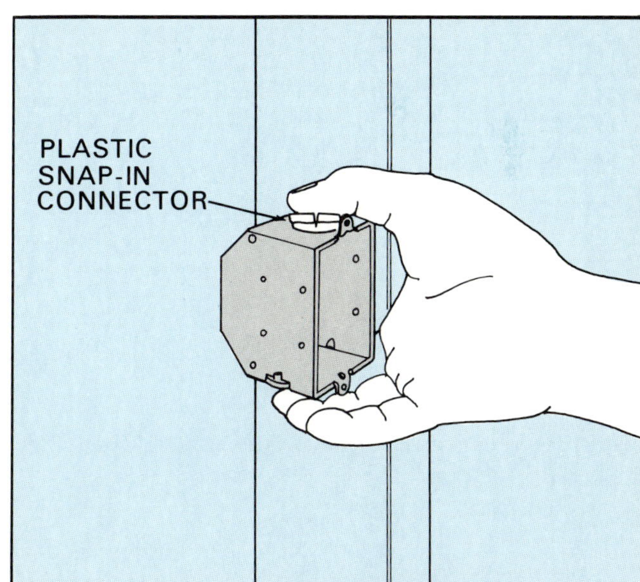

FIGURE 142. Plastic snap-in connector.

small locking wedge fastens the cable when squeezed with pliers (Figure 143). A third type snaps in place and locks the cable with the turn of a screwdriver (Figure 144). Still another snap-in type uses the screwdriver as a lever to lock a wedge against the cable (Figure 145).

(b) To install cable using clamps, proceed as follows:

(1) *Insert cable under clamp leaving 15 to 20 cm (6 to 8 inches) beyond the clamp for connections and tighten bolt (Figure 146).*

FIGURE 143. Small locking wedge fastens cable when clamped firmly with pliers.

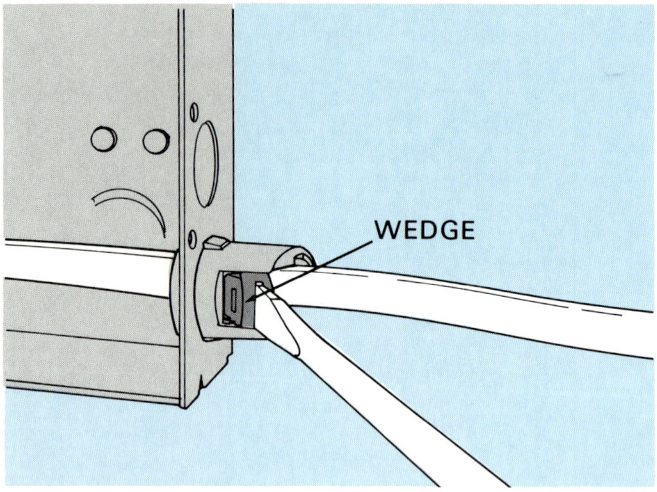

FIGURE 145. Screwdriver used as a lever to lock the wedge against the cable.

FIGURE 144. A turn of the screwdriver snaps wedge into place.

FIGURE 146. Pull 15 to 20 cm (6 to 8 inches) of cable beyond the clamp and tighten the bolt or screw.

(2) *If "quick-clamp" type, pry up edge of clamp and pull cable through (Figure 147).*

7. *Pull cable for remaining boxes in the circuit as follows (Figure 148):*
   (1) *Move roll or pull cable end to first box.*
   (2) *Pull cable through holes, notches or guard strips to second box.*
   Pull cable through knockout and fasten.
   (3) *Cut cable at the roll.*
   (4) *Insert cable end through a knockout in first box and fasten with connector or clamp.*
   (5) *Repeat for each box until circuit is complete.*

NOTE: The bathroom receptacle in Figure 148 is installed in a circuit with the receptacle in the adjacent bathroom and may be connected to a GFCI in the SEP or individual GFCI's may be installed.

8. *Pull cable to switch boxes (Figure 149).*

The Code calls wiring from light to switch box the **switch loop**. Two-wire cable **with ground** should be used for most switch loops. Run the cable to each switch connection from the ceiling light outlet box through the plate to the switch box.

FIGURE 147. Quick-clamp type cable clamp.

NOTE: Some switch installations require the use of a cable with more than two conductors. Switch installations are discussed under Section III-E, "Connecting Switches and Circuits."

9. *Identify the branch circuit at the SEP.*

Service entrance panel covers usually have a numbered chart inside the door on which each circuit can be identified. Label each circuit as you pull the cable to simplify the completion of the chart later.

Attach masking tape to cable end in the SEP and label the circuit. For example, Circuit No. 2, Bedroom No. 1, and Bathroom.

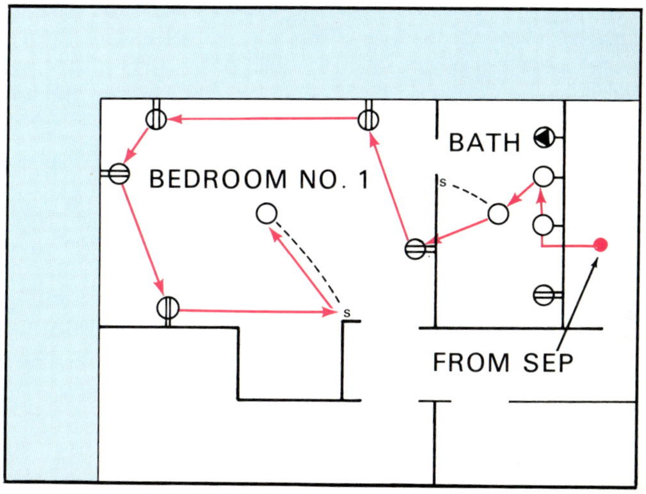

FIGURE 148. Pull cable to lighting outlets, receptacles and switches indicated for circuit No. 2.

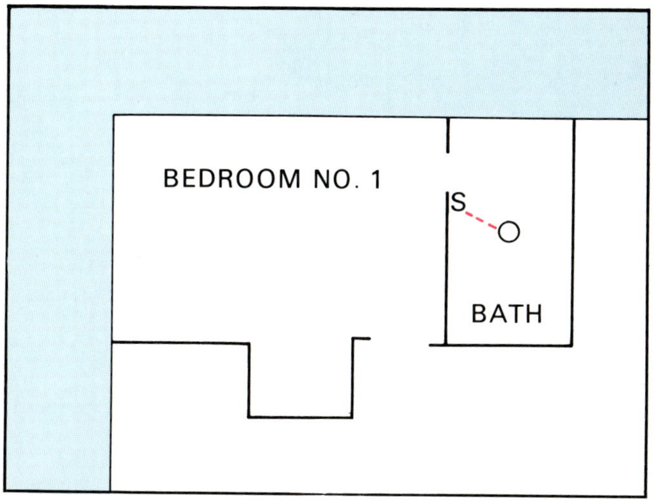

FIGURE 149. Switch loop for bath.

G. Anchoring Cable

Cable must be anchored or fastened as specified by the Code on surfaces over which it runs. Anchoring cable is discussed as follows:

1. Types of Anchors.
2. Strapping or Stapling Cable.

1. TYPES OF ANCHORS

The Code requires cable to be anchored by staples, straps or similar fittings so designed and installed as not to injure the cable (Figure 150). The use of staples is permitted but they must be installed carefully. The cable sheathing and wire insulation will likely be damaged if a staple is bent or driven too far.

2. STRAPPING OR STAPLING CABLE

Where cable is run through holes drilled in framing members, you are not required to fasten nonmetallic sheathed cable except within 30 cm (12 inches) of the box (Figure 151).

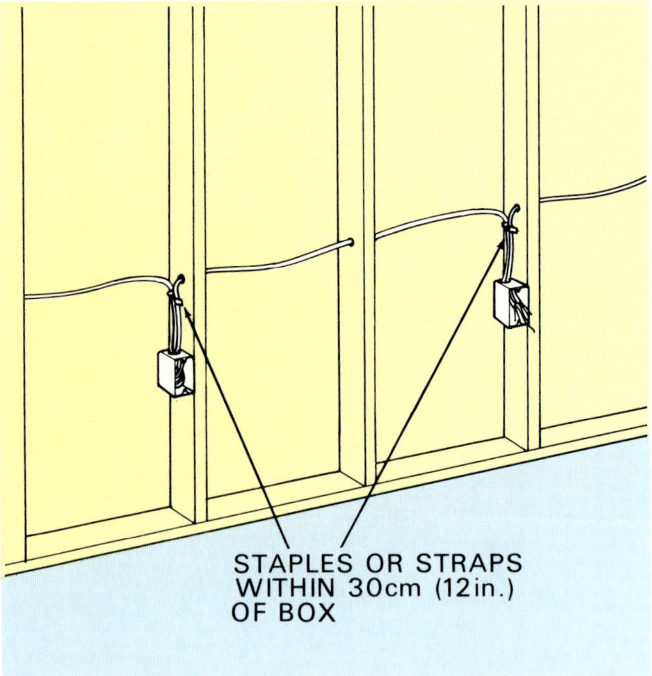

FIGURE 151. Nonmetallic sheathed cable must be fastened within 30 cm (12 inches) of every metal box.

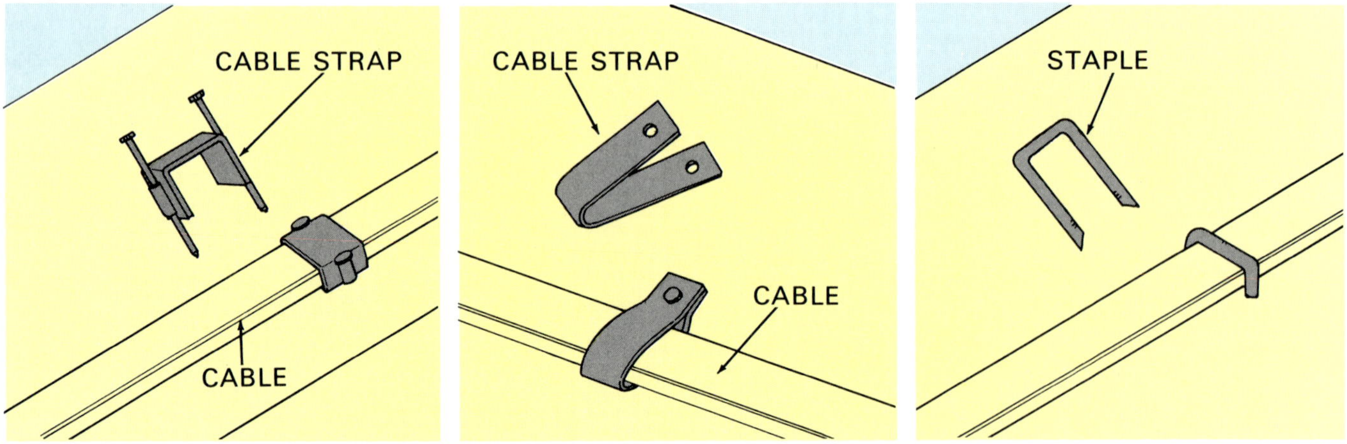

FIGURE 150. Means of anchoring cable.

All cable on studs or run across the face of framing members and at right angles to joists, rafters and sills must be fastened at least every 1.4 m (4½ feet) and within 30 cm (12 inches) of every metal box (Figure 152). Cable must be fastened **within 20 cm (8 inches)** of nonmetallic boxes if cable connectors or clamps are not used at the box. If multi-gang (more than one) nonmetallic boxes are installed, you must fasten the cable at the box.

Where notches in framing are covered with steel plates, the cable is considered secured. But you must still fasten the cable within 30 cm (12 inches) of every metal box and within 20 cm (8 inches) of nonmetallic boxes (Figure 153).

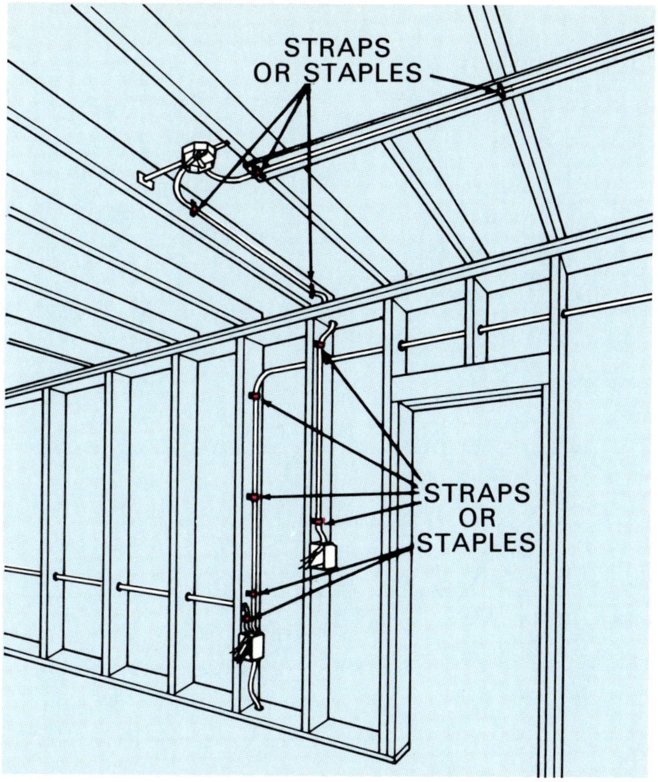

FIGURE 152. Cable fastened to studs run across framing members and at right angles to framing.

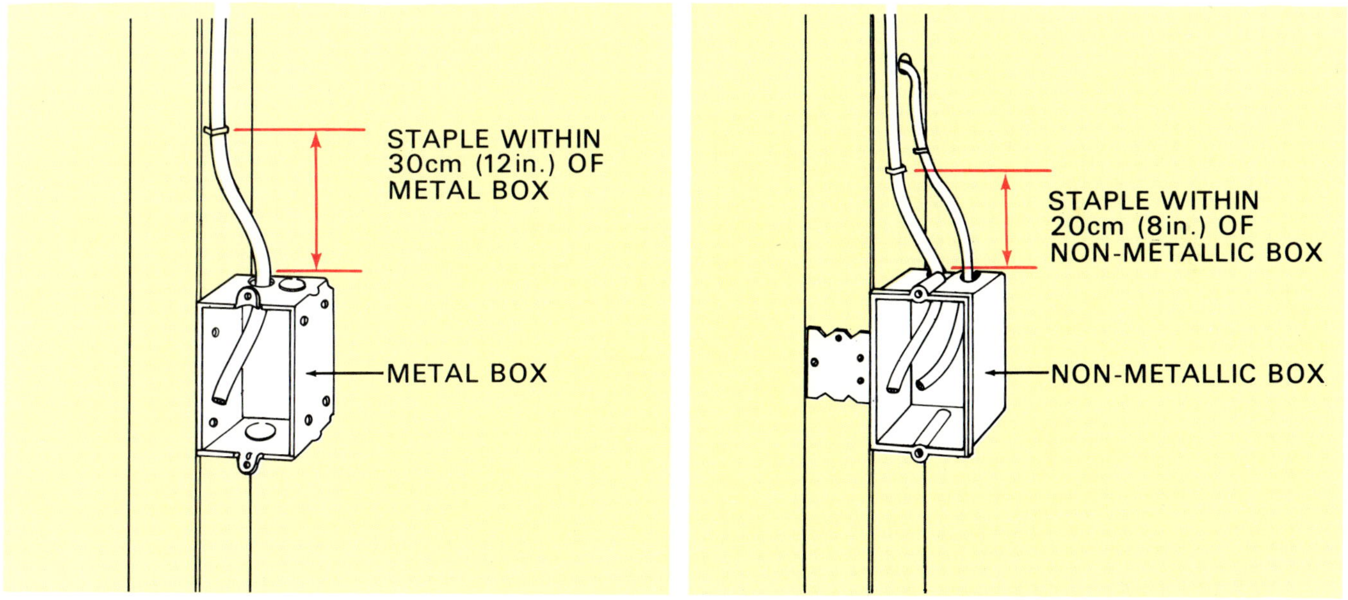

FIGURE 153. Cable fastened within 30 cm (12 inches) of every metal box and within 20 cm (8 inches) of nonmetallic boxes.

NOTES

III. Connecting Circuits

The service entrance panel box or cabinet should have the cable ends neatly in place around the sides of the box after cable has been pulled to the outlets (Figure 154). The open space in the panel box will be occupied by the service entrance wires, the neutral bar, and the panel for the snap-in circuit breakers or fuses (Figure 155). After all wallboard and plaster are in place, your next task is to connect the wiring to the device in each box.

Upon successful completion of this section, you will be able to **use correct procedures to connect all circuits required for wiring a house, including receptacles, switches, lighting outlets, and the required grounding.** Connecting circuits is discussed under the following headings:

A. Connecting Conductors.
B. Grounding the Electrical System and Equipment.
C. Connecting Circuits and Receptacles.
D. Installing Lighting Fixtures.
E. Connecting Switches and Circuits.
F. Installing Doorbells, Chimes, and Smoke Detectors.

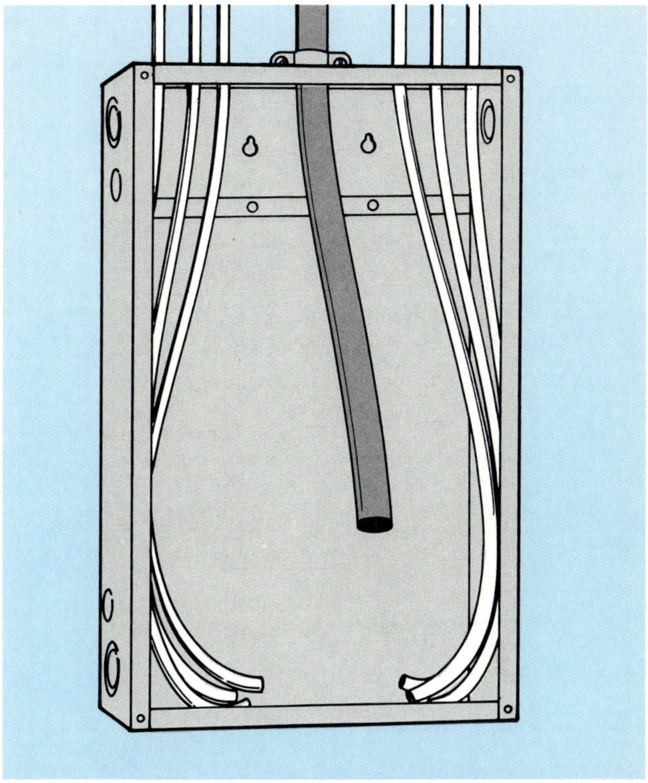

FIGURE 154. Ends of cable for each circuit pulled into the SEP box (cabinet).

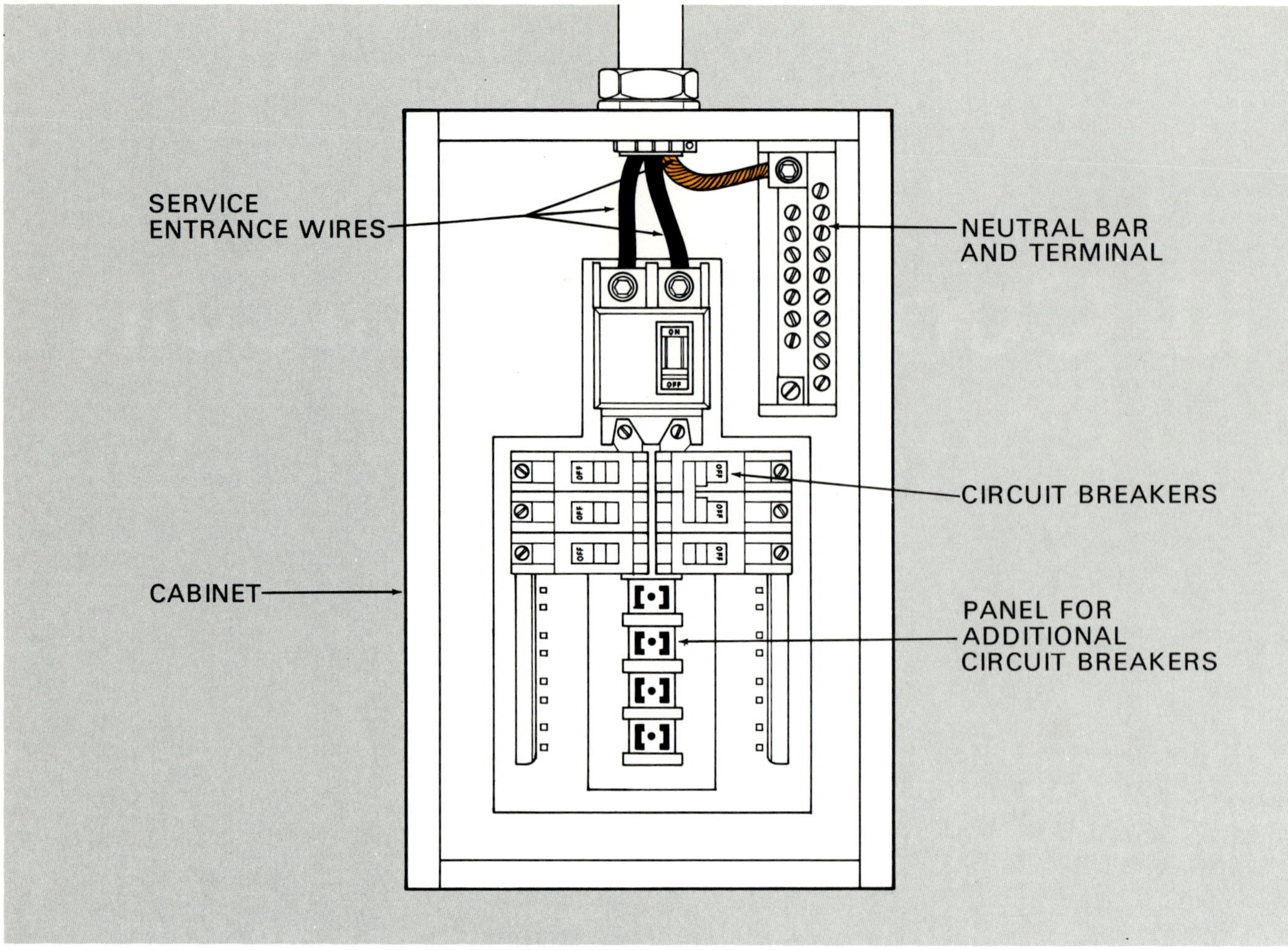

FIGURE 155. SEP box with service entrance wires, neutral bar and circuit breaker panel installed.

A. Connecting Conductors

A dependable wiring job depends to a great extent on good connections throughout the system. Wherever wires are attached to terminals or spliced together, take time to make the connection correctly and prevent problems in the future.

Connecting conductors is discussed as follows:

1. Preparing Cable for Connections.
2. Types of Splices and Connectors.
3. Splicing Wires.

1. PREPARING CABLE FOR CONNECTIONS

The first step in preparing cable for the connection in boxes is to remove the outer covering called a sheath, if not removed before installing in the box. A cut down the center of the cable will open the sheath making it easy to separate from the insulated wires inside. A knife blade will do the job but may cut through the insulation on the wires. Many prefer to use a cable ripper because it does not usually penetrate deep enough to cut the wire insulation (Figure 156).

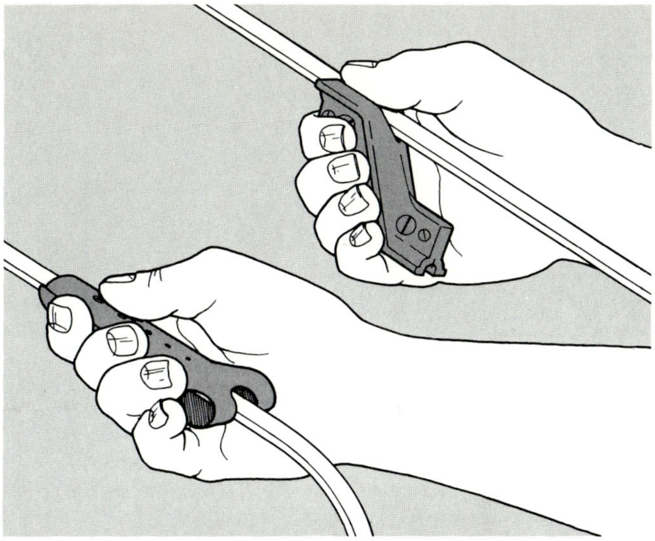

FIGURE 156. Cable ripper used to open the sheath on nonmetallic sheathed cable.

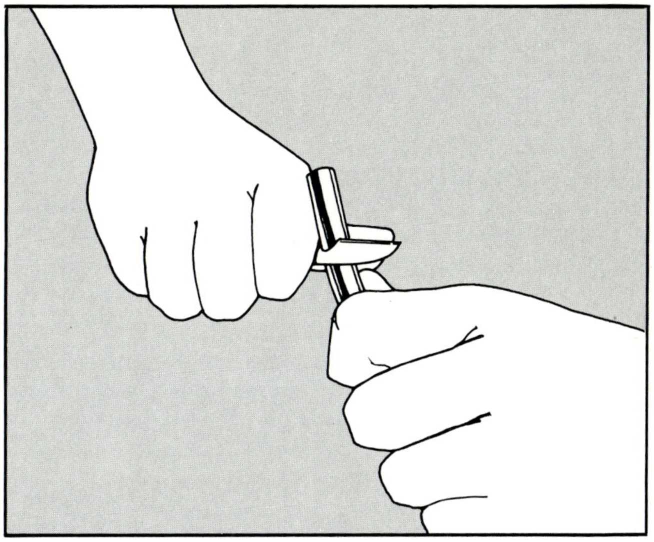

FIGURE 157. To remove insulation with a knife, start the cut at angle to prevent a nick in the wire.

Preparing cable for connections is discussed under the following headings:

a. Removing Cable Sheath.

b. Removing Insulation from Wires.

a. Removing Cable Sheath

Proceed as follows to remove cable sheath with cable ripper:

1. *Place the cutting edge of the tool where you want the cut to start on the cable.*
2. *Squeeze and pull to the end of the cable.*
3. *Peel the sheathing and cut it loose with knife or side cutter.*
4. *Remove any remaining paper and other filler.*

b. Removing Insulation from Wires

Insulation may be removed with a knife or a wire stripping tool. Proceed as follows:

1. *If you use a knife, start the cut on the insulation at an angle to prevent cutting into the wire (Figure 157).*

   A nick in the wire will weaken it and reduce current carrying capacity. Remove about 1.5 to 2 cm (5/8 to 3/4 inch), depending on type of connection to be made.

2. *If you use a wire stripper, insert the wire into the correct wire gage hole size on the tool and squeeze the handles.*

   Be sure the hole size is correct or you will damage the wire. Squeeze to make the cut through the insulation and strip the insulation off the end of the wire (Figure 158). This method does not cut into the wire.

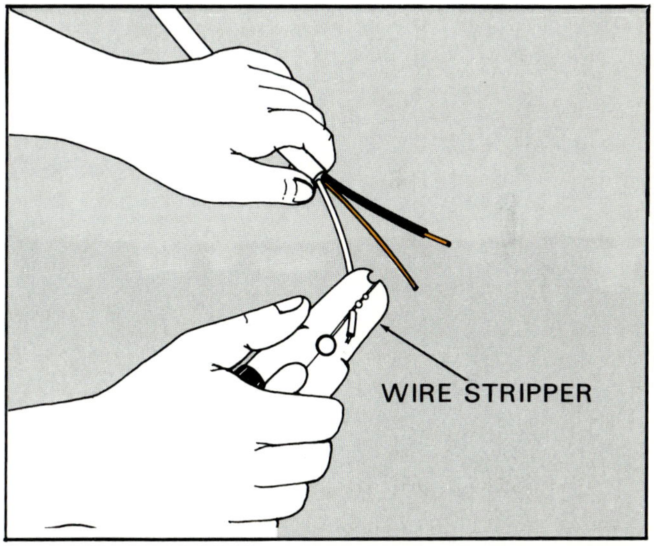

FIGURE 158. To remove insulation with a wire stripper, match stripper notch to wire size, insert the wire, squeeze the handle and strip the insulation off the wire.

2. TYPES OF SPLICES AND CONNECTORS

Splicing is the joining together of two or more wires to make a connection. Splices must conduct electricity as well as an uncut wire to be sure of good electrical connections. Loose connections may allow arcing which could cause a fire or damage equipment when the system is placed in service.

Soldering was once the accepted method of splicing wires but is seldom used in modern house wiring. Soldering is **not** permitted for connecting the grounding electrode conductor to the grounding electrode.

Solderless connectors are available for most types and sizes of splices (Figure 159). They are easy to use and make good connections quickly. There are several brands but they are usually called **wire nuts,** which is the name used in this book (Figure 160). Solderless connectors are used for practically all splices in house wiring circuits. They are available in sizes to fit all house wiring. The size of wire must be matched with correct wire nut size given on the package. Another type of solderless connector is the crimp-type made of plastic covered metal. **Split-bolt** connectors of various types and sizes are used to splice heavier wires (Figure 161).

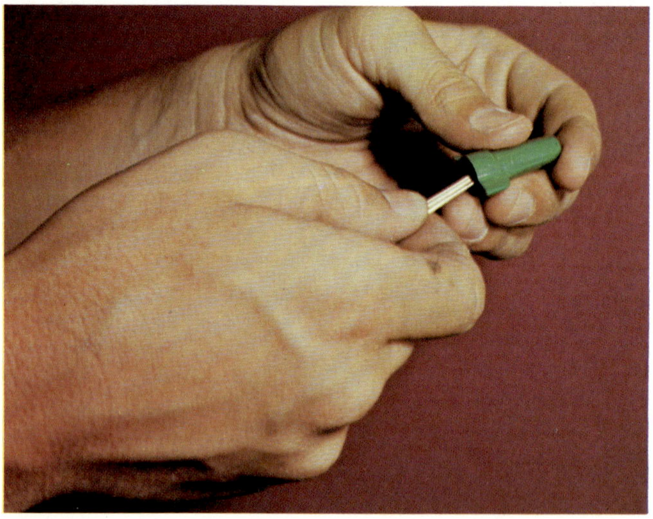

FIGURE 159. Solderless connector for quick and easy splicing.

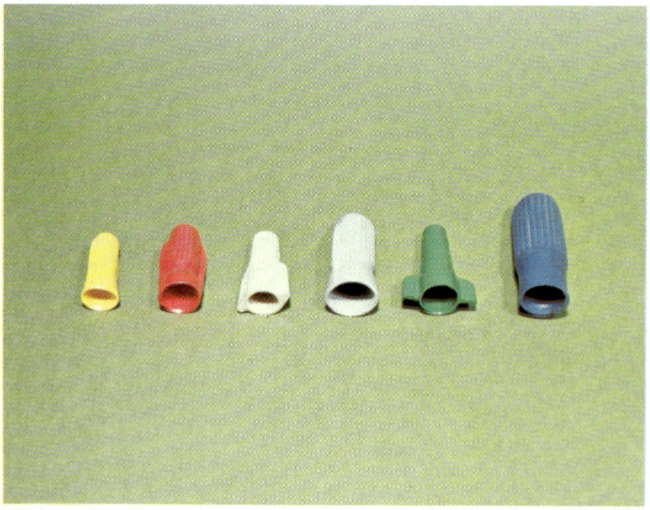

FIGURE 160. Wire nuts of various types, size and color.

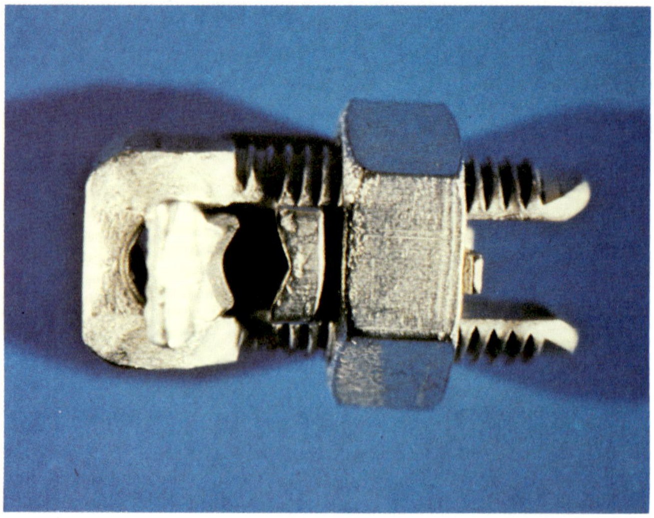

FIGURE 161. Split-bolt for splicing heavy conductors.

3. SPLICING WIRES

Remember, all splices must be enclosed in a box. There are two methods for splicing wires:

a. Solderless.

b. Soldered.

a. Solderless

Solderless splices may be made by the use of **three** different types of connectors. They are described as follows:

(1) WIRE NUT CONNECTORS are the most satisfactory. Proceed as follows:

1. *Remove insulation from wire ends.*

   Use a wire stripper.

   Remove just enough insulation for the connector to cover the bare wire when tight (Figure 162). Follow instructions on the box.

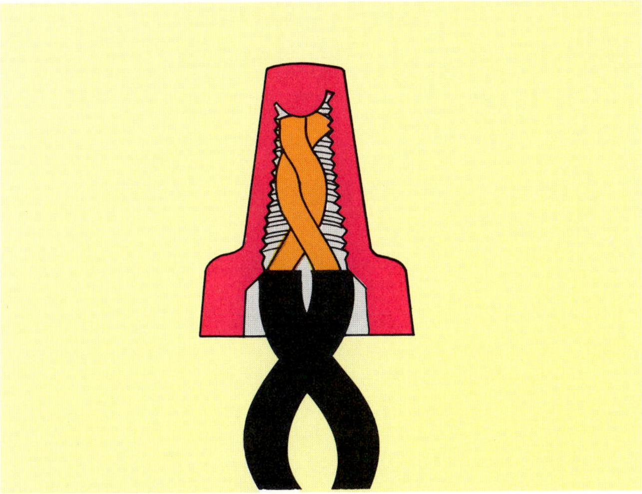

FIGURE 162. The amount of insulation to remove depends on size of wire and wire nut.

2. *Hold wires side by side.*
   It is not necessary to twist the wires together.
3. *Place connector over wire end and turn as you would a nut until tight (Figure 163).*
   Be sure both wires are locked in and no bare wires are exposed.

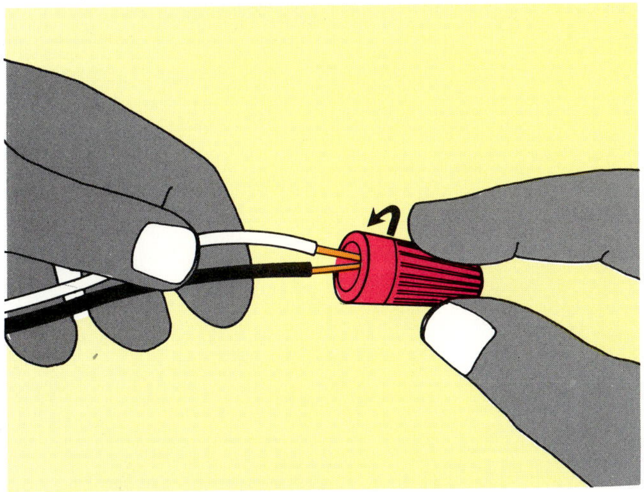

FIGURE 163. Attaching a wire nut (solderless connector).

(2) CRIMPED-CONNECTORS are used but may not be approved by some local codes. The wire ends are stripped, placed in a connector and crimped with a crimping tool (Figure 164).

(3) SPLIT-BOLT CONNECTORS are normally used for large wires (Figure 165). The wire ends are stripped, placed in the connector and the nut is tightened. The connection is then wrapped with layers of plastic electrical tape at least the same thickness as the original insulation.

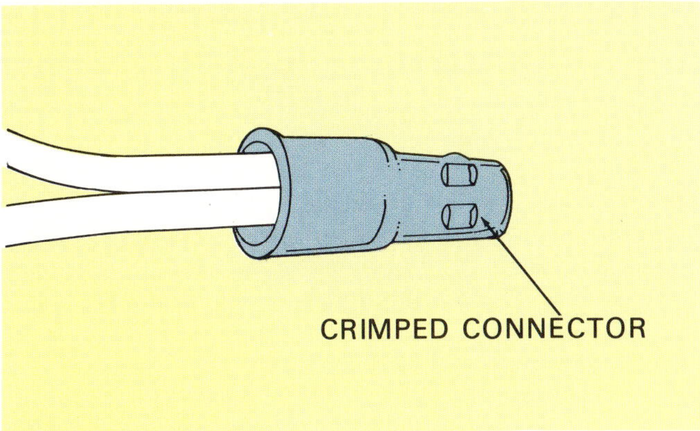

FIGURE 164. A crimp-connector splice.

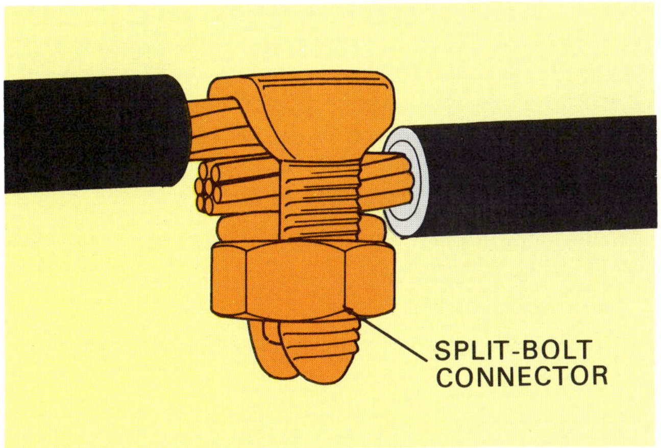

FIGURE 165. A split-bolt connector splice. Wrap with tape to same thickness as wire insulation.

b. Soldered

Wires may be spliced with solder. Wire ends are stripped, cleaned and twisted together. The wire ends are heated with a soldering iron and rosin-core solder is melted by the heated wire (Figure 166).

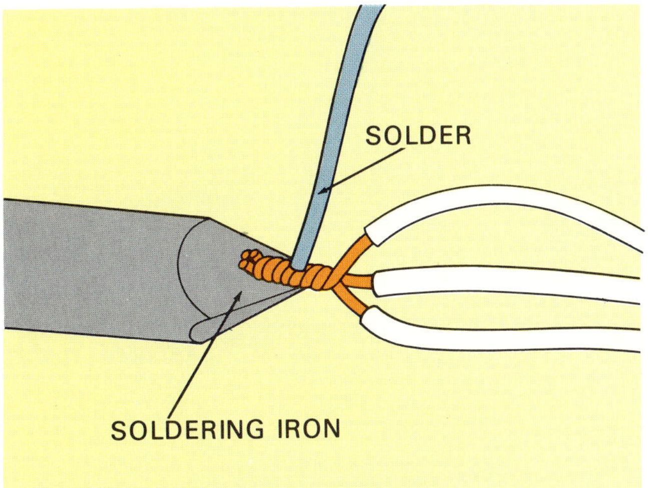

FIGURE 166. Soldering a splice. Heat wires with soldering iron underneath to melt solder held on top.

B. Grounding the Electrical System and Equipment

A **ground** is an intentional connection between an electrical system and the earth. The preferred way to ground a house wiring system prior to the 1978 Code was to connect a conductor from the neutral bar in the SEP to a metal cold water pipe in direct contact with the earth for at least 3.1 m (10 feet) (Figure 167). This type of installation must now be supplemented by one or more additional grounding electrodes. Details on conductor sizes and other methods of grounding are given under Section IV, "Installing Service Entrance Equipment."

Grounding makes an electrical system safer for people to use and prevents damage to the system and property. If circuits and equipment are not correctly grounded, you may get a severe or even fatal shock. Such items as appliances, power tools and electric motors may be damaged or destroyed by a ground fault. A ground fault is an insulation failure between an energized (hot) conductor and ground.

Grounding the electrical system and equipment is discussed as follows:

1. Grounding the System and Circuits.
2. Grounding the Equipment and Conductor Enclosures.

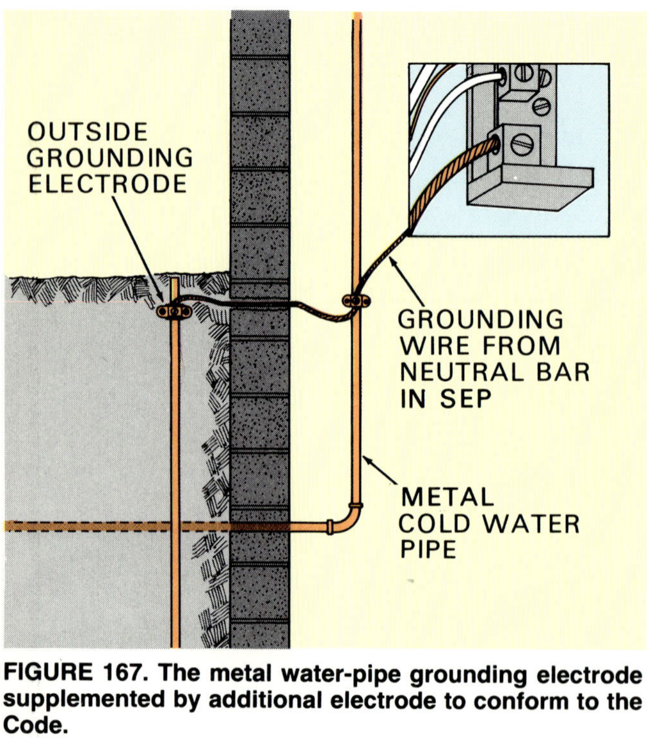

FIGURE 167. The metal water-pipe grounding electrode supplemented by additional electrode to conform to the Code.

1. GROUNDING THE SYSTEM AND CIRCUITS

Grounding the system and circuits covers all current-conducting parts of the electrical installation, including the necessary wiring in the SEP and the circuits. In this discussion, you are concerned primarily with grounding the house wiring circuits from the SEP to all parts of the house.

System and circuit grounding is accomplished by connecting the SEP neutral bar to ground as discussed earlier (Figure 167) and by connecting the white neutral wire from each circuit to the neutral bar (Figure 168).

In 120-volt installations there must be two conductors, one black and one white. The black wire is commonly referred to as the "**hot**" wire, but its technical name is the **ungrounded conductor.** The white wire is commonly called the **neutral wire,** but its technical name is **grounded neutral conductor.** Both wires carry current, but the white neutral wire is connected to ground through the neutral bar in the SEP (Figure 168). The white neutral wire must be continuous throughout the system. How can this be done when the wire must be cut at every box or outlet to make spliced connections?

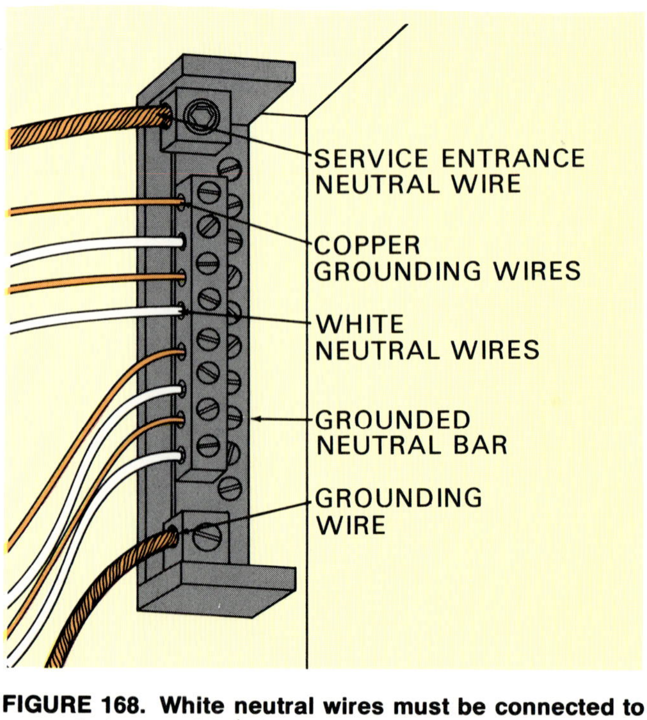

FIGURE 168. White neutral wires must be connected to grounded neutral bar in the service entrance panel for each circuit.

It can be made electrically continuous by joining the cut ends with good solid connections at every box (Figure 169). By joining white wire to white wire at every connecting point, the effect is the same as having one continuous white wire. It runs from the SEP to every outlet in the circuit. **You must not install a switch, a fuse or a circuit breaker in the neutral wire.** The rule to remember is, **the neutral is always white but the white wire is not always a neutral.** The white wire is not a neutral if it is connected to a switch. In an exception to the code requirements, the Code allows a white wire to be connected to a switch in some instances. The exception is explained in detail in Section E, "Connecting Switches and Circuits."

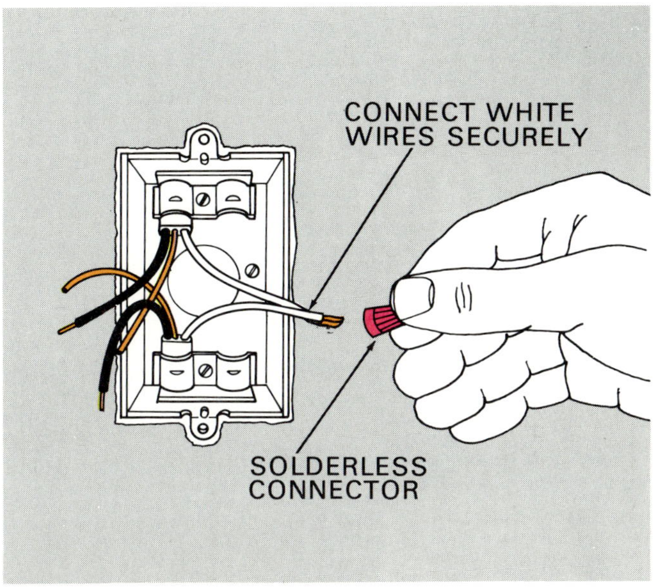

FIGURE 169. The white neutral wire must be made continuous by spliced connections at each box or outlet in the circuit.

2. GROUNDING THE EQUIPMENT AND CONDUCTOR ENCLOSURES

Equipment grounding covers the non-current-carrying parts of the system. Included are metal frames on equipment, appliances, and motors, and all metal outlet boxes, switch boxes, receptacles and metal conduit.

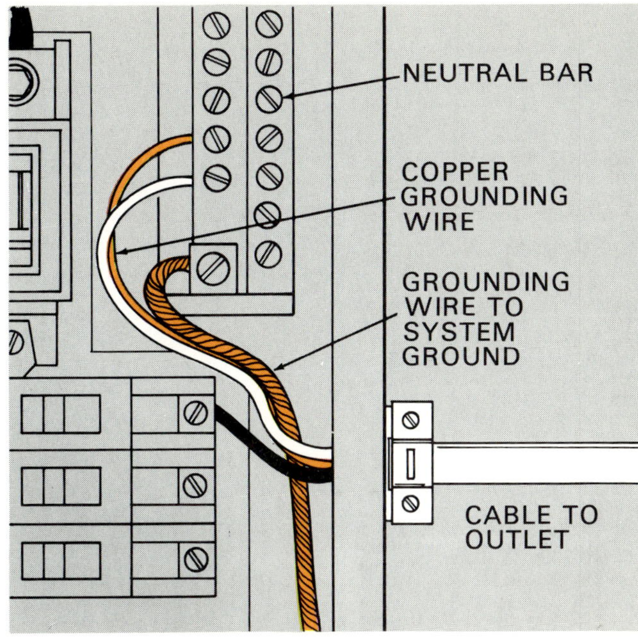

FIGURE 170. The grounding wire must be connected to the neutral bar in the SEP.

According to the Code, equipment is connected to ground for the following purposes:

- To **limit the voltage to ground** on equipment frames and on metal enclosures of the wiring system.
- To **assure operation of over-current devices** in case of ground faults.

The grounding wire is there to help protect a person from a dangerous shock or to protect equipment from damage in case of insulation failure in the circuit. If a ground fault occurs and the current in the line exceeds its overcurrent protection, the circuit breaker trips or the fuse opens, disconnecting the circuit.

In 120-volt house wiring using nonmetallic sheathed cable w/ground, **equipment grounding** is accomplished by means of the bare grounding wire in the cable. It must be connected to the neutral bar in the SEP (Figure 170). The bare grounding wire does not normally carry current. From the neutral bar, it must be connected to metal outlet boxes, to every outlet in the circuit and to the frames of all equipment. The grounding wire may sometimes be insulated. If so, it must be green or green with yellow stripes.

By connecting the white neutral wire and the bare grounding wire to the neutral bar in the SEP, you have grounded both the circuit and the equipment. The following example illustrates grounding the clothes washer and its circuit (Figure 171). Note the terminal connections on the back of the washer. The black and white wires of the cord carry current to operate the washer. When the cord is plugged into the receptacle, the neutral wire is connected to ground on the neutral bar through the white wire in the circuit cable. An accidental contact of the non-grounded current-carrying wires to circuit ground will result in a short circuit, and the circuit breaker or fuse will open to break the circuit.

♣ The green grounding wire in **Figure 171** is fastened to the **frame** of the washer by means of a metal screw terminal. When plugged into the receptacle, the grounding wire is connected by the grounding circuit to the neutral bar at the SEP. This means that the **equipment** is **grounded** through the grounding wire to the neutral bar. Thus, if a non-grounded circuit conductor should make contact with the frame or machinery of the washer, a short circuit fault will occur and open the fuse or trip the circuit breaker. Grounding the outlet boxes and receptacles has the effect of grounding all appliances and equipment that may be plugged into any receptacle on each circuit, provided the appliance cord has a grounding conductor and plug.

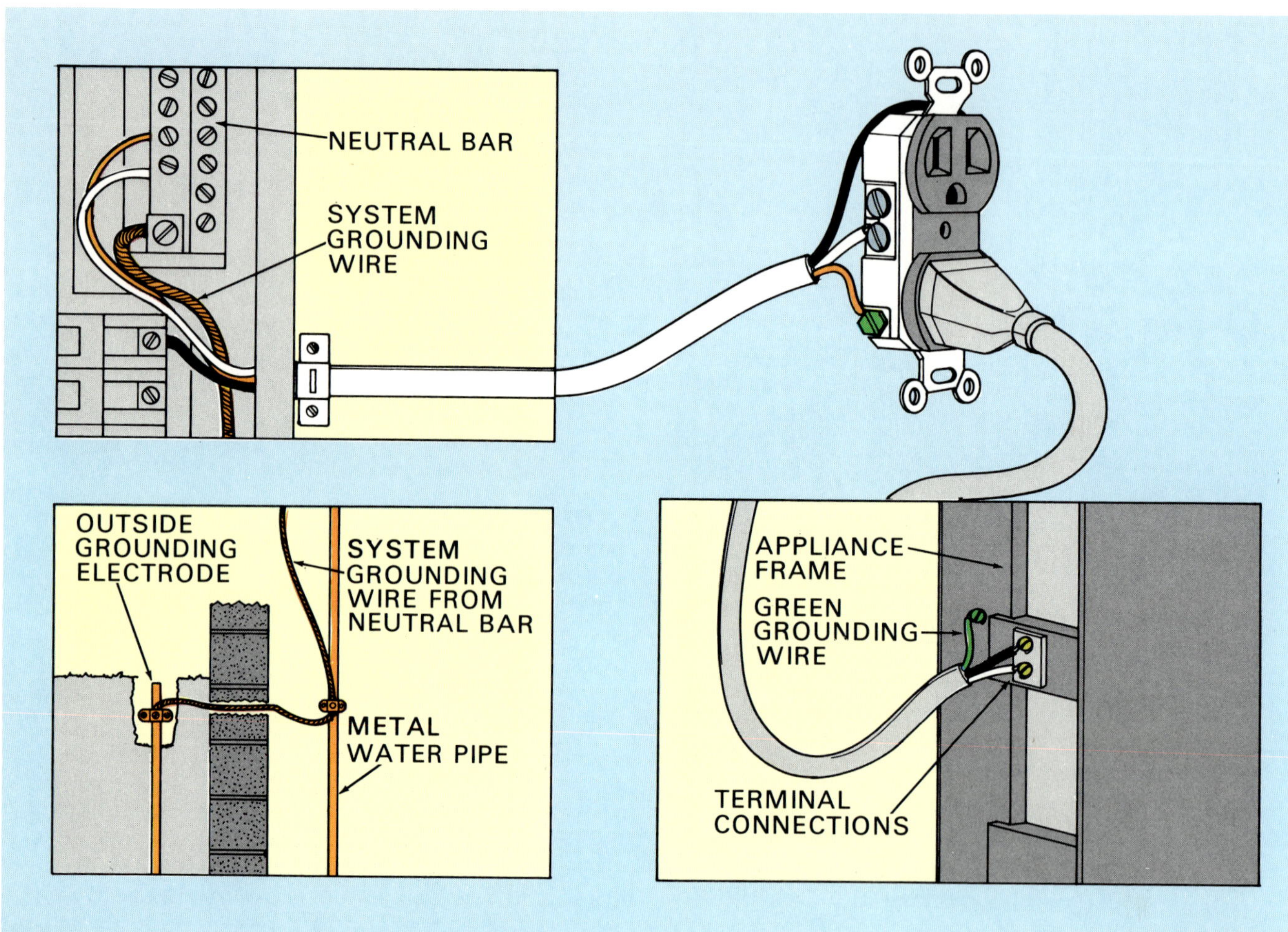

♣ **FIGURE 171. The grounding wire connects to the metal frame of the washer, to the receptacle and receptacle box and to the netural bar in the service entrance panel.**

To ground the system and circuits and the equipment, proceed as follows (Figure 172):

1. *In the SEP, locate the end of the cable marked "clothes washer."*
2. *Remove the sheath and filler from cable clamp to the ends of the cable.*
3. *Remove 2 cm (¾ inch) of insulation from each insulated wire.*
4. *Insert the white neutral wire under a terminal screw on the neutral bar and tighten. On most neutral bars the wire-end is inserted under the terminal screws without a loop.*
5. *Insert the bare grounding wire under a separate terminal screw on the neutral bar and tighten the screw.*

   To complete the circuit wiring in the SEP, connect the black wire to the circuit breaker. Most have push-in set-screw type connections. Snap the circuit breaker into position. Be sure it is the correct size for the conductor.

6. *Connect the grounding wire to the metal box and to the receptacle grounding screw.*

   The bare grounding wire in the cable must be connected to the metal outlet box and to the green grounding screw on the receptacle (Figure 173A). The usual way to connect the grounding wire to a metal box is with a "jumper" wire, sometimes called a "pigtail," and a screw (Figure 173A). The "jumper" is a short piece of wire, 10–15 cm (4–6 inches) long. Another method of grounding the box is by means of a clip. Insert the end of the jumper wire into the clip and force the clip onto the edge of the box (Figure 173B). Use of clips is not allowed by some local codes.

   In single-gang plastic boxes, the grounding wire is not required to be fastened to the box. The grounding wires are fastened to each other and to the grounding terminal on the receptacle (Figure 174).

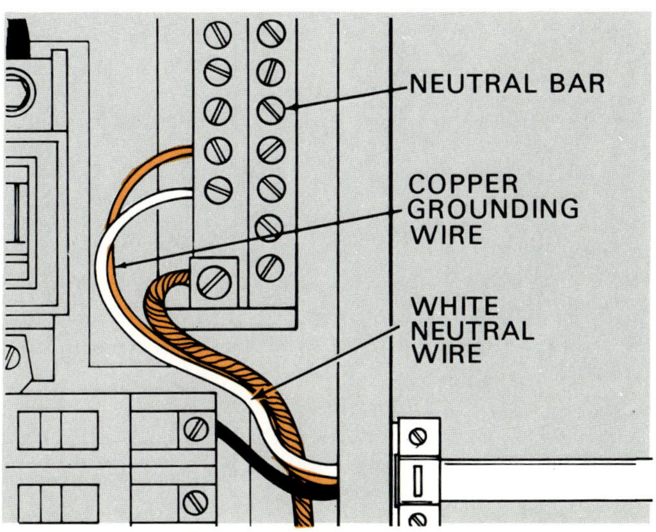

FIGURE 172. Connect the black wire in the cable to the circuit breaker.

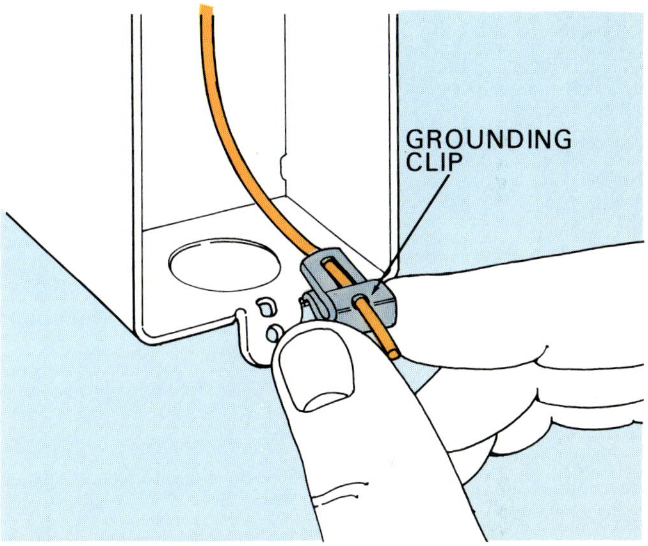

FIGURE 173B. A grounding clip fastens the jumper wire to the edge of the box.

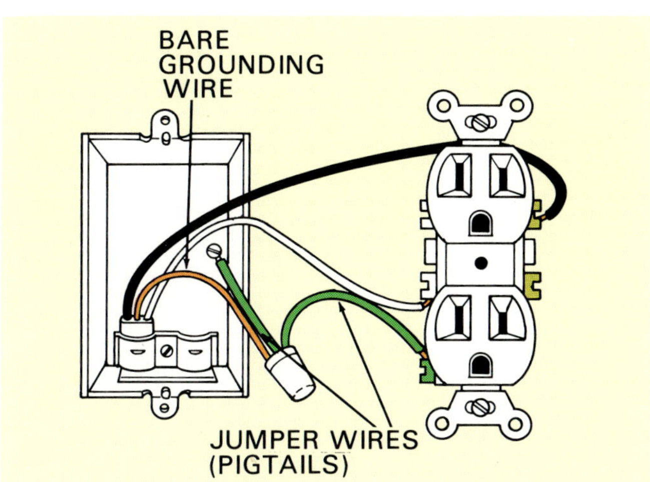

FIGURE 173A. Connect the bare grounding wire from the cable to the receptacle grounding screw and to the box, using two jumper wires and a wire nut solderless connector.

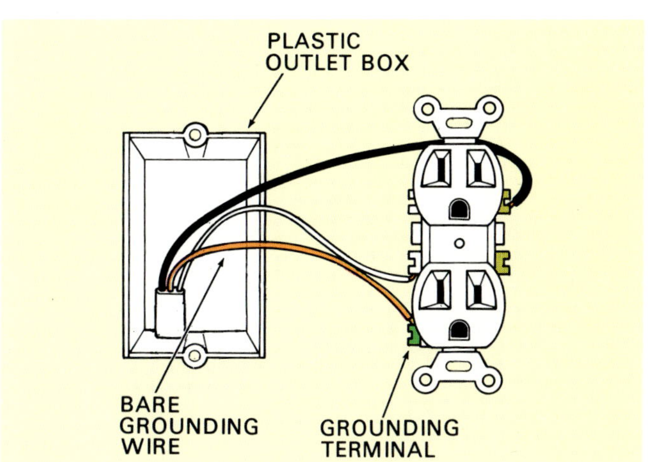

FIGURE 174. For single-gang plastic box, a grounding wire to the box is not required.

The Code permits the following exceptions to the required jumper wire for receptacle grounding:

(a) **Surface mounted boxes.**

Where the box is surface mounted, the metal to metal contact between the metal receptacle yoke and the box is permitted for grounding the box to the yoke (Figure 175).

(b) **Flush-mounted boxes.**

Contact receptacles or yokes designed and **listed** for the purpose are permitted for grounding the circuit between the yoke and flush-type boxes (Figure 176). Several manufacturers offer specially constructed receptacles which provide a ground between the receptacle yoke and box, thus eliminating the grounding screw and jumper wire (Figure 176).

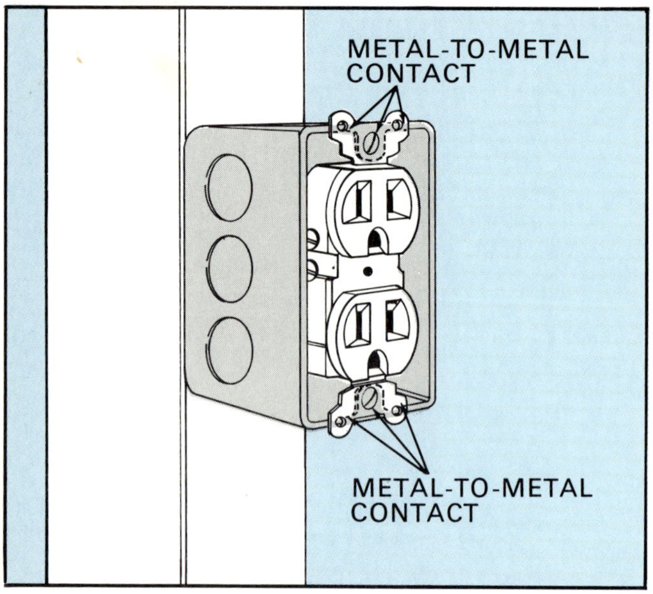

FIGURE 175. Surface-mounted boxes having metal-to-metal contact between the receptacle and box are considered to be grounded.

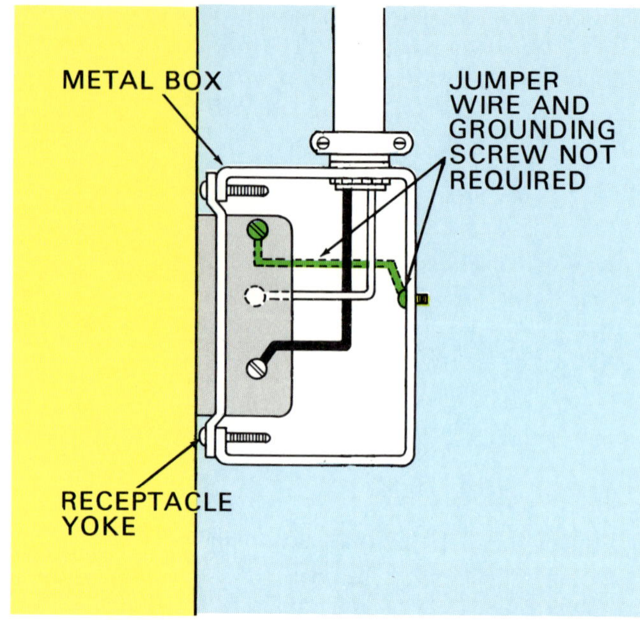

FIGURE 176. Receptacles with yokes that are designed and listed for the purpose may be installed without a grounding wire to the box.

C. Connecting Circuits and Receptacles

In this section, you will connect circuits for each type of receptacle. Connecting circuits and receptacles is discussed as follows:

1. Types, Sizes and Colors of Receptacles.
2. Connecting Receptacles.

1. TYPES, SIZES AND COLORS OF RECEPTACLES

The most common type of receptacle in house wiring is the standard grounding type duplex receptacle. It is available in 15- and 20-ampere, 120-volt rating with the U-shaped opening as the grounding terminal (Figure 177).

The same receptacle is available with a decorator face (Figure 178) and with cover plates in decorator colors including white, brown, grey, ivory, pink, blue, yellow and black. You may also purchase receptacles designed for 2-circuit installation. Another type allows one outlet to be wired for 15-ampere, 125 volts and one for 15-ampere, 250 volts (Figure 179). Receptacles may be rated for **125** and **250** volts even though they are connected to **120** and **240** volts.

Other types and sizes of receptacles are the single receptacle type, single receptacle with a switch, single receptacle with pilot light, and a duplex receptacle with weatherproof cover (Figure 180).

Self-contained kitchen ranges normally require a heavy duty 50 ampere, 125/250-volt receptacle. The flush-mounted type is most often used in new construction (Figure 181). A similar 250-volt receptacle is used for clothes dryers that operate on 240 volts but with different configurations (arrangement of openings) because of different ampere ratings (Figure 182).

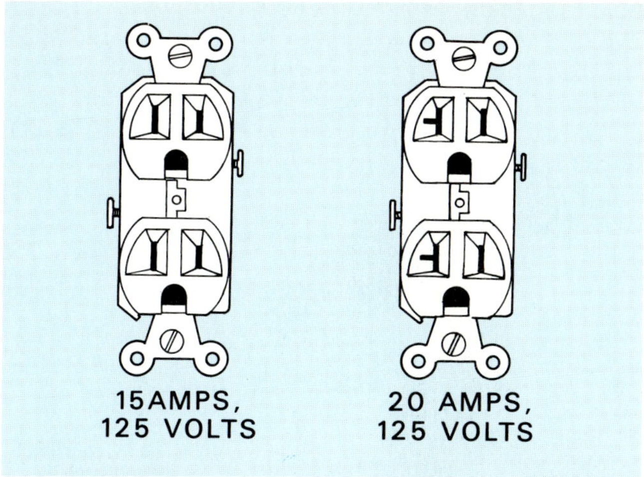

FIGURE 177. Standard grounding duplex receptacles.

FIGURE 178. Grounding duplex receptacle with decorator face.

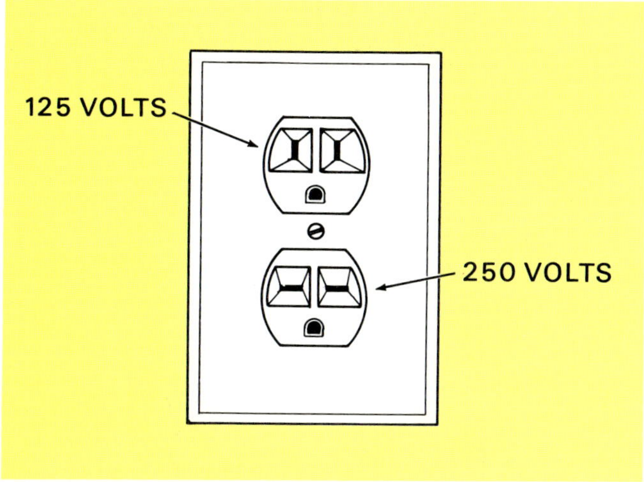

FIGURE 179. Grounded duplex receptacle rated at 15 amperes, 125 volts in upper outlet and 15 amperes, 250 volts in lower outlet. Note the arrangements of openings, paralleled for 125 volts and tandem for 250 volts.

FIGURE 180. Duplex receptacle with weatherproof cover.

FIGURE 181. Flush-mounted 250-volt, 50-ampere kitchen range receptacle and plug. Also used for some arc welding machines.

FIGURE 182. Dryer receptacle rated at 30 amperes, 250 volts.

Table VII shows the configurations for general purpose, non-locking receptacles and plugs. Each type is also available as a locking plug which provides an extra safety factor. These configurations are approved as the standard by the "National Electrical Manufacturers Association" (NEMA) and are endorsed by Underwriters Laboratories and the electrical industry generally. Note that each configuration is designated by numbers and letters. The range outlet shown in Figure 182 is No. 10-50 R on the chart and the dryer is 10-30 R.

You may also note that most terminals in Table VII are identified by letters. The letter "G" marks the terminal to which an equipment grounding wire is attached. The letter "W" identifies the white neutral wire. If the terminal is unmarked, it is connected to an ungrounded (hot) circuit conductor. On devices containing two or three terminals for hot circuit conductors, the terminals are marked "X," "Y" and "Z."

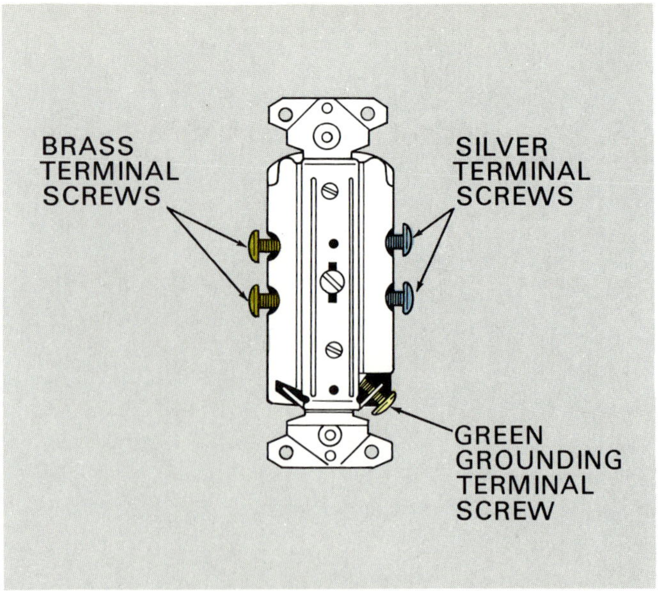

FIGURE 183. Brass-colored terminals for black wires, silver-colored terminals for white wires and green terminal for grounding wire.

2. CONNECTING RECEPTACLES

Connecting receptacles is discussed as follows:

a. Connecting 120-Volt Duplex Receptacles.
b. Connecting Split-Wired Duplex Receptacles.
c. Connecting Individual Appliance Circuits.

a. Connecting 120-Volt Duplex Receptacles

Examine the duplex receptacle before you start wiring. Looking at an example from the back (Figure 183), note the two **brass terminal** screws on one side. Always connect a **black** wire or current-carrying wire to a brass terminal (black wire to dark terminal). On the opposite side are the two silver terminal screws. Connect only white wires to silver terminals (white wires to light terminals). The green hexagonal grounding screw is at the bottom on the right. (It may be at the top on some receptacles). Connect the bare or green grounding wire to the green screw.

Some receptacles do not have different colored terminals. If not, the Code requires the word **"white"** to be located next to the terminals for the white neutral wires. Also, if the grounding terminal is not colored green, it must have the word **"green"** located next to it.

At the top and bottom center are the mounting screws. They fit the holes on the front of the box. Use these to mount the receptacle after you complete the wiring. Plaster ears at bottom and top hold the receptacle in correct position if the box is set too deep.

Assume you are wiring receptacles in bedroom No. 1 in the standard house plan. No. 12-2 w/g cable is installed and clamped inside the box. You should have about 15 to 20 cm (6 to 8 inches) of cable to work with. Mount the duplex receptacle with U-shaped ground opening at the bottom. Proceed as follows to connect the first receptacle in the circuit:

1. *Remove cable sheath and filler and strip about 1.6 cm (⅝ inch) of insulation from the exposed insulated wires.*

2. *Make a loop of the wire ends of the black and white wires as shown in Figure 125.*

   Leave an opening at the end of each loop so it can be easily placed under the screw terminal.

TABLE VII. NEMA CONFIGURATIONS FOR GENERAL-PURPOSE NONLOCKING PLUGS AND RECEPTACLES

			15 AMPERE		20 AMPERE		30 AMPERE		50 AMPERE		60 AMPERE	
			Receptacle	Plug	Receptacle	Plug	Receptacle	Plug	Receptacle	Plug	Receptacle	Plug
2-POLE 2-WIRE	1	125 V	1-15R	1-15P								
	2	250 V		2-15P	2-20R	2-20P	2-30R	2-30P				
2-POLE 3-WIRE GROUNDING	5	125 V	5-15R	5-15P	5-20R	5-20P	5-30R	5-30P	5-50R	5-50P		
	6	250 V	6-15R	6-15P	6-20R	6-20P	6-30R	6-30P	6-50R	6-50P		
	7	277 V AC	7-15R	7-15P	7-20R	7-20P	7-30R	7-30P	7-50R	7-50P		
3-POLE 3-WIRE	10	125/250 V			10-20R	10-20P	10-30R	10-30P	10-50R	10-50P		
	11	$3\phi\Delta$ 250 V	* 11-15R	* 11-15P	* 11-20R	* 11-20P	* 11-30R	* 11-30P	* 11-50R	* 11-50P		
3-POLE 4-WIRE GROUNDING	14	125/250 V	* 14-15R	* 14-15P	14-20R	14-20P	14-30R	14-30P	14-50R	14-50P	14-60R	14-60P
	15	$3\phi\Delta$ 250 V	* 15-15R	* 15-15P	15-20R	15-20P	15-30R	15-30P	15-50R	15-50P	15-60R	15-60P
4-POLE 4-WIRE	18	$3\phi Y$ 120/208 V	18-15R	18-15P	18-20R	18-20P	* 18-30R	* 18-30P	* 18-50R	* 18-50P	18-60R	18-60P

Configurations approved as NEMA Standard.

3. *Connect the black (hot wire) to one of the brass colored terminal screws on the receptacle (Figure 184).*

   Place the loop on the terminal so it will tighten as the screw is turned clockwise. With pliers, pull the loop snug around the screw and tighten with a screwdriver. When tight, the insulation should run right up to the screw so bare wire doesn't show.

4. *Connect the white (neutral) wire to a silver-colored terminal opposite the black wire connection (Figure 184).*

   Install the wire loop under the silver terminal and tighten.

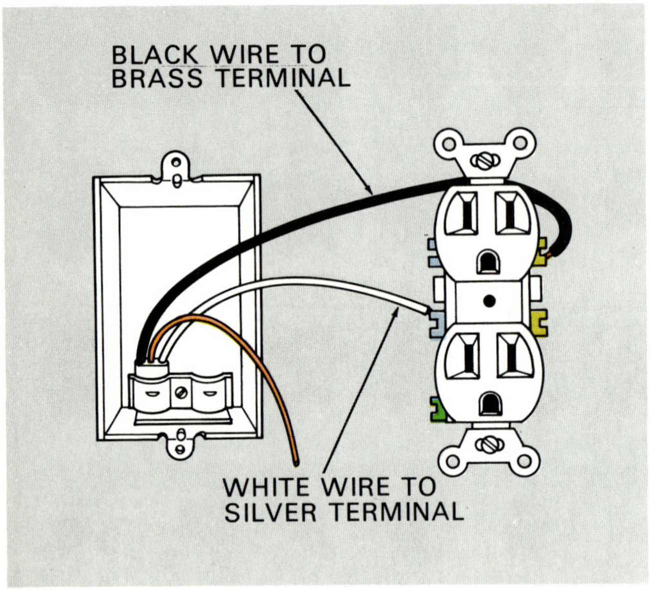

FIGURE 184. Connect black wires to brass terminals and white wires to silver terminals on the receptacle. Place the loop under the screw terminal so it will tighten as the screw is turned.

5. *Connect the grounding wire.*

   You can connect the wires to the receptacle in any order you like. Many people prefer to make the grounding connections first. The grounding wire in the cable is bare or, if insulated, green. The grounding wire must always be connected to every metal box. The grounding wire must be connected to both the box and the receptacle unless the receptacle is the type with special grounding yoke or the box is plastic. Assume you are wiring a receptacle that must be grounded to the box. Two jumper wires connect to the grounding wire(s) in the circuit cable, one from the box and the other from the receptacle.

   Proceed as follows:

   (1) *Connect the jumper wire to the grounding screw in the box.*

   Cut a 15- to 20-cm (4- to 6-inch) length of green insulated or bare wire and connect one end to the grounding screw in the box (Figure 185).

   (2) *Cut a second 15- to 20-cm (4- to 6-inch) length of green insulated or bare grounding wire and connect one end to the green grounding terminal on the receptacle (Figure 185).*

   (3) *Complete the grounding wire connections.*

   Three wire ends must be brought together, the bare wire from the cable and the two jumper wires. Place the three ends of the wires together, attach a wire nut and tighten (Figure 185).

   The receptacle is now completely wired if only one cable enters the box. However, in this instance, you must connect a second outgoing cable to extend the run to other receptacles.

6. *Connect second cable.*

   To connect the second cable, attach the black wire to the second brass terminal screw and

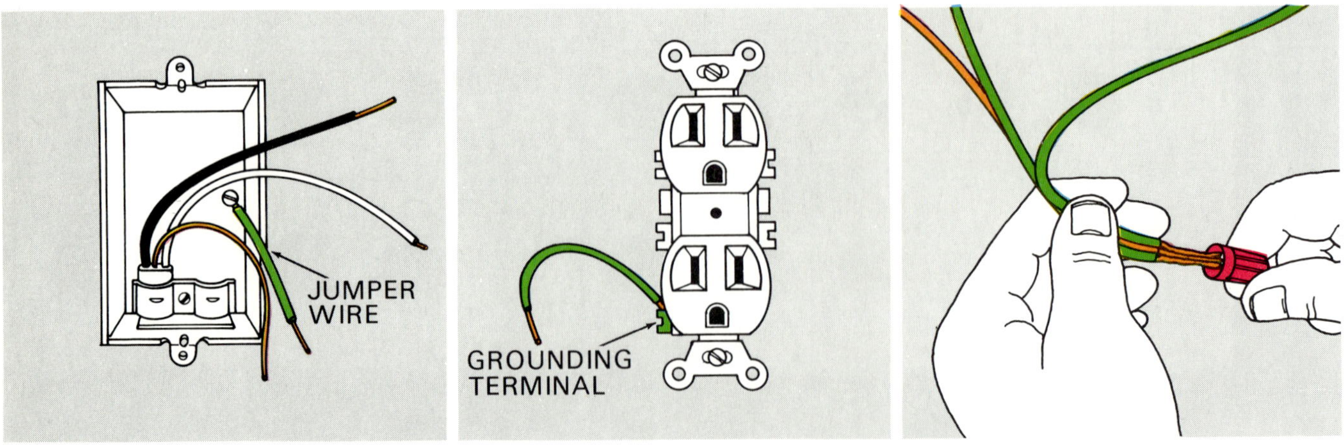

FIGURE 185. Connect one jumper wire to the box grounding screw and one to the receptacle grounding screw. Place the ends of the jumper wires together with the grounding wire from the cable and fasten together with a wire nut.

the white wire to the second silver-colored terminal screw (Figure 186).

Since the grounding connections have already been made to the receptacle and to the box, you need not repeat these steps.

7. *Connect the bare grounding wire from the second cable to the grounding wire from the first cable.*

   Instead of three grounding wire ends fastened with a wire nut, you will now have four. Remove the wire nut and place the end of the grounding wire with the ends of the first three. Replace the wire nut and tighten (Figure 186). (On the job you would connect all four grounding wires at once, knowing that the outgoing cable must be connected). This completes connections to the receptacle and box for the incoming cable and the outgoing cable.

8. *Place the receptacle in the box.*

   Bend the wires slightly so they will fold in accordion fashion. Push the receptacle into the box with wires folded behind (Figure 187).

9. *Attach the receptacle to the box.*

   The receptacle should be mounted so it is straight up and down. The box may sometimes tilt right or left. A wide slot is provided for the mounting screws at top and bottom of the receptacle to permit adjustments as needed (Figure 188).

10. *Attach cover plate.*

    After all finish work is completed on wall surfaces, attach the cover plate. Insert center screw to hold cover plate (Figure 189).

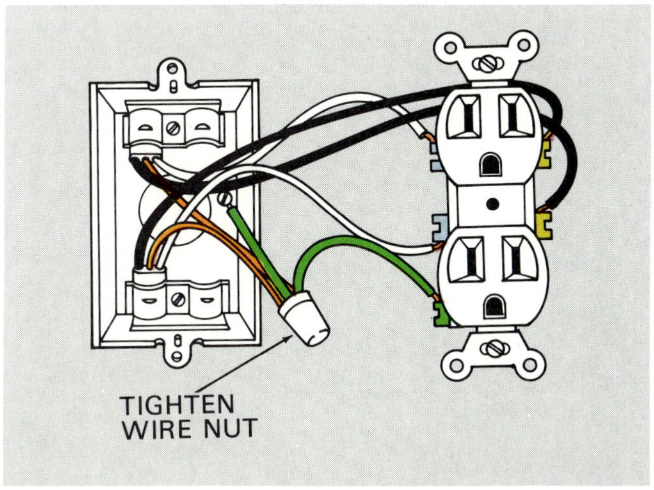

FIGURE 186. Where you have two cables entering the box, place the ends of the two cable grounding wires together with ends of the two jumper wires and fasten with a wire nut.

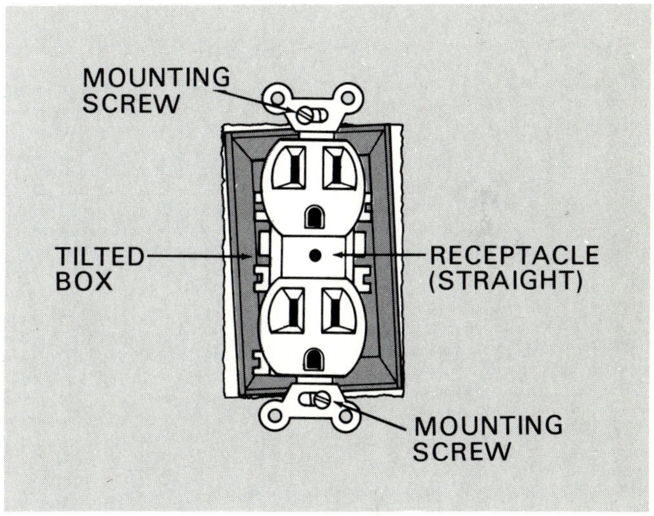

FIGURE 188. Wide slot in strap permits adjustment of receptacle left or right as required for correct position.

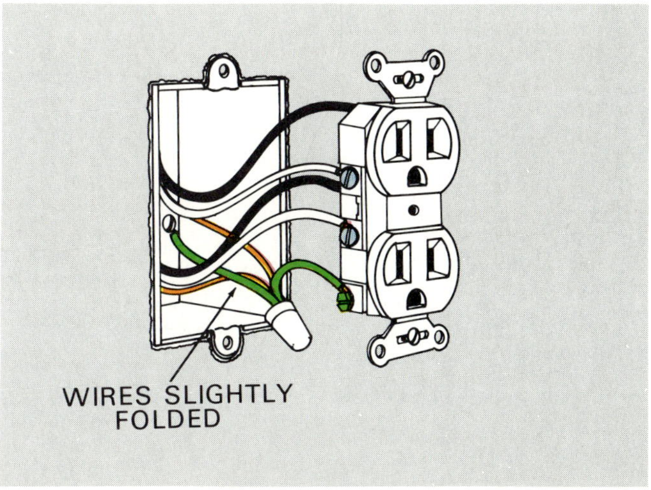

FIGURE 187. Fold wires accordion style to push them into the box more easily.

FIGURE 189. Attach cover plate after all plastering and painting is completed.

11. *Connect back-wired receptacles.*

Many receptacles are made so wiring can be connected to either side terminals or to back-wiring holes. Connections to back-wired receptacles are similar to side screw terminal connections. The difference is that wires are inserted into holes on the back of the receptacle (Figure 190). Back-wired receptacles have a strip (depth) gage on the receptacle to indicate the depth of the hole. The length of insulation to be removed from the wires must match the length of the strip gage (Figure 190). If you are installing back-wired receptacles, proceed as follows:

(1) *Strip the correct length of insulation from the black and white wires.*

(2) *Insert the wires into the holes. (Figure 190).*

With many back-wired devices you need only to insert the wire into the hole near the terminal screw to complete the connection. The wire is clamped when it is inserted. With other back-wired devices, you must first loosen a screw near the hole, insert the wire and tighten the screw against the wire. Follow instructions on the carton or packing slip.

Insert the black wire into the hole near the brass terminal and the white wire into the hole near the silver terminal.

If the screw terminals are not identified by color, the hole for each white neutral wire must have a white ring around it or it must have the word white printed next to the hole.

If an outgoing cable for another receptacle is to be connected, repeat the procedure for black and white wires. Insert the wires in the holes provided near the second set of terminals.

A hole may not be provided for the bare grounding wire. Use the green grounding terminal screw if there is no terminal marked green. Make all grounding connections in the same way as with screw terminal receptacles.

12. *Mount the receptacle in the box.*

Follow the procedures described above to connect any number of receptacles on a circuit. Connect the remaining receptacles in the bedroom. Be sure you understand the grounding wire connections. The details of that procedure will be omitted from some later circuit drawings to simplify the illustration.

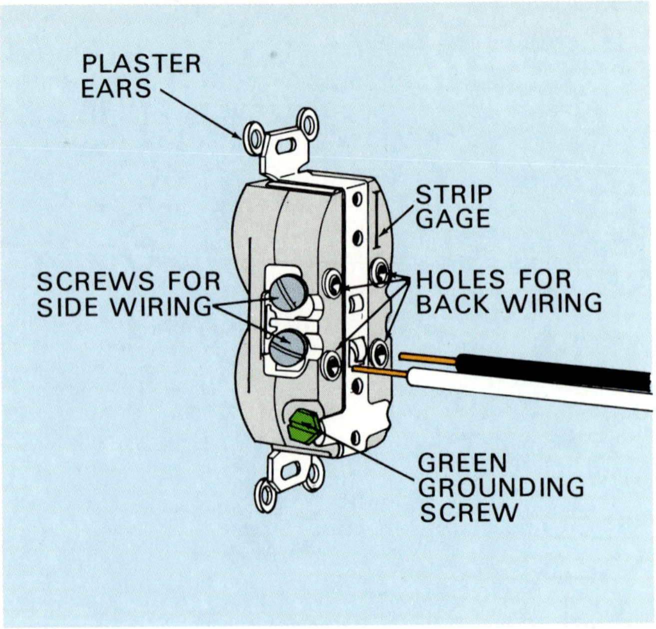

FIGURE 190. Strip insulation as required and insert wire into holes of back-wired receptacle.

b. Connecting Split-Wired Duplex Receptacles

The standard duplex receptacle has two outlets and two sets of terminals. Both outlets are on the same circuit when wired normally. If you wish to connect a separate circuit to each outlet, you may do so. Most manufacturers attach a **break-off connection tab** between the brass terminals on the receptacle. When the connector tab is removed (Figure 191), the two outlets are no longer electrically connected. That is, they are split. Each outlet can then be connected to a separate circuit. There is seldom any need to remove the tab on the neutral (silver screws) side of the receptacle.

Where do you need a split-wired duplex receptacle? There may be several instances where separate circuits on a duplex receptacle are desirable. One example is in the kitchen appliance circuits. Dividing the load between the upper and lower receptacle outlets prevents possible overloading of a single circuit (Figure 192). In addition, **it is less expensive since two 120-volt circuits are constructed using a 3-wire with ground cable correctly installed to a 120/240-volt circuit.** Only one cable is needed instead of two to obtain two 120-volt circuits. Some workshop circuits may also need split receptacles for shop equipment.

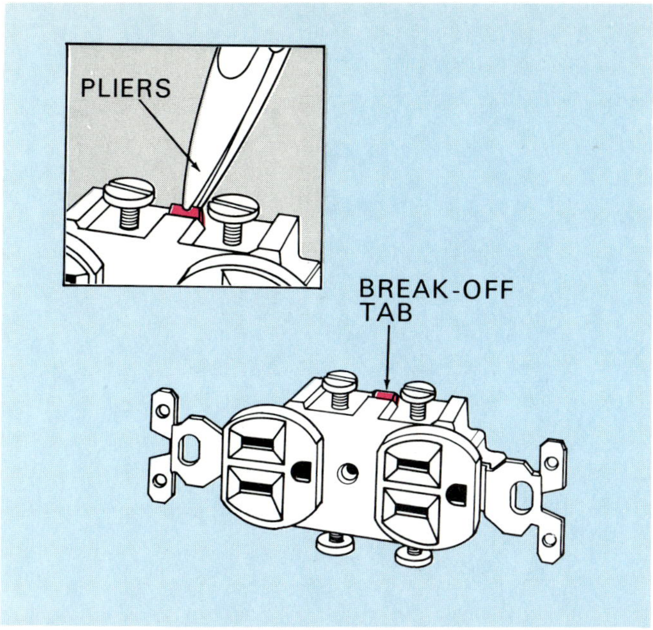

FIGURE 191. Remove the connector tab to wire the duplex receptacle with a separate circuit on upper and lower outlets.

All receptacles wired to this point have been connected to 120-volt, 15- or 20-ampere circuits using 12–2 w/g or 14–2 w/g. To wire the duplex receptacle outlets in the kitchen on separate 120-volt circuits would require two 12–2 w/g cables. In this section, you will follow procedures for wiring split receptacles on separate circuits using one 12–3 w/g cable for a 120/240-volt circuit.

When **two** circuits are wired to the receptacle on a split 240-volt circuit using **one 12–3 w/g cable,** the load must be balanced to avoid problems. To balance

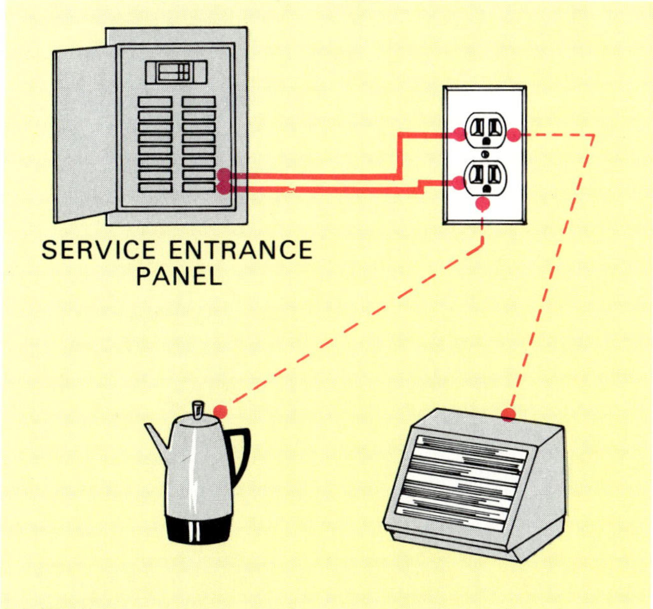

FIGURE 192. Separate (split) circuits for each outlet prevent overloading a duplex receptacle.

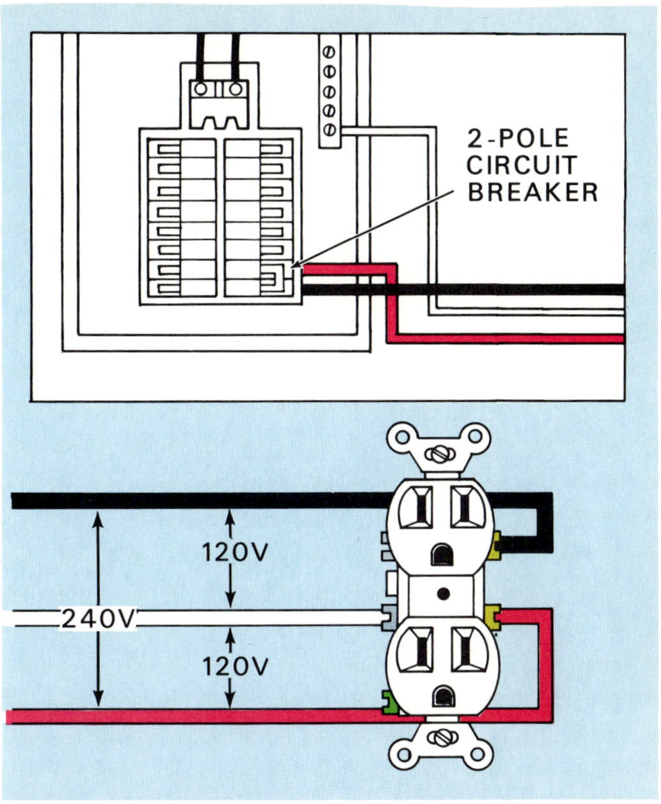

FIGURE 193. A split 240-volt circuit must have a means to disconnect both hot wires at the same time such as a two-pole circuit breaker.

the load, the black wire and red wire must be connected to electrically **opposite legs** of the SEP. On most panels, this is done by placing one above the other (Figure 193).

The Code requires the circuit for a split-wired receptacle to have a double-pole disconnect at the SEP. This may be either a double-pole circuit breaker or two single-pole circuit breakers with a handle tie (Figure 193). The purpose of this requirement is to be sure both hot conductors are disconnected at the same time. If only one of the two single circuit breakers protecting the 120/240-volt circuit is turned off, the other hot wire could cause severe shock to a person who is working on the receptacle if contact is made. Where the circuit is protected by fuses, install a pullout fuse block to insure that both hot wires are disconnected at once.

When the two hot wires of the circuit are connected to a double-pole circuit breaker or to two single-pole circuit breakers with a handle tie or to a fuse block, the load is divided between two opposite legs of the 120/240-volt supply in the SEP (Figure 193). The load is balanced between the two circuits and the danger of an overload on the neutral wire is avoided.

To connect the circuit breaker for the split-wired receptacle, place the two hot wires in the two breaker openings and tighten the screws. Run the grounding wire and the neutral wire to the neutral bar and fasten to separate terminals.

Figure 194 shows a split receptacle wired for two 120-volt circuits using two 12-2 w/g cables. When wired in this manner, you have two cables practically side by side from the SEP to the receptacle.

Figure 195 shows the same split receptacle wired with **one cable** (12-3 w/ground) and a **split 120/240-volt circuit.**

This is a less expensive way to wire the split receptacle. It saves both material and labor, since four wires will do the work of six, counting grounding wires. Proceed as follows to wire the split duplex receptacle using 12-3 w/g cable:

1. *Remove the break-off tab between the two brass terminals.*
   Assume you are installing a standard duplex receptacle with a break-off tab. Remove the tab between the **brass** terminals with pliers or screwdriver (Figure 191). Remember, the break-off tab between silver terminals is not removed, since one white neutral wire will serve both circuits.
2. *Connect the two hot wires (Figure 195).*
   Assume the receptacle box has a 12-3 w/g cable installed. One hot wire is red and one is black. The neutral is white and the grounding wire is bare.
   Connect the red wire to one brass terminal and the black wire to the other brass terminal. Where several duplex receptacles are to be wired, as over the kitchen counter, it is best to place the top receptacle on one circuit, for example, red, and the bottom receptacle on the other.

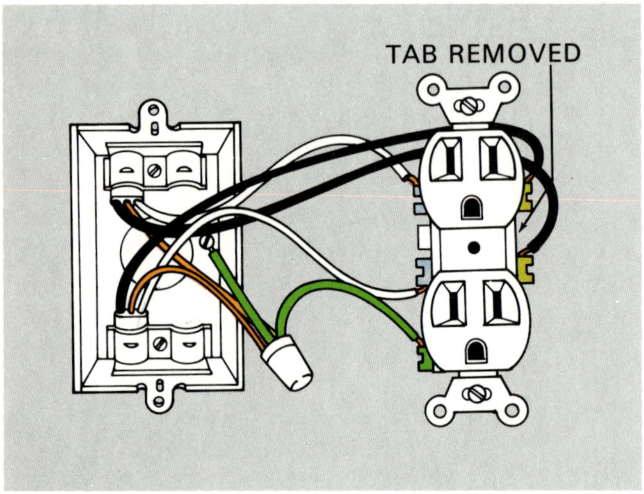

FIGURE 194. Split receptacle wired with two 12-2 with ground cables.

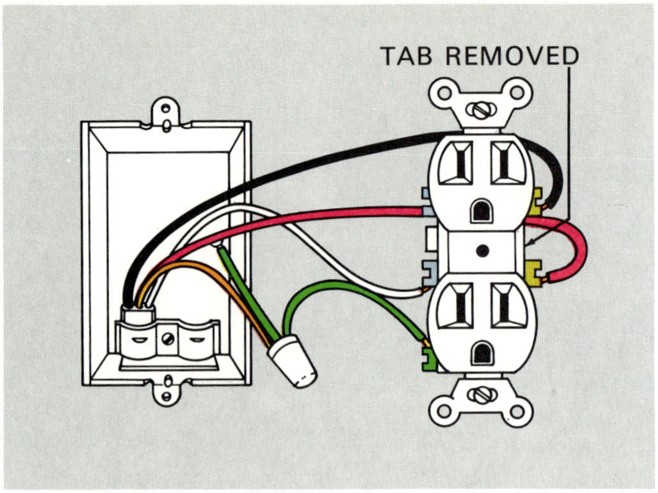

FIGURE 195. Split receptacle wired with 12-3 with ground cable.

3. *Connect the white neutral wire.*
   Connect the white neutral wire to one of the silver terminals. This neutral serves both hot wires. The tab should still be in place between the silver terminals.
4. *Connect the bare grounding wire.*
   Place the jumper wire ends from receptacle and box together with the end of the bare grounding wire from the cable. Fasten with a wire nut (Figure 196).

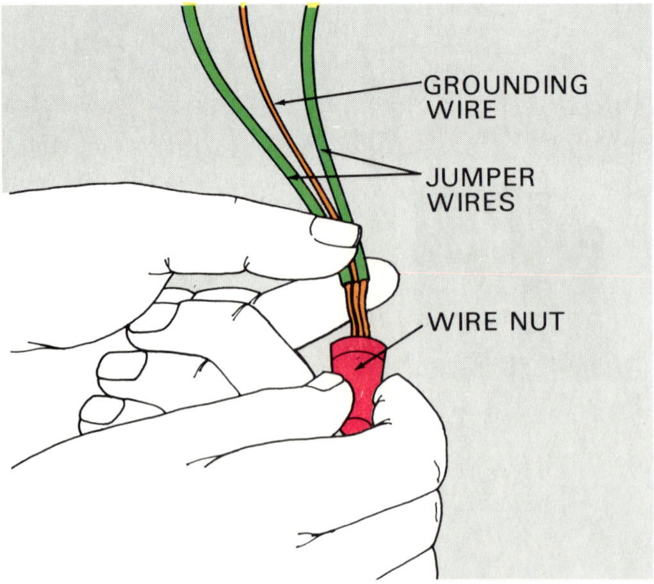

FIGURE 196. Connect grounding wire from cable to box grounding wire.

c. Connecting Individual Appliance Circuits

Appliances may be classified in three groups. They are:

—Portable.

—Stationary.

—Fixed.

—**PORTABLE APPLIANCES** such as coffeemakers, mixers and blenders may be plugged into the 20-ampere, 120-volt, kitchen, dining room and den circuits. They do not require special wiring and may be disconnected by pulling the plug from the receptacle.

In this section you are most concerned with stationary and fixed appliances which are wired on individual circuits.

—**STATIONARY APPLIANCES** are those that the homeowner usually moves with him from one house to another. Examples are self-contained ranges, clothes dryers and window air conditioners.

—**FIXED APPLIANCES** are those that are permanently installed such as built-in ovens and counter cooking tops, trash compactors, garbage disposals and water heaters.

Stationary and fixed appliances are generally wired on separate circuits. The Code requires that each appliance be protected by a **circuit breaker** at the SEP, or, if fuses are in a **pullout block,** no other disconnect is required. However, if the appliance is connected to a SEP containing only fuses, a separate disconnecting means is required. For example, the water heater is usually wired directly to connections at the tank. If you have a fused SEP without a pullout block for the water heater, a separate disconnect switch must be installed (Figure 197). If connected to a circuit breaker in the SEP, the special disconnect is not required.

Any water heater with capacity of 120 gallons or less must have fuse or circuit breaker protection at least 125% of the nameplate current. For example, a water heater with a nameplate rating of 30 amperes would require 40-ampere protection. ($30 \times 1.25 = 37.5$ Install next highest common size, 40 ampere).

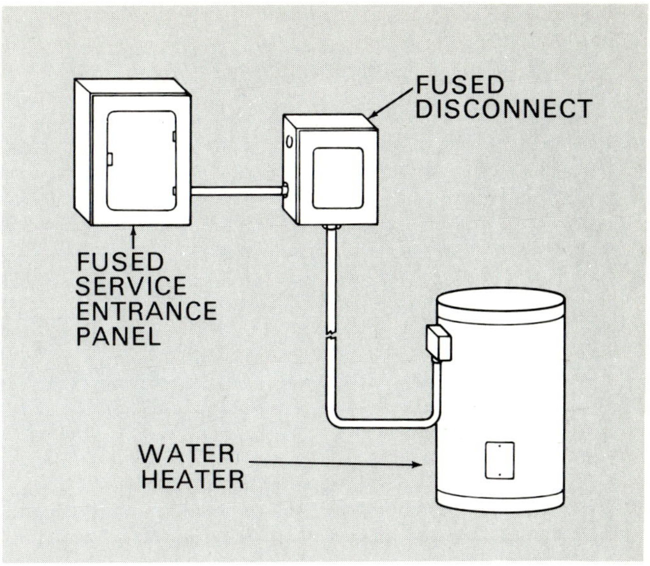

FIGURE 197. Disconnect required between a SEP with fuses and without a pullout block for the water heater.

Table VIII shows typical wire sizes and circuit breaker or fuse sizes for home appliances and equipment. Check your local code and select the correct materials for appliances.

Connecting individual appliance circuits is discussed as follows:

(1) Connecting Self-Contained Range Receptacles.

(2) Connecting Separate Ovens and Countertop Cooking Units.

(3) Connecting Clothes Dryers.

(4) Connecting Water Heaters.

(5) Connecting Clothes Washers.

(6) Connecting Electric Heaters.

TABLE VIII. TYPICAL WATTAGE, VOLTAGE, WIRE SIZE AND BREAKER OR FUSE RATINGS FOR APPLIANCES

Load	Typical Wattage	Volts	Wire	Circuit Breaker Rating**	Comments
KITCHEN					
Range	12000	120/240	3 No. 6	50A-2P	Direct to Service Entrance
Range (double oven)	21000	120/240	3 No. 4	70A-2P	Direct to Service Entrance
Oven (built-in)	5000	120/240	3 No. 10	30A-2P	Direct to Service Entrance
Range Top	6800	120/240	3 No. 8	40A-2P	Direct to Service Entrance
Dishwasher	1200	120	2 No. 12	20A-1P	Recommend Separate Circuit*
Waste Disposer	300	120	2 No. 12	20A-1P	May be combined with Dishwasher
Broiler-Roaster	1500	120	2 No. 12	20A-1P	Kitchen Receptacle Circuits
Fryer	1300	120	2 No. 12	20A-1P	Kitchen Receptacle Circuits
Coffeemaker	1100	120	2 No. 12	20A-1P	Kitchen Receptacle Circuits
Toaster	1100	120	2 No. 12	20A-1P	Kitchen Receptacle Circuits
Grill—(Portable)	1000	120	2 No. 12	20A-1P	Kitchen Receptacle Circuits
Exhaust Fan	300	120	2 No. 12	20A-1P	Combine with Light Circuits
Kitchen Receptacles	1920	120	2 No. 12	20A-1P	Minimum of two circuits Consider 120/240 V.—3 wire Split Receptacle Circuit
Refrigerator	300	120	2 No. 12	20A-1P	Recommend separate circuit for
Freezer	350	120	2 No. 12	20A-1P	Refrigerator and Freezer
LAUNDRY					
Washer	1200	120	2 No. 12	20A-1P	Separate Laundry Circuit
Dryer	5000	120/240	3 No. 10	30A-2P	Direct to Service Entrance
Dryer (High Speed)	9000	120/240	3 No. 6	50A-2P	Direct to Service Entrance
Hand Iron	1000	120	2 No. 12	20A-1P	Use Laundry Circuit
Water Heater	5000	240	2 No. 10	30-40 2P	Consult Utility

*May be combined with waste disposer
**1P = 1-pole, 2P = 2-pole

(1) CONNECTING SELF-CONTAINED RANGE RECEPTACLES. Assume you are wiring a flush mounted 50-ampere 125/250-volt receptacle (No. 10-50R, Table VII) in the kitchen. No. 6, 3-wire cable is installed in the receptacle box.

The range operates at 120 volts on low heat and on 240 volts at higher settings. It requires a cable with three wires. The **white neutral wire** in a 3-wire cable may be used as a **ground wire** because of an exception in the Code. The exception is allowed for **ranges and clothes dryers only.** Also, the same section of the Code allows the use of service entrance cable with a bare neutral wire for grounding if the range or dryer is on a separate circuit and it originates at the SEP. Proceed as follows to connect a range receptacle:

1. *Remove the receptacle cover if required.*

   Some receptacles do not require this step.

2. *Note the terminal markings on the receptacle (Figure 198) and make the connections.*

   Attach the hot wires (usually black and red) to the terminals marked "y" and "x." Connect the white wire to the terminal marked "W."

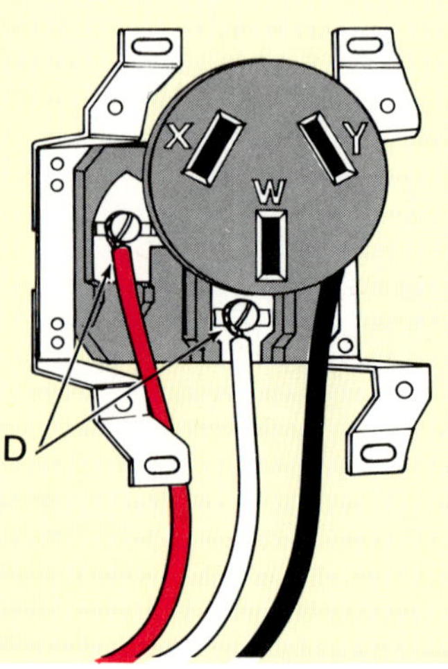

FIGURE 198. Connect the hot wires to terminals marked "x" and "y". Connect the white wire to the "W" terminal.

3. *Mount the receptacle in the box.*

NOTE: The grounding wire from the frame of the range connects to the grounded terminal of the receptacle when the connection (cord) from the range is plugged into the receptacle.

(2) CONNECTING SEPARATE OVENS AND COUNTERTOP COOKING UNITS. Except for different wire size, follow the same procedures to connect separate ovens and countertop cooking units as were used for connecting the self-contained range (Figure 198). Use the wire sizes shown in Table VIII or as recommended by the manufacturer.

NOTE: Receptacles are not required for the connection of separate ovens or countertop cooking units. They may be directly connected by bringing a properly sized cable from the SEP to the unit if a disconnect is provided in the SEP. The disconnect can be a circuit breaker or pullout fuse block.

(3) CONNECTING CLOTHES DRYERS. Procedures for connecting 240-volt clothes dryers are the same as for ranges. The wire size depends on the type of dryer. Most dryers are wired with No. 10 wire in a 30-amp circuit. If the dryer is a **high speed type,** one cycle may operate at up to 9000 watts. If you have this type, you must install a 50-amp receptacle and use No. 6 wire.

Proceed as for connecting ranges (Figure 198).

(4) CONNECTING WATER HEATERS. The water heater operates at 240 volts. It may require either No. 10-2 w/g for a 30-amp circuit or 12-2 w/g for a 20-ampere circuit, depending on local codes. Wiring is usually direct to the conductors in the water heater. However, some local codes may require a "junction" (outlet) box located 30 or more centimeters (12 or more inches) above the heater. Procedures are given as follows:

Insert the cable. Connect as in Figure 200.

The cable contains one black wire, one white wire and a bare grounding wire. The white wire must be used as a hot wire to provide a 240-volt circuit. The Code requires the white wire to be identified as hot wherever it is accessible. This may be done by wrapping it with black tape at the splice in the box and at the breaker terminal in the SEP. Black paint may be used if preferred (Figure 200).

(b) If the **junction box is not required,** make the connections directly to the conductors in the water heater and connect the grounding wire to the metal case (Figure 201).

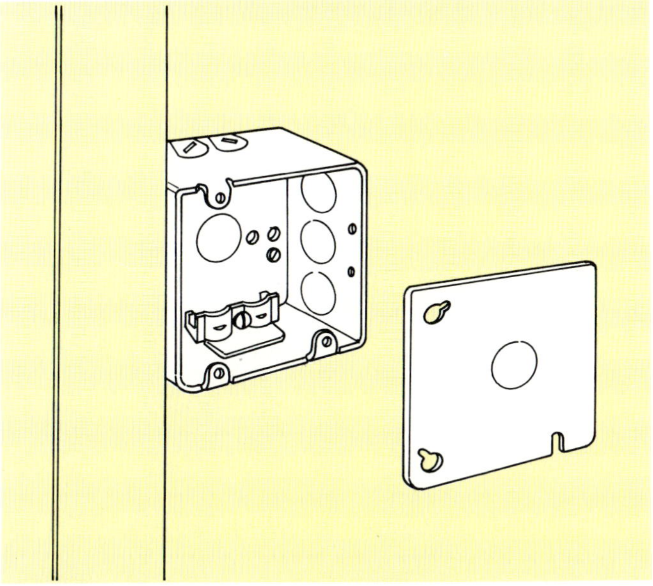

FIGURE 199. Junction box used only for making wiring connections. No device is installed.

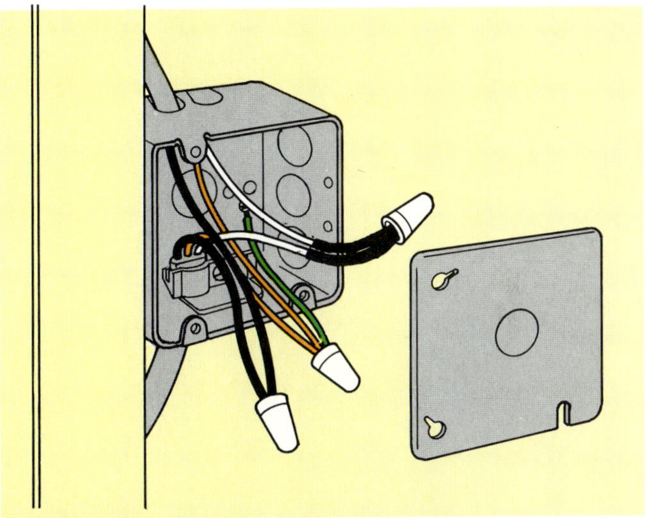

FIGURE 200. Install the cable and make required connections in the box.

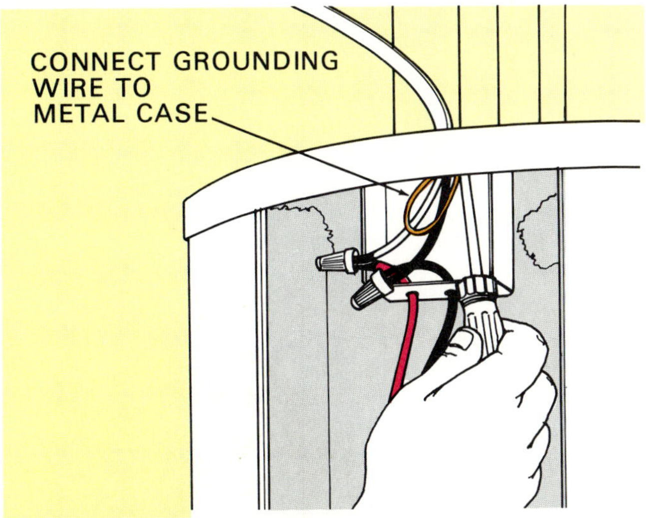

FIGURE 201. Connect the 240 volt conductors directly to heater conductors and the grounding wire to the metal case of the heater.

In the interest of energy conservation, some electric power suppliers offer the homeowner lower rates for off-peak electrical consumption. A switching device installed by the power supplier controls the time-of-day operation of the heater. The device allows the heater to operate during the periods when demand for electricity on the lines is lower. Other types may cycle the heater operation, turning off electricity for a few minutes each hour. Power supplier personnel will connect this type of device (Figure 202).

(5) CONNECTING CLOTHES WASHERS. Most clothes washers can be plugged into a 120-volt receptacle and do not require special wiring. However, they should always be plugged into a receptacle on a separate circuit and be grounded.

(6) CONNECTING ELECTRIC HEATERS. Electric heat for residences includes sizes from electric furnaces large enough to heat an entire house, to small bathroom wall heaters. Where electric heat is installed, central whole house heating with electric furnace or heat pump is a common choice. However, a large number of individual room units of various kinds are installed. One example is the electric baseboard heater. They may be either 120-volt or 240-volt units. Baseboard heaters may have a built-in thermostat or may be controlled by a wall-type thermostat.

In most central heating installations, the electrician will provide the rough-in wiring (Figure 203) and the installer of the equipment will make the connections.

The electrician may be required to install electric baseboard heaters, wall heaters and ceiling bathroom heaters. Proceed as follows:

1. *Determine type and size branch circuit required.*

   Check the unit to determine whether a 120-volt or 240-volt circuit is required. You must also know the load demand of the heating unit to determine the correct wire size to install.
2. *Install cable if required.*

   Most heaters are installed on individual branch circuits. Run the cable to the outlet if not already installed. The branch circuit wires must have an ampacity of not less than 1.25 times the total ampere load of the heater (and fan, if any).

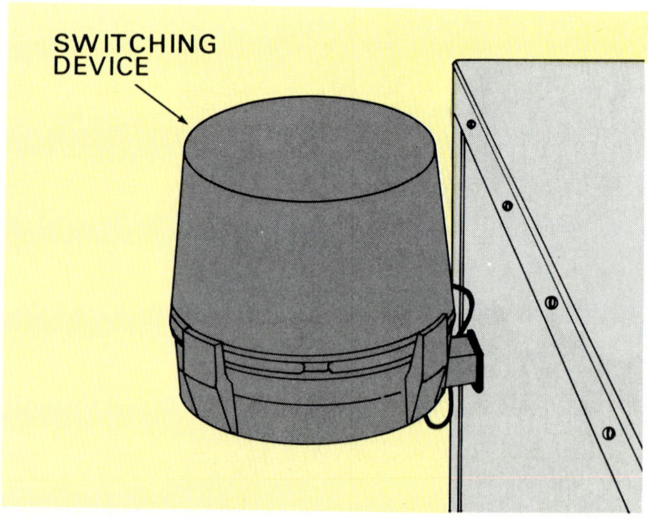

FIGURE 202. Switching device installed by power supplier to cycle the operation of water heaters and air conditioners during peak demand periods.

FIGURE 203. Rough-in wiring for central heating and air conditioning provided by the electrician.

3. *Connect the heater.*

(a) If you are installing a **ceiling heater,** as often used in bathrooms, mount a standard outlet box. Install the heater according to the instructions and connect the wiring. Ceiling heaters are usually prewired. Connect black to black and white to white wires or follow manufacturer's instructions. Install grounding wires as usual. (Instructions for installing switches are given in Section E. "Connecting Switches and Circuits.")

(b) If you are installing a **baseboard heater** or a wall heater (Figure 204), an outlet box is not normally required. Install the heater on the wall according to instructions included with the unit. Install a connector clamp through the opening in the heater box and make the connections to terminals on the unit or to wires attached to the unit (Figure 204).

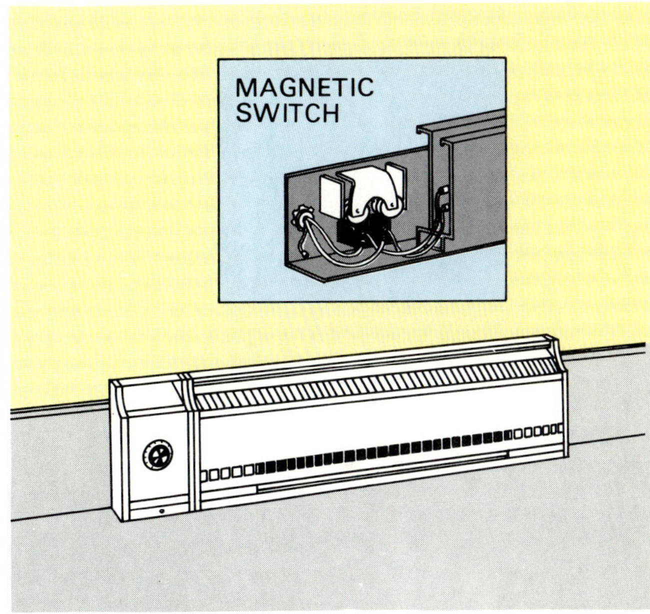

FIGURE 204. Wiring connections for baseboard heater.

D. Installing Lighting Fixtures

In this section, you will mount the lighting fixtures on walls and ceilings. Wiring connections will be made to the fixtures only. Switching connections are covered in detail in Section E.

Lighting fixtures are available in many sizes, styles and shapes. As installed in the home, they are either ceiling or wall fixtures. However, of more interest to you during installation is the method of mounting for the different types of fixtures. The following four mounting methods cover most installation procedures:

1. Mounting Direct-to-Box.
2. Mounting Strap-to-Box.
3. Mounting Fixture-to-Stud.
4. Mounting Recessed Fixtures.

1. MOUNTING DIRECT-TO-BOX

♣ Most of the lighting fixtures you will install will have fixture wiring factory installed. The fixture wires are connected to the wires from the circuit cable (source) by wire nuts. However, the fixture mounted direct-to-box may have either lead wires or two terminal screws. Assume you are mounting the type with terminal screws. This type of fixture with a pull switch may be installed in a bedroom closet as in bedroom No. 1 in the standard house plan. This is a simple connection. Since it is at the end of a circuit, it requires only two terminals and a grounding connection. Fixtures with exposed metal now must have a grounding terminal or screw for the bare grounding wire connection. **Incandescent closet fixtures must have at least 46 cm (18 inches) of clearance between the fixture and the storage area where combustible material may be stored.** A surface-mounted fluorescent fixture must have at least six inches clearance horizontally between the fixture and the shelf.

1. *Connect circuit cable wires to fixture.*

   Attach the black wire to the brass screw, white wire to the silver screw and the bare ground to the box grounding screw (Figure 205).

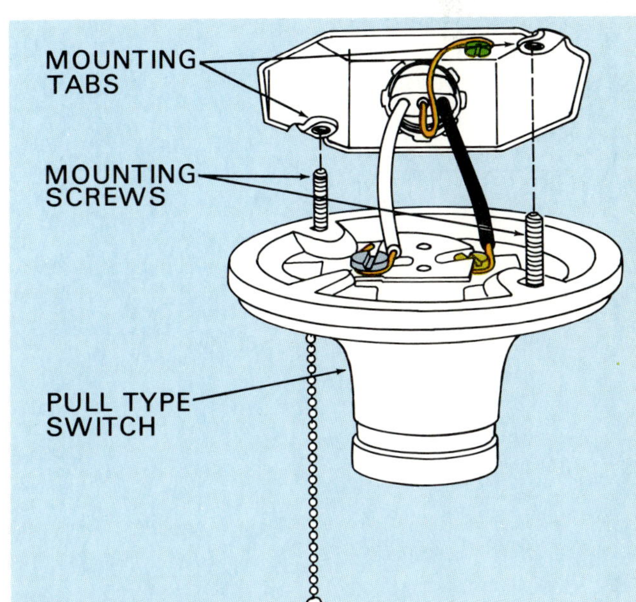

FIGURE 205. Pull-switch fixture mounted directly to the box.

2. *Attach fixture to box.*

   Fold wires slightly and place fixture against the bottom of the box. Insert bolts and tighten (Figure 205).

   If the fixture is porcelain, do not overtighten as you may break the porcelain.

2. MOUNTING STRAP-TO-BOX

Assume you are mounting a wall fixture in a box in bathroom No. 1 of the standard house plan. Wires from the incoming cable (source) and from the outgoing cable (to following outlet) must be connected to the fixture wires. Proceed as follows:

1. *Connect circuit cable wires to fixture wires.*

   Connect three black wires with wire nut. Connect one each from **source cable** (top) from **fixture** (right) and from **outgoing cable** (bottom) (Figure 206). Connect three white wires with wire nut, one each from source, fixture and outgoing cable. Connect three bare grounding wires, one from each cable, one from the fixture, plus pigtail wire to grounding screw (4 wire ends total) (Figure 206). Fold wires into the box.
2. *Attach fixture to box.*

   Bolt strap to box and insert headless bolts. Attach fixture to headless bolts using ornamental nuts (Figure 207).
3. *If you have a ceiling fixture with bolts, attach to strap, as in Figure 207.*

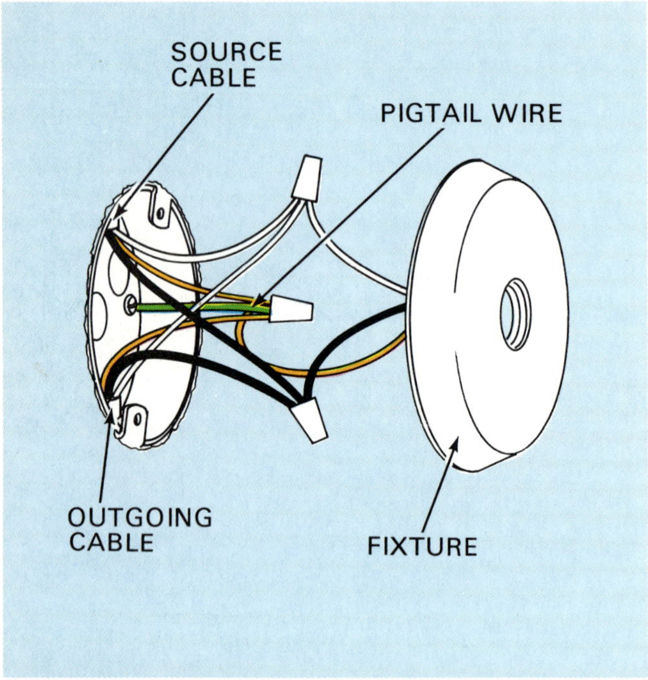

FIGURE 206. Connect wires from source cable, outgoing cable and fixture.

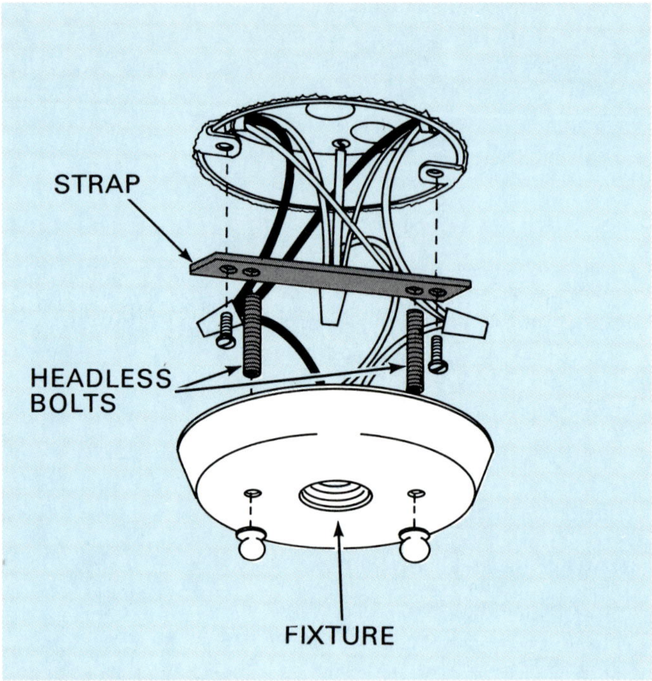

FIGURE 207. Attach strap to box and fixture to headless bolts.

3. MOUNTING FIXTURE-TO-STUD

Heavy suspended fixtures, such as large chandeliers, require more support than those described above. Many fixtures of this type require a **stem** in addition to the stud. The stud and stem are joined together by a fixture coupling sometimes called a hickey. (Figure 208). Proceed as follows:

1. *Screw coupling onto stud on outlet box.*

   Pull wires from fixture stem through coupling (Figure 208).
2. *Screw fixture stem into coupling and connect wires (Figure 208).*
3. *Push canopy against ceiling and tighten holding screw (Figure 208).*

A Code rule effective April 1, 1982 requires that lighting fixtures with exposed metal parts have a grounding terminal. You must connect the grounding wire to the terminal if you are using nonmetallic cable after the effective date (Figure 208).

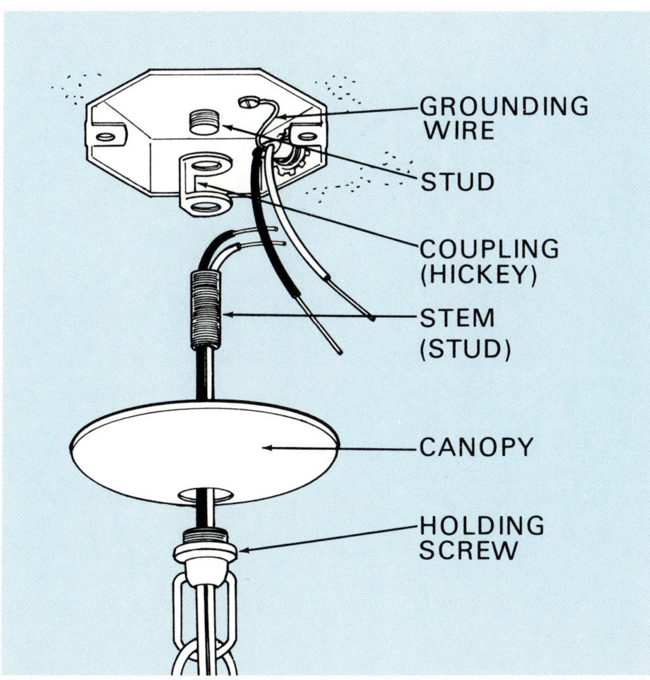

FIGURE 208. Screw coupling to stud and stem, connect wiring and attach canopy.

4. MOUNTING RECESSED FIXTURES

Recessed fixtures are self-contained. They should be mounted before plaster or dry wall is in place. Most are prewired and can be quickly mounted using adjustable brackets (Figure 209). Wiring connections are in the attached junction boxes. The Code requires that no thermal insulation may be installed within 8 cm (3 inches) of the recessed fixture enclosure, wiring compartment or ballast. Free circulation of air must be allowed around the fixture. Also, recessed portions of the enclosure must be spaced at least 1.3 cm (½ inch) from combustible material.

Proceed as follows:

1. *Mount the fixture according to procedures in Section II, "D. Installing Device Boxes and Outlet Boxes."*
2. *Connect wiring as required.*

FIGURE 209. Recessed fixtures are prewired and are mounted with screws or nails. They must have 1.3 cm (½ inch) clearance between the fixture and the joist.

E. Connecting Switches and Circuits

The information contained in this section on connecting switches and circuits is very important in your training program. Upon successful completion of procedures included in the material, you will be able to connect switches to fit most situations you will find in common wiring jobs.

Connecting switches and circuits is discussed under the following headings:

1. Types and Sizes of Switches.
2. Connecting Single-Pole Switches.
3. Connecting Switched-Duplex Receptacles.
4. Connecting Three-Way Switches.
5. Connecting Four-Way Switches.
6. Connecting Combination Devices.
7. Connecting Double-Pole Switches.

NOTE: Before installing switches described in this section, be sure to read the installation instructions included in or on the carton.

1. TYPES AND SIZES OF SWITCHES

In house wiring, the switches you install will likely be AC switches for control of lighting circuits. Most of those used in lighting circuits are toggle switches. Other kinds available are the rocker-type and push-button switches (Figure 210). Older type AC-DC switches are seldom used because the AC type is quieter and lower in cost.

Switches have their safe ampere and voltage ratings stamped on the metal yoke and, if horsepower rated, their maximum rating. A switch may be rated for both 120-volt and 240-volt operation. For example, 10 amperes at 120 volts or 5 amperes at 240 volts. In others the switch may be rated only for 120-volt use. Check your switches before mounting to be sure the ratings are suitable for the circuit.

FIGURE 210. Types of switches used in the home.

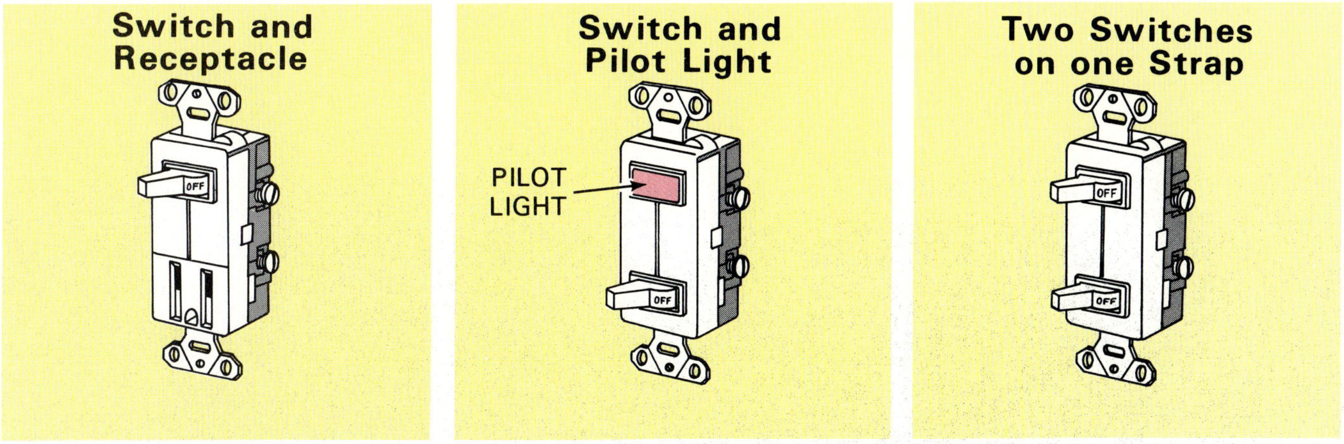

FIGURE 211. Examples of combination switches.

Switches are available in various combinations such as switch and receptacle, switch and pilot light or with two switches on one strap (Figure 211).

As with receptacles, switches may be either side-wired or back-wired or both. Other useful variations are the time-delay switch and the rotary-dimmer switch (Figure 212).

The **time-delay** switch turns lights on as an ordinary switch. However, when turned to "delay," which is the "off" position, the lamp remains "on" for a convenient period of time. In the garage in the standard house plan, for example, it will allow time to enter your car comfortably before the lamp turns "off."

A **rotary-dimmer** switch turns the lamp to any level of lighting desired, from "off" to full brightness. Dimmer switches are often used to control lighting levels in living and dining rooms.

Both time-delay and rotary-dimmer switches are mounted in standard switch boxes. Both connect in the circuit in the same manner as standard single-pole switches.

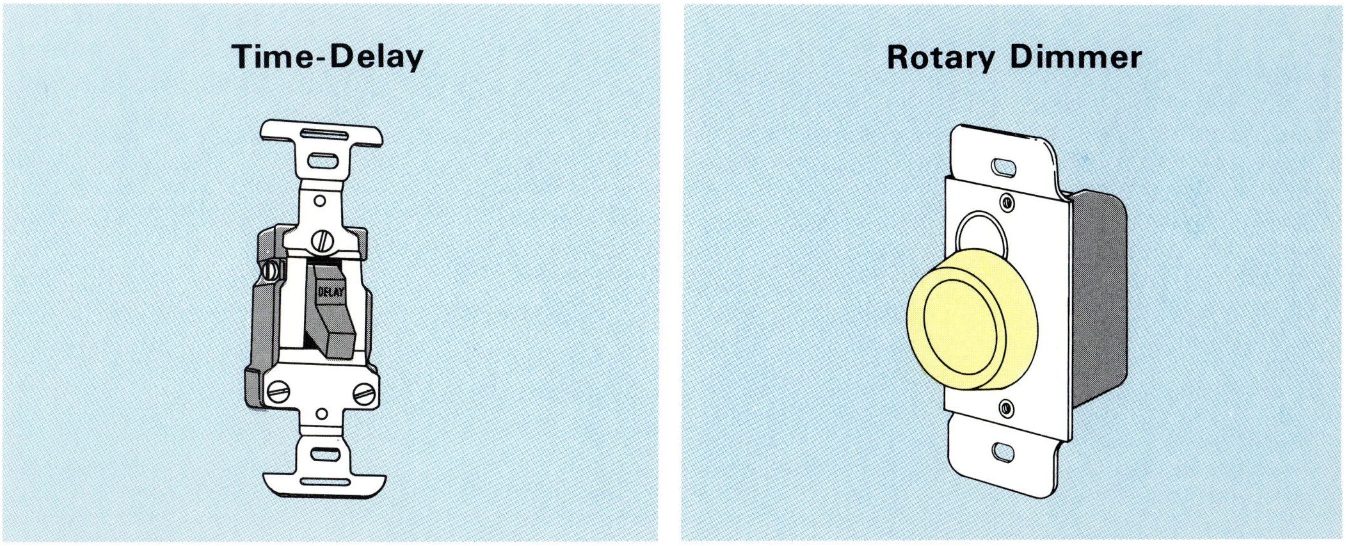

FIGURE 212. Time-delay and dimmer switches.

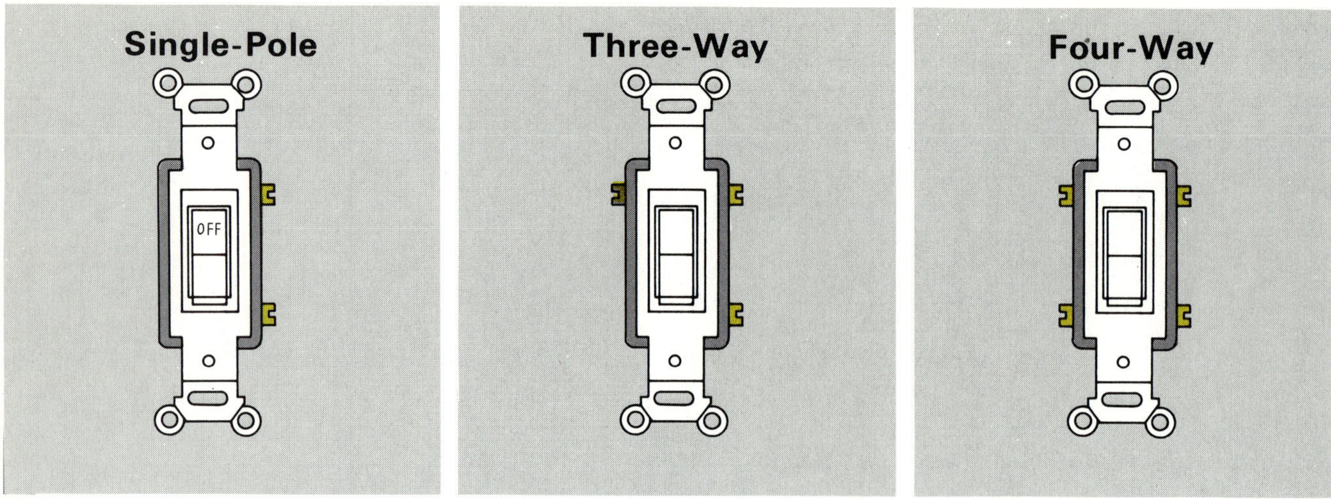

FIGURE 213. Single-pole, three-way and four-way switches.

Switches also differ in their internal wiring. You are likely to work with three different types in house wiring. Terminals may be located on ends, front or sides. The three types are (Figure 213):

—Single-Pole Switches
—Three-Way Switches.
—Four-Way Switches.

2. CONNECTING SINGLE-POLE SWITCHES

A single-pole switch controls lighting from one position. But one single-pole switch may control several lights or receptacles on the same circuit. Wiring connections could be different for one or two or more lights on a circuit. You will also find different connections depending on whether the power source enters the circuit at the switch or through an outlet. The term **source** is used to indicate the power source from the SEP.

The Code requires that metal switch **boxes** be grounded. But the switches themselves do not usually have a grounding terminal unless used for a special purpose. So, unlike receptacles, the pigtail grounding wire is not normally required to be connected from box to switch.

A jumper wire is required to connect the grounding wire at the lighting outlet box, **only if the circuit continues to another outlet.** If the outlet is at the end of the circuit, the bare wire connects directly to the grounding screw in the outlet box. Make all required connections at each outlet or switch before moving to another.

Both of the above grounding situations are seen in the circuit in bedroom No. 1 in the standard house plan (Figure 214). A single-pole switch controls the light. Current from the SEP enters at the switch box (source). The bare grounding wire is connected to the grounding screw in the switchbox by a jumper wire. The ends of the bare wire from the cables are connected to the jumper wire with a wire nut.

At the outlet box for the light, the bare grounding wire is connected directly to the grounding screw on the metal box. No jumper is required because the circuit does not continue beyond the lighting outlet.

Connecting single-pole switches is discussed as follows:

a. Connecting Single-Pole Switch with Switch Source.
b. Connecting Single-Pole Switch with Lighting Outlet Source.
c. Connecting Single-Pole Switch for Two Lamps Using Lighting Outlet Source.
d. Connecting Single-Pole Switch for Two Lamps with Switch Source and Unswitched Line.

a. Connecting Single-Pole Switch with Switch Source

Single-pole switches control lights from one position, as in a bedroom. The lever is marked "off" and "on" (Figure 213).

1. *Connect switchbox wiring (bedroom No. 1, circuit No. 2) in standard house plan (Fig. 214).*
   (1) *Connect the black wire from the source cable to one terminal on the switch.*
   (2) *Connect the black wire from the outgoing cable to the other switch terminal.*

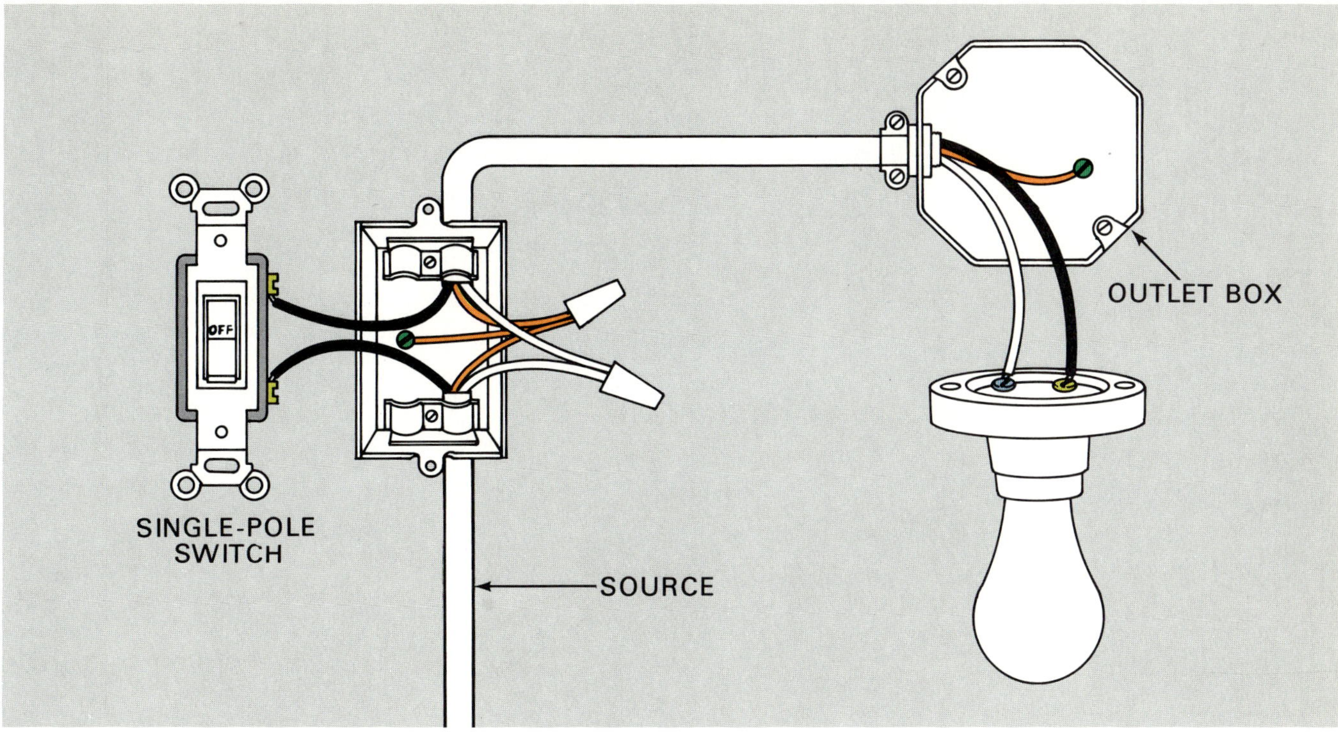

FIGURE 214. Single-pole switch and lighting outlet with switch-source wiring.

(3) *Connect the white neutral wire from each cable with a wire nut.*

Note that the white neutral wire does not connect to the switch. **Neutral wires do not connect to switches.**

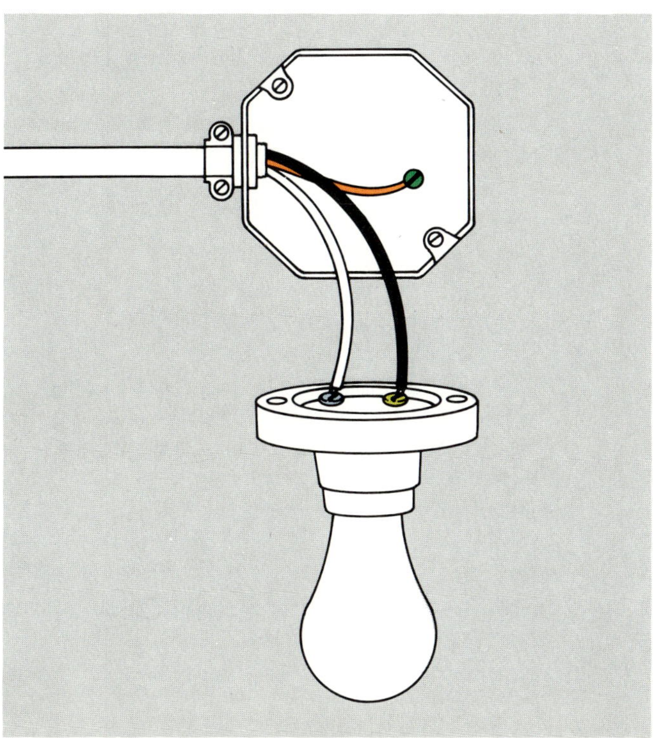
FIGURE 215. Outlet-box wiring for single-pole switch control.

(4) *Connect the bare grounding wire.*

Connect the jumper wire to the screw in the switch box. Place the ends of the bare wire from each cable together with the jumper wire and fasten with a wire nut.

2. *Connect outlet box wiring (Figure 215).*

(1) *Connect the grounding wire to the grounding screw in the outlet box.*

(2) *Connect cable wires to fixture terminals.*

Connect black wire to brass and white wire to silver terminal.

To simplify illustrations, bare grounding wire connections are not shown in some of the remaining circuits. You may refer to Figures 214 or 216A if necessary for bare grounding wire connections. If the circuit continues beyond the outlet box at the fixture, connect the grounding wires from cable ends to the box by means of a jumper wire. Use the procedure shown at the switch box in Figure 216A.

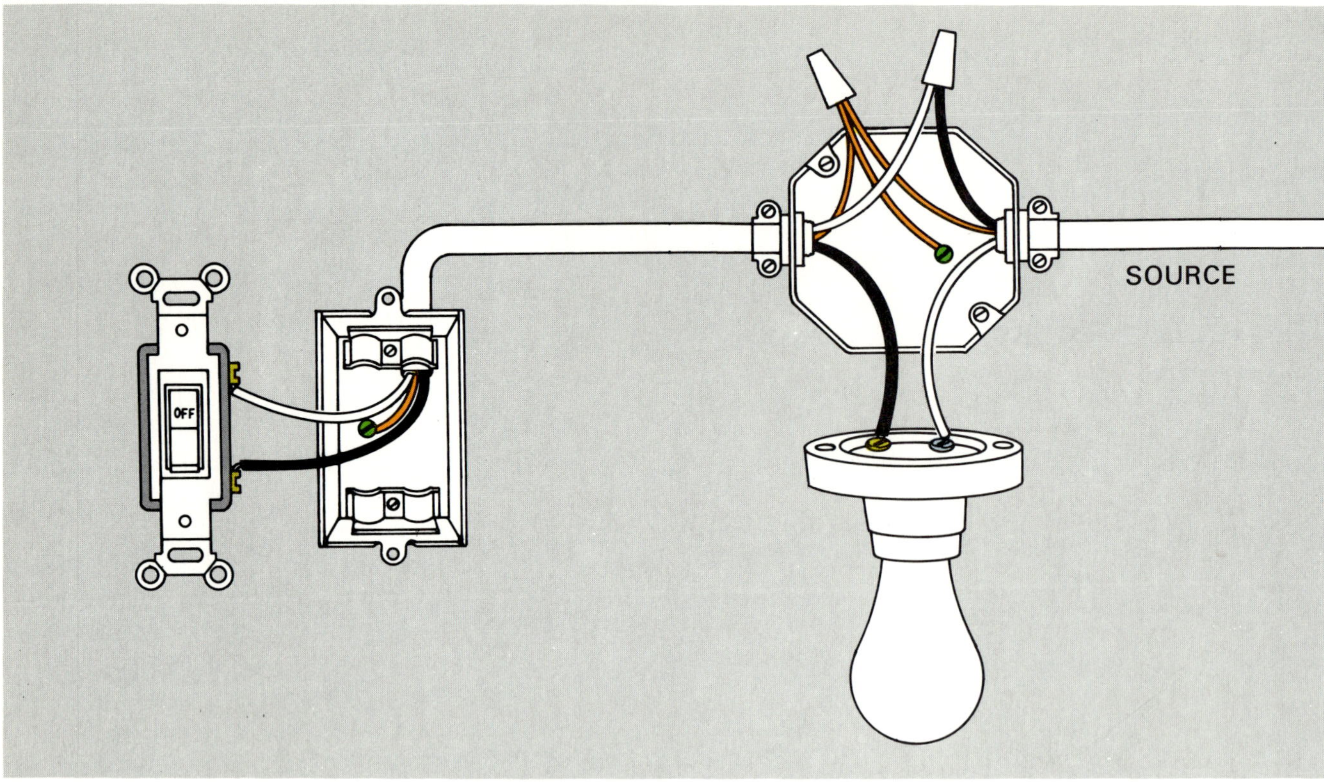

FIGURE 216A. Where the white wire is installed on the supply side of the switch loop, it is not necessary to identify as a hot wire.

b. Connecting Single-Pole Switch with Lighting Outlet Source

You may want to connect either (1) one single-pole switch to one light outlet or (2) two single-pole switches to two different light outlets.

(1) ONE SINGLE-POLE SWITCH TO ONE LIGHT OUTLET (Figure 216A). An example of the single-pole switch with lighting outlet source is in bedroom No. 3, circuit No. 3 of the standard house plan.

1. *Connect wiring at the light outlet.*

   (1) *Connect white wire from the switch to black wire from source cable.*

   Remember that an **exception** was noted in the first discussion of Code requirements regarding the **white wire** in a cable (Section B, 1, "Grounding the System and Circuits"). The exception is as shown in a **switch loop,** sometimes called a switch leg. A switch loop is required when the power source enters the lighting outlet box first.

   The Code permits connecting the white wire in a cable to a switch, **provided it is used for the supply to the switch but not as a return conductor from the switch to the switched outlet.**

   In this instance and all others **where permitted, the white wire is used as if it were a black wire.** The reason it is permitted is that two-wire cable is made only with one black and one white wire, not two black wires. This white wire is considered an unidentified or "hot" wire.

   To connect a single-pole switch with lighting outlet source, proceed as follows (Figure 216A):

   (2) *Connect black wire from switch to brass terminal on the fixture.*

   (3) *Connect white wire from source to silver terminal on fixture.*

   (4) *Connect bare grounding wires.*

   Connect pigtail grounding wire to ground screw in switch box. Connect wire ends from pigtail wire, source cable and switch-box cable with wire nut.

2. *Connect wiring at the switch.*

   (1) *Connect the black wire to one switch terminal.*

   (2) *Connect the white wire to the other switch terminal.*

   (3) *Connect bare grounding wire to box grounding screw.*

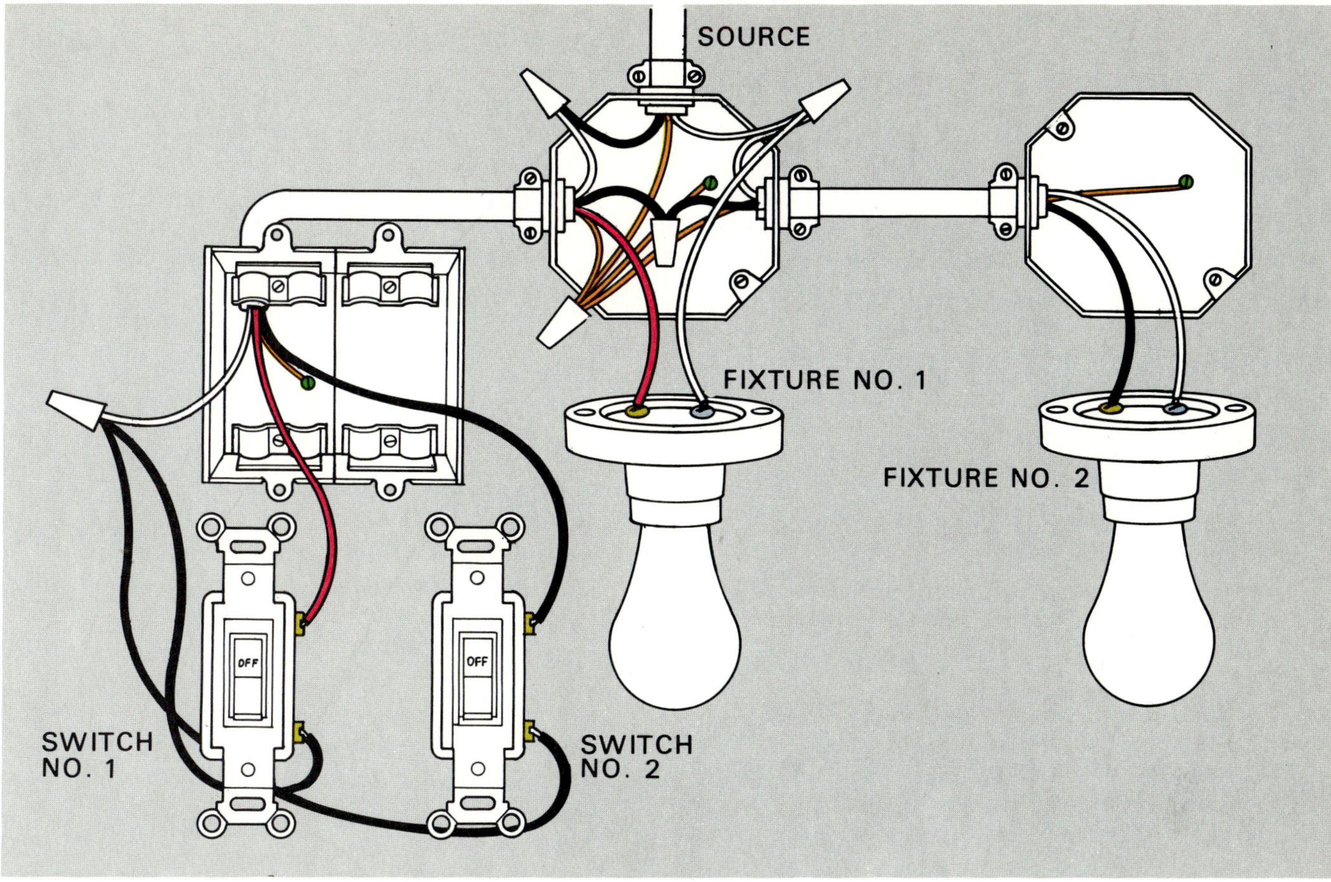

FIGURE 216B. Two single-pole switches control two different light outlets.

(2) TWO SINGLE-POLE SWITCHES TO TWO DIFFERENT LIGHT OUTLETS (Figure 216B). The grounding wires are not shown in this illustration so the other wiring can be shown more clearly. Connect grounding wires as in Figure 216A.

1. *Connect wiring at fixture No. 1.*

   This circuit requires 3-wire cable from the source to the switch and 2-wire to fixture No. 2.

   (1) *Connect black wire from source to white wire from switch box.*

   (2) *Connect white wire from source to a white jumper wire and to white wire going to fixture No. 2.*

   (3) *Connect black wire going to the switch to black wire from fixture No. 2.*

   (4) *Connect red wire to brass-colored light terminal and white jumper wire to silver light terminal.*

2. *Connect wiring at Fixture No. 2.*

   (1) *Connect black wire to brass terminal and white wire to silver terminal.*

3. *Connect wiring at switches.*

   (1) *Connect black wire from box at fixture No. 1 to a terminal on switch No. 2.*

   (2) *Connect white wire from box at fixture No. 1 to two black jumper wires.*

   (3) *Connect red wire from fixture No. 1 to a terminal on switch No. 1.*

   (4) *Connect one black jumper wire to Switch No. 1 and the other black jumper wire to switch No. 2.*

c. Connecting Single-Pole Switch for Two Lamps Using Lighting Outlet Source

Assume you are wiring the wall lighting outlets in bathroom No. 1 of the standard house plan on one switch. Grounding wire connections are omitted for greater clarity of circuit wire connections.

The circuit begins at the lighting fixture on the left. Connections are similar to those for Figure 216 except that the outgoing cable to the second outlet must be included.

Proceed as follows (Figure 217):

1. *Connect wiring at the switch.*

   (1) *Connect the black wire to one switch terminal.*

   (2) *Connect the white wire to the other switch terminal.*

   (3) *Connect bare grounding wires as in previous section (not shown).*

2. *Connect wiring at first outlet box.*

   (1) *Connect white wire from the switch to black wire from source cable.*

   (2) *Connect black wire from switch to black wires from fixture and cable going to next lighting outlet.*

   (3) *Connect white wire from source cable to white wires from fixture and cable going to next lighting outlet.*

   (4) *Connect grounding wire as in previous sections (not shown).*

3. *Connect wiring at second outlet box.*

   (1) *Connect black wire to brass terminal.*

   (2) *Connect white wire to silver terminal.*

   (3) *Connect grounding wire as in previous section (not shown).*

   Circuit No. 1 ends at this outlet.

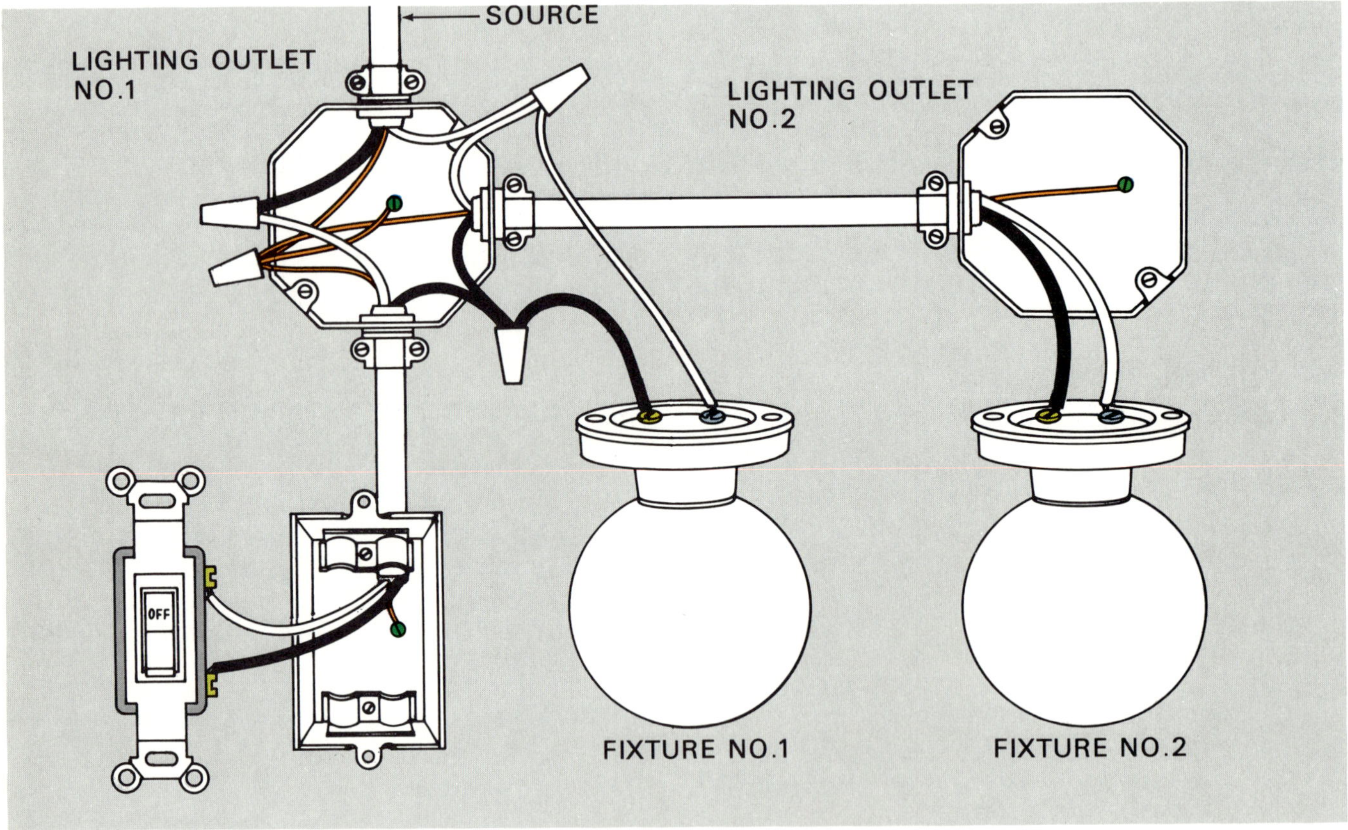

FIGURE 217. Single-pole switch controls two lighting outlets with lighting outlet source.

d. Connecting Single-Pole Switch for Two Lamps with Switch Source and Unswitched Line

In the standard house plan, the circuit often continues from outlet to outlet. **Three-wire** cable is required if an *unswitched* cable is to be continued beyond the light fixtures. The 3-wire cable contains a white, a black, and a red wire (Figure 218). The black and red wires are both hot. Figure 219 shows an example of a switch source circuit, two wall lamps with two lights and an unswitched cable continuing beyond the lights.

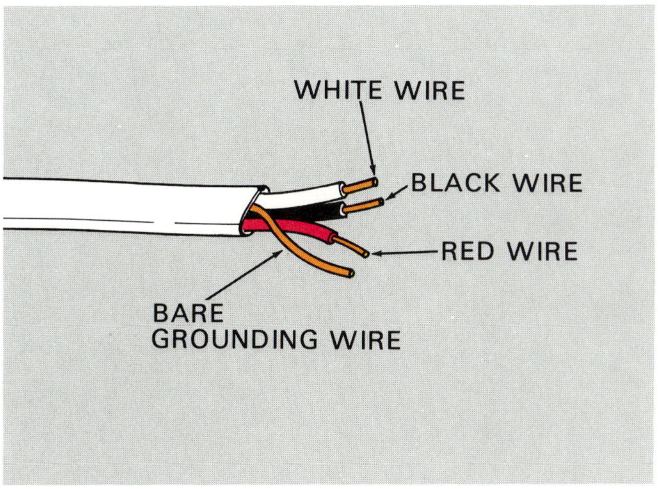

FIGURE 218. Three-wire with ground cable.

🍁 Proceed as follows using 14–3 w/g or 12–3 w/g cable (Figure 219).

1. *Connect wiring at the switch.*
   (1) *Connect white wire from source to white wire of outgoing cable.*
   (2) *Connect black wire from source to black wire of outgoing cable and to 15 cm (6-inch) black jumper. Connect black jumper to either switch terminal.*
   (3) *Connect red wire of outgoing cable to other switch terminal.*
   (4) *Connect bare grounding wires as in previous section (not shown).*
2. *Connect wiring at first outlet box.*
   (1) *Connect white wire from switch to white wire in outgoing cable and white wire from fixture.*
   (2) *Connect red wire to red wire in outgoing cable and to black wire from fixture.*
   (3) *Connect black wire to black wire.*

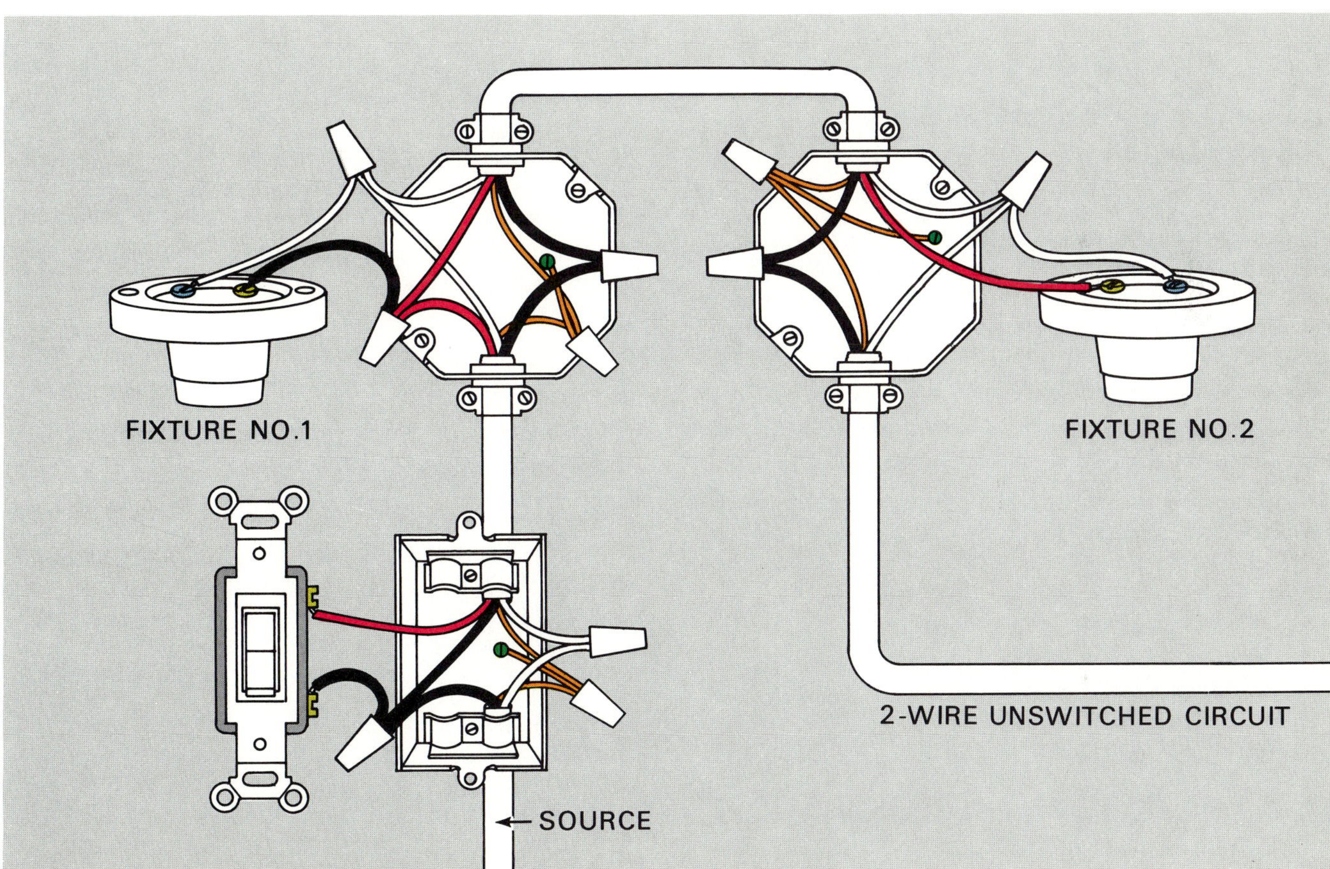

FIGURE 219. Single-pole switch with two lamps and unswitched line.

(4) *Connect grounding wires as in previous sections (not shown).*

3. *Connect wiring at second outlet box.*

   (1) *Connect white wire to white wire from both cables and to silver terminal on fixture.*

   (2) *Connect black wire to black wire from each cable.*

   (3) *Connect red wire from first outlet to bronze terminal on fixture.*

   (4) *Connect grounding wires as in previous section (not shown).*

   As shown (Figure 219), a two-wire unswitched circuit extends beyond the second lighting outlet to the convenience outlet. If desired, one or more additional light outlets could be added, using 3-wire cable from outlet to outlet, until the point where the unswitched line is needed.

3. CONNECTING SWITCHED DUPLEX RECEPTACLES

Connecting switched-duplex receptacles is discussed as follows:

a. Connecting Switched-Duplex Receptacles.
b. Connecting Switched-Split-Duplex Receptacles.

a. Connecting Switched-Duplex Receptacles

Switched-duplex receptacles are often installed in living areas of the home. Table lamps supplying part, or all of the lighting, are more convenient to use if connected to a switched receptacle that you can turn on when entering a room. The switch may control both outlets on the receptacle or only one.

To connect the switch to both halves of a duplex receptacle outlet, proceed as follows:

Assume you are wiring a bedroom circuit in the standard house plan. The source is at a wall receptacle. The switch is near the bedroom door. It controls certain bedroom receptacles (Figure 220).

1. *Connect wiring at the receptacle.*

   Connect black wire from switch to either brass terminal. Connect white wire from switch to black wire from source cable. Connect white wire from source cable to silver terminal on receptacle.

2. *Connect cable going to next receptacle.*

   Connect black wire to brass terminal and white wire to silver terminal.

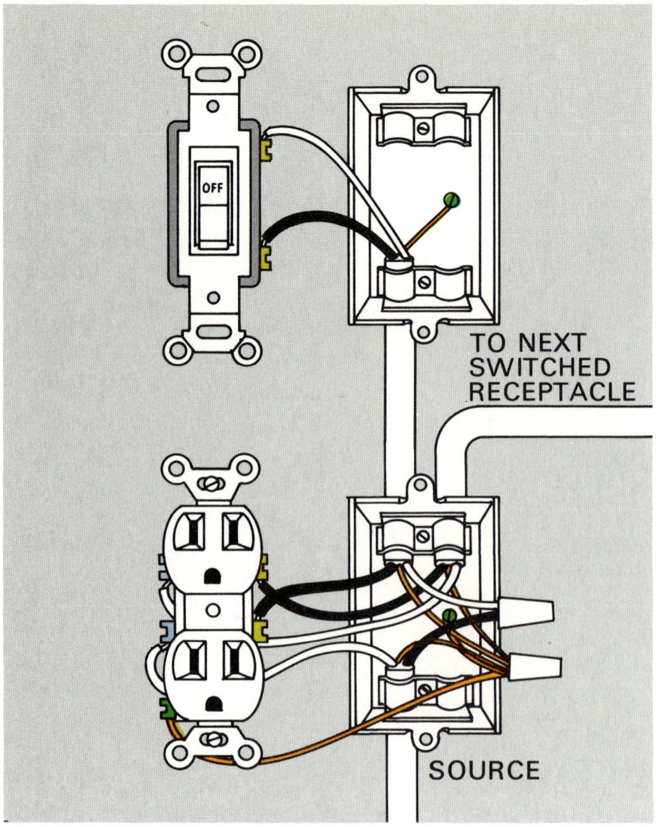

FIGURE 220. Wiring for a switched-duplex receptacle.

3. *Connect grounding wires (not shown).*

   Connect grounding wires as in previous sections.

4. *Connect wiring at the switch.*

   Connect the black wire to one terminal and white wire to the other terminal.

b. Connecting Switched-Split-Duplex Receptacles

It is often desirable to have one half of the receptacle controlled by a switch and have the other half remain constantly hot. In the standard house plan, assume you are wiring the living room receptacle on the wall next to the bedroom with switch in the hall.

Proceed as follows:

1. *Remove connecting tab between two brass terminals of the receptacle.*

   Use screwdriver or pliers to remove tab. The outlets are no longer electrically connected on the hot side when tab is removed.

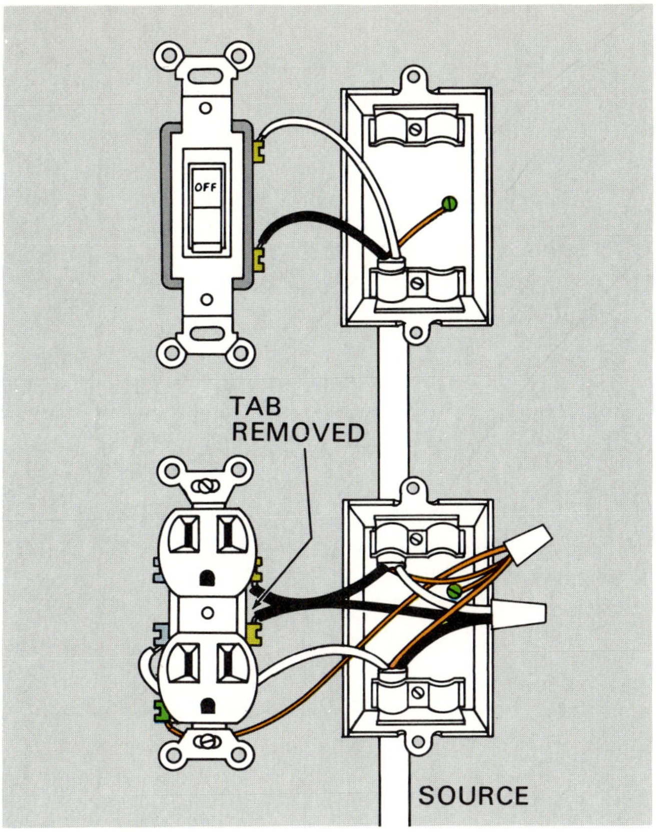

FIGURE 221. Switched-split-duplex receptacle wiring.

2. *Connect wiring at receptacle (Figure 221).*

   Procedures are the same as for a single circuit except for one black jumper wire to a brass terminal. Connect black source wire to a black jumper and to white wire from switch. Connect black jumper to lower brass terminal. The jumper wire makes the lower outlet constantly hot (not controlled by switch). Connect black wire from switch to top brass terminal. This outlet is controlled by the switch. Connect source white wire to silver terminal.

3. *Connect wiring at the switch.*

   Connect black wire to one terminal and white wire to other terminal (Figure 221). Connect grounding wire to grounding screw in box (not shown).

4. *Connect grounding wires as in previous sections.*

4. CONNECTING THREE-WAY SWITCHES

Three-way switches make it possible to switch lighting fixtures on or off from either of **two** locations. Why not three locations if they are three-way switches? The three-way description refers to the **number of terminals** on the switch, not the number of locations. Three-way switches have three terminals instead of two, as on single-pole switches. With the use of three-way switches, it is convenient to turn lights on or off without retracing steps. You may turn the light on at the switch nearest the door where you enter the room and turn it off at the other switch as you leave (Figure 222). Three-way switches are also useful in long halls, stairways and between house and garage.

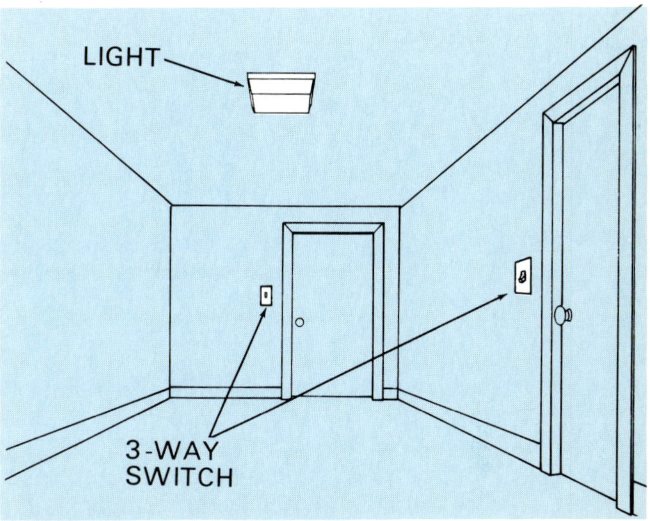

FIGURE 222. Three-way switches turn light on or off at two locations.

Three-way switches are used only in pairs. The toggle (handle) on a three-way switch is not marked for "on-off" position (Figure 223). Either up or down position may switch the light on or off, depending on the position of the toggle on the other switch.

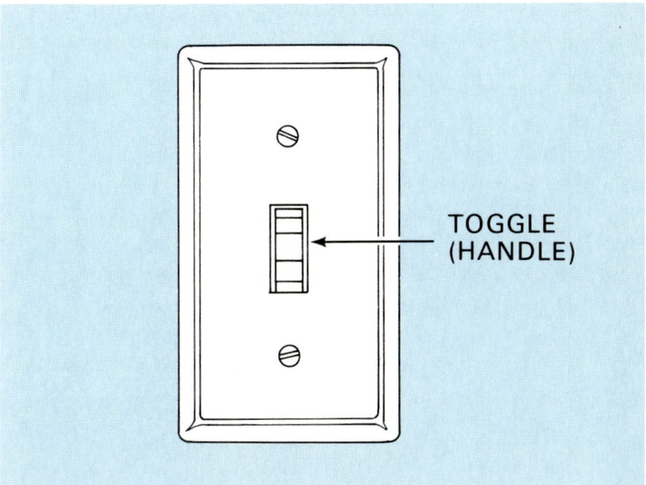

FIGURE 223. The toggle on a three-way switch is not marked to indicate "on-off."

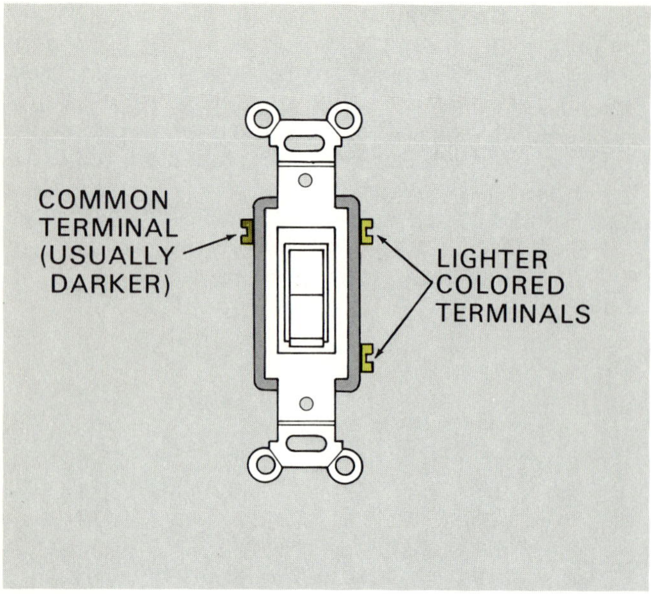

FIGURE 224. The common terminal screw on a three-way switch is usually, but not always, darker than the other two terminals.

On a three-way switch, one of the three terminals is a **common** or pivot terminal. You must identify the common terminal to wire the switch correctly. Usually it is not hard to locate. The common terminal may be a different color than the other two (Figure 224), usually darker. But on some switches the word "common" is printed next to the common terminal. It is often alone or on one side, but this varies according to the manufacturer. The carton or packing slip will indicate its location on the switch if not otherwise identified.

There are several possible combinations of lights and switches in 3-way switch wiring. As with single-pole switches, connections are different depending on where the power source enters the circuit. Also, the position of the light fixture in relation to the switches affects the wiring procedures.

The **power source** of the 3-way switch circuits is **two-wire** cable (14–2 w/g or 12–2 w/g). For wiring *between* the switches, you must use **three-wire cable**.

Connecting three-way switches is discussed as follows:

a. Rules for Wiring Every Three-Way Switch.
b. Wiring Three-Way Switches with Lighting Outlet Ahead of Switches Using Light Outlet Source.
c. Wiring Three-Way Switches with Lighting Outlet Between Switches Using Switch Source.
d. Wiring Three-Way Switches with Light at End of Run Using Switch Source.

a. Rules for Wiring Every Three-Way Switch

Wiring procedures for three-way switch circuits can be confusing, especially in the beginning. However, there are rules for wiring these circuits that will simplify the wiring procedures if carefully observed. When wiring a three-way switch circuit using 2 and 3 wire cable, the rules are as follows (Figure 225).

NOTE: Figure 225 is intended only to demonstrate the application of the basic rules shown below. Detailed procedures for actual wiring situations are shown in figures 226, 227 and 228.

1. *Connect white wire from source to silver terminal on the light fixture.*

   It will be spliced on some circuits but not connected at any other terminal. Note in figure 225, the white wire is shown as a spliced wire.
2. *Connect black wire from source to common terminal of either switch.*
3. *Connect black wire from brass terminal on light fixture to common terminal of the other switch.*

   These three steps complete the connection to the common terminals on each switch and to both terminals on the light fixture.
4. *To complete the circuit, connect the two remaining terminals of one switch to the two remaining terminals of the other switch.*

   The remaining wires (shown as a red wire and a black wire spliced to white) are known as **switching** or **traveler** wires. These wires always connect the remaining two lighter colored terminals of one switch to the remaining lighter colored terminals of the other switch.

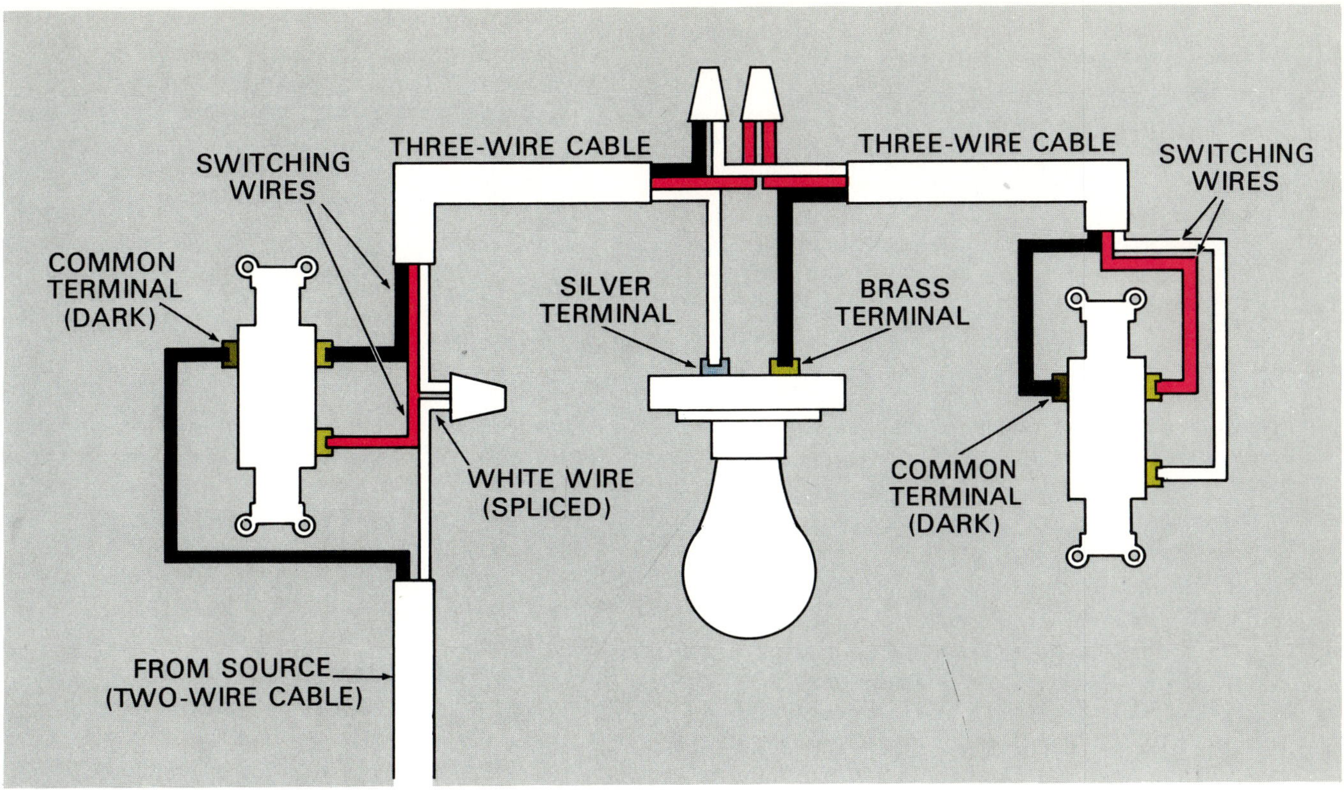

FIGURE 225. Basic wiring layout for wiring a three-way switch.

b. Wiring Three-Way Switches with Lighting Outlet Ahead of Switches Using Light Outlet Source

Wiring from the source to the first switch box on the circuit is 14–2 w/g or 12–2 w/g cable (Figure 226). Three-wire w/g cable is used between the switches. Insulation on wires in three-wire cables is usually **black, white** and **red.** Proceed as follows:

1. *Connect wiring in lighting outlet box (Figure 226).*
   (1) *Connect white wire from the source cable to silver terminal on lighting fixture.*
   (2) *Connect black wire from common terminal of switch No. 1 to brass terminal on light fixture.*
   (3) *Connect black wire from source to white wire from switch No. 1.*
   This white wire is spliced at switch box No. 1 to the white wire in the three-wire cable and continues to switch No. 2 common terminal. As in the case of previous switched circuits, the white wire serves as a hot wire for the supply to switch. When used on the supply side of the switch circuit, the Code does not require it to be identified (painted black).
   (4) *Connect grounding wires as in previous sections (not shown).*

2. *Connect wiring to switch No. 1 and box (Figure 226).*
   (1) *Splice white wire from lighting outlet cable to white wire from cable to second switch (used as hot wire to switch No. 2).*
   (2) *Connect black wire from lighting outlet cable to common terminal on switch No. 1.*
   (3) *Connect black wire from three-wire cable to a light-colored terminal on switch No. 1 (switching wire).*
   (4) *Connect red wire from three-wire cable to other light-colored terminal (switching wire).*
   (5) *Connect grounding wires and ground the switch box (not shown).*

3. *Connect wiring to switch No. 2 and box (Figure 226).*
   (1) *Connect white wire to common terminal of switch No. 2.*
   (2) *Connect black wire to a light-colored terminal of switch No. 2.*
   (3) *Connect red wire to a light-colored terminal of switch No. 2.*
   (4) *Connect grounding wire (not shown).*

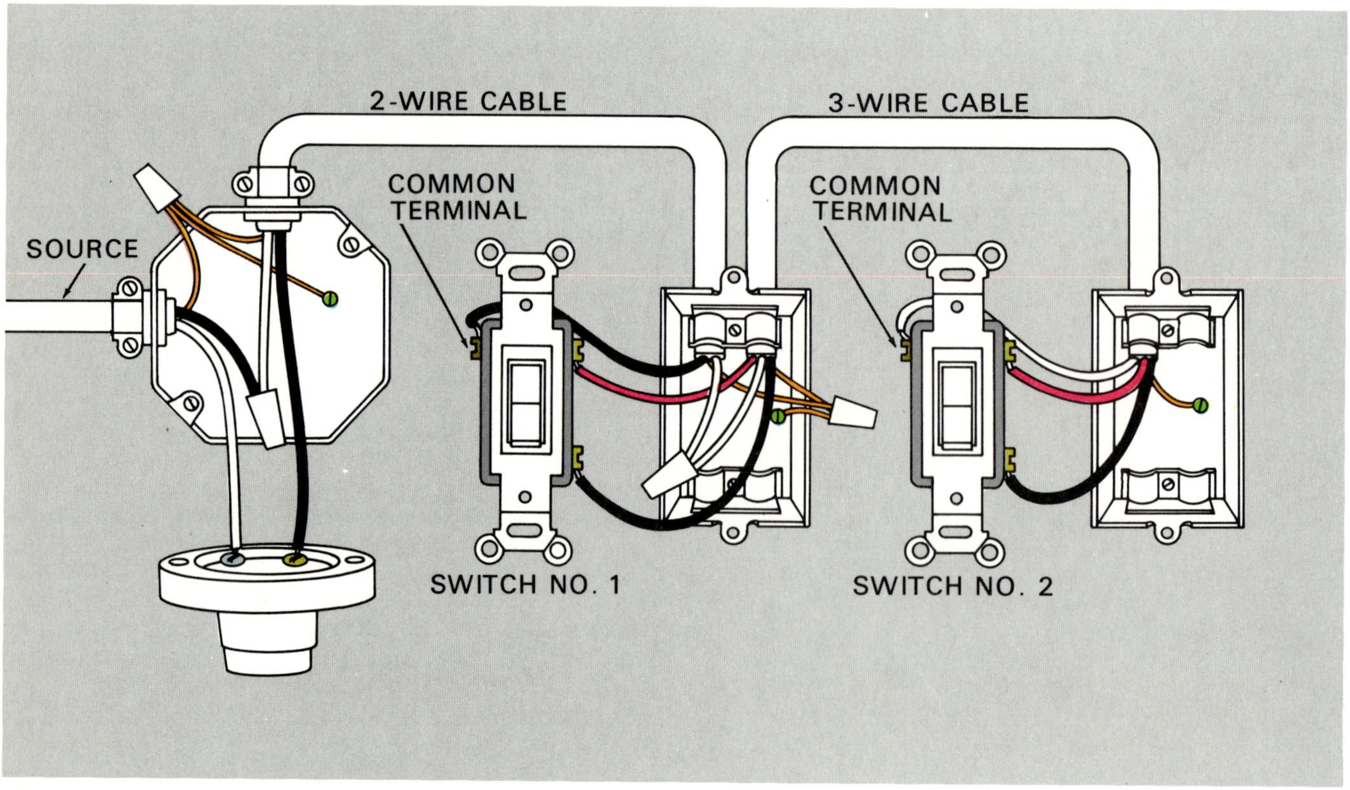

FIGURE 226. Wiring for three-way switch with lighting outlet source.

c. Wiring Three-Way Switches with Lighting Outlet Between Switches Using Switch Source

Assume you are wiring switches for a **hall light** (Figure 227):

1. *Connect wiring at switch No. 1.*
   (1) *Connect black wire from source cable to common terminal of switch.*
   (2) *Splice white wire from source cable to white wire from outgoing cable to lighting outlet box.*
   (3) *Connect black wire from outgoing cable to light-colored terminal on switch No. 1.*
   (4) *Connect red wire from outgoing cable to other light-colored terminal on switch No. 1.*
   (5) *Connect grounding wires (not shown).*
2. *Connect wiring to lighting outlet box (Figure 227).*
   (1) *Connect white wire from switch No. 1 cable to silver terminal on light fixture.*
   (2) *Splice black wire from switch No. 1 cable to white wire from switch No. 2 cable.*
   (3) *Splice red wires from each cable together.*
   (4) *Connect black wire from switch No. 2 cable to brass terminal on light fixture.*
   (5) *Connect grounding wires (not shown).*
3. *Connect wiring to switch No. 2 (Figure 227).*
   (1) *Connect black wire to common terminal.*
   (2) *Connect red wire to light-colored terminal.*
   (3) *Connect white wire to other light-colored terminal.*
   (4) *Connect grounding wire (not shown).*

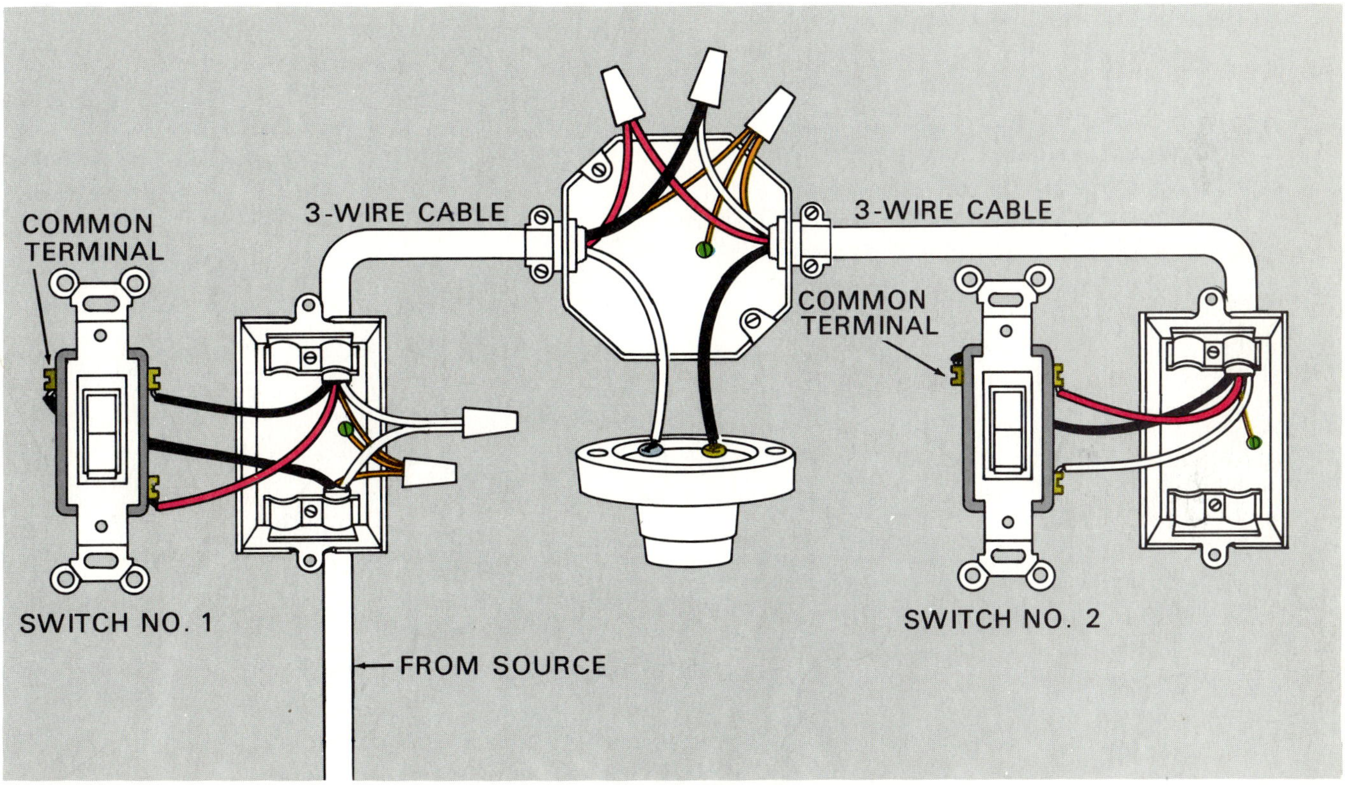

FIGURE 227. Wiring for three-way switches with lighting outlet between switches.

d. Wiring Three-Way Switches with Light at End of Run Using Switch Source

Assume you are wiring a lighting fixture with switches ahead of the light at the end of a run (Figure 228).

1. *Connect wiring in switch No. 1.*
   (1) *Connect black source wire to common terminal of switch No. 1 and white source wire to white wire in three-wire cable going to switch No. 2. Connect red and black wires (switching wires) from three-wire cable to light-colored terminals.*
   (2) *Connect grounding wires (not shown).*
2. *Connect wiring in switch No. 2 (Figure 228).*
   (1) *Connect black wire from outgoing cable to switch No. 2 common terminal.*
   This wire also connects to brass terminal of fixture.
   (2) *Splice white wire to white wire.*
   (3) *Connect black wire and red wire from switch No. 1 to other two switch No. 2 terminals.*
   (4) *Connect grounding wires (not shown).*
3. *Connect wiring to lighting fixture (Figure 228).*
   (1) *Connect black wire to brass terminal and white wire to silver terminal.*
   (2) *Connect grounding wire to box (not shown).*

5. CONNECTING FOUR-WAY SWITCHES

Four-way switches are **used only in combination with two three-way switches** to control lighting from three or more locations. A circuit containing two three-way switches and one four-way switch gives control from three locations (Figure 229). By adding four-way switches to the circuit, you can have as many control locations as desired for the same lighting outlet(s). Four-way switches are so called because they have **four terminals**. Installation of four-way switches is easier than the three-way. You will use three-wire cable between all switches and two-wire cable between the light and the first switch.

Assume you are installing the switches in the kitchen of the standard house plan. The source enters at the lighting outlet. Install 14–3 w/g cable between switches and 14–2 w/g cable between lighting outlet and first switch.

Steps for connecting four-way and three-way switches with source at the lighting outlet are as follows (Figure 230):

1. *At the fixture, connect the black wire from switch No. 1 to the brass terminal on the fixture. Connect the white wire from the source to the silver terminal on the fixture. Splice black wire from source to white wire going to the box at switch No. 1.*

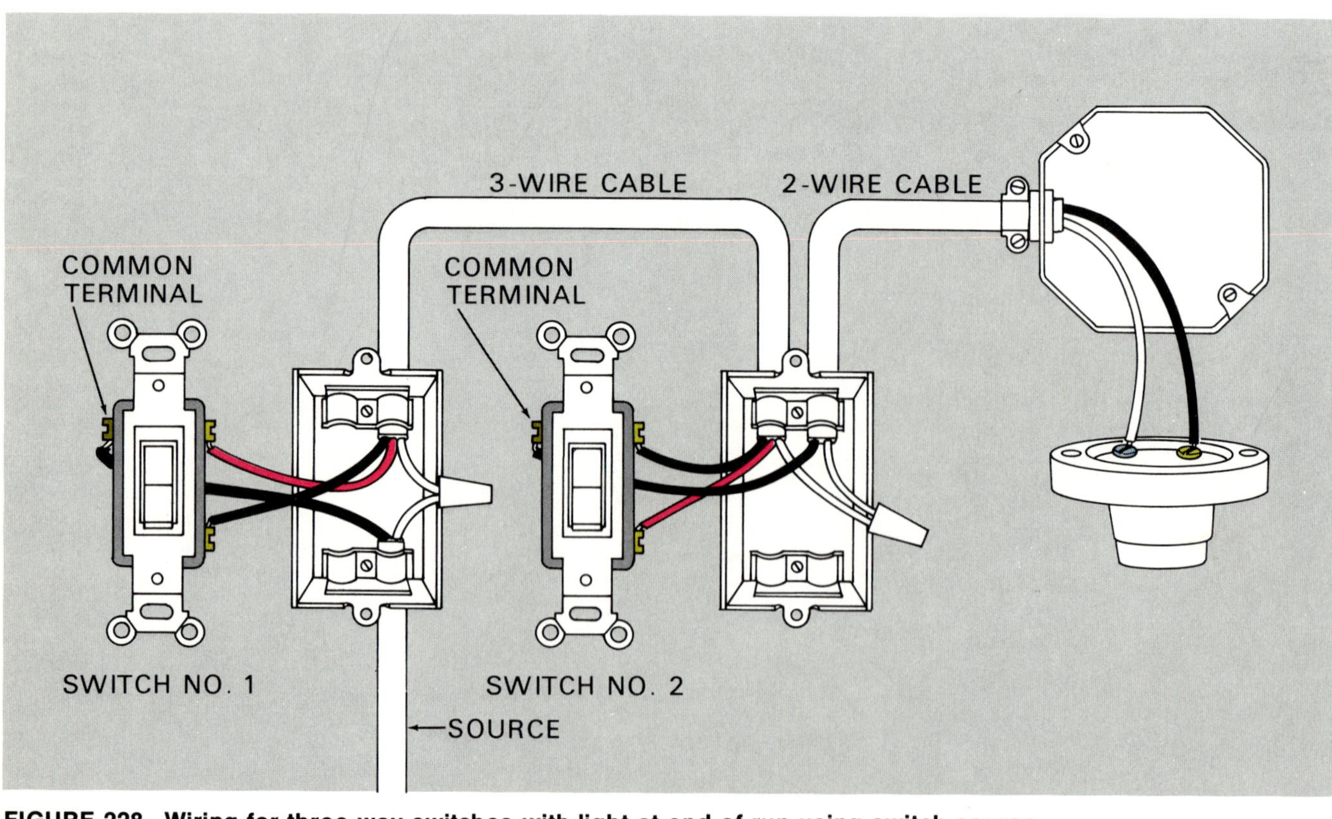

FIGURE 228. Wiring for three-way switches with light at end of run using switch source.

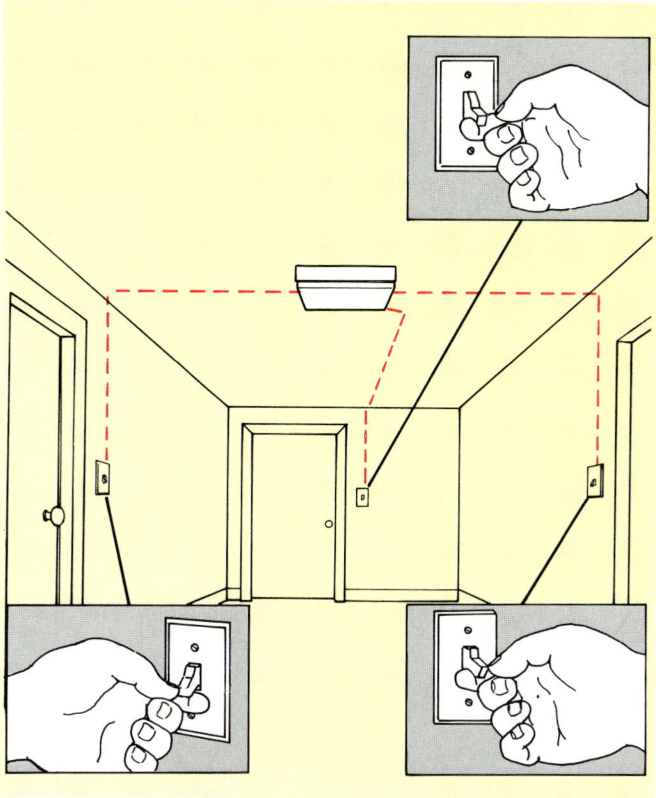

FIGURE 229. A circuit containing one four-way switch and two three-way switches allows control from three locations.

2. *At the box for switch No. 1, splice black wire from switch No. 2 to the white wire going toward the fixture. Connect white and red wires from switch No. 2 to the light-colored terminals. Connect black wire from light fixture to common terminal on switch No. 1.*
3. *At the box for switch No. 2, the four-way switch, splice together the black wires from incoming and outgoing cable. The black wire does not connect to switch No. 2. Connect the switching wires, white wires from both cables, to terminals on one side. Connect red wire from switch No. 3 to terminal opposite white wire from switch No. 3. Connect second red wire to remaining terminal.*
4. *At switch No. 3, connect the black wire to the common terminal, white wire to a light-colored terminal and red wire to the other light-colored terminal.*
5. *Connect grounding wires at each box (not shown).*

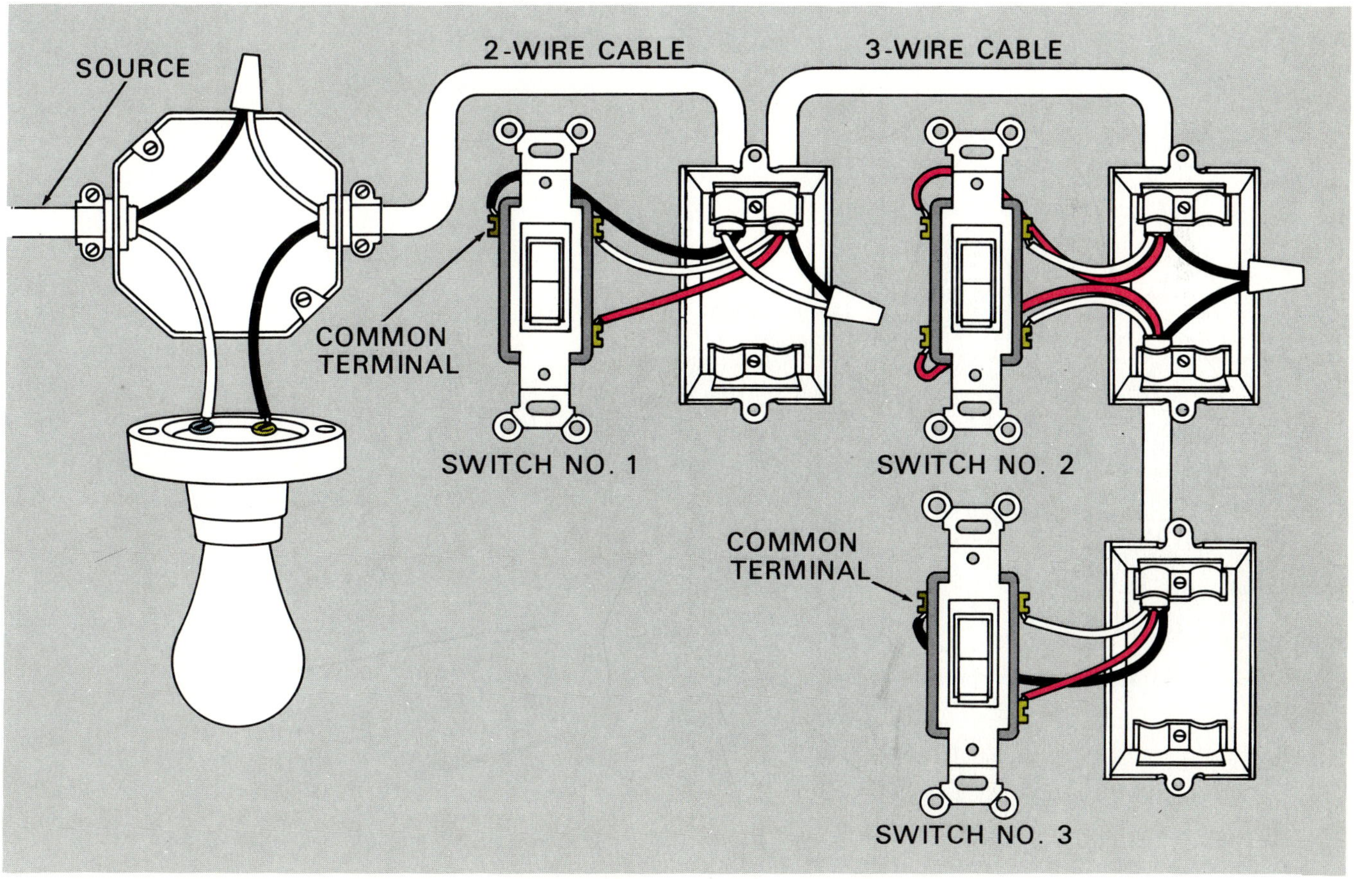

FIGURE 230. Wiring for four-way switch and two three-way switches.

6. CONNECTING COMBINATION DEVICES

Connecting combination devices is discussed as follows:

a. Connecting Combination Switch and Receptacle.
b. Connecting Switch with Pilot (Locator) Light.

a. Connecting Combination Switch and Receptacle

Many options are available for wiring a combination switch and receptacle. Directions for wiring various kinds of circuits are usually given on the carton or on a packing slip included with combination devices. An example is seen in Figure 231.

In Figure 231 the switch is connected to a light only. It does not control the outlet. The receptacle is unswitched, always hot. In Figure 232 the switch controls only the receptacle. In Figure 233A, the switch and outlet are on split-wired 120/240-volt circuits. The switch controls a light and the receptacle is available for other use.

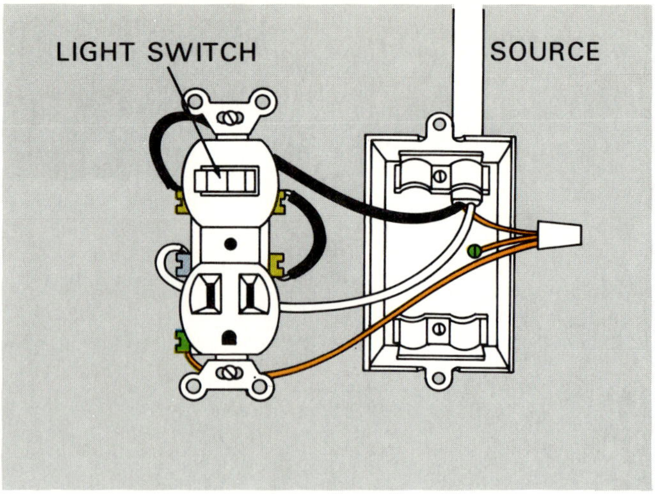

FIGURE 232. Switch controls only the receptacle outlet.

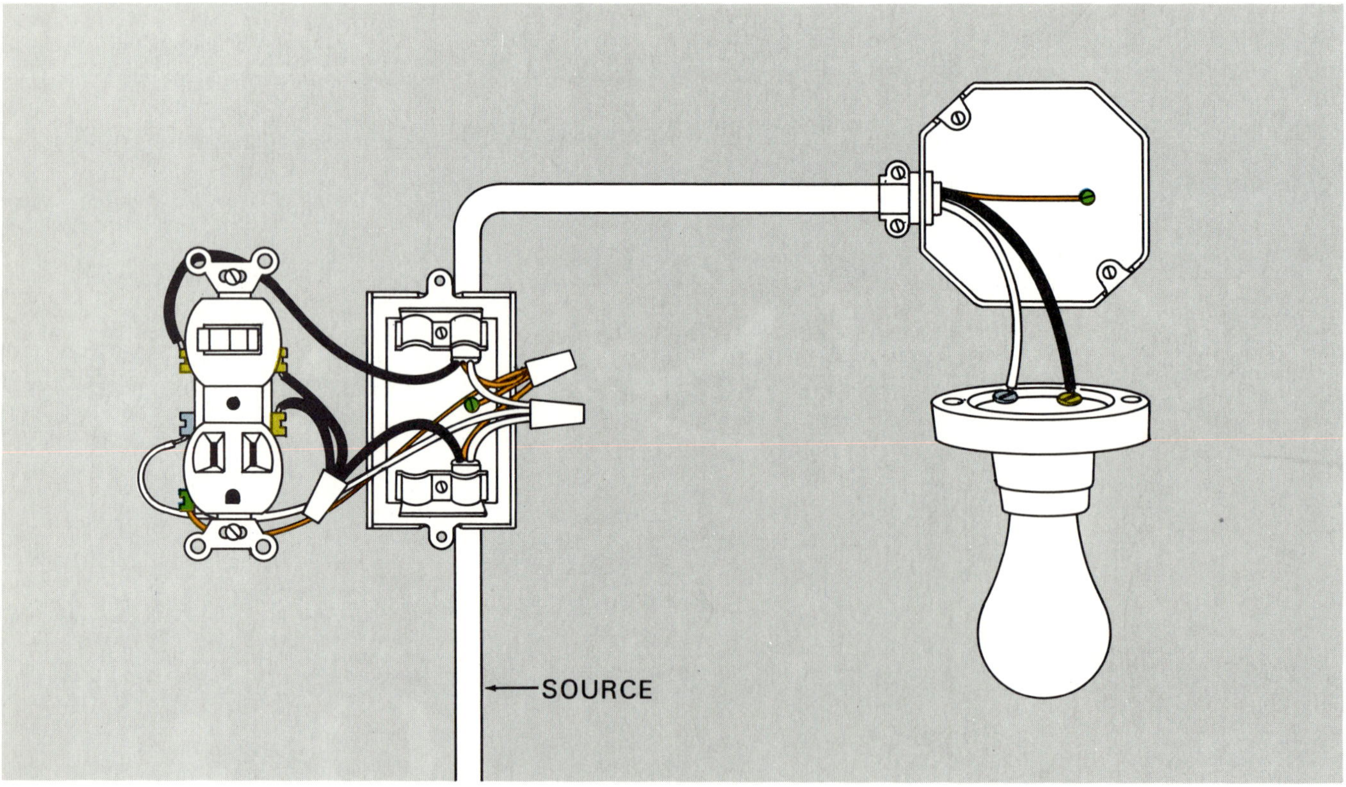

FIGURE 231. Combination switch and receptacle where switch controls a light but not the receptacle.

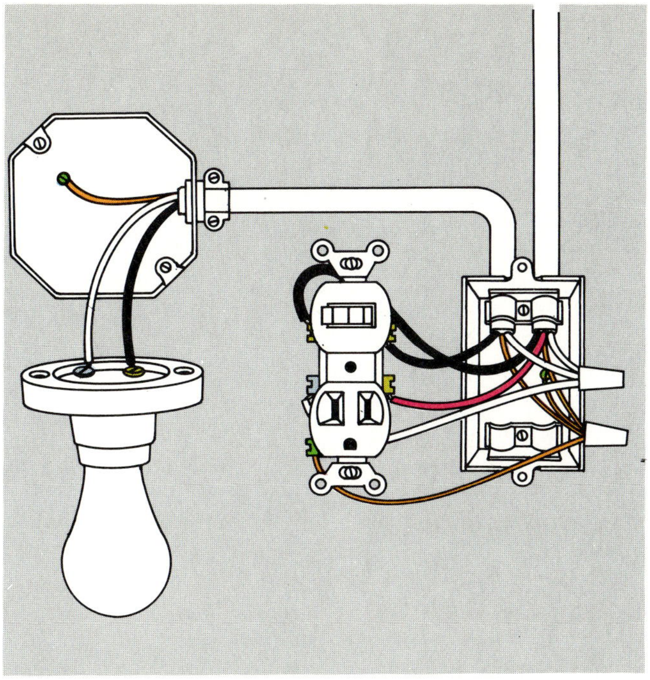

FIGURE 233A. The switch and outlet are on split-wired circuits.

b. Connecting Switch with Pilot (Locator) Light

A switch with pilot light may be connected either of two ways. When connected one way, the pilot light glows when the switch is **off.** This locates the switch for you in the dark. When wired another way, the pilot light glows when the switch is **on.** This last wiring method is useful when a light or equipment is out of sight of the switch position. An example is a basement light controlled by a switch beyond the basement door. Connections are different for different manufacturers. Follow directions included with the pilot light.

7. CONNECTING DOUBLE-POLE SWITCHES

A 240-volt circuit has two hot wires. The switch for a 240-volt circuit must be a double-pole type so both hot wires will be broken when the switch is turned off. Wire the switch and receptacle connections as shown in Figure 233B. Wrap white wires with black tape at each terminal to show they are hot.

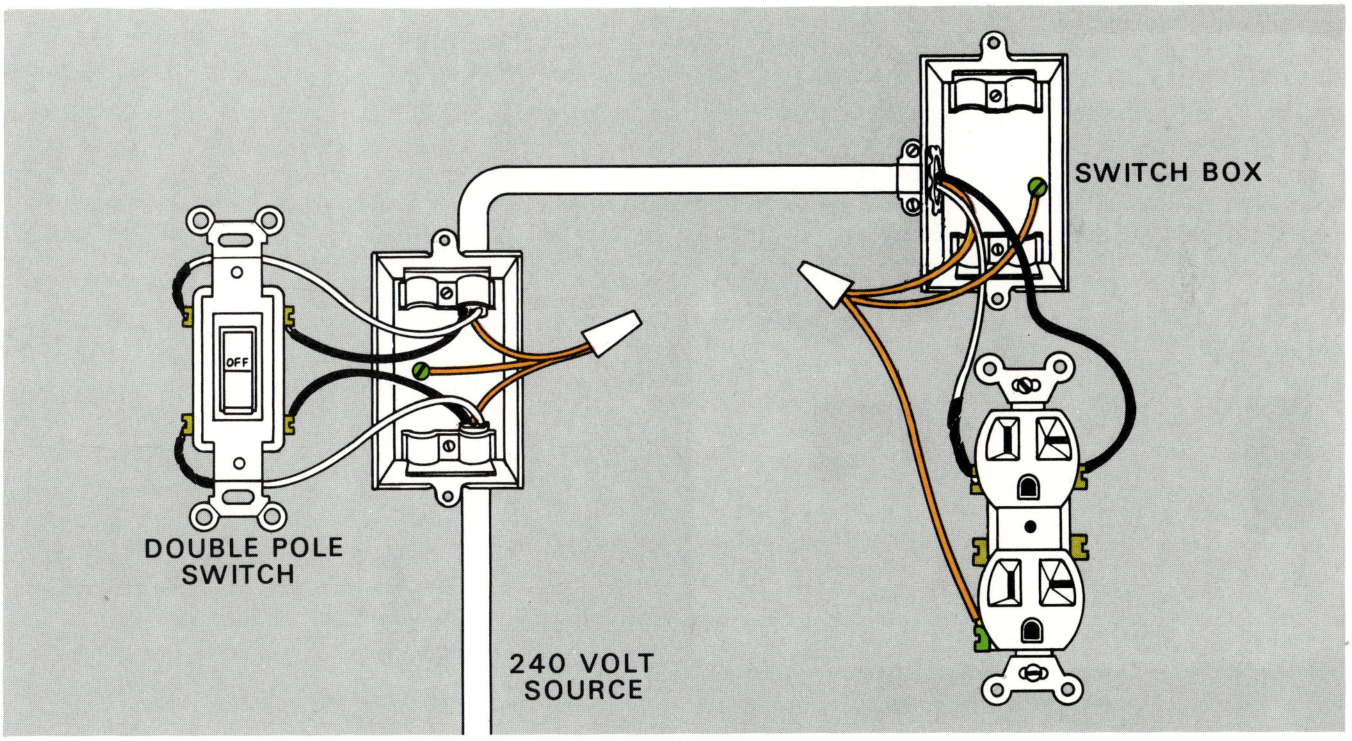

FIGURE 233B. Double-pole switch required for 240-volt circuit.

F. Installing Doorbells, Chimes and Smoke Detectors

Installing doorbells, chimes and smoke detectors is discussed under the following headings:

1. Installing Doorbells and Chimes.
2. Installing Smoke Detectors.

1. INSTALLING DOORBELLS AND CHIMES

In new homes, doorbells have largely been replaced by chimes. Most homes have pushbuttons at the front and rear doors and sometimes at a third door. The pushbuttons may each have a different note that indicates from which door the signal comes.

Doorbells and chimes operate at low voltage, usually 16 to 24 volts AC. A transformer must be installed to reduce the voltage as required. For doorbells and chimes, the transformer reduces the voltage from 120 volts to the 16 to 24 volts required. Number 18 – 2 or 20 – 2 insulated wire may be used for connections between the transformer, the chimes and pushbuttons. Assume you are installing a front and rear door chime system.

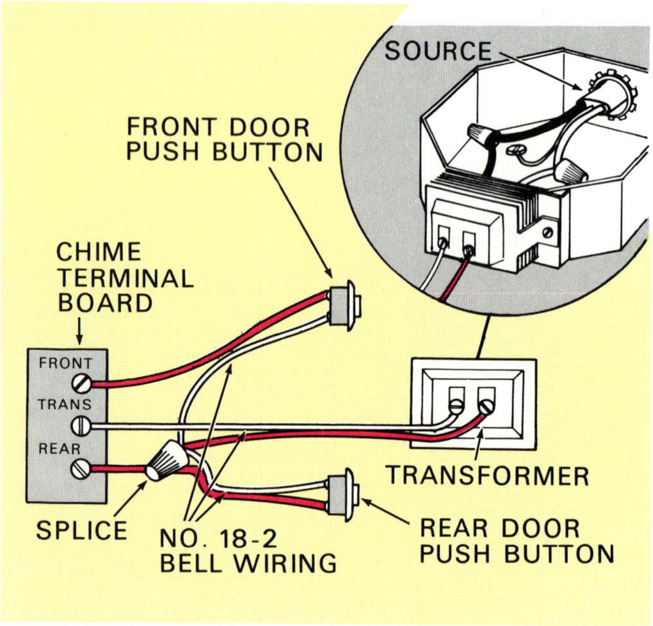

FIGURE 234. Circuit wiring for a front and rear door (two-note) chime system.

Proceed as follows (Figure 234):

1. *Install the transformer.*

   Mount the transformer on an outlet box. Most will have a bracket that can be attached to a knockout in the box. Mount according to manufacturer's instructions. Make line voltage connections inside the box.

2. *Connect 120-volt circuit to transformer.*

   Connect white wire to silver terminal and black wire to brass terminal or white wire to white wire, and black wire to black wire. Connect the bare wire to a grounding screw or clip on the outlet box.

3. *Connect low voltage wiring pushbuttons and chimes (Figure 234).*

   (1) *Using 18–2 cable, connect red and white wires at transformer.*

   (2) *Connect white wire from transformer to center terminal on chime terminal board (staple wires for each run).*

   (3) *Connect red wire from front door button to front door terminal at the chime.*

   (4) *Connect red wire from rear door button to rear door terminal at the chime.*

   (5) *Splice white wires from each doorbell to red wire from transformer.*

2. INSTALLING SMOKE DETECTORS

Smoke detector units are now available that are sensitive to very small amounts of smoke and/or gases produced by combustion. An alarm sounds to warn of danger. These units are relatively low in cost. Many building codes require that they be installed in all new homes. Two or more units may be required in a 2-story house.

Installing smoke detectors is discussed as follows:

a. Installing Battery-Operated Smoke Detectors.
b. Installing 120-Volt Smoke Detectors.

a. Installing Battery-Operated Smoke Detectors

Battery-operated types are self-contained and do not require any wiring. They may be mounted on the wall or ceiling. Follow instructions included with the product to select the best location. Fasten the unit to wall or ceiling as follows:

1. *Place template on mounting surface (Figure 235).*

   A template is usually supplied with unit. Tape the template on the ceiling or wall and drill holes for the plastic screw anchors if required. Tap the anchors into the holes with a hammer, then remove the template.

2. *Attach unit to wall or ceiling (Figure 235).*

   Insert screws through the base of the alarm unit and tighten them into the plastic anchors to complete the job.

   An alarm device warns when the battery needs to be replaced. In Figure 236, a small flag pops out. Other types have audio signals. A battery usually lasts about one year before requiring replacement.

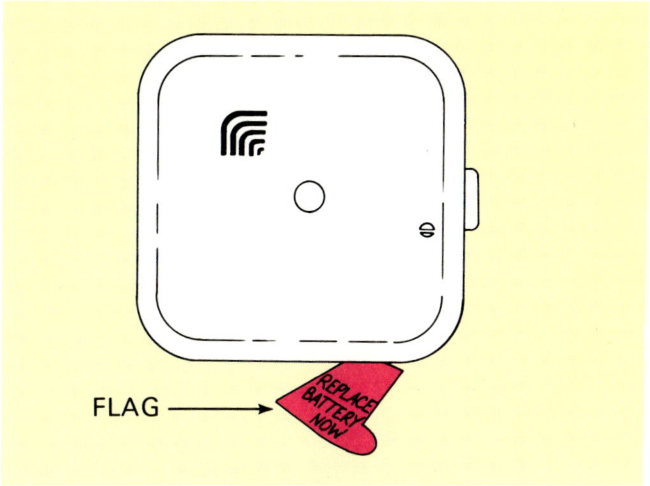

FIGURE 236. Small flag pops out of unit to indicate a weak battery. Other types have audio signals.

b. Installing 120-Volt Smoke Detectors

Follow instructions of the manufacturer to locate the unit correctly. Remember, the circuit must have power at all times and cannot be switched off.

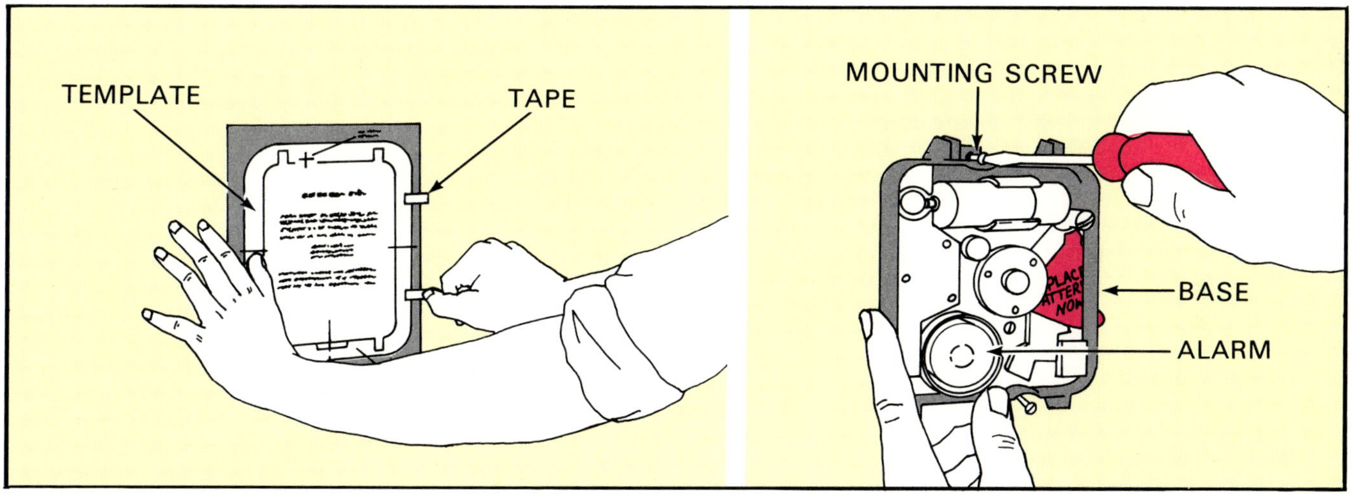

FIGURE 235. Smoke detector attached to wall.

FIGURE 237. Permanently installed 120-volt unit.

The 120-volt type may be plugged into a convenient receptacle if desired. However, most people prefer not to have a cord running down the wall. Also, a plug-in unit may be accidentally disconnected.

To permanently connect the unit to the wiring system, proceed as follows:

1. *Locate the unit according to directions.*
2. *Mount the unit on wall or ceiling (Figure 237).*
3. *Connect to outlet box.*

   Install outlet box if necessary and run cable. Connect white to white and black to black wires with wire nuts. Connect the bare wire to the box with screw or clip.
4. *Test for operation as directed by manufacturer's instructions.*

IV. Installing Service Entrance Equipment

Service entrance equipment has been discussed briefly and in general terms in previous sections. However, to install the equipment, much more information is required.

Upon completion of this section, you will be able to **select the type of service entrance equipment required and determine the size needed.** You will be able to **install the parts of the service entrance equipment and connect the necessary wiring.** Installing service entrance equipment is discussed as follows:

A. Determining Type and Size of Service Entrance Panel and Equipment.
B. Installing the Service Entrance Panel Cabinet.
C. Locating and Mounting the Service Drop Eyebolt or Insulator Rack.
D. Installing Service Entrance Cable or Conduit and Attachments.
E. Connecting Service Entrance Wiring.
F. Installing Ground Fault Circuit Interruptors (GFCI).
G. Connecting Circuits in Service Entrance Panel.

A. Determining Type and Size of Service Entrance Panel and Equipment

Determining types and sizes of SEP and equipment is discussed under the following headings:

1. Types of Service Entrance Panels and Equipment.
2. Determining Size of Service Entrance Panel and Equipment.

1. TYPES OF SERVICE ENTRANCE PANELS AND EQUIPMENT

Service entrance panels may be one of two types:

a. Circuit-Breaker-Type Service Entrance Panel.
b. Fuse-Type Service Entrance Panel.

a. Circuit-Breaker-Type Service Entrance Panel

More circuit-breaker-type SEP's than the fuse type are installed in new homes.

Circuit-breaker entrance panels have several advantages over fuse types (Figure 238). The switch on

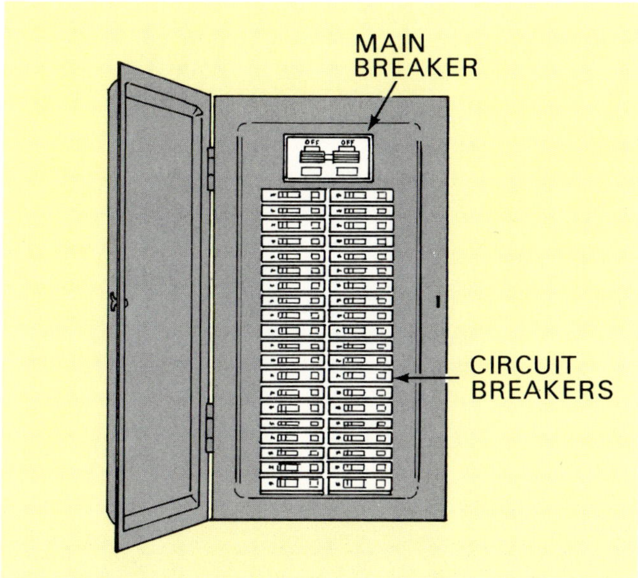

FIGURE 238. Service entrance panel equipped with circuit breakers.

each breaker, when tripped, may be reset after the problem has been corrected without replacing it. Circuit breakers are easy to switch off when necessary to make repairs or for other reasons. Circuit breakers are available in sizes to protect any circuit desired. They are rated in amperes and will carry 80 percent of their rated loads continuously. They will take short periods of overload without tripping but provide protection from continued overloads or short circuits. Also, they cannot be tampered with and it is not easy to install a larger size accidentally or intentionally. The circuit breaker *main* is a heavy duty 2-pole on-off switch which can be used to turn off all power to the panel (Figure 239). It will trip when its rated current is exceeded.

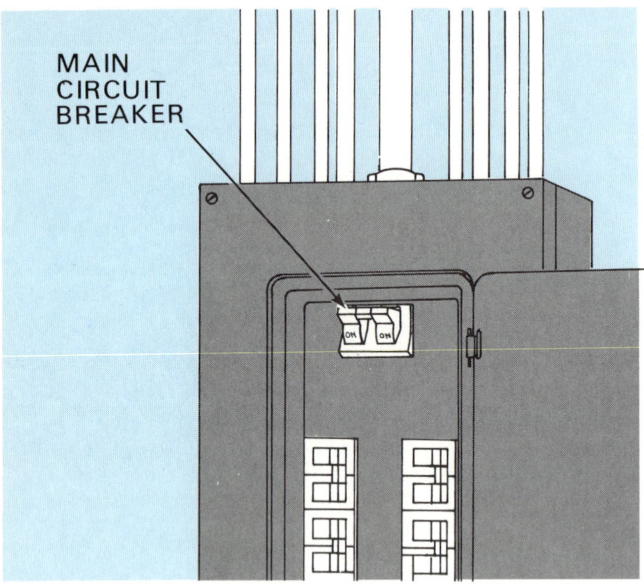

FIGURE 239. Main circuit breaker switches off all current to the service entrance panel.

b. Fuse-Type Service Entrance Panel

♣ Fuse-type service entrance panels cost less to purchase but are less convenient to use (Figure 240). When a fuse "blows," it must be replaced unless equipped with special reset-type fuses.

Both fuse and circuit breaker SEPs are available with **main fuses** and **main breakers** that allow disconnecting of power to all circuits. The main fuses are mounted in a pull-out block (Figure 241).

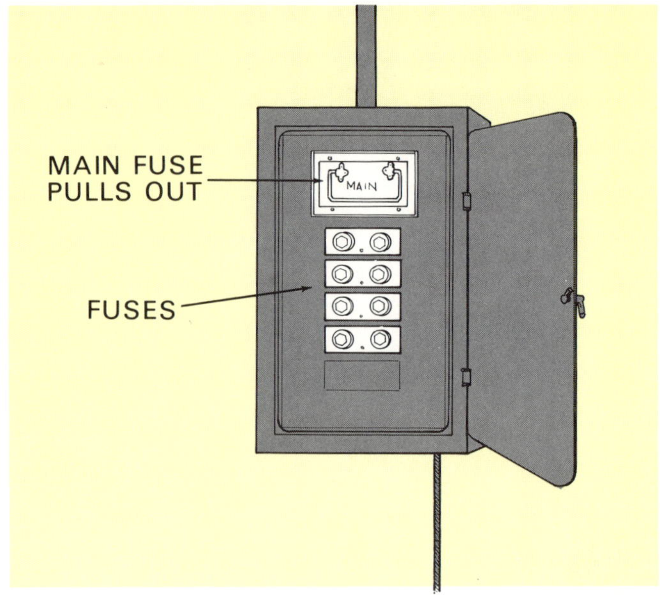

♣ **FIGURE 240. Fuse-type service entrance panel.**

♣ **FIGURE 241. Main fuses are mounted in a pullout block.**

For use with mobile homes, a combination of **meter base** and circuit breakers in a single cabinet is allowed in some areas (Figure 242).

FIGURE 242. Combination meter base and circuit breaker service entrance panel in a single cabinet for mobile home installations.

2. DETERMINING SIZE OF SERVICE ENTRANCE PANEL AND EQUIPMENT

SEPs with 100, 150 or 200 amperes are commonly installed in homes. In larger homes, two panels are often used, for example, one 150-ampere and one 100-ampere.

> **NOTE: Article 220 of the code has been revised to replace the use of "watts" with "volt-amperes".**

The Code requires residential service entrance equipment to be rated for at least **100-ampere capacity** unless (1) **the computed load for the house is less than 10kVA (10000 VA)** or (2) **there are less than six 2-wire branch circuits in the house**. If either (1) or (2) applies, a 60-ampere service is allowed. Very few houses would qualify for either of the exceptions for 100-ampere service.

From a practical standpoint, you may consider 100 amperes as the minimum requirement for service entrance equipment rating in a home. If you install electric heat or air conditioning and an electric water heater, 150-ampere is the minimum and 200-ampere may be required. Power supplier representatives are usually available for advice on size of service entrance.

The Code provides guidelines for determining the minimum size of service entrance equipment for each individual home, including examples in Chapter 9 of the Code.

Demand factors are considered in computing the total load for a home. Use of the demand factor takes into account the difference between the maximum capacity of the system and the actual load it is likely to carry. **Appliances** have been tested and assigned a percentage demand factor to be used in computing the total load. Some **circuits** are also rated at less than their maximum capacity. For example, **each of the two small appliance circuits shown in the standard house plan as required by the code has a 2400 volt ampere capacity (120 volts × 20 amperes = 2400 volt-amperes). To convert to kVA, divide by 1000 = 2.4 kVA. The code allows them to be rated at 1500 volt-amperes (1.5 kVA) each because it is not likely that both will be fully loaded at the same time.**

To determine the size of service entrance required for the standard house plan by one method shown in the Code, proceed as follows:

The Code permits use of optional methods in computing the total load. Optional method from Code Section 220–30 may be used for single family dwellings with an ampacity of 100 amperes or greater. This would include most new homes.

Compute the size of service entrance required in the 215 square meter (2000 square foot) standard house plan.

Assume the 215 square meter (2000 square foot) house has the following:

—**Two** 20-ampere **small appliance** circuits.
—**One** 20-ampere **laundry** circuit.
—**Two** 5 kVA **wall mounted ovens**.
—**One** 5.8 kVA **counter-top cooking unit**.
—**One** 1.2 kVA **dishwasher**.
—**One** 4.5 kVA **water heater**.
—**One** 8.5 kVA **clothes dryer** (high speed).
—**Two** 1.2 kVA **built-in bathroom heaters**.
—**One** 9.6 kVA **central air conditioner**.

Follow the procedures from optional method **demand factors** in the Code and from Table IX.

CALCULATE "OTHER LOAD"

1. *Include 1500 volt-amperes for each two-wire, 20 ampere small appliance branch circuit.*

   Rated by the code at 1500 volt-amperes each; 2 × 1500 = 3000 or = 3 kVA.

2. Three volt-amperes per 0.09 square meter (square foot) for general lighting and general-use receptacles.

   Multiply 28 times the number of square meters (3 times the number of square feet) in the house, exclusive of open porches and garages.

   3 volt-amperes × 2000 square feet
   = 6000 or = 6 kVA
   Total steps 1 and 2 = 9 kVA

3. *List the nameplate ampere rating of all fixed appliances, ranges, wall-mounted ovens and counter-mounted cooking units.*

Check the nameplates of each item included and list the volt-amperes of each.

Fixed Appliances	Volt Amperes
Laundry circuit (rated by Code at 1500 volt-amperes) or	1.5 kVA
Two wall-mounted ovens (5000 volt amperes each) 10,000 or	10.0 kVA
Countertop cooking unit 5800 or	5.8 kVA
Dishwasher 1200 or	1.2 kVA
Water Heater 4500 or	4.5 kVA
Total "Other Load"	**23.0 kVA**
9 kVA plus 23 kVA =	**32.0 kVA**

4. *List the nameplate ampere or kVA rating of all motors and of all low-power-factor loads.*

   Include here only those motors and other loads not included in steps 3 and 5. If the motor rating is given in amperes, convert to volt-amperes by multiplying volts × amperes. For example, assume a motor rated 9 amperes, 240 volts: 240 × 9 = 2160 volt-amperes or 2.16 kVA.

   NOTE: Code Section 220–21 states that where it is unlikely that two dissimilar loads will be in use at the same time, it is permissible to omit the smaller of the two in computing the total load. This applies to heating and air conditioning loads described in Step 5 as follows:

5. *Use the largest of the following:*
   (a) 100 percent of the nameplate rating(s) of the air conditioning and cooling, including heat pump compressors.
   (b) 65% of the nameplate rating(s) of the central electric space heating, including integral supplemental heating in heat pumps.
   (c) 65% of the nameplate rating(s) of electric space heating if less than four separately controlled units.
   (d) 40% of the nameplate rating(s) of electric space heating of four or more separately controlled units.

TABLE IX. OPTIONAL CALCULATION FOR DWELLING UNIT LOAD IN kVA (Code Table 220–30)

Largest of the following four selections.

(1) 100 percent of the nameplate rating(s) of the air conditioning and cooling, including heat pump compressors.

(2) 65 percent of the nameplate rating(s) of the central electric space heating including integral supplemental heating in heat pumps.

(3) 65 percent of the nameplate rating(s) of electric space heating if less than four separately controlled units.

(4) 40 percent of the nameplate rating(s) of electric space heating of four or more separately controlled units.

Plus: 100 percent of the first 10 kVA of all other load.
40 percent of the remainder of all other load.

Computations for the standard house plan are as follows:

Air conditioning included in total at 9.6 kVA.

6. *Compute calculated load.*

   Refer to Table IX.

100% of nameplate rating of the air conditioner	9.6 kVA
100% of the first 10 kVA of all other load	10.0 kVA
40% of the remainder of all other load (22 kVA × 40%)	8.8 kVA
Total calculated load	28.4 kVA

   28.4 kVA = 28400 VA ÷ by 240 volts = 117 ampere service rating. Use 150-ampere service rating.

   Allowances should always be made for some future expansion of the load.

   This home may be served by 150-ampere service, which provides enough capacity for future needs. The service entrance cable or wires, the main disconnect switch and the service entrance panel must be sized for 150 amperes.

   Some local codes specify minimum wire sizes for a service entrance of a given capacity. For example, No. 2 THW wire with a 100-ampere service, No. 1/0 for 150-ampere and No. 3/0 for 200-ampere.

   The Code in "Notes to Tables 310–16 through 310–19" provides a simple method to determine required conductor size for the service entrance of residences, as follows:

TABLE X. CONDUCTOR TYPES AND SIZES IN COPPER AND ALUMINUM FOR SERVICE ENTRANCE RATINGS LISTED (Code Notes 310–19, Number 3)

Conductor Types and Sizes RH-RHH-RHW-THW-THWN-THHN-XHHW

Copper	Aluminum and Copper-Clad AL	Service Rating in Amperes
AWG	AWG	
4	2	100
3	1	110
2	1/0	125
1	2/0	150
1/0	3/0	175
2/0	4/0	200

B. Installing the Service Entrance Panel Cabinet

The **cabinet** is sometimes called the box (Figure 243). It is the steel cabinet with hinged door in which are mounted the service entrance terminals, circuit terminals and circuit breakers or fuses.

To mount the SEP, proceed as follows:

1. *Remove cover.*

   Remove cover mounting screws (usually four), and replace cover in carton for protection.

2. *Mount the cabinet.*

   For **flush mount,** line up front edge of cabinet making allowance for the finished wall surface. Screw (preferred) or nail to stud through small knockouts or holes on each side. For **surface mount,** screw or nail through keyhole slots in back of box.

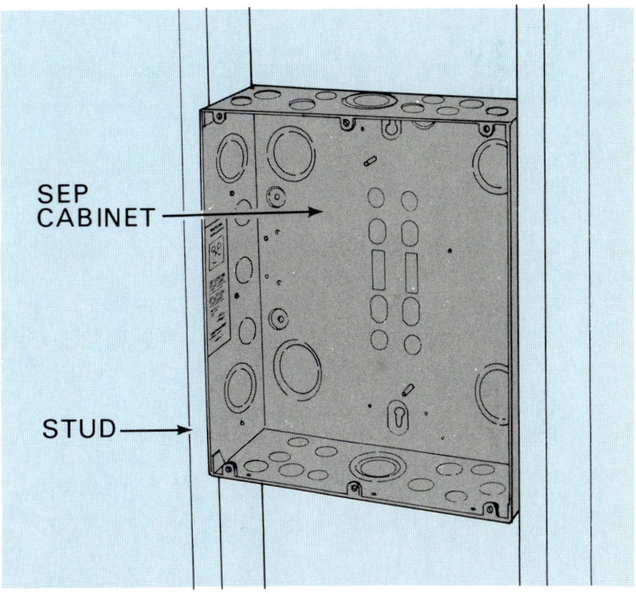

FIGURE 243. Service entrance panel cabinet encloses the service entrance terminals, circuit terminals and circuit breakers or fuses.

C. Locating and Mounting the Service Drop

Locating and mounting the service drop eyebolt or insulator rack are discussed under the following headings:

1. Locating the Service Drop Insulator or Insulator Rack.
2. Mounting the Service Drop Insulator or Insulator Rack.

1. LOCATING THE SERVICE DROP INSULATOR OR INSULATOR RACK

Before you install the service entrance equipment, you must know the location of the **service drop** (wires from the power source to the house). Location of the service drop depends to some extent on the location of the service entrance panel. Instructions for locating the service entrance panel were given in Section II, "Installing Wiring."

The **service drop** is usually located and installed by the power supplier. It includes the service conductors required to reach from the street main, transformer or meter pole, to the house (Figure 244). Some power suppliers furnish and install the insulator, or **insulator rack**, required for the service drop. In most areas, it is installed by the electrician. Locate the service drop attachment at the location agreed upon with the power supplier.

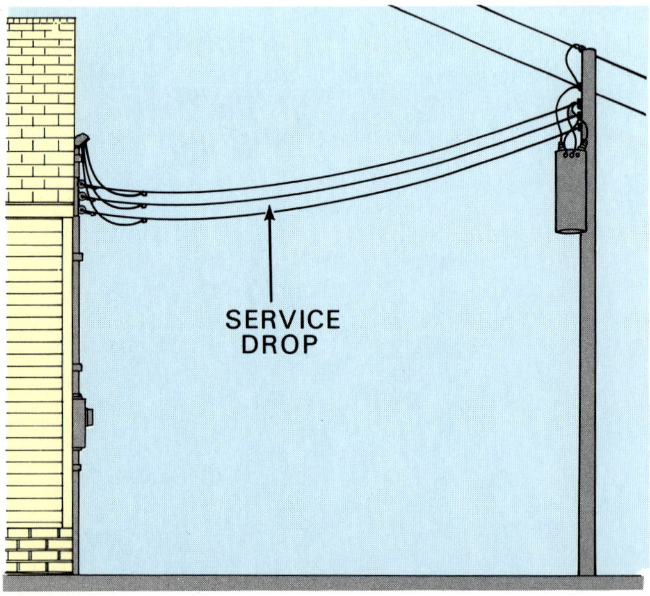

FIGURE 244. The service drop is the cable or wires that extend from the power company transformer to the house or building. It is usually located and installed by the power supplier.

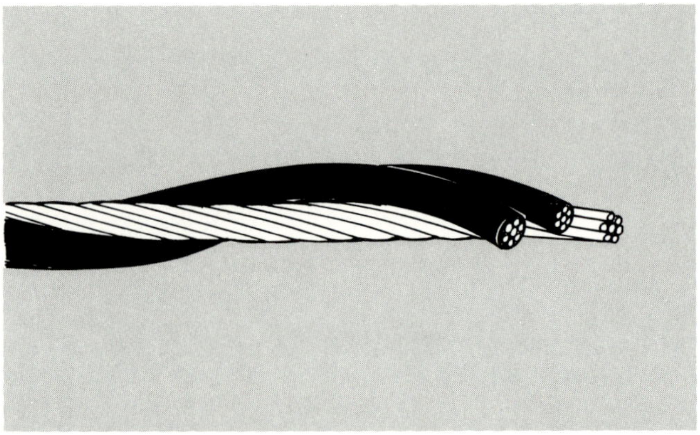

FIGURE 245. Triplex cable, made up of a bare neutral wire with two insulated cables, is used for most service drops to new homes.

The service drop conductors for a 120/240-volt installation may be three separate wires including two black insulated wires and a neutral wire (Figure 244). The neutral may be a white insulated wire but usually it is bare.

Most service drop installations to new homes are made by use of triplex cable, which is made up of a bare neutral wire with the two insulated cables wrapped around it for support (Figure 245). Proceed as follows:

1. *Select tentative location for insulator or insulator rack.*
2. *Check ground clearance (Figure 246).*

The Code requires that service drop conductors, where not in excess of 600 volts, nominal, shall have the following minimum clearance from final grade.

3.05 m (10 feet)—at the electric service entrance to buildings, or at the drip loop of the building electric entrance, measured from final grade or other accessible surface only for service-drop cables supported on and cabled together with a grounded bare messenger and limited to 150 volts to ground.

3.66 m (12 feet)—for those areas listed in the 4.57 m (15 feet) classification when the voltage is limited to 300 volts to ground.

4.57 m (15 feet)—over residential property and driveways, and those commercial areas not subject to truck traffic.

5.49 m (18 feet)—over public streets, alleys, roads, parking areas subject to truck traffic, driveways on other than residential property, and other land traversed by vehicles such as cultivated, grazing, forest, and orchard.

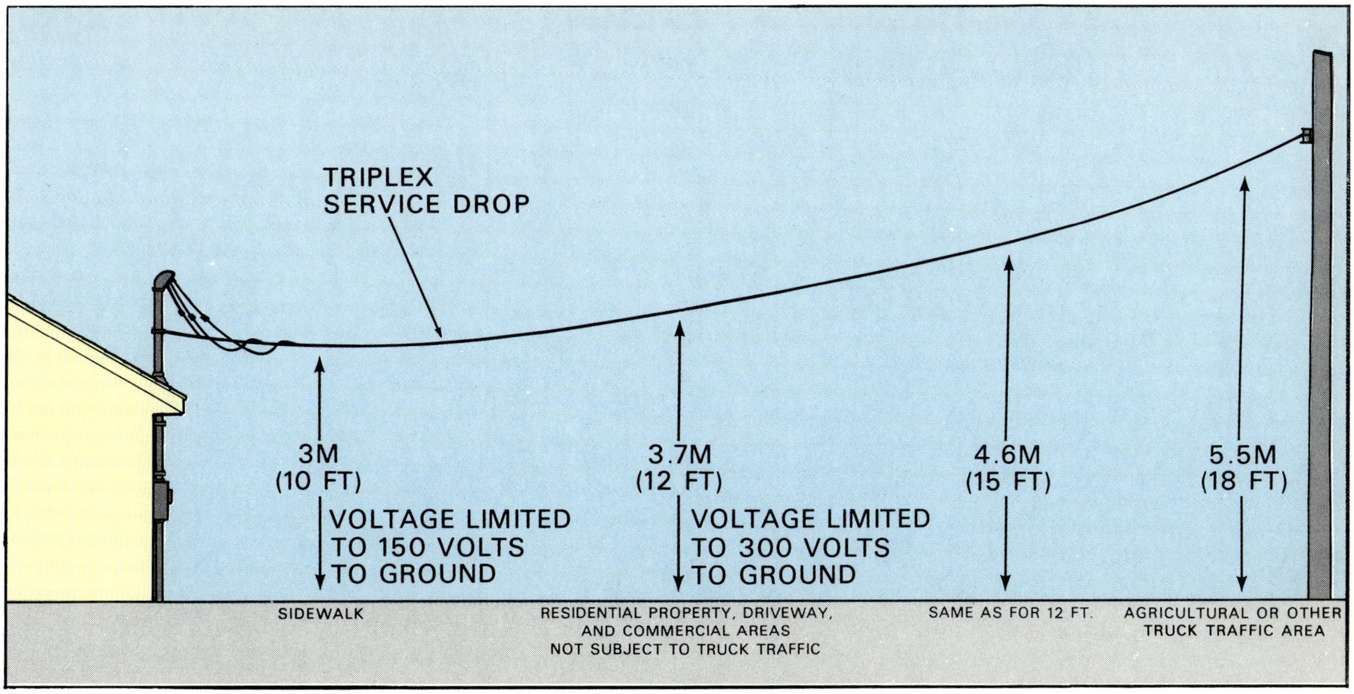

FIGURE 246. Service drop conductors have minimum clearance requirements, depending on the sidewalks, or traffic areas to be crossed.

3. *Check clearance over roof.*

   If the service drop conductors pass over a roof, the minimum clearance above the roof must be as follows:

   If the roof slope is **not less than 4 in 12 inches,** and voltage between conductors is not over 300, a 0.9 m (3 feet) clearance is permitted.

   —**0.46 meters (18 inches) if terminated at a mast** and not more than 1.2 meters (four feet) of the overhang portion of the roof is crossed by the service drop.

4. *Check clearance for windows and other items.*

   Conductors shall have a clearance of not less than 0.9 meters (3 feet) from windows, doors, porches, fire escapes, or similar locations.

   Conductors run above the top level of a window shall be considered out of reach from that window.

5. *Check power supplier requirements.*

   Some power suppliers have requirements that are different from the Code for service drop mounting. Check with your power supplier for instructions.

2. MOUNTING THE SERVICE DROP INSULATOR OR INSULATOR RACK

Proceed as follows:

1. *At the selected location, fasten the cable to the side of the house by means of an insulator rack (Figure 248A) or to a mast through the roof, usually made of 5.1 cm (2 inch) conduit (Figure 247).*

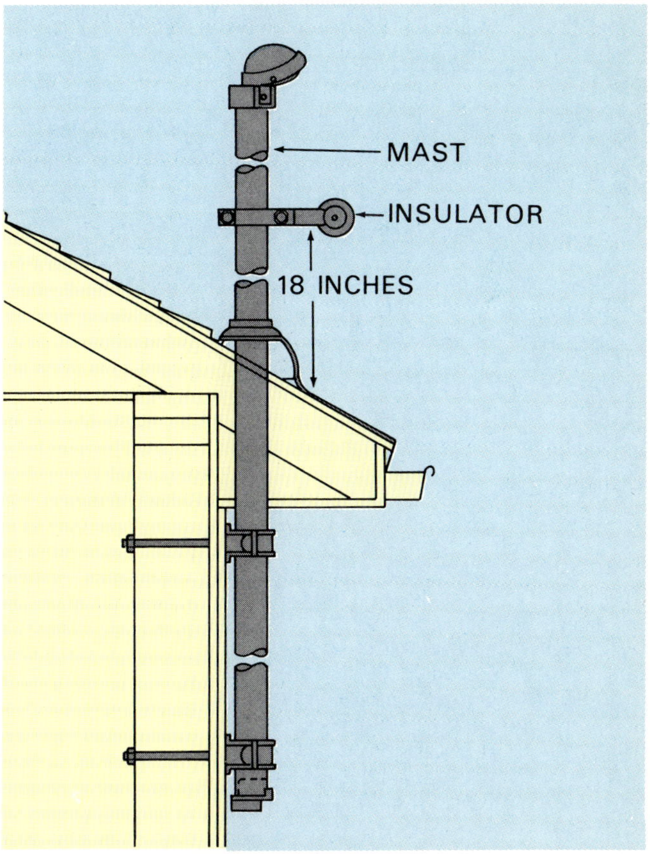

FIGURE 247. Two-inch conduit mast through the roof.

2. *Locate the entrance head above the service drop to allow room for the drip loops (Figure 248A).*

   The drip loops prevent water from entering the entrance head. Proceed as follows:

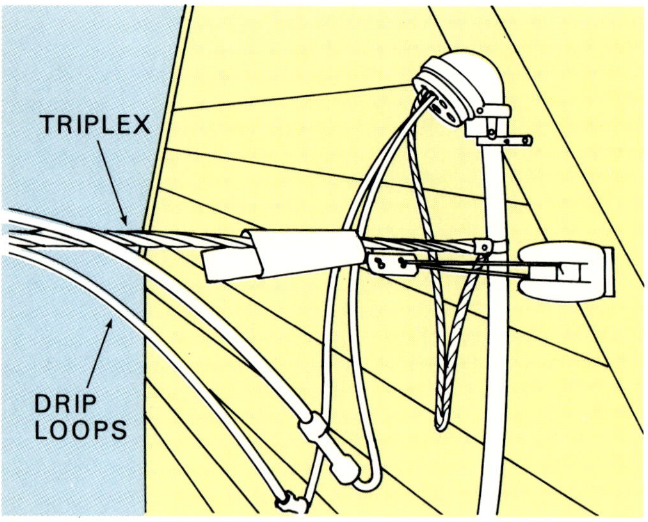

FIGURE 248A. The entrance head must be mounted higher than the service drop to allow room for drip loops.

If mounting a screw-type insulator, screw it directly into a stud, plate or other reinforcement in back of the siding (Figure 248B).

If mounting an insulator rack, fasten the insulator rack with lag screws.

Both types must be fastened securely to withstand the strain of the service drop during extreme weather conditions.

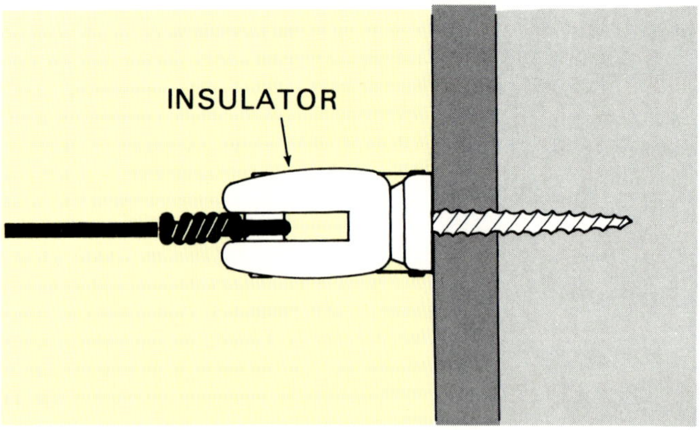

FIGURE 248B. Fasten screw-type insulator directly into a stud or other framing member.

D. Installing Service Entrance Cable or Conduit and Attachments

Your local code may determine the type of installation you will use for the service entrance wiring and attachments. Four methods may be used for the installation. They are discussed as follows:

1. Installing Service Entrance Cable and Meter Base.
2. Installing Service Entrance Conductors in Conduit.
3. Installing a Mast Type of Service Entrance.
4. Installing Underground Service Entrance Cable.

1. INSTALLING SERVICE ENTRANCE CABLE AND METER BASE

Wire used for service entrance cable may be:

a. Aluminum.

b. Copper.

a. Aluminum

Aluminum wire or cable is often installed for the service entrance because it costs less than copper. The following procedures are recommended when connecting aluminum wire:

1. *Be sure the service entrance terminals are labeled Al-Cu.*
2. *Strip the insulation, being careful not to nick the wires.*
3. *Wire-brush the conductor strands.*
4. *Thoroughly coat the stripped conductor with a suitable anti-oxidant compound such as ALNOX or PENETROX A13.*
5. *Insert conductor and tighten connector screw very tightly.*

b. Copper

Copper is very satisfactory for service entrance installation but in larger sizes, it is expensive. To install copper, use care when stripping insulation so you don't nick the wires. Also, be sure to tighten the connector screw very tightly.

Service entrance cable is used for this installation. The bare neutral wire of the cable is made up of many small wires that are wound around the insulated wires (Figure 249). Proceed as follows to install the service entrance cable and meter base:

1. *Measure and cut cable for installation between entrance head and meter base.*
   (1) *Determine from power supplier or other authority the locations of the entrance head and meter base.*
   (2) *Mark the location of each.*
   (3) *Measure the distance between the two marks and add 1.1 meter (3½ feet) for connections at the entrance head and meter base.*
   (4) *Cut the measured length of cable.*
2. *Remove the outer jacket from cable.*
   The outer layer of heavy plastic jacket and an inner layer of lighter material are both removed for a length of about 0.9 m (3 feet).

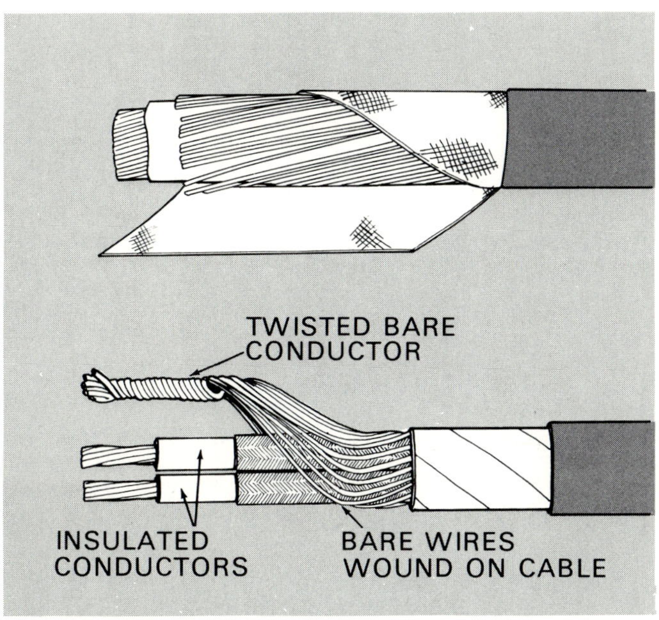

FIGURE 249. Service entrance cable with strands of bare neutral wire wound around the insulated conductors. Unwrap the separate strands and twist to form the neutral wire.

3. *Prepare neutral wire for installation.*
   Unwrap the separate strands, pull to one side and twist together to form a bare conductor (Figure 249).
   NOTE: If the neutral wire is insulated, it must be unmistakably identified at each end.
4. *Install wires in entrance head (Figure 250).*
   Remove screws from entrance head (weatherhead) to disassemble. Insert the insulated wires through the bushing holes provided. Insert the neutral wire through the notch below the bushing and assemble the entrance head. Cable covering must be under the clamp of the entrance head to prevent damage to conductors. Do not clamp too tight; insulation may be damaged. You should have about one meter (three feet) of wire extended from the entrance head. These wires are later connected to the service drop wires by the **power supplier** when the installation is complete.

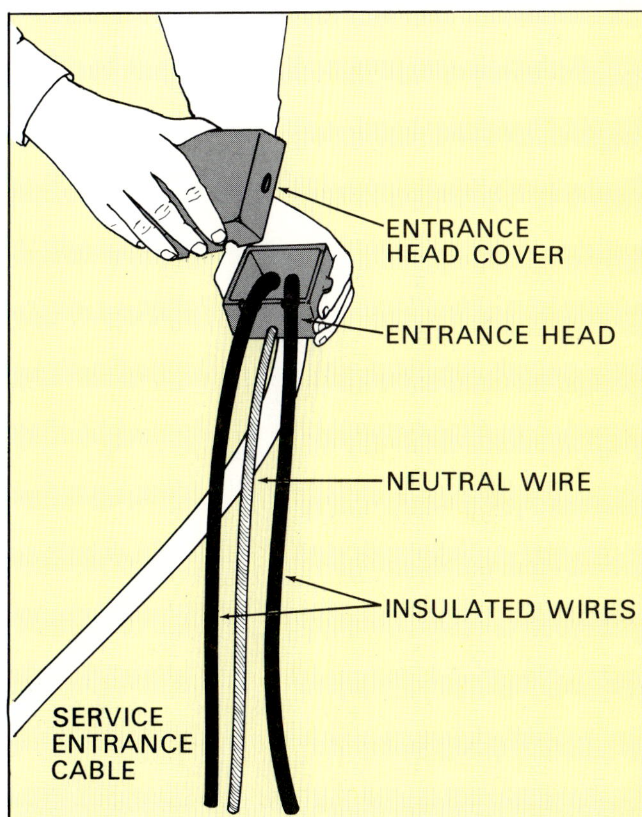

FIGURE 250. Insulated wires are inserted through the holes provided in the entrance head and the neutral wire through the notch or other type opening.

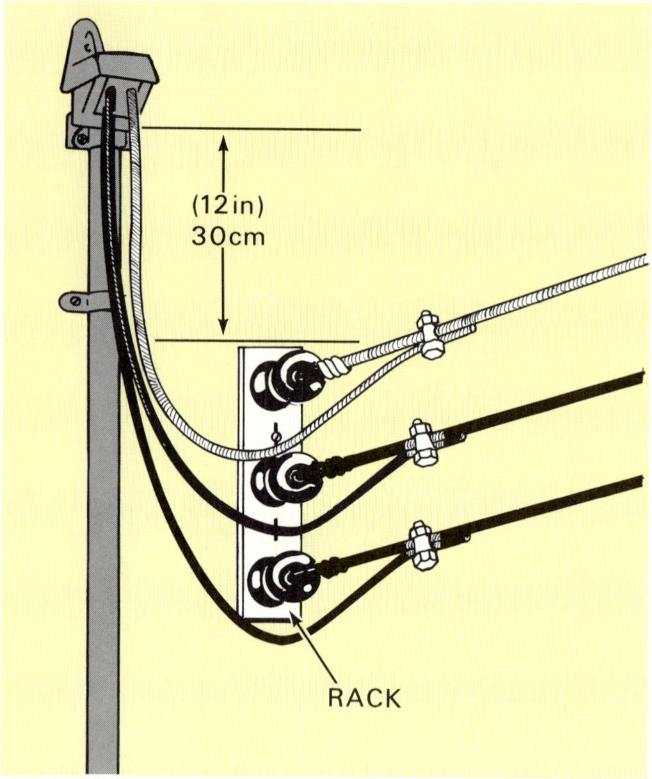

FIGURE 251. Entrance head should be 30 cm (12 inches) above the rack. Each connection must be wrapped with tape for insulation.

5. *Install the entrance head.*

   Mount the entrance head 30 cm (12 inches) higher than the rack and in line with the meter location (Figure 251). Fasten in place with screw(s).

6. *Install the meter base.*

   Check with the power supplier for recommendations on meter height.

   If none, mount the meter base 1.4 to 1.7 m (4 ½ to 5 feet) above ground at the marked location. Fasten screws in holes in the back. The meter is later installed by the power company (Figure 252).

7. *Attach connectors to meter base (Figure 253).*

   Install watertight connector at top and regular connector at bottom outlets of the meter base. Do not tighten the part holding the cable.

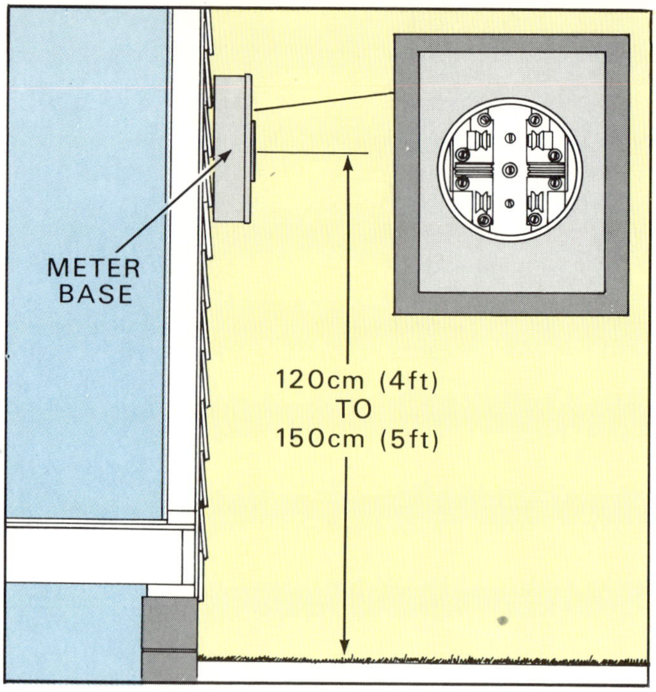

FIGURE 252. Locate meter base at height of 1.4 to 1.7 m (4 ½ to 5 feet) or as recommended by your power supplier. Fasten screws in holes provided in the back.

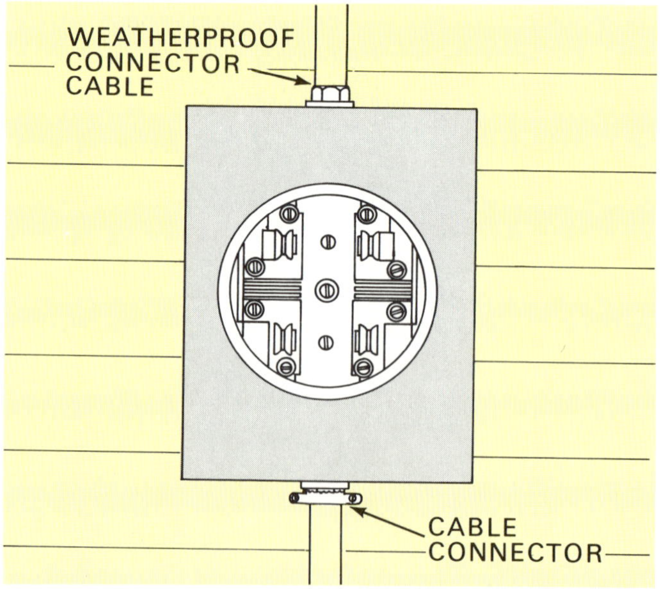

FIGURE 253. Watertight connector installed at upper opening in the meter base.

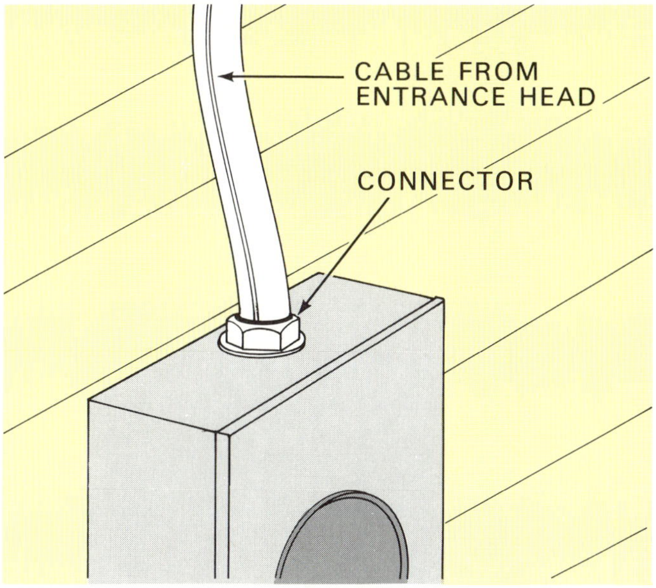

FIGURE 254. Cable from entrance head inserted through top of meter base.

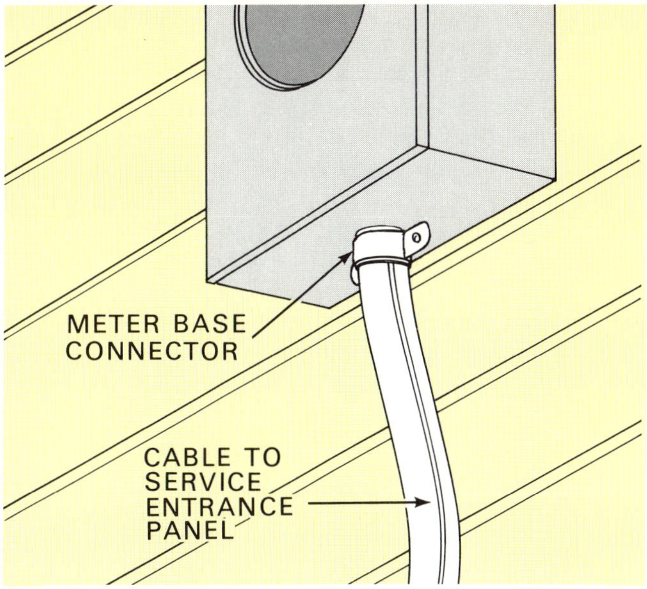

FIGURE 255 Cable installed in meter base to be extended to service entrance panel.

8. *Insert the end of cable from entrance head through the top meter base opening and tighten connector on the cable (Figure 254).*

   You should have about 15 to 20 cm (6 to 8 inches) of cable inside the meter base for terminal connections.

9. *Drill hole through wall for cable.*

   Locate the hole for the shortest or most convenient route to reach the SEP inside the house.

10. *Measure and cut cable length to reach service entrance panel.*

    Add 35 to 60 cm (14–24 inches) to the measured distance for meter connections and SEP connections.

11. *Install cable between meter base and SEP.*

    (1) *Insert wires in bottom outlet of meter base.*

    Strip cable covering from insulated conductors.

    (2) *Insert cable into meter base connector and tighten (Figure 255).*

    (3) *Insert cable through the wall and pull to SEP (Figure 256).*

    Strip outside cable covering from wires that go into the SEP.

    Remove knockout, install connector, and insert cable. Approximately 15 to 35 cm (6 to 14 inches) of cable should extend inside the panel box for terminal connections. Exact length depends on location of knockout opening and position of lugs in the panel.

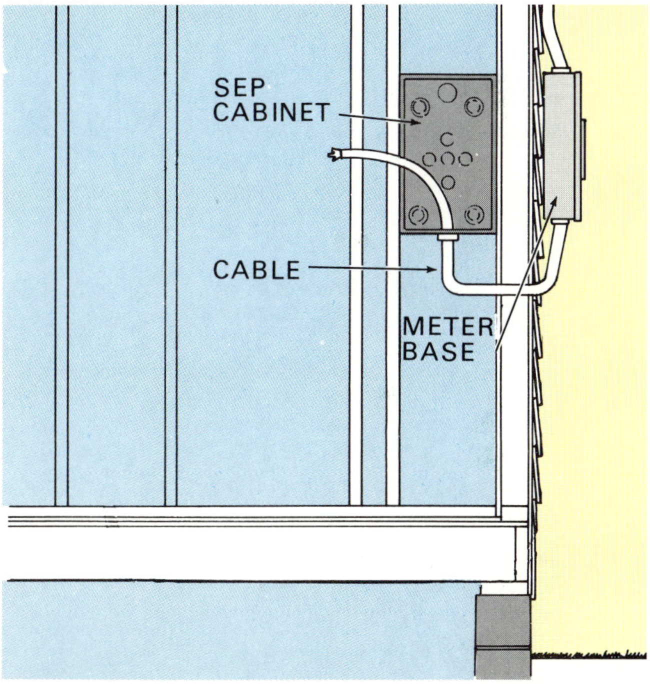

FIGURE 256. Cable run from meter base to service entrance panel.

If cable goes through a masonry wall, do not concrete around the cable. Use a mastic-type sealer.

(4) *Install sill plate (if required) and sealer (Figure 257).*

Many localities do not require a sill plate. At the point where the cable enters the building, install the sill plate where required to seal the hole. Apply a sealing compound to any space between the cable and sill plate, and where cable enters the building if no sill plate is used.

(5) *Install cable straps (Figure 258).*

Service entrance cable must be supported by straps or other means within 30 cm (12 inches) of every service head, gooseneck or connection and at least every 1.4 m (4½ feet) in between. Install straps within 30 cm (12 inches) of the meter connections and SEP and at least every 1.4 m (4½ feet) of cable between.

12. *Connect cable to meter base terminals (Figure 259).*

(1) *Strip cable covering from the conductors and strip the insulation from the ends of the two insulated conductors.*

(2) *Connect one insulated conductor to an upper (line) terminal and the second to the other upper (line) terminal.*

(3) *Connect the bare neutral conductor to the center neutral terminal.*

Some power suppliers prefer not to have the neutral conductor cut for this connection.

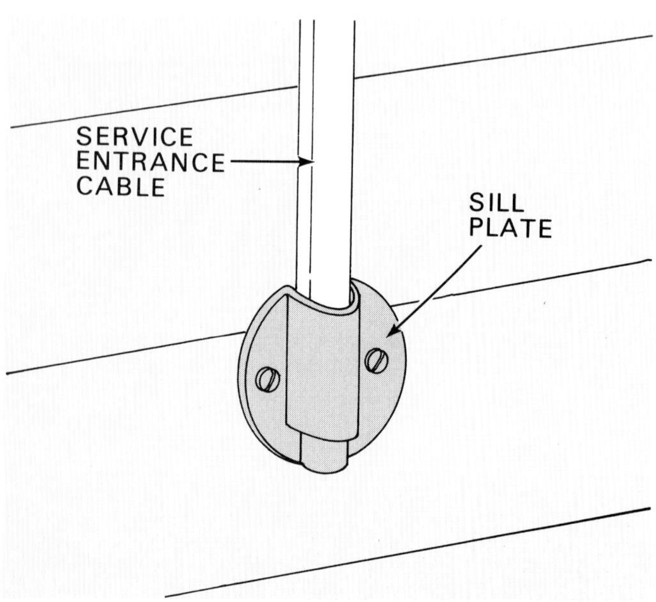

FIGURE 257. Sill plate installed in wall opening.

(4) *Repeat steps 1–3 for conductor connections to lower (output) terminals and to the neutral terminal.*

(5) *Force the conductors against the back of the meter base.*

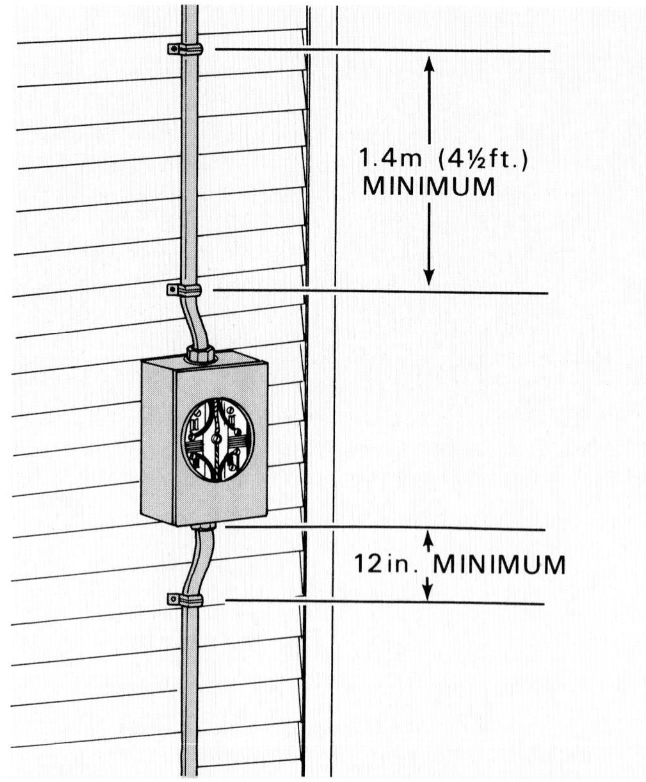

FIGURE 258. Service entrance cable supported by straps or other support every 1.4 m (4½ feet) and within 30 cm (12 inches) of every entrance head, gooseneck or other connection.

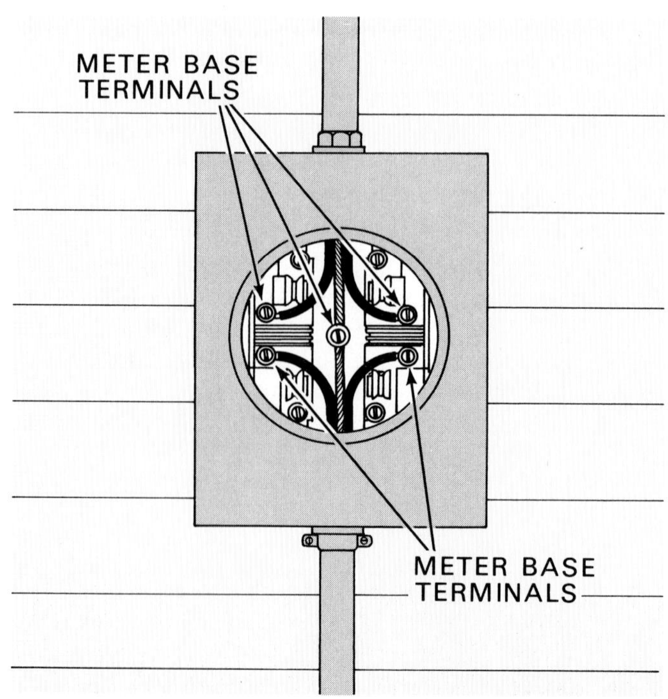

FIGURE 259. Meter base terminals completely wired.

2. INSTALLING SERVICE ENTRANCE CONDUCTORS IN CONDUIT

Some details of conduit installations are given in Section V., "Installing Metallic Conduit." However, you can assemble the required conduit and connections for rigid conduit, IMC (intermediate metal conduit) or EMT (electrical metallic tubing) service entrance requirements using wrenches, a hacksaw with 13 teeth per centimeter (32 teeth per inch), a pipe reamer and threaders for the rigid conduit. The Code now requires the use of **standard conduit cutting dies** with a 0.63 mm taper per centimeter (¾-inch taper per foot) for threading rigid conduit. EMT conduit cannot be threaded on the job because it is very thin-walled. Use connectors and couplings for EMT.

Other than assembling the conduit, procedures for conduit service entrance installation are similar to cable, but there are some notable differences.

First, there is a difference in the wire used inside the conduit. Every wiring installation you have learned to this point has used cable with 2 or more wires. **In conduit wiring, separate insulated wires are pulled into the conduit (Figure 260).** The conduit provides protection against weather and mechanical damage so outer cable covering is not required. Other differences in installation are pointed out in procedures.

Wires used inside service entrance conduit may be Type TW, THW, RHW or other types approved for that use. The neutral wire may be bare, but if insulated it should be white. Colored or white wires may not be available in every size. If not, you must identify the neutral wire by painting it white, wrapping with tape or with a tag at each terminal point. The other two wires may be black or one red and one black.

Installing service entrance wiring in conduit is discussed as follows:

NOTE: The Code now permits the use of intermediate metal conduit (IMC) wherever rigid conduit is required in residential wiring. The wall thickness of IMC is thinner than rigid conduit but thicker than EMT. Procedures for installing IMC are the same as for rigid conduit.

a. Installing Service Entrance Wiring in Rigid Conduit.
b. Installing Service Entrance Wiring in Thinwall Conduit (EMT).

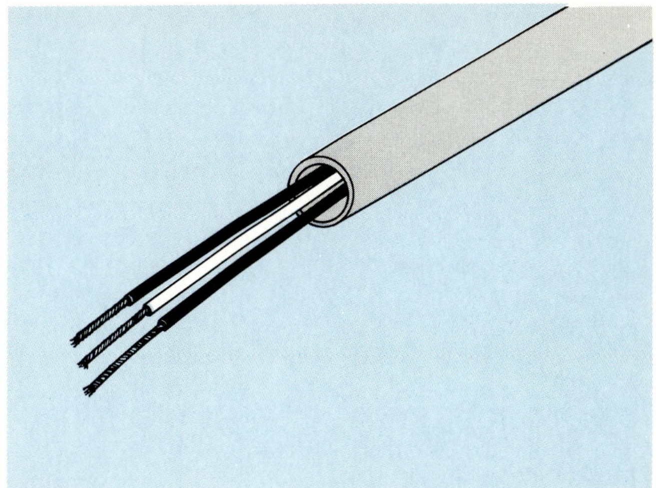

FIGURE 260. Separate insulated wires are installed in conduit.

a. Installing Service Entrance Wiring in Rigid Conduit

Rigid metal conduit is steel pipe that looks very much like water pipe (Figure 261). But water pipe may never be substituted for conduit. Rigid metal conduit has a rust resistant finish inside and out. The inside finish is smooth to prevent damage to wires as they are pulled through, which is not the case with water pipe. Metal conduit must carry the underwriters label to meet Code requirements. It is soft enough to bend as required.

For service entrance installations and other small jobs, you may use fittings instead of bending the conduit.

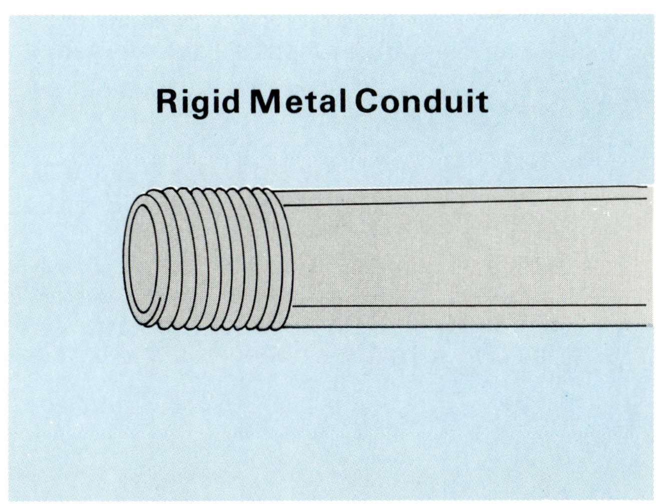

FIGURE 261. Rigid metal conduit looks much like common water pipe but there are important differences.

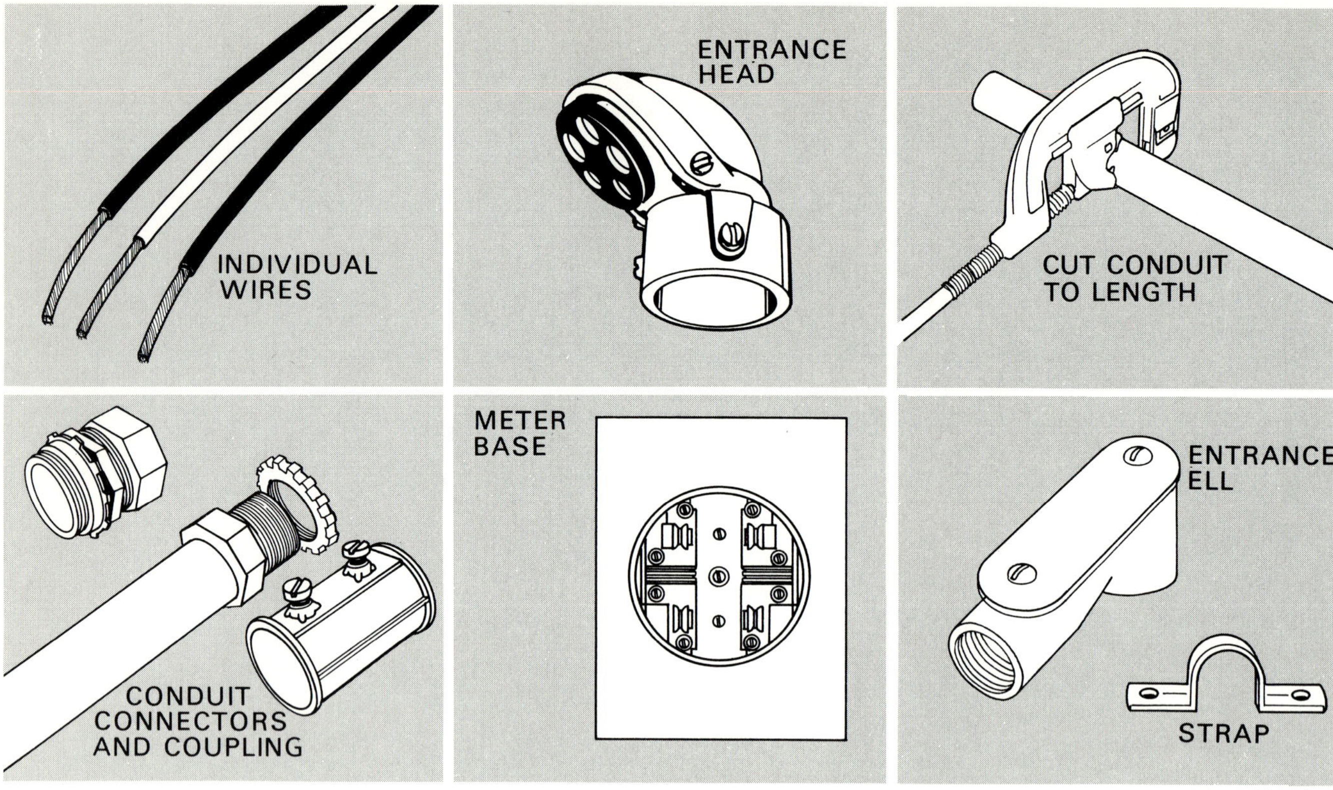

FIGURE 262. Materials required to install service entrance wiring in rigid conduit.

The materials required for installing 120/240-volt entrance wiring in rigid conduit include (Figure 262) three individual wires, entrance head, conduit cut to measured length, conduit connectors, meter base, entrance ell and straps. The entrance **head** must be the type made for conduit (Figure 263). The entrance **ell** is installed where the conduit enters the house. It has a removable back to make it easier to install the wires.

The conduit is assembled first and wires are pulled through after the assembly is mounted on the wall. Proceed as follows:

1. *Determine size conduit to install.*

   Table XI shows maximum number of wires that can be installed in conduit of given sizes. Select conduit of correct size for your service entrance.

   NOTE: Where at least one heavy appliance such as a dryer or water heater will be in use, the neutral wire usually may be one size smaller than the other two service entrance wires.

FIGURE 263. Entrance head with fitting for rigid conduit installation.

TABLE XI. MAXIMUM NUMBER OF CONDUCTORS IN TRADE SIZES OF CONDUIT OR TUBING
(Code Tables 3A and 3B)
(Based on Code Table 1, Chapter 9)

Conduit Trade Size (Inches)		½	¾	1	1¼	1½	2
Type Letters	**Conductor Size AWG, MCM**						
TW, T, RUH,	14	9	15	25	44	60	99
RUW,	12	7	12	19	35	47	78
XHHW (14 thru 8)	10	5	9	15	26	36	60
	8	2	4	7	12	17	28
RHW and RHH	14	6	10	16	29	49	65
(without outer	12	4	8	13	24	32	53
covering),	10	4	6	11	19	26	43
THW	8	1	3	5	10	13	22
TW,	6	1	2	4	7	10	16
T,	4	1	1	3	5	7	12
THW,	3	1	1	2	4	6	10
RUH (6 thru 2),	2	1	1	2	4	5	9
RUW (6 thru 2),	1		1	1	3	4	6
FEPB (6 thru 2),	0		1	1	2	3	5
RHW and	00		1	1	1	3	5
RHH (with-	000		1	1	1	2	4
out outer	0000			1	1	1	3
covering)	250			1	1	1	2
	300			1	1	1	2
	350				1	1	1
	400				1	1	1
	500				1	1	1
	14	13	24	39	69	94	154
THWN,	12	10	18	29	51	70	114
	10	6	11	18	32	44	73
	8	3	5	9	16	22	26
THHN,	6	1	4	6	11	15	26
FEP (14 thru 2),	4	1	2	4	7	9	16
FEPB (14 thru 8),	3	1	1	3	6	8	13
PFA (14 thru 4/0)	2	1	1	3	5	7	11
PFAH (14 thru 4/0)	1		1	1	3	5	8
Z (14 thru 4/0)							
XHHW (4 thru 500MCM)	0		1	1	3	4	7
	00		1	1	2	3	6
	000		1	1	1	3	5
	0000		1	1	1	2	4
	250			1	1	1	3
	300			1	1	1	3
	350			1	1	1	2
	400				1	1	1
	500				1	1	1
XHHW	6	1	3	5	9	13	21
	600				1	1	1
	700					1	1
	750					1	1
	14	3	6	10	18	25	41
	12	3	5	9	15	21	35
RHW,	10	2	4	7	13	18	29
	8	1	2	4	7	9	16
RHH	6	1	1	2	5	6	11
	4	1	1	1	3	5	8
(with	3	1	1	1	3	4	7
outer	2		1	1	3	4	6
covering)	1		1	1	1	3	5
	0		1	1	1	2	4
	00			1	1	1	3
	000			1	1	1	3
	0000			1	1	1	2

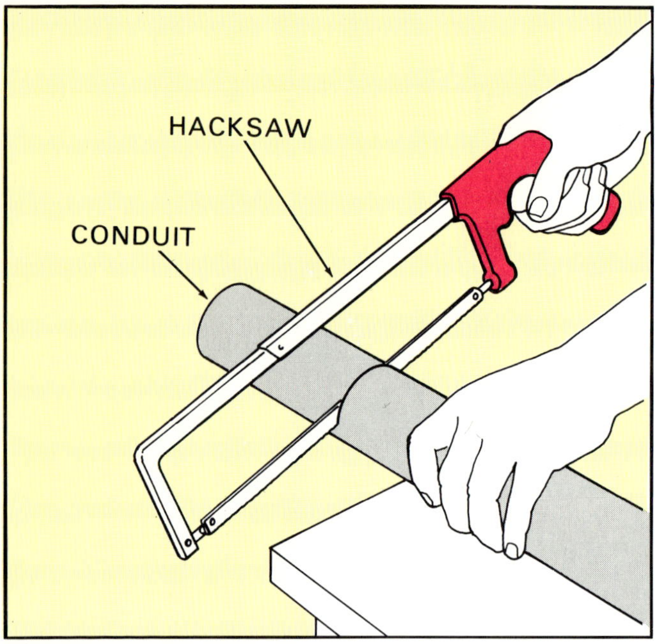

FIGURE 264. Cut rigid conduit with pipe cutter or fine-tooth hacksaw.

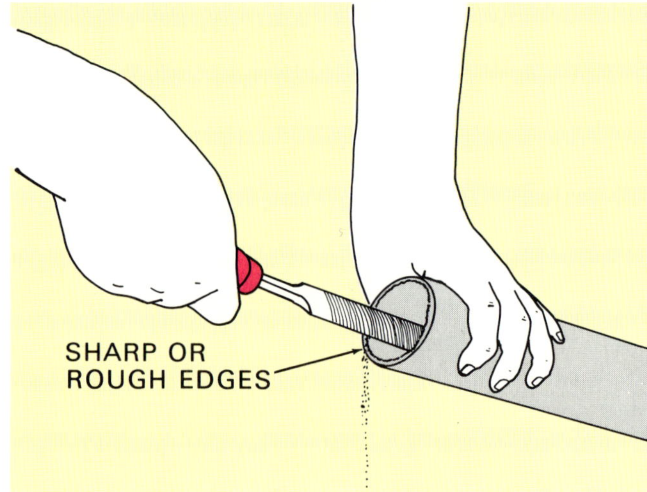

FIGURE 265. File or ream ends of the conduit to remove sharp or rough edges that might cut wiring insulation.

The number of conductors allowed in Table XI for various sizes of conduit applies to all conduit wiring installations. They are not intended to apply to short sections of conduit used to protect exposed wiring from physical damage according to the Code. A general recommendation is to install 3.8 cm or 5 cm (1½ or 2 inch) conduit for the house service entrance.

2. *Install service-drop insulator rack as for cable.*
3. *Locate entrance head as for cable.*
4. *Measure space between locations of service entrance head and meter, between meter and entrance ell and from entrance ell to service entrance cabinet.*
5. *Cut measured lengths of conduit with a pipe cutter or hacksaw (Figure 264). File or ream ends of each piece to remove rough edges (Figure 265).*

   This *must* be done to prevent damage to insulation as wires are pulled through.
6. *Thread conduit where required with tapered conduit dies.*

   The top end of the entrance head may be threaded or not depending on the type of entrance head used.

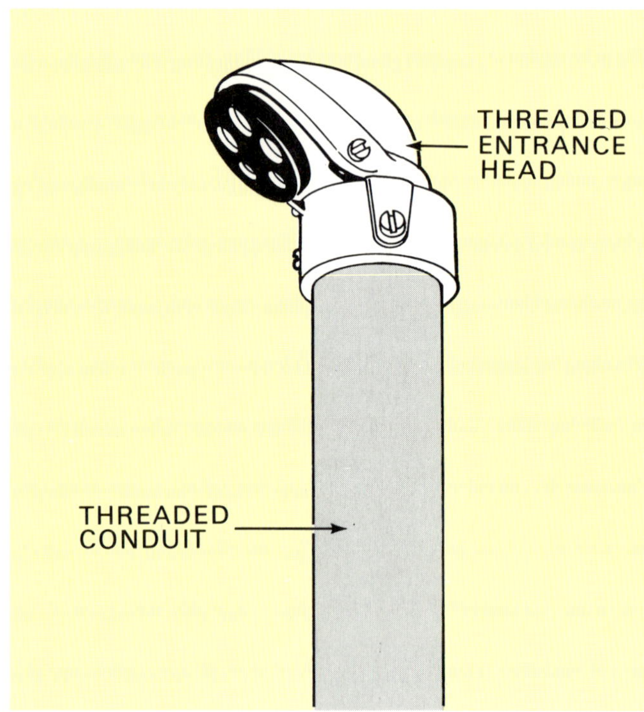

FIGURE 266. If entrance head is threaded, screw it on the conduit end.

7. *Assemble parts on the ground.*
   (1) *Attach entrance head.*

   If the entrance head is threaded, screw it on the threaded end of the conduit (Figure 266). If not threaded, place on end of conduit and tighten bolts or set screw(s) on each side of clamp (Figure 267).

(2) *Connect conduit to meter (Figure 268).*

Most power suppliers require a threaded hub on top of the meter base to make the connection water tight. Attach the hub to the meter base, usually with four bolts. Insert threaded conduit and tighten. If conduit enters the meter base at bottom, or side, use two locknuts, one inside and one outside. Attach locknut to threaded conduit end. Screw the locknut on as far as it will go. Place conduit through opening in meter base, attach inside locknut and turn until snug. Turn outside locknut until very tight.

(3) *Connect entrance ell.*

Attach entrance ell to threaded conduit length from meter base (Figure 269).

(4) *Connect short length of conduit between entrance ell and service entrance cabinet.*

This completes the assembly of all outside connections (Figure 269).

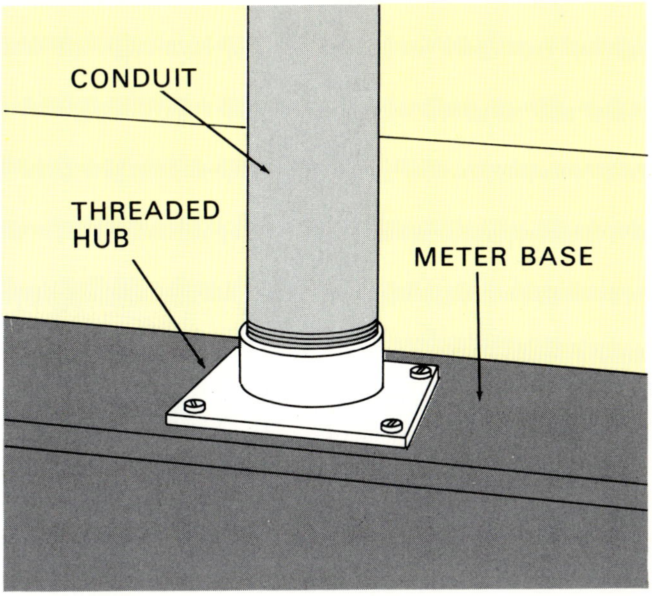

FIGURE 268. Attach hub to meter base and screw conduit into the threaded hub.

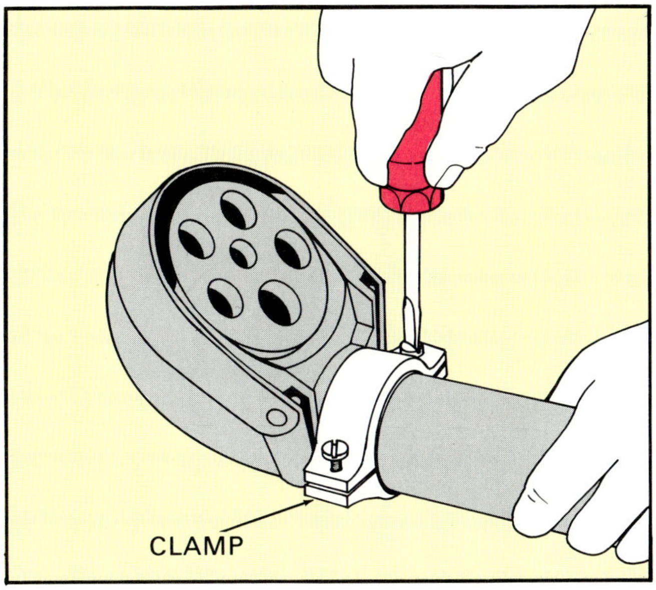

FIGURE 267. For bolt-on entrance heads, tighten the bolts that clamp the conduit.

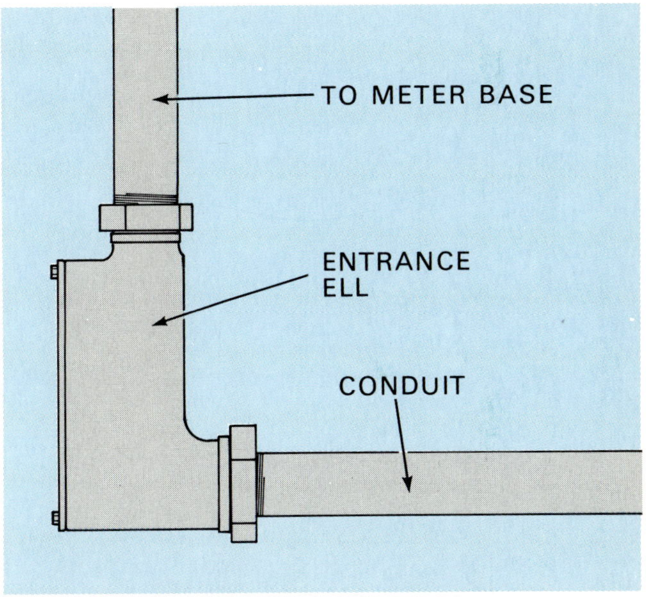

FIGURE 269. Install entrance ell and connect short length of conduit to reach service entrance cabinet.

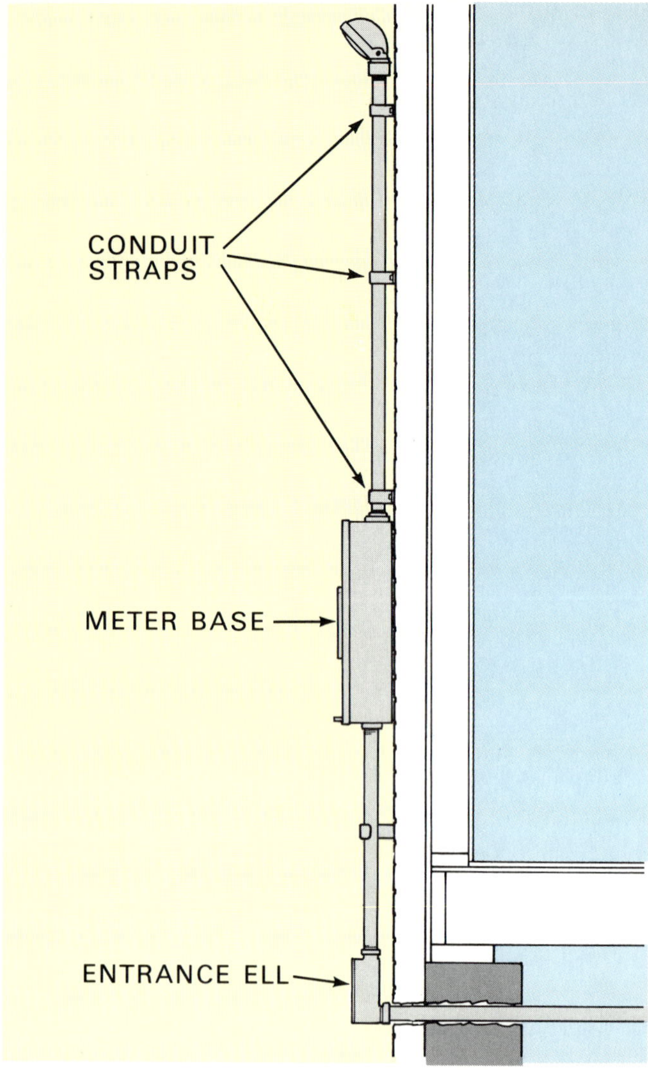

FIGURE 270. Install the assembly on the wall and fasten in place with straps.

8. *Install the assembly on the wall.*

   Insert the short length of conduit through the wall with the assembly held upright. Fasten the conduit to the building with conduit straps (Figure 270). Install support brackets behind the conduit where mounted away from the wall as in Figure 270.

   It must be supported within 0.9 m (3 feet) of every connection, plus additional supports not over 3.1 m (10 feet) apart for 1.3 and 1.9 cm (½ and ¾ inches), not over 3.7 m (12 feet) for 2.5 cm (1 inch), 4.3 m (14 feet) for 3.2 and 3.8 cm (1 ¼ and 1 ½ inches) and 4.9 m (16 feet) for 5.1 and 6.4 cm (2 and 2 ½ inches).

9. *Connect conduit to service entrance cabinet.*

   Attach locknut and grounding bushing. Twist locknut tight to insure solid grounding connection. Run a jumper wire from one bushing to the neutral bar (Figure 271). (Refer to Code Table 250–94 for correct size conductor.)

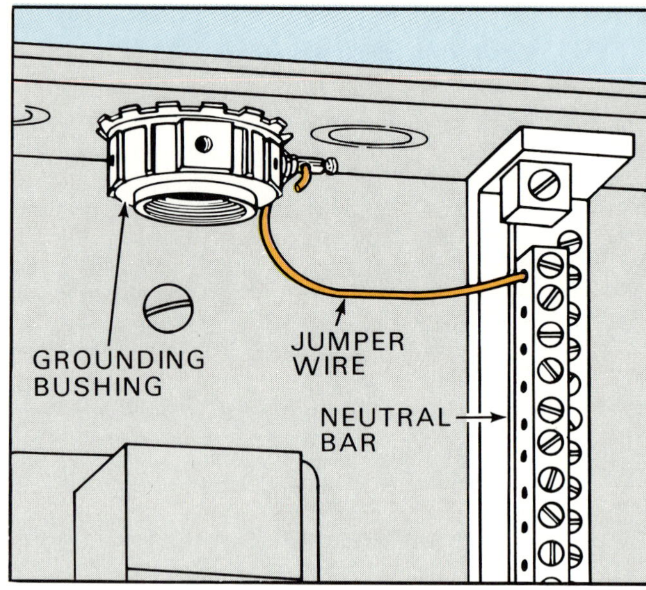

FIGURE 271. Jumper wire connects the grounding bushing and neutral bar.

10. *Install the wires.*

    Measure and cut the first set of wires between entrance head and meter base so you will have 0.9 m (36 inches) extending from the entrance head and 20 to 25 cm (8 to 10 inches) inside the meter base.

    Remove front cover from meter base and push wires up through meter opening through conduit to entrance head. Remove top of entrance head and thread wires through bushing holes. Replace top on entrance head and leave 0.9 m (3 feet) (or length recommended by your power supplier) of wire extended for connection to service drop (Figure 270).

    Measure and cut second set of wires to reach from bottom terminals of meter base to terminals inside the SEP. Push wires through bottom opening of meter base toward the entrance ell. Remove cover on entrance ell to push wires through more easily. Pull wires into service entrance cabinet. Replace cover on entrance ell (Figure 270).

11. *Connect wiring to meter base.*

    Follow procedures in Step 12 Under IV. D. 1., "Installing Service Entrance Cable and Meter Base."

b. Installing Service Entrance Wiring in Thinwall Conduit (EMT)

The Code refers to this type of conduit as Electrical Metallic Tubing (Figure 272). It is popularly known as Thinwall or EMT. Sizes of EMT used in residential wiring are the same as for rigid conduit. As the name indicates, the walls are very thin and cannot be threaded. Instead, special pressure type or setscrew connectors are used to fasten the tubing to the boxes. Setscrew connectors cannot be used outside where exposed to weather. Couplings are used to join lengths of EMT (Figure 273).

Procedures for installing EMT for the service entrance are practically the same as for rigid. **The major difference is in the type of connectors and couplings used.** Proceed as follows to install EMT using weathertight pressure connectors (Figure 274).

1. *Install the unthreaded end of the pressure connector on the end of the conduit.*
2. *Insert the connector into the meter base opening.*
3. *Screw the locknut on the connector inside the box until tight.*
4. *If using a setscrew coupling indoors to join two pieces of conduit, insert each end of conduit into the coupling to the stops. Turn the setscrews down firmly to lock the conduit to the coupling (Figure 275).*

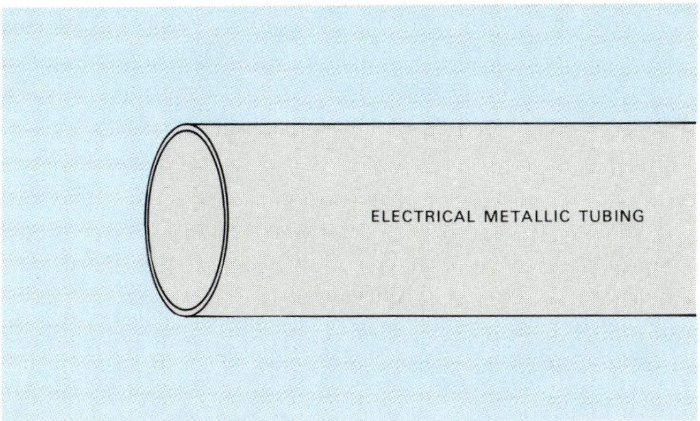

FIGURE 272. Electrical metallic tubing, commonly called thinwall or EMT.

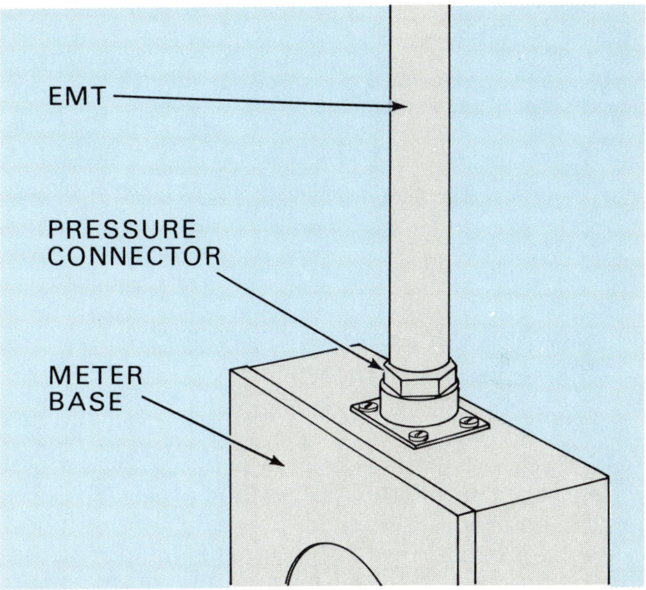

FIGURE 274. Weathertight pressure connectors are used for EMT service entrance installations.

FIGURE 273. EMT connectors and couplings.

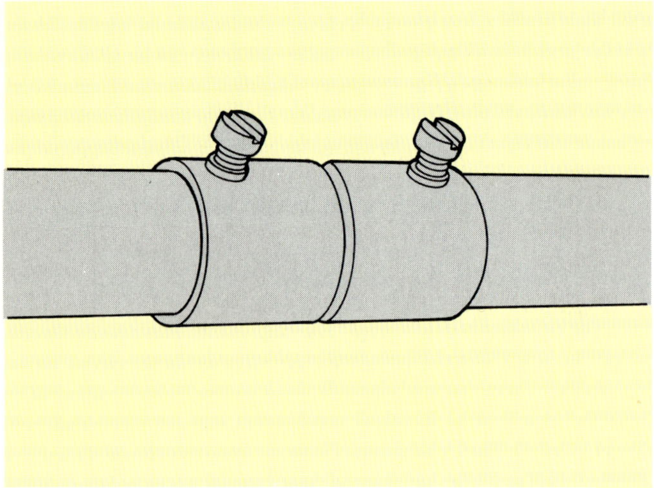

FIGURE 275. Setscrew EMT couplings may be used inside where protected from weather.

5. *Proceed as for rigid conduit to assemble conduit and install service entrance wiring.*

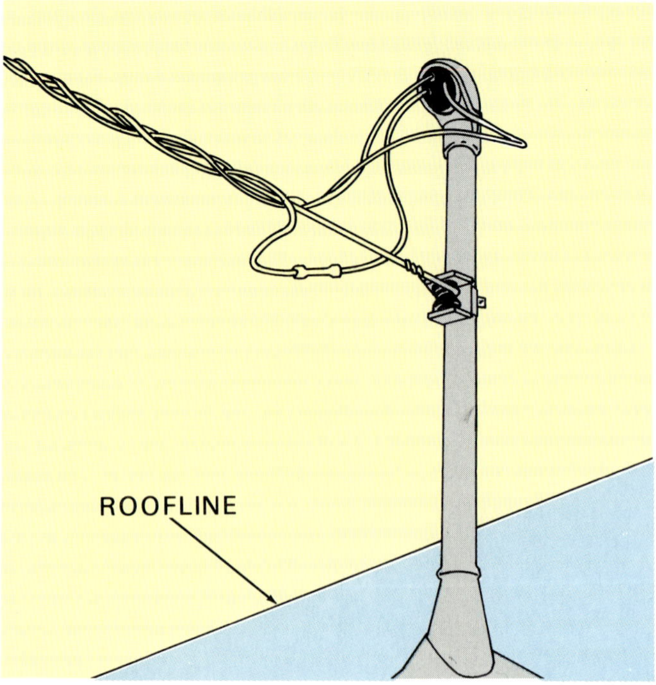

FIGURE 276. The conduit mast-type service entrance is preferred for single-story homes with low roof lines.

3. INSTALLING A MAST TYPE OF SERVICE ENTRANCE

On single-story houses with low roof lines, a rigid conduit mast is the preferred method of installing the service entrance. The rigid conduit mast extends through the roof to give the required clearance above ground (Figure 276). Service drop clearance above the roof itself may be only 46 cm (18 inches) if no more than 1.2 m (4 feet) of roof area has to be crossed by the service drop (Figure 277). Conduit size should be at least 5.1 cm (2 inches) to provide the necessary rigidity.

Masts are available in kit form which includes conduit, mounting brackets, roof flashing, entrance head and a stand-off insulator that clamps to the mast. Mast height extending more than 0.9 m (3 feet) above the roof should be supported by guy wires.

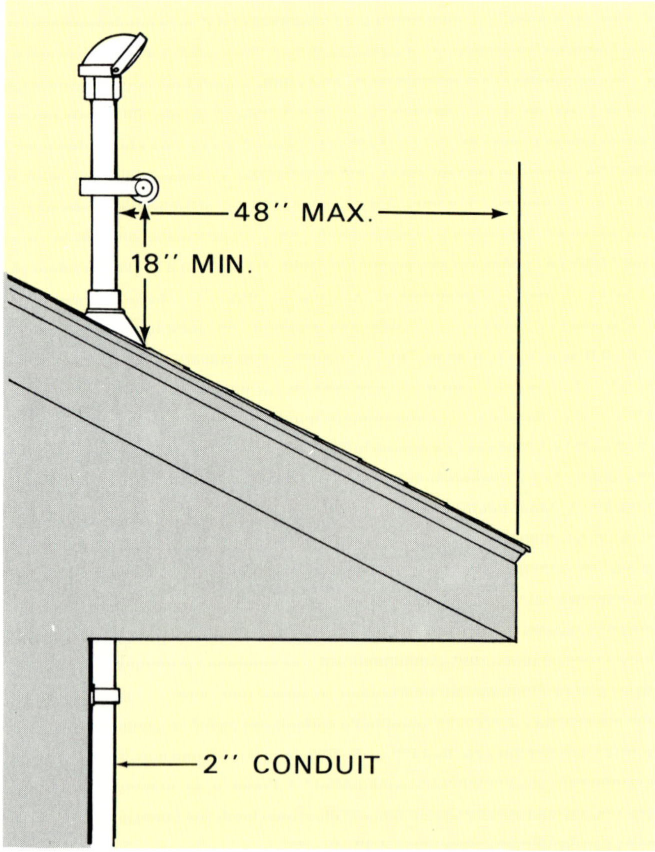

FIGURE 277. Clearance for the service entrance may be as little as 46 cm (18 inches) if no more than 1.2 m (four feet) of roof area is crossed by the service drop.

Proceed as follows to install (Figure 278):

1. *Locate meter at 1.4 to 1.5 m (4½ to 5 feet) height.*
2. *Measure to determine length of conduit needed.*

   Measure from top of meter box to required height above roof line. Assume 0.8 to 0.9 m (2½ to 3 feet), for example, above roof line to allow clearance for drip loop and weatherhead.
3. *Cut conduit to measured length.*
4. *Cut necessary holes through boards and roofing.*
5. *Install necessary bolts for bracket supports.*
6. *Insert conduit through holes and raise to position.*
7. *Fasten in place with brackets.*

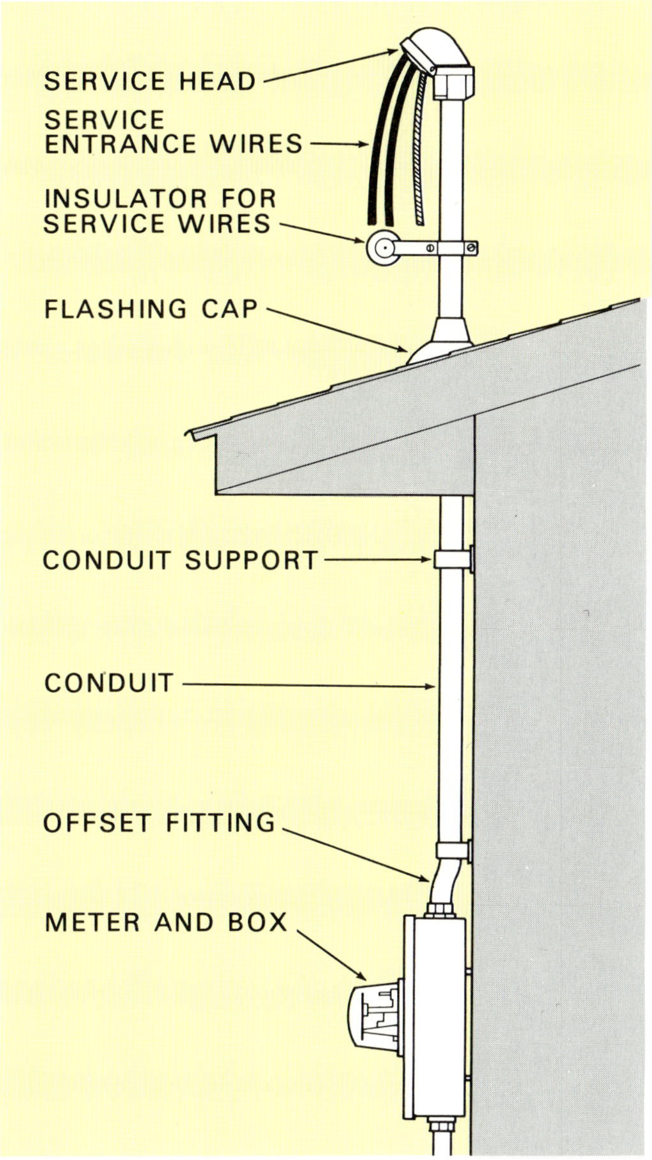

FIGURE 278. Complete mast-type service installation. An off-set fitting is installed on the conduit end to be connected to the meter base.

8. *Install roof flashing and seal.*
9. *Install insulator for service drop (May be furnished and installed by power supplier).*
10. *Install service head as in previous procedures.*
11. *Install meter base as for conduit.*

    Install offset fitting on the conduit end to connect conduit to meter base (Figure 278). Adjust the meter base to correct position and fasten in place on the wall. Connect meter base to conduit fitting.

4. INSTALLING UNDERGROUND SERVICE ENTRANCE CABLE

Underground service is gaining favor as a means of service entrance installation. Many subdivisions now require it and more farmers are using it.

Among the advantages of underground installation are reduced danger of damage from lightning, wind, ice and falling branches. On farms it eliminates danger of tall moving equipment coming in contact with overhead wires.

In some localities, the entire distribution system is underground and the transformers are mounted on a concrete pad or fiberglass mounting unit (Figure 279). In others, only the wiring from the power sup-

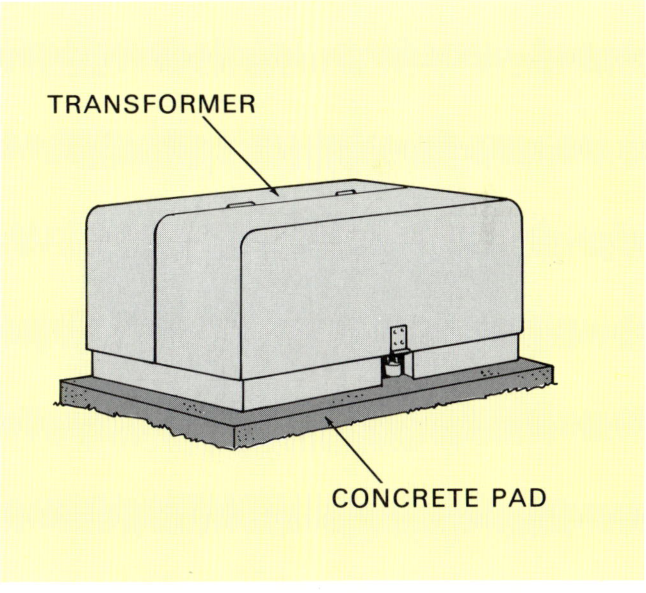

FIGURE 279. Pad-mounted transformers are used in many subdivisions for underground distribution.

plier overhead lines to the house is underground (Figure 280). For both types, cable is laid loosely in a trench and covered 60 to 120 cm (24–48 inches) deep, depending on local codes. If there is danger that the cable might be damaged, run it through conduit.

The cable to be installed is called Underground Service Entrance cable. The identification on the cable is Type USE (Figure 281). The conductors inside the cable are insulated with moisture-resistant insulation and the cable covering is a tough moisture-resistant material. If cable is supplied by the power supplier, as is sometimes done, it may not be USE. Most popular in some areas is single-conductor cable having rubber insulation and no outer covering.

♣ Both recessed and surface-mounted meter installations may require a standard 90° sweep elbow with bushing, in addition to the usual conduit fittings. The bushing on the end of the elbow prevents damage to Underground Service Entrance cable (Figure 282).

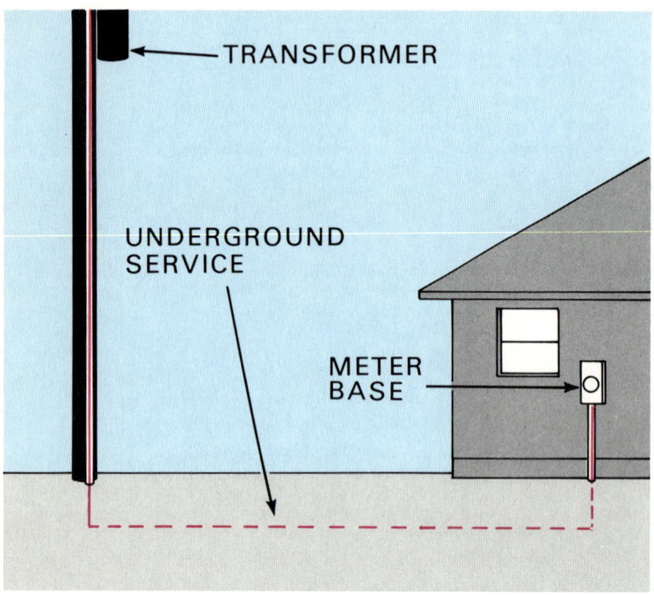

FIGURE 280. In some localities, only the wiring from the power supplier overhead lines to the house is underground.

Two methods of mounting the meter and conduit for underground service entrance are:

a. Installing Recessed Meter and Conduit.
b. Installing Surface-Mounted Meter and Conduit.

NOTE: Some power suppliers may not allow recessed meter installation. Check specifications before installing. Extra space may be required on each side and below the meter base to work on the meter.

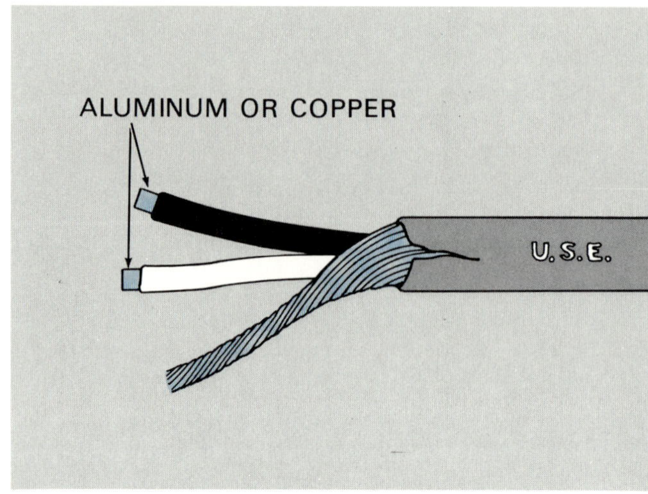

FIGURE 281. Underground service entrance cable (USE) may be buried directly in the earth.

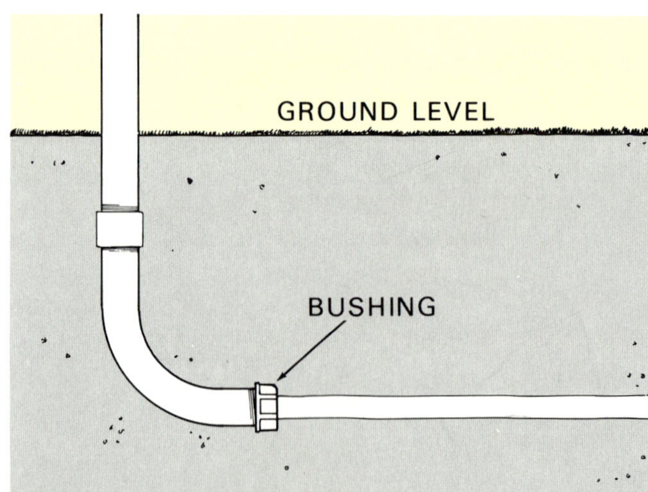

♣ **FIGURE 282. Bushings on the end of the elbow or conduit prevent damage to cable insulation.**

a. Installing Recessed Meter and Conduit

Only rigid galvanized conduit is allowed by most codes. It is usually 5.1 cm (2-inch size). The recessed system requires installation before construction is complete, because the conduit must be mounted inside the wall space.

Proceed as follows to install conduit for underground service entrance with recessed meter and conduit (Figure 283):

1. *Dig hole for conduit installation.*

   The conduit, when installed, extends underground, so you must dig the hole to the required depth to meet the trench from the power supplier transformer. Required depth may vary from 60 to 120 cm (24 to 48 inches), depending on local codes.

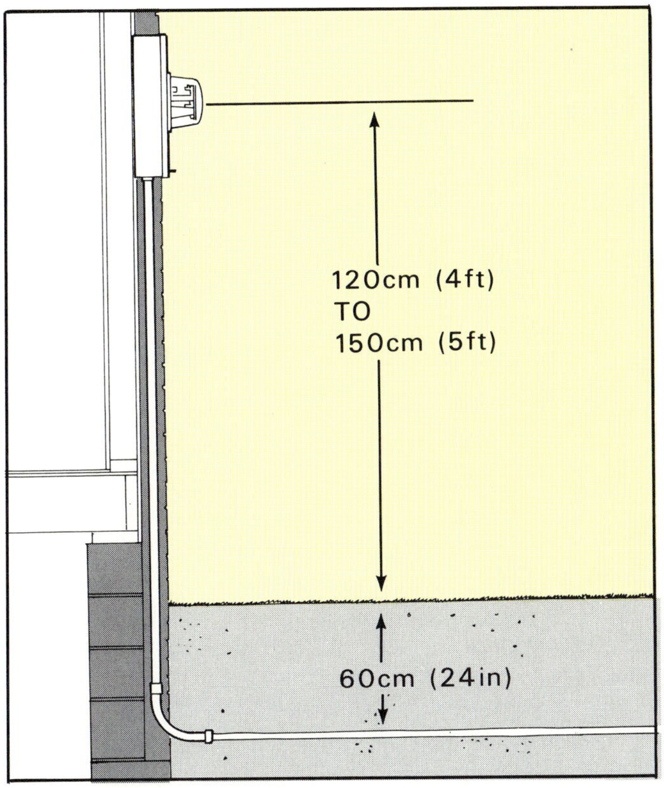

FIGURE 283. Recessed meter and underground installation.

2. *Mount the meter base in the wall space on the inside wall.*

   The **base** should be flush with the outside wall. The **meter** will extend from the wall when mounted.

3. *Measure from bottom of meter base to required depth underground, including the standard elbow (Figure 283).*
4. *Cut conduit to measured length, ream and thread each end.*
5. *Attach conduit to meter base and screw coupling on bottom end of conduit.*

   Use grounding type insulating bushing on end of conduit in the meter base to insure grounding of conduit (Figure 284).
6. *Screw elbow to coupling and attach bushing. Turn elbow so it is perpendicular to building when tight.*

   Power supplier personnel will install underground cable and connect to meter base terminals.

FIGURE 284. Insulating bushing with grounding lug at conduit connection in meter base.

b. Installing Surface-Mounted Meter and Conduit

Procedures for installing the surface-mounted meter base are practically the same as for recess mounted. Proceed as follows (Figure 285):

1. *Dig hole as for recess mount.*
2. *Mount meter base on outside wall surface.*
3. *Follow steps 3–6 for recess mount.*
4. *Attach mounting straps (Figure 285).*

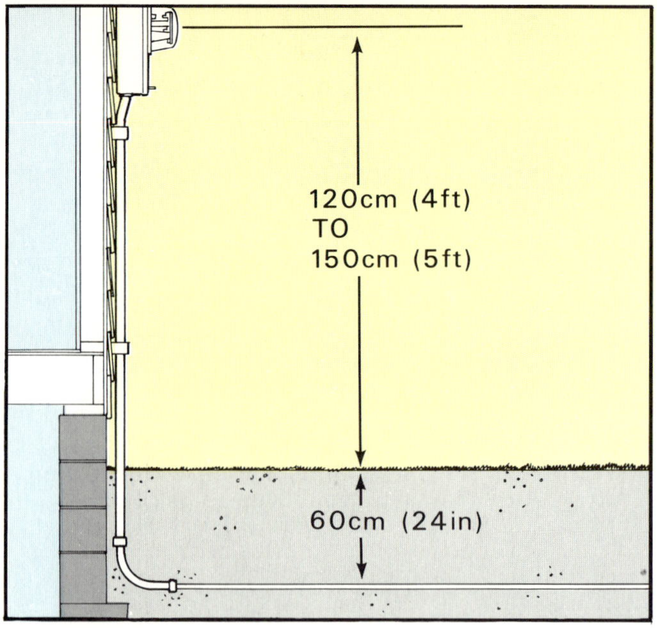

FIGURE 285. Surface-mounted meter and underground installation.

E. Connecting Service Entrance Wiring

The interior of the service entrance cabinet includes the main breaker or fuse pullout, the neutral bar and the circuit breaker panel (Figure 286). Instructions for installing parts of the SEP are provided by the manufacturer. Follow instructions to mount the main breaker, the neutral bar and circuit breaker panel in the cabinet. Two to four bolts usually fasten each piece. This is usually done after all wiring has been installed in the house and circuit wires pulled into the empty cabinet.

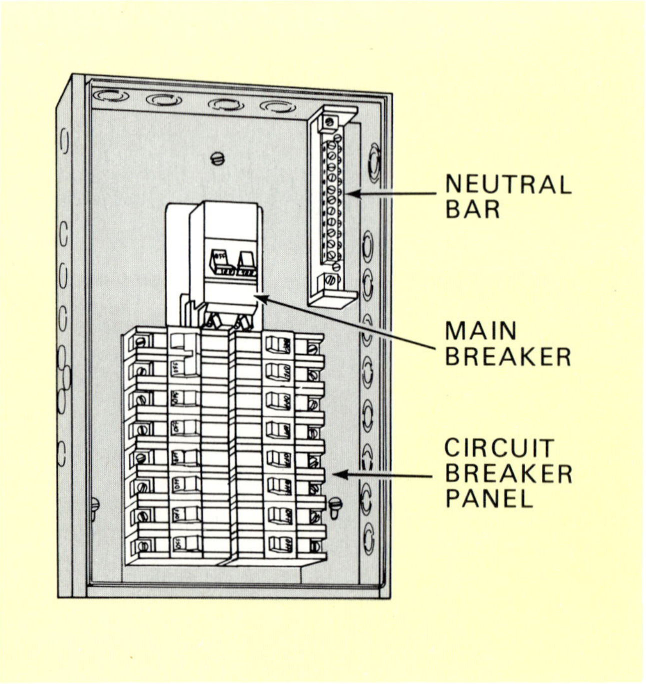

FIGURE 286. Service entrance cabinet contains main breaker, neutral bar and circuit breaker panel.

The neutral bar may be mounted below the circuit breaker panel, on either side or above it, depending on the manufacturer (Figure 286). The main breaker, or fuse pullout, which disconnects all electricity to the house, is usually mounted above the circuit breaker panel (Figure 286). Or, it may be an integral part of the panel.

According to electrical equipment manufacturers, **overheating** may be a problem in SEP's after they are completely wired and placed in service. You can reduce the amount of heat generated by making SEP wiring connections in a neat, orderly manner. Excess wire length and poor connections are said to be major causes of heat generation. Cut off any excess wire before connecting to the panel and make sure all connections are tight. Without the excess length, it is easier to route circuit wires around the edges of the cabinet, allowing better air circulation (Figure 287). Shorter wires also reduce the amount of resistance. Lower resistance means less heat will be produced.

Compare the two panels in Figure 287. Wiring on the left panel is the ideal. On the right is a poor wiring job. Each panel has the same number and ratings of breakers.

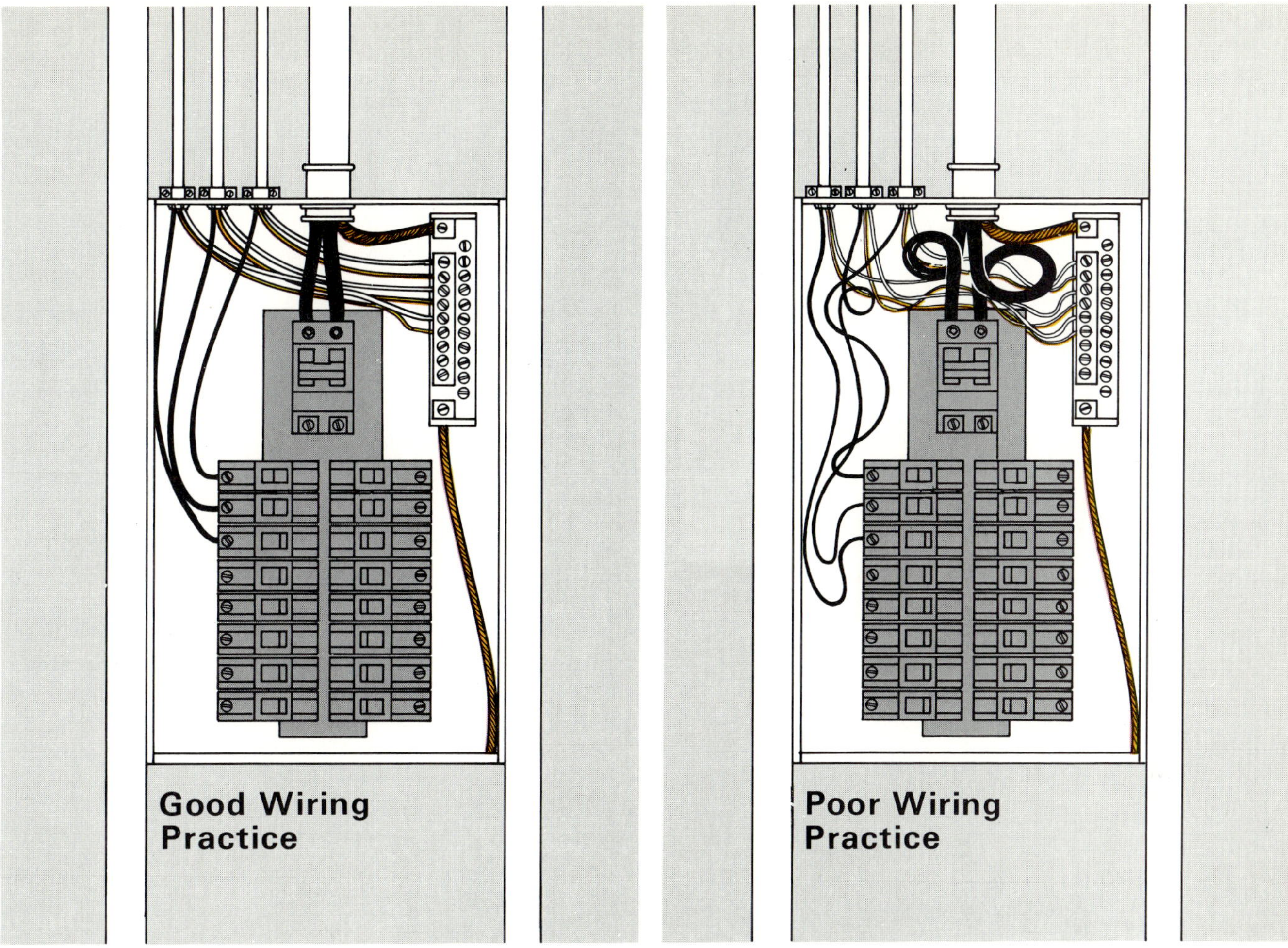

FIGURE 287. Remove excess wire length for easier installation and better air circulation.

Connecting SEP wiring is discussed as follows:

1. Connecting Service Entrance Conductors to Service Entrance Panel Terminals.
2. Installing Wiring System Ground.

1. CONNECTING SERVICE ENTRANCE CONDUCTORS TO SERVICE ENTRANCE PANEL TERMINALS

Proceed as follows:

1. *Connect the two hot conductors (Figure 287).*

   The hot conductors usually are either two black wires or one red and one black. Blue and other colors may be used for hot wires. Cut off any excess length on the service entrance wires. Insert each wire into the connector provided in the main breaker terminal. Tighten the terminal screws. Many are hexagonal and require use of an "Allen" wrench of the correct size. Others are slotted for regular screwdrivers. **If using aluminum wire, apply inhibitor paste to each terminal and conductor. Be sure the terminals are labeled Al-Cu. Make every connection as tight as practicable.**
2. *Connect the neutral wire to the neutral bar terminal.*

   Apply inhibitor and tighten securely with an Allen wrench, or screwdriver.

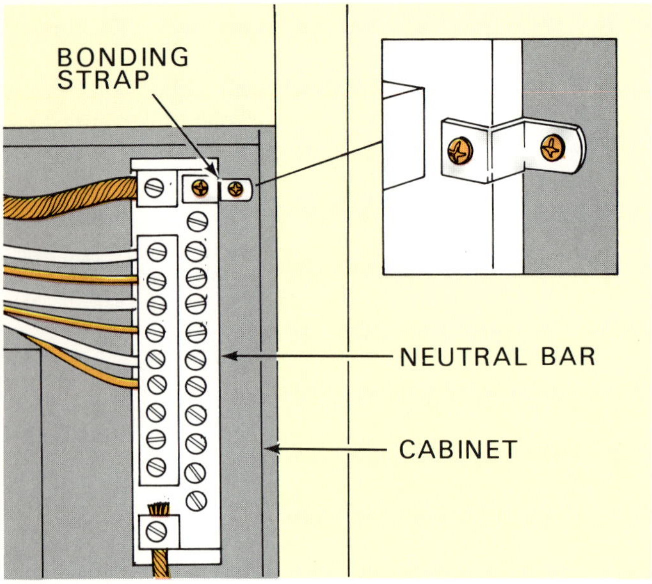

FIGURE 288. Bonding strap must be bolted to neutral bar and to cabinet.

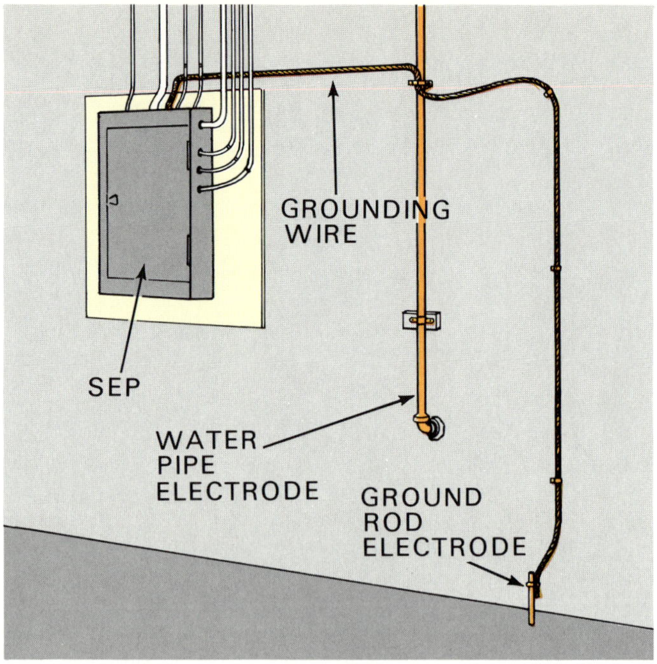

FIGURE 289. If metal water pipe is used as a grounding electrode, it must be supplemented by one or more additional electrodes.

3. *Ground and bond neutral bar to the panel cabinet.*

   Use bolt or strap provided (Figure 288). Insert bonding screw through the hole provided in the neutral bar and tighten. If a bonding strap is provided, bolt one end to the neutral bar and the other to the metal cabinet. Bonding the cabinet to the neutral bar provides additional grounding protection against ground faults that may occur on the inside or outside surface of the cabinet.

2. INSTALLING WIRING SYSTEM GROUND

To ground the house wiring system, you must connect a **ground wire** (grounding electrode conductor) from the neutral bar on the SEP to an approved grounding electrode, such as a metal water pipe and/or to other approved grounding electrode(s). **If metal water pipe is used, it must be supplemented by one or more additional electrodes to meet Code requirements** (Figure 289).

Installing system ground is discussed as follows:

a. Sizes and Types of Grounding Conductors.
b. Grounding Electrodes.
c. Connecting Grounding Conductors and Grounding Electrodes for System Ground.

a. Sizes and Types of Grounding Conductors

Sizes and types of grounding conductors are discussed under the following headings:

(1) Sizes of Grounding Conductors.
(2) Types of Grounding Conductors.

(1) SIZES OF GROUNDING CONDUCTORS. The size of the grounding conductor depends on the size of your service entrance conductors. **You may use No. 8 copper wire if your service entrance conductors are No. 2 or smaller.** However, you may want to use No. 6 instead, because it is simpler to install. For example, a No. 8 ground wire must be enclosed in conduit or armored cable. A No. 6 or larger wire may be stapled rigidly to the building surface without being covered unless exposed to severe physical damage (Figure 290). Select grounding conductor size from Table XII (taken from Code Table 250–94).

The grounding electrode conductor must be of copper, aluminum or copper-clad aluminum.

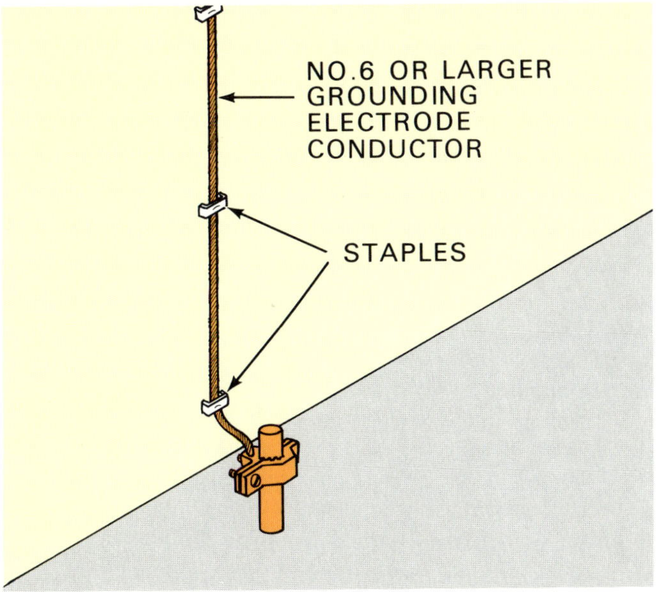

FIGURE 290. Number six or larger grounding electrodes may be stapled rigidly to the wall without being covered unless it is exposed to severe physical damage.

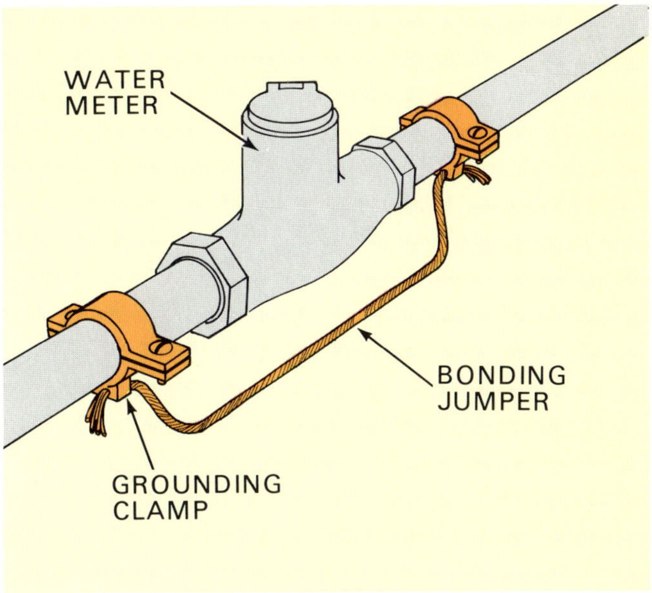

FIGURE 291. Run a bonding jumper around a water meter or insulated sections of pipe to assure grounding continuity.

♣**(2) TYPES OF GROUNDING CONDUCTORS.** Aluminum or copper-clad aluminum may not be used in contact with masonry or the earth or in corrosive conditions. If used outside, it may not be installed within 46 cm (18 inches) of the earth.

♣ **TABLE XII. SIZES OF GROUNDING ELECTRODE CONDUCTORS FOR AC SYSTEMS (From Code Table 250–94)**

Reprinted by permission from NFPA 70, National Electrical Code® Copyright 1984, National Fire Protection Association, Boston, MA

Size of Largest Service-Entrance Conductor		Size of Grounding Electrode Conductor	
Copper	Aluminum or Copper Clad Aluminum	Copper	Aluminum or Copper Clad Aluminum
2 or smaller	0 or smaller	8	6
1 or 0	2/0 or 3/0	6	4
2/0 or 3/0	4/0 or 250 MCM	4	2
Over 3/0 thru 350 MCM	Over 250 MCM thru 500 MCM	2	0

b. Grounding Electrodes

Bonding is important in installing grounding electrodes. It is especially important in meeting Code requirements.

Bonding means the permanent joining of metallic parts to form a conductive path for electrical current that will assure electrical continuity and capacity to conduct safely any current likely to be imposed on it. The bonding jumper around the water meter in Figure 291 is an example.

♣ Whether used as part of the grounding electrode system or not, the interior metal water piping system must always be bonded to either the service entrance cabinet, the grounding electrode conductor or the grounding electrode. Also, other metal piping which may become energized, such as gas pipes, must be bonded. The size of the bonding conductor must be the same as the grounding electrode conductor.

♣ Grounding electrodes are discussed under the following headings:

(1) Bonded Electrode Systems.
(2) Made and Other Electrodes.

(1) BONDED ELECTRODE SYSTEMS. Bonded electrode systems are discussed as follows:

(a) Four-Electrode Bonded Systems.
(b) Three-Electrode Bonded Systems.

(a) Four-electrode bonded systems are discussed as follows:

Under the Code, each of the following four items, if available, must be bonded together to form the grounding electrode system.

- —A **metal underground water pipe** in direct contact with the earth for at least 3.1 m (10 feet).
- —The **metal frame of the building** (not applicable to most homes).
- —An **electrode encased in at least 5.1 cm (2 inches) of concrete** at the bottom of the footing consisting of at least 6.1 m (20 feet) of one or more 1.3 cm (½-inch) steel bars or rods, or bare copper at least No. 4 AWG.
- —A **ground ring** of bare copper not smaller than No. 2 AWG encircling the building, buried at least 76 cm (2½ feet) deep.

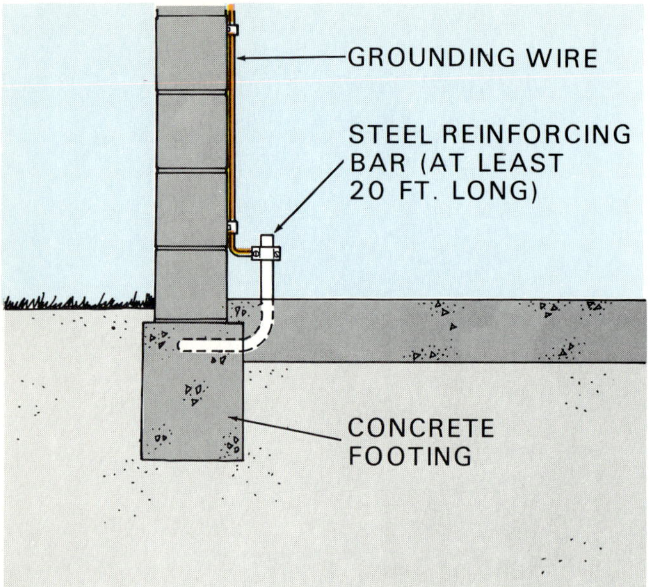

FIGURE 292. Half-inch (1.3 cm) steel reinforcing bars or No. 4 copper wire at least 6.1 m (20 feet) long may be used as supplemental grounding electrodes if encased by 5.1 cm (2 inches) of concrete in direct contact with the earth.

Note that the code does *not* state that all of the systems described **must be present,** only that they must be bonded together if available.

If you connect to the metal water pipe for the grounding electrode, the pipe must be in direct contact with the earth for at least 3.1 m (10 feet) (a metal well casing may be used). If there is a water meter or insulated joints or sections in the line that may be removed for repairs or replacement, you must run a jumper around it, using the same size conductor as the ground wire (Figure 291). Where used as a grounding electrode, a metal underground water pipe must be supplemented by an additional electrode as previously described.

(b) Three-electrode bonded systems are discussed as follows:

The metal underground water pipe must be bonded to both of the following electrodes **where available:**

—An **electrode** encased by at least 5.1 cm (2 inches) of concrete in direct contact with the earth and at least 6.1 m (20 feet) long (Figure 292). If steel reinforcing bars are used, they must be at least 1.3 cm (½ inch) in diameter. Bare solid or stranded copper wire No. 4 or larger may be used.

—A **ground ring** encircling the building in direct contact with the earth buried at least 76 cm (2½ feet) deep, made of bare copper wire No. 2 or larger, at least 6.1 m (20 ft.) long (Figure 293A).

Where none of the above electrodes are available, one or more of the following types may be used.

(2) MADE AND OTHER ELECTRODES. Made electrodes are those installed using metal rod, pipe or plate. Made electrodes must be free of paint or enamel and buried below permanent moisture level where practicable. They are discussed as follows:

—**Rod and pipe made electrodes** must not be less than 2.4 m (8 feet) in length and pipe shall be 1.9 cm (¾ inch) trade size or larger. The grounding

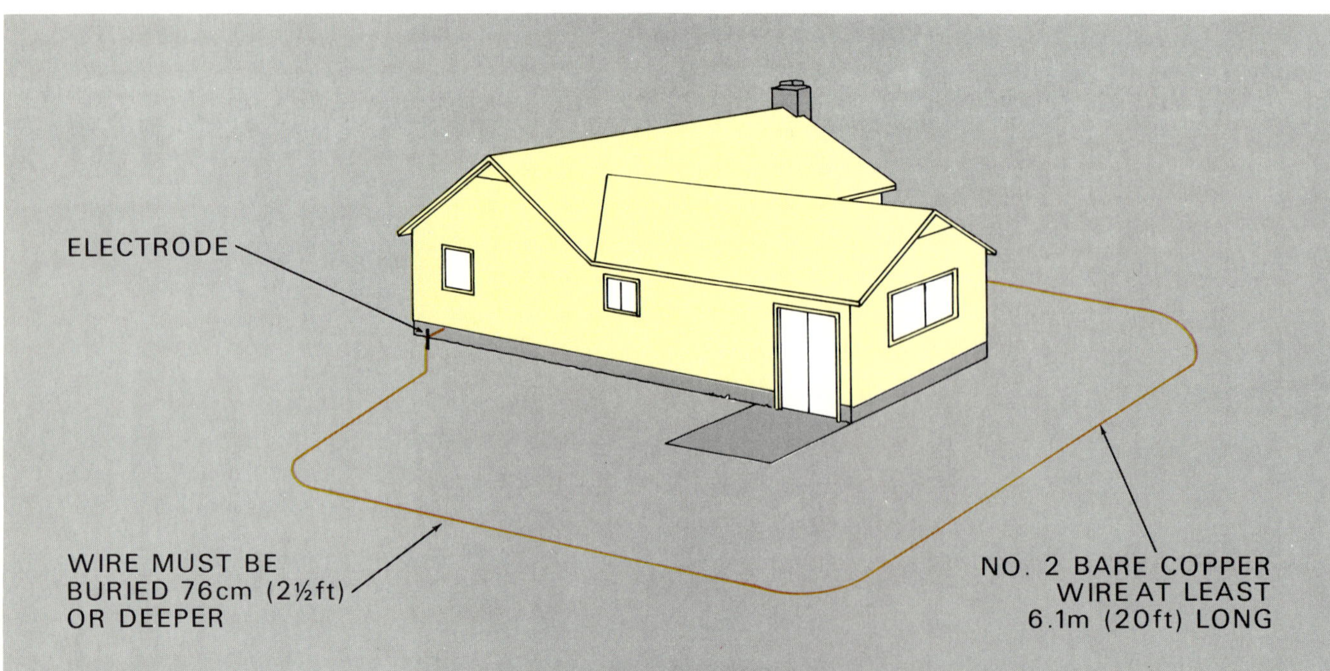

FIGURE 293A. Ground ring of bare copper wire, No. 2 or larger, at least 6.1 m (20 feet) long must be buried at least 76 cm (2½ feet) deep.

electrode conductor for made electrodes must be No. 6 copper or No. 4 aluminum. They must be galvanized or metal coated for corrosion protection (some local codes do not permit the use of galvanized pipe).

The electrode must be installed so that at least 2.44 m (8 feet) of length is in contact with the soil.

The electrode, rod or pipe, shall be driven to a depth of 2.4 m (8 feet) unless rock bottom is encountered. If you hit rock bottom at less than 1.2 m (4 feet), you must bury the electrode in a trench 762 mm (2 ½ feet) deep, or drive it at an oblique angle not to exceed 45 degrees from vertical. The top end of the electrode must be flush with, or below ground level unless it and the grounding electrode conductor are both protected against physical damage. Where multiple rod, pipe or plate electrodes are installed, they must be at least 1.8 m (6 feet) apart (Figure 293B).

—**Steel or iron rods** must be at least 1.6 cm (⅝ inch) diameter. Non-ferrous rods shall be listed and not less than 1.3 cm (½ inch) diameter.

—**Plate electrodes** of iron or steel must expose at least 0.19 square meters (2 square feet) to the soil and must be at least 6 mm (¼-inch) thick. Non-ferrous plates shall be at least 1.5 mm (0.06-inch) thick.

—**Underground metal gas piping systems** may be used for grounding electrodes if electrically continuous and without interruptions. Its use must be approved by the gas supplier and local codes.

—Other **metal piping systems** and **underground tanks** may be approved by some local codes.

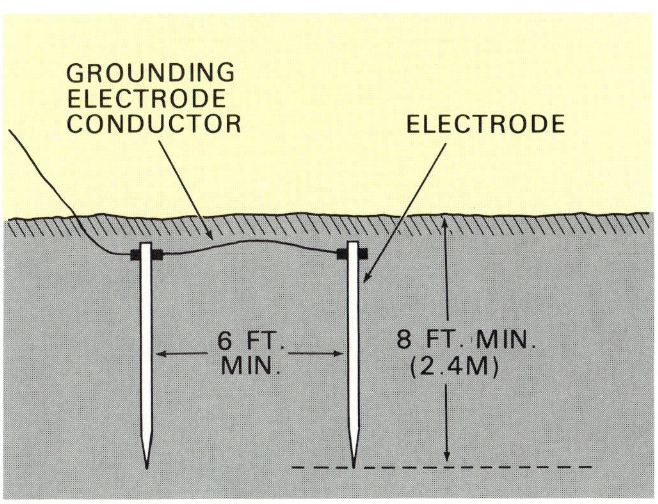

FIGURE 293B. Top end of electrode must be flush with ground level or below, if unprotected. Two or more rods must be at least six feet apart.

c. Connecting Grounding Conductors and Grounding Electrodes for System Ground

The connection of the grounding conductor to the grounding electrode must be accessible and made in a manner that will provide a permanent and effective ground.

Proceed as follows to connect the system ground:

1. *Connect the grounding conductor at the neutral bar.*

   Insert the grounding conductor under the terminal provided on the neutral bar and tighten securely (Figure 294).

2. *Connect the grounding conductor at the water pipe or other grounding electrode.*

   The ground wire must be clamped tightly to the water pipe or other grounding electrode. Use bronze, brass, plain or malleable iron clamps (Figure 295).

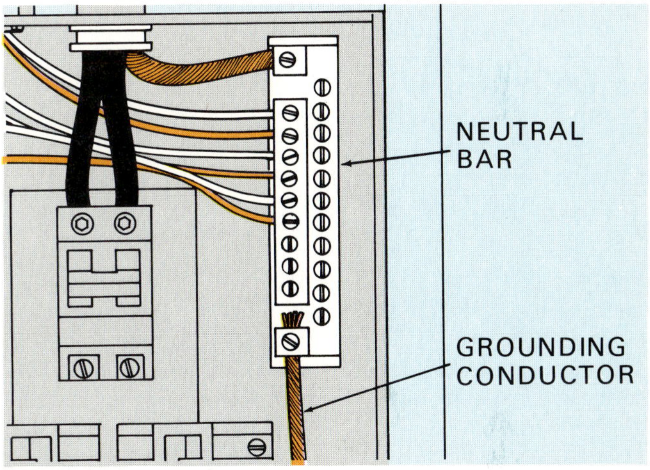

FIGURE 294. Grounding conductor connected to neutral bar.

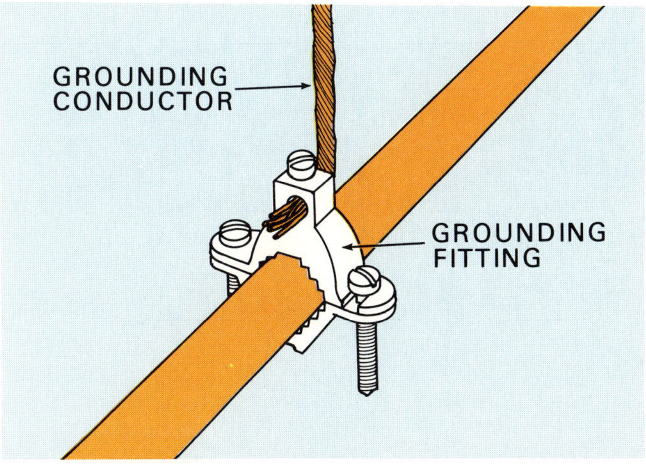

FIGURE 295. Grounding fitting clamped tightly to water pipe.

Two steps are involved to make the connection. **First**, connect the clamp to the pipe using the two bolts provided. Tighten the bolts until the ridges in the clamp bite into the water pipe. **Second**, insert the ground wire in the slot on top of the clamp and draw the screw(s) or bolt(s) tight against the wire. The metal water piping bonding jumper must be accessible.

3. *Bond the first grounding electrode to at least one other grounding electrode.*

   Follow procedures as in step two.

   NOTE: Some power suppliers and public service authorities require the ground wire to be installed differently. It does not go through the SEP. Instead, it connects to the neutral overhead wire at the weatherhead or at the neutral bar in the meter base. It comes down the side of the house and fastens directly to the water pipe and/or other type of ground (Figure 296).

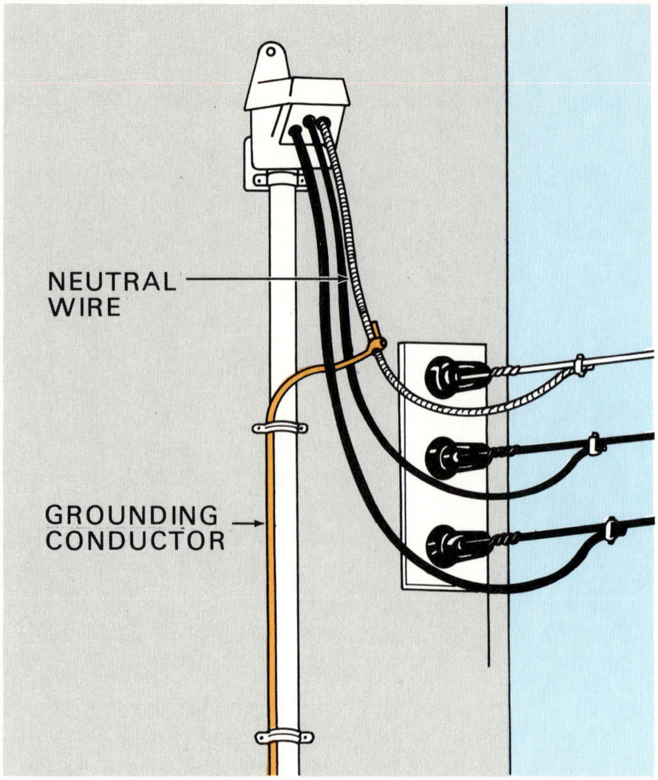

FIGURE 296. Grounding conductor connected to the overhead neutral wire and directly to a grounding electrode is required by some Codes.

F. Installing Ground Fault Circuit Interrupters (GFCI)

GFCI's are installed to protect people from shock hazards. As discussed in the previous sections, system and equipment grounding provides protection for the electrical system and equipment and considerable protection against shock in most situations. However, there is a need for a more sensitive device to protect people from shock in the more hazardous locations in and around the home. The GFCI's are designed for this purpose.

Ground faults **can** damage equipment. This costs money. But should a person be a part of this unintended path, the result could be injury or death.

Only 50 milliamperes, or 0.0033 of the current required to open a 15-ampere circuit breaker or fuse, can be fatal.

A GFCI can help prevent harm to people in situations like these.

All 125-volt single phase, 15- and 20-ampere construction site receptacles which are not part of the permanent structure and which ARE IN USE BY EMPLOYEES must have GFCI protection. This is interpreted to mean that employees may use a portable GFCI that can be moved from one receptacle to another on the job (Figure 297).

The Code requires GFCI protection for all 120-volt, **15- and 20-ampere receptacles** installed in **bathrooms, garages** with some exceptions, and all **outdoor receptacles** of homes. Also, any **receptacle within 4.6 m (15 feet) of a swimming pool** must be controlled by a GFCI. A garage receptacle in space provided for a plug-in appliance such as a freezer,

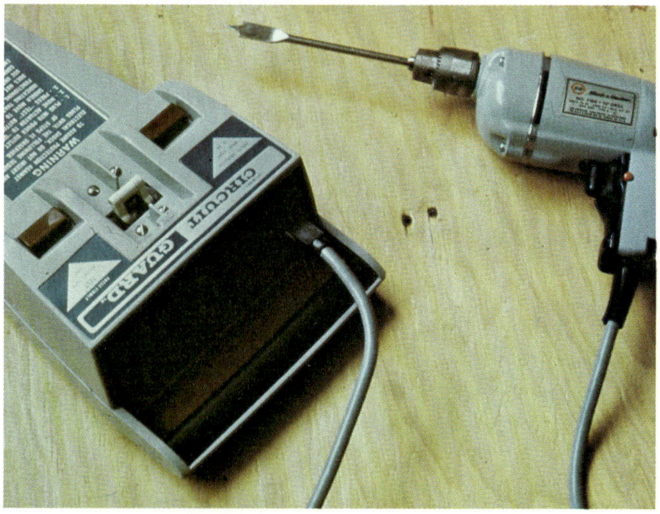

FIGURE 297. Portable GFCI's can be located where needed on construction sites.

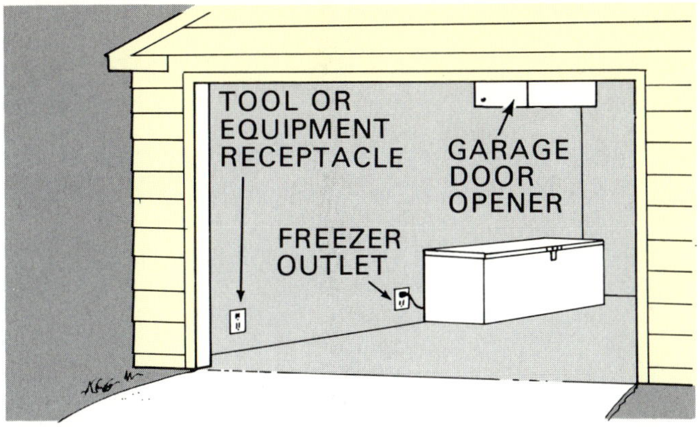

FIGURE 298A: A GFCI is not required for a garage receptacle in a space provided for a plug-in appliance or a garage door opener.

does not require a GFCI (Figure 298A). Also, a plug-in receptacle for a garage door opener is excepted (Figure 298A).

A ground fault exists when an unintended current path is established between an ungrounded conductor and ground. This situation can occur not only in defective electrical equipment but also from accidental misuse of equipment in good working order.

GFCI's protect people from shock by monitoring the current level in a circuit at all times. If an accidental ground fault occurs in an electric drill, for example, the GFCI interrupts (turns off) the circuit before any harm can be done (Figure 298B). If as little as five thousandths of an ampere flows accidentally to ground, the unit interrupts the circuit in less than a second. A higher current flow, which might endanger a life; will be turned off in twenty-five thousandths of a second.

Installing GFCI's is discussed as follows:

1. Types and Sizes of GFCI's.
2. Connecting GFCI's.
3. Testing GFCI's.

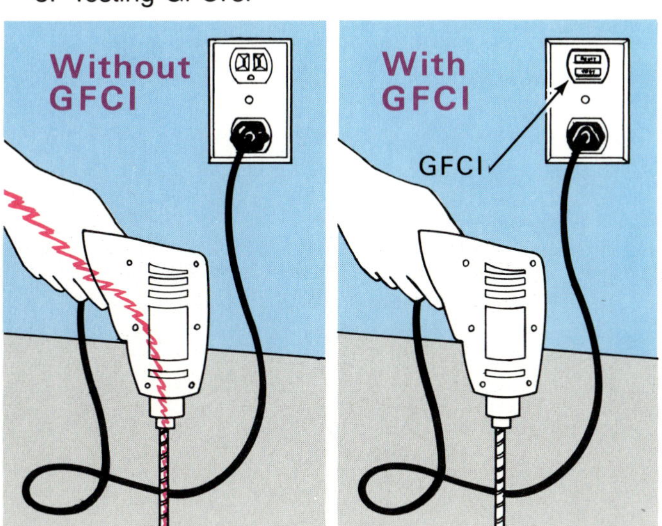

FIGURE 298B: GFCI's protect people from fault currents as small as five thousandths of an ampere.

1. TYPES AND SIZES OF GFCI's

Three types of GFCI's are available. A portable type plugs directly into an existing receptacle (Figure 299). A second type can be permanently installed to replace a standard receptacle (Figure 300). For outdoor use, weatherproof types are available (Figure 301). The third type is a combination circuit breaker-GFCI. It is installed instead of a standard circuit breaker in the SEP (Figure 302). This type protects all outlets on that circuit.

The GFCI rating in volts and amperes must be the same as the circuit it protects.

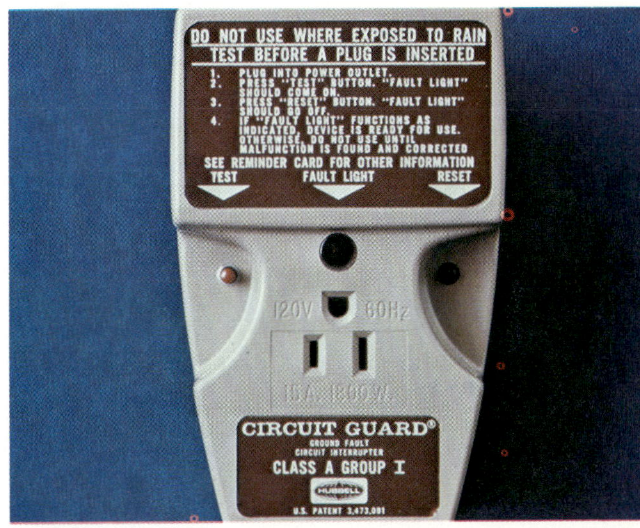

FIGURE 299. The portable GFCI plugs directly into an existing receptacle.

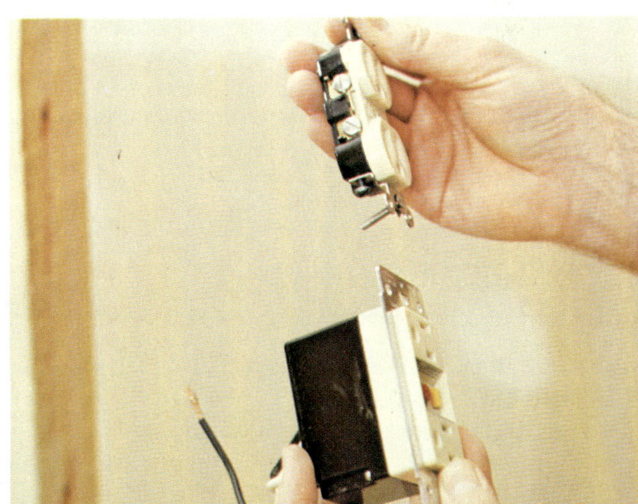

FIGURE 300. A GFCI (bottom) can be permanently installed in place of a receptacle.

FIGURE 301. Weatherproof GFCI can be installed outdoors.

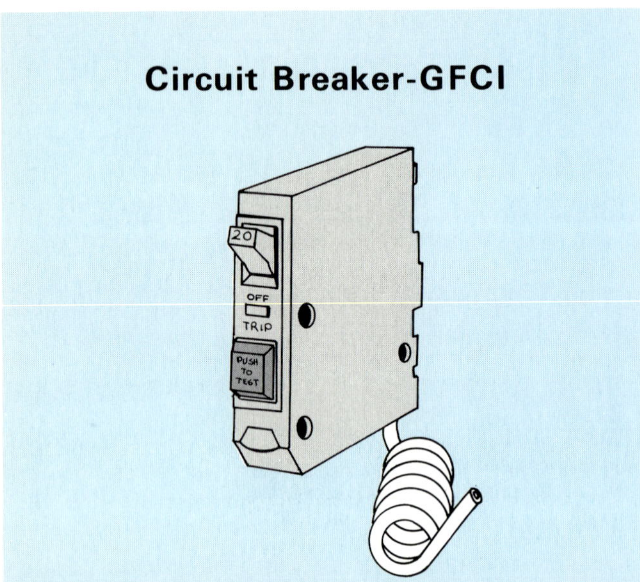

FIGURE 302. The combination circuit-breaker GFCI is installed in the service entrance panel. It protects all outlets on that circuit.

2. CONNECTING GFCI'S

Connecting GFCI's is discussed under the following headings:

a. Connecting the Plug-In Type GFCI.
b. Connecting Permanently Installed Receptacle Type GFCI.
c. Connecting the Combination Circuit Breaker Type GFCI.

a. Connecting the Plug-In Type GFCI

1. *To connect the plug-in type, plug the device directly into the receptacle (Figure 303A).*

FIGURE 303A. The plug-in GFCI fits a standard receptacle outlet.

b. Connecting the Permanently Installed Receptacle Type GFCI

To connect the permanently installed receptacle-type GFCI in a circuit, follow manufacturer's directions. Proceed as follows (Figure 300):

1. *Install the GFCI in a standard switch (receptacle) box.*
2. *Connect wiring as for standard receptacle, white to white terminal, black to brass terminal and ground to ground (some types are self-grounding).*
3. *If desired, make feed-through connections to other receptacles (Figure 303B).*

   Most types have terminals for continuing to other outlets, which then are also protected. They provide GFCI protection at the point of installation and at all other receptacles connected beyond the unit on the same circuit.
4. *Mount the outdoor type as directed by manufacturer's instructions.*

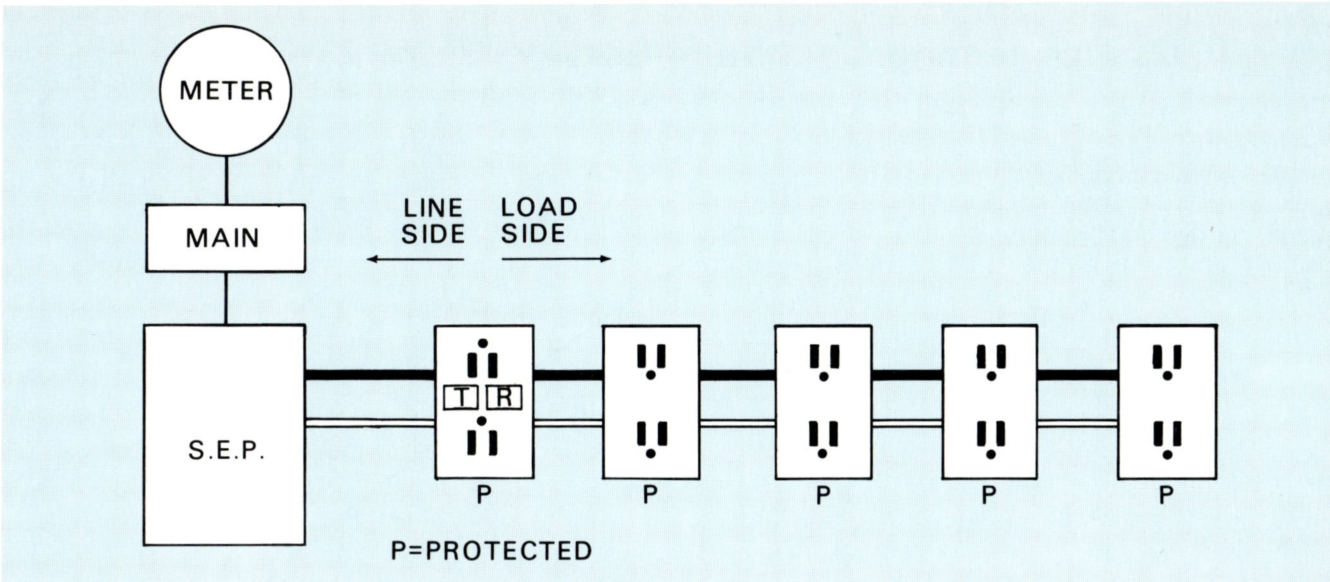

FIGURE 303B. A GFCI installed as the first receptacle in a circuit provides protection to all other receptacles beyond it on the circuit.

c. Connecting the Combination Circuit Breaker Type GFCI

To connect circuit breaker type GFCI in service entrance panel, follow manufacturer's instructions. Most will be connected as illustrated.

Proceed as follows to connect the combination circuit breaker and GFCI (Figure 304):

1. *Connect the black wire to terminal marked load power.*

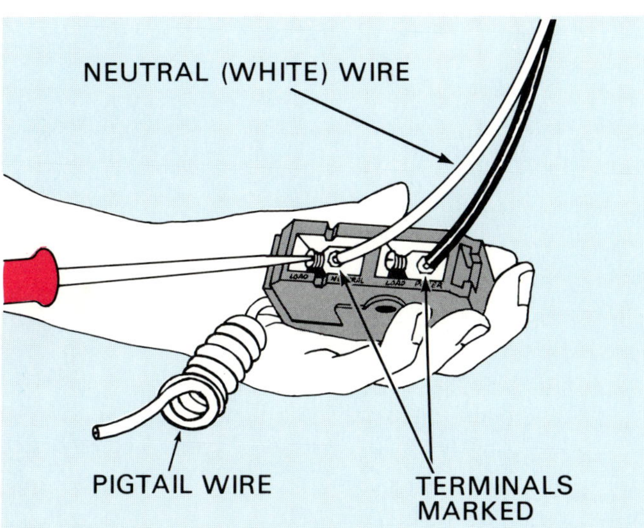

FIGURE 304. Connect white circuit wire to the neutral terminal and black circuit wire to other terminal on circuit breaker GFCI.

2. *Connect white neutral wire from circuit to screw terminal on GFCI marked load neutral.*
3. *Plug the device into the SEP clips.*
4. *Connect the white pigtail wire to the neutral bar (Figure 305).*

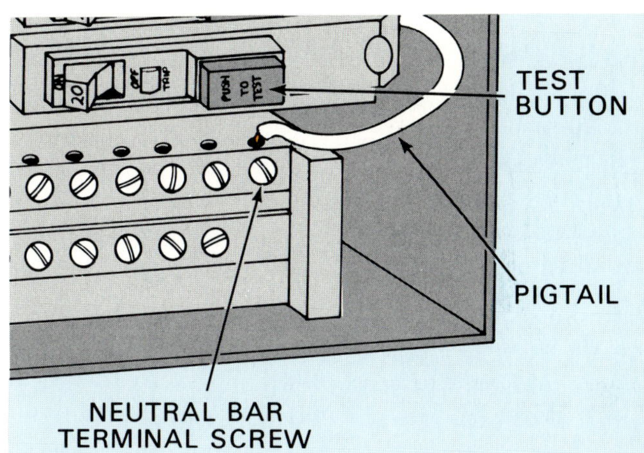

FIGURE 305. Test the GFCI monthly according to directions on the units.

3. TESTING GFCI'S

Each GFCI has a test button that you can use to determine if the device is working (Figure 305). Owners should test each unit monthly by pushing the test button. If the unit is working, it will trip the circuit breaker on the circuit. Reset the circuit breaker after the test.

G. Connecting Circuits in Service Entrance Panel

As illustrated in Figure 287, an orderly wiring job in the SEP allows better air circulation, reducing heat buildup. Another advantage is that it is much easier to identify circuit wiring and breakers when trouble-shooting and for adding circuits at a later date if desired.

You will install a single-pole circuit breaker for each 120-volt circuit and a double-pole one for the 240- or 120/240-volt circuits (Figure 306). The capacity of the circuit breaker is rated in amperes and is marked on the lever. You must match the rating of the circuit breaker to the amperage of the circuit it protects.

The ends of each circuit cable should have already been pulled into the service entrance cabinet and identified. Check each cable before you make the connection. Mark the identification of the circuit on the chart inside the cabinet door.

For each cable connection, strip the sheath from the cable. Extend the wires to the terminals to which they connect and cut off any excess length. Then remove about 1.3 cm (½-inch) of insulation from each individual wire.

Connecting circuits in the SEP is discussed as follows:

1. Connecting 120-Volt Circuits with Circuit Breakers.
2. Connecting 240-Volt Circuits with Circuit Breakers.
3. Connecting 120/240-Volt Circuits with Circuit Breakers.
4. Connecting 120-Volt Circuits with Fuses.
5. Connecting 240-Volt Circuits with Fuses.
6. Connecting 120/240-Volt Circuits with Fuses.

1. CONNECTING 120-VOLT CIRCUITS WITH CIRCUIT BREAKERS

Proceed as follows:

1. *Extend the white wire and the bare grounding wire around the edge of the cabinet to the neutral bar (Figure 307).*

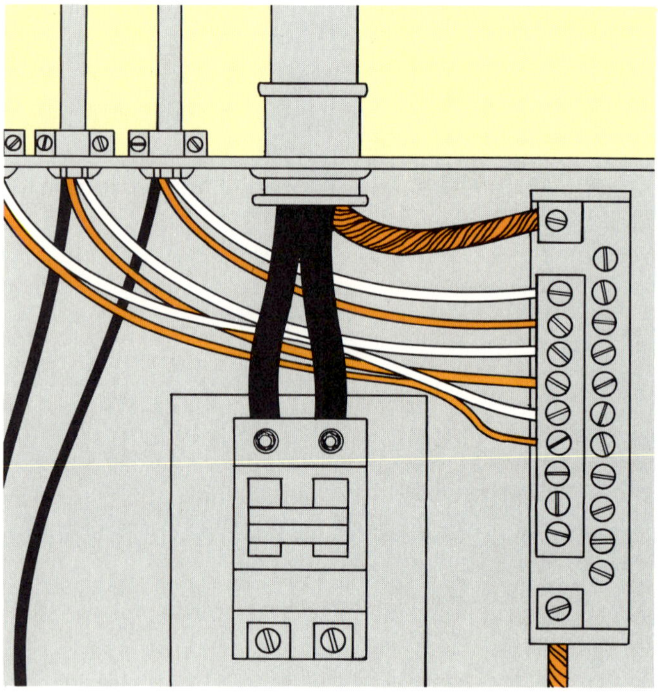

FIGURE 307. White neutral wire and grounding wire attached to terminals on neutral bar.

FIGURE 306. Single-pole and double-pole circuit breakers.

2. *Insert the white wire under a neutral bar terminal and tighten the screw firmly.*
3. *Insert the bare grounding wire under a neutral bar terminal and tighten the screw.*
4. *Insert the black wire from the cable under the terminal screw on the circuit breaker and tighten the screw (Figure 308).*
5. *Insert the circuit breaker into the slot at one end and push the blade into the clip terminal until it stops (Figure 309).*

   Procedures for some makes may vary slightly, but most will plug into the terminal.

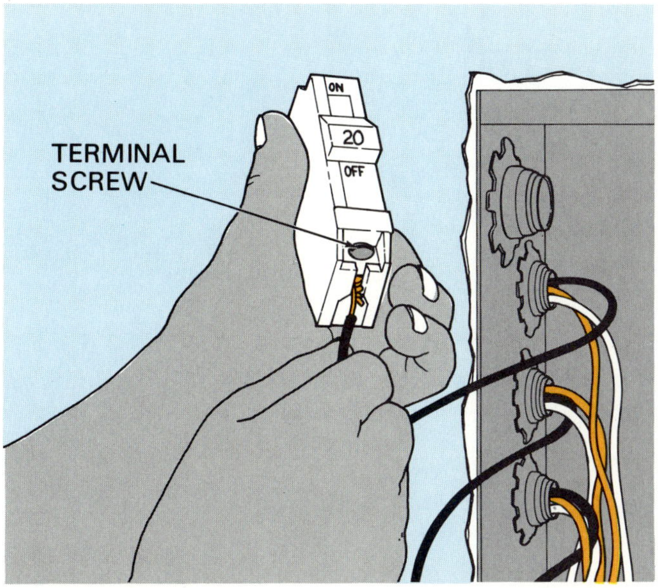

FIGURE 308. Black ("hot") wire inserted into circuit breaker terminal.

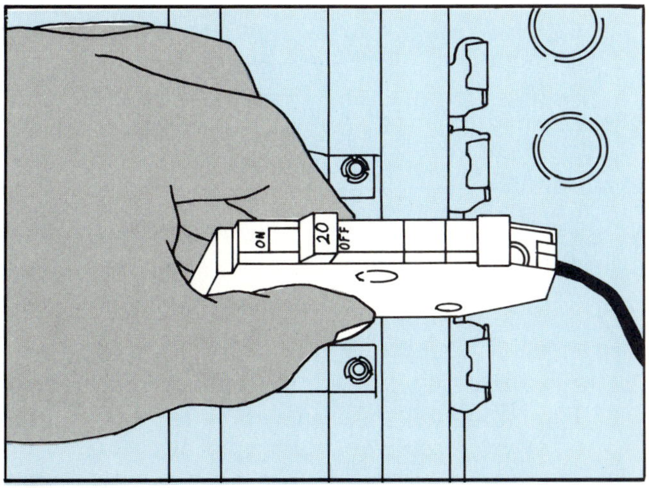

FIGURE 309. Circuit-breaker snaps into position on the panel.

To remove a circuit breaker after installation, insert a screwdriver blade under the end opposite the wire connection and pry up until the clip releases.

CAUTION: When removing a circuit breaker from a SEP where power is turned on, be sure to turn off power at main disconnect before removing breaker.

2. CONNECTING 240-VOLT CIRCUITS WITH CIRCUIT BREAKERS

Certain equipment such as water heaters, 240-volt air conditioners and 240-volt heaters are installed on straight 240-volt circuits, not 120/240-volt as with a range or dryer. These 240-volt circuits are wired with 2-wire w/g cable as follows (Figure 310):

1. *Connect black wire to a terminal on the 240-volt circuit breaker.*
2. *Connect white wire to the other breaker terminal (paint the end of white wire black, or wrap with tape to identify as hot wire).*
3. *Connect the grounding conductor to the neutral bar.*

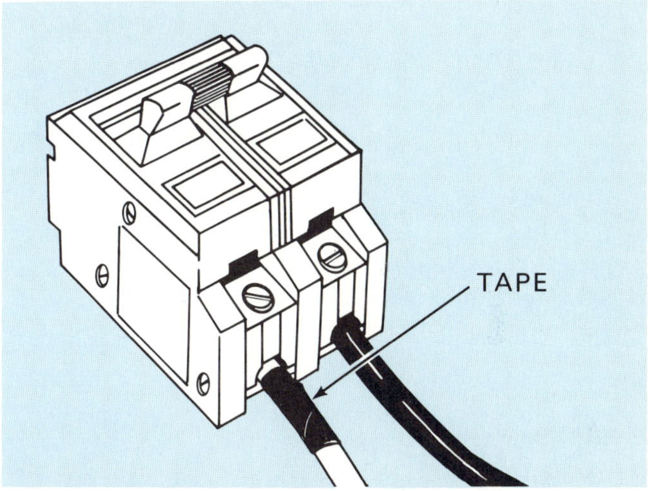

FIGURE 310. Double-pole circuit breaker with 240 circuit wires attached.

3. CONNECTING 120/240-VOLT CIRCUITS WITH CIRCUIT BREAKERS

1. *Extend the white neutral wire from the cable around the edge of the cabinet to the neutral bar.*
2. *Insert the white neutral wire under a neutral bar terminal and tighten the screw.*
3. *Insert one hot wire (black) under a terminal on a double-pole circuit breaker and tighten (Figure 311).*
4. *Insert the other hot wire (red) under the other terminal on the double-pole circuit breaker and tighten.*
5. *Insert grounding wire under a neutral bar terminal and tighten screw.*

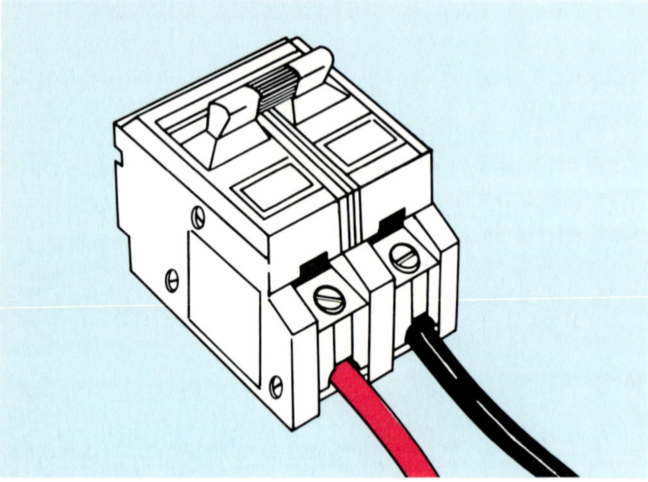

FIGURE 311. Double-pole circuit breaker with 120/240 circuit wires attached.

4. CONNECTING 120-VOLT CIRCUITS WITH FUSES

Procedures are similar to those for circuit breaker wiring. Proceed as follows (Figure 312):

1. *Extend the white neutral wire and the bare grounding wire to the neutral bar.*
2. *Insert the white neutral wire under a neutral bar terminal and tighten.*
3. *Repeat for the bare grounding wire.*
4. *Extend the black wire to a fuse terminal.*
5. *Insert the black wire under the terminal screw and tighten.*

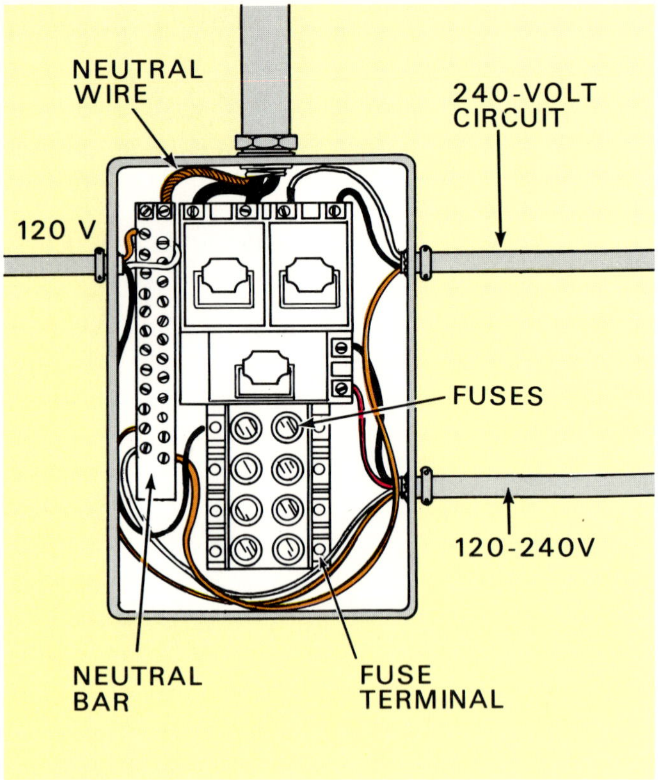

FIGURE 312. Fuse panel with 240-volt circuit, 120/240 volt circuit and 120-volt circuit installed.

5. CONNECTING 240-VOLT CIRCUITS WITH FUSES

Proceed as follows (Figure 312):

1. *Connect the black wire to a 240-volt fuse terminal.*
2. *Connect the white wire to the other 240-volt terminal and identify as a hot wire with tape or paint.*
3. *Connect grounding wire to neutral bar.*

6. CONNECTING 120/240-VOLT CIRCUITS WITH FUSES

Proceed as follows (Figure 312):

1. *Extend the white neutral wire to the neutral bar.*
2. *Insert the white neutral wire under a neutral bar terminal screw and tighten.*
3. *Extend one hot wire (black) to a fuse connection (Figure 312).*
4. *Insert the hot wire under the fuse-terminal screw and tighten.*
5. *Repeat steps 3 and 4 for the second hot wire (red).*
6. *Insert grounding wire under a neutral bar terminal and tighten screw.*

V. Installing Metallic Conduit

Three types of metallic conduit were described in Section IV, "Installing Service Entrance Equipment." The three types are **rigid conduit, intermediate metallic conduit,** and **EMT (thinwall) conduit** (Figure 313). All three come in 3.1 m (10 feet) lengths and must be UL approved. Other types are **flexible metallic conduit,** often called **Greenfield liquidtight** and **flexible metal** conduit (Figure 313). The use of flexible metallic conduit is limited to dry locations and it must be protected from physical damage if not covered. In certain areas it is used extensively, but in the United States as a whole, it is not widely used in house wiring. It is not discussed in this section.

Rigid conduit and IMC are seldom used in house wiring, except as described for service installations, and are not discussed in this section.

EMT is lighter and easier to handle and is approved for most requirements in house wiring installations. Many local codes require the use of EMT for all residential wiring. EMT is available in sizes from 1.3 to 10.2 cm (½-inch to 4 inches).

Rigid nonmetallic conduit (PVC) is economical and may be used in walls, floors and ceilings (Figure 313). Liquidtight flexible metal conduit is similar to flexible metallic conduit but has a jacket that is liquidtight, nonmetallic and sunlight resistant (Figure 313).

Upon completion of this section, **you will be able to install metallic conduit and the necessary wiring.**

Installing metallic conduit and wiring is discussed as follows:

A. Installing Electrical Metallic Tubing (EMT).
B. Installing Wiring in Metallic Conduit.

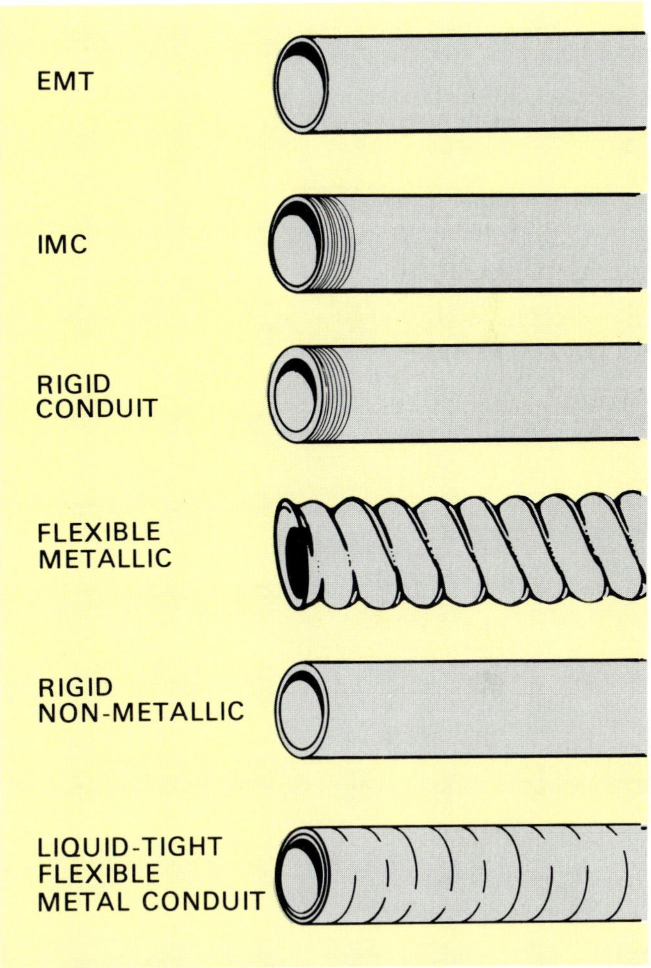

FIGURE 313. Types of conduit installed for house wiring.

A. Installing Electrical Metallic Tubing (EMT)

Installing EMT is discussed under the following headings:

1. Cutting Electrical Metallic Tubing.
2. Bending Electrical Metallic Tubing.
3. Installing Electrical Metallic Tubing.

1. CUTTING ELECTRICAL METALLIC TUBING

Cut the measured length of EMT with either a fine blade hacksaw (13 teeth to the centimeter or 32 teeth to the inch) (Figure 314) or with a tubing cutter. Use a reamer to remove sharp edges after cutting.

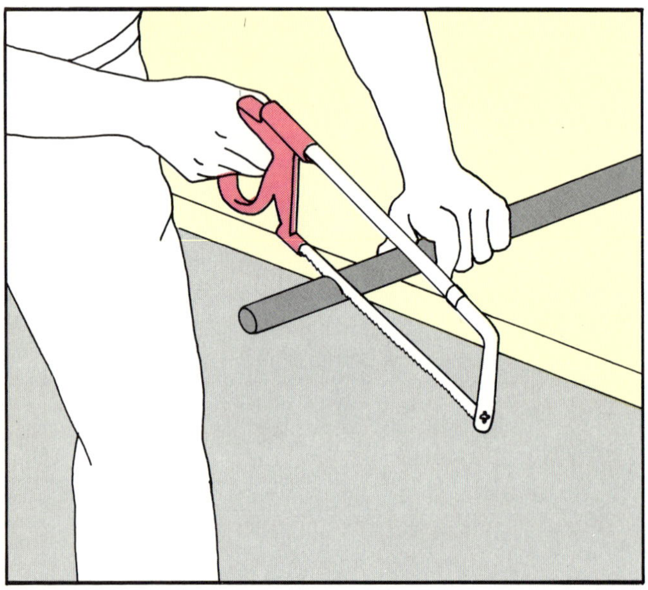

FIGURE 314. Cutting EMT with a fine-tooth hacksaw.

2. BENDING ELECTRICAL METALLIC TUBING

You must be very careful when bending EMT to avoid kinking or crushing the pipe, which would reduce the usable inside area. A run of EMT between outlets, or between fittings or between outlet and fitting must not contain more than the equivalent of four quarter bends total (360 degrees). Bend the conduit first, then cut the length required.

Thinwall conduit benders are available which are marked to help you make the type and length of bend required (Figure 315). The foot step helps to apply steady pressure during bending (Figure 315). Always work on a hard surface when bending EMT. Follow manufacturer's directions.

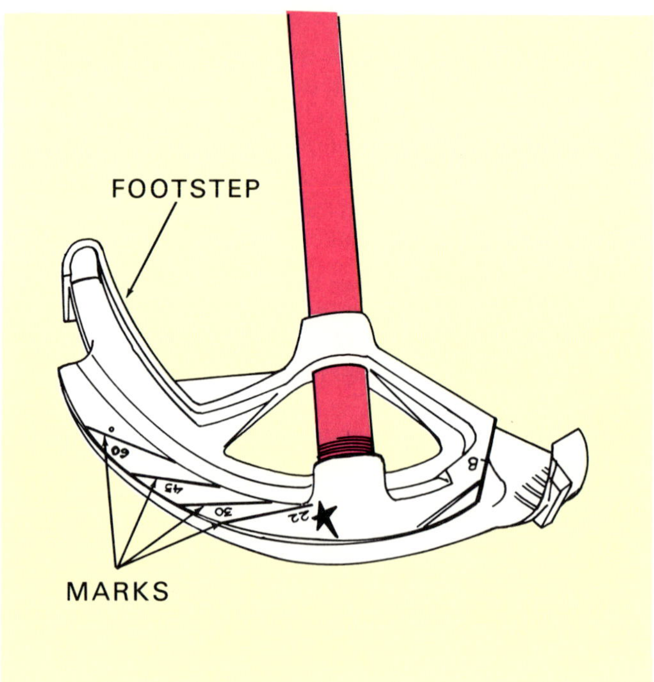

FIGURE 315. Marks on bender serve as a guide.

Assume you want a 90° bend, 28 cm (11 inches) high. Proceed as follows to bend 1.3 cm (½-inch) EMT using one type of bender:

1. *Place the bender at the correct distance from the end of the conduit.*

   The correct distance is **equal to the desired height of the bend, minus an amount usually called the "take-up" height.** The take-up height to be applied for each bend is 12.7 cm for 1.3 cm (5 inches for ½-inch) conduit, 15.2 cm for 1.9 cm

(6 inches for ¾-inch) and 20.3 cm for 2.5 cm (8 inches for 1-inch) conduit. For the 27.9 cm (11-inch) height, subtract the take-up height of 12.7 cm (5 inches) for a remainder of 15.2 cm (6 inches).

2. *Set the mark "B" of the bender 15.2 cm (6 inches) from the end (Figure 316).*
3. *Pull on the handle with a steady pressure and press the foot step at the same time.*

   Keep the conduit tight against the floor throughout the bend.

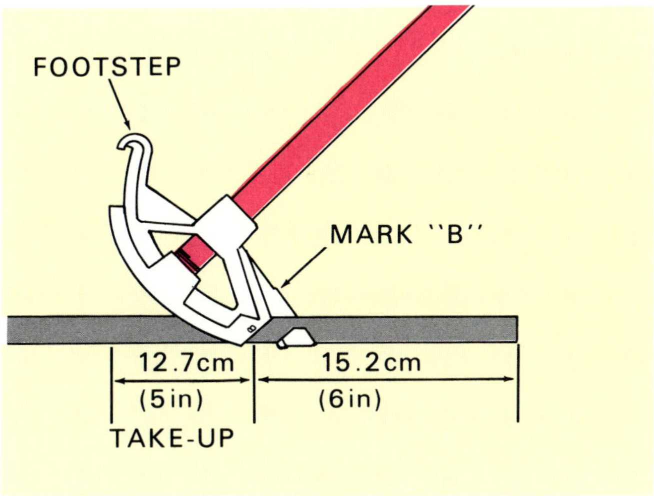

FIGURE 316. Set mark on bender as indicated.

Continue heavy foot pressure until the stub end of the pipe is at right angles to the floor (Figure 317).

NOTE: When handle of the bender is pulled to the vertical position, you have a 45° bend. For a 90° bend, continue the pressure until the stub end of the pipe is vertical and the handle is at 45° with the floor (Figure 318).

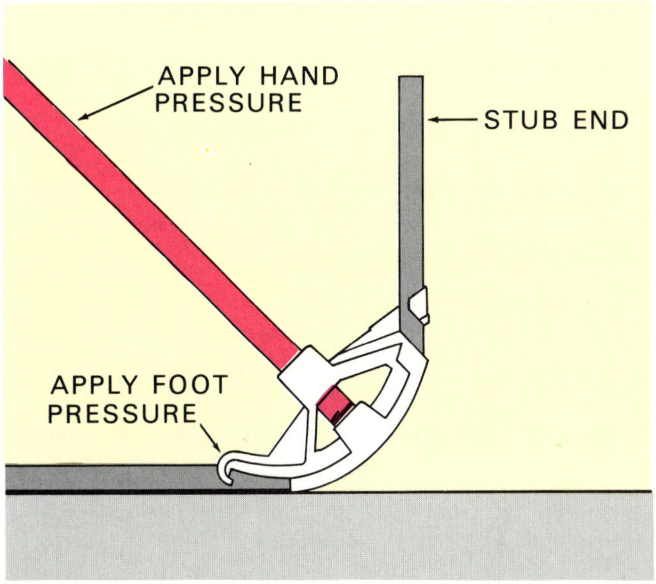

FIGURE 317. Apply hand and foot pressure until desired position is reached.

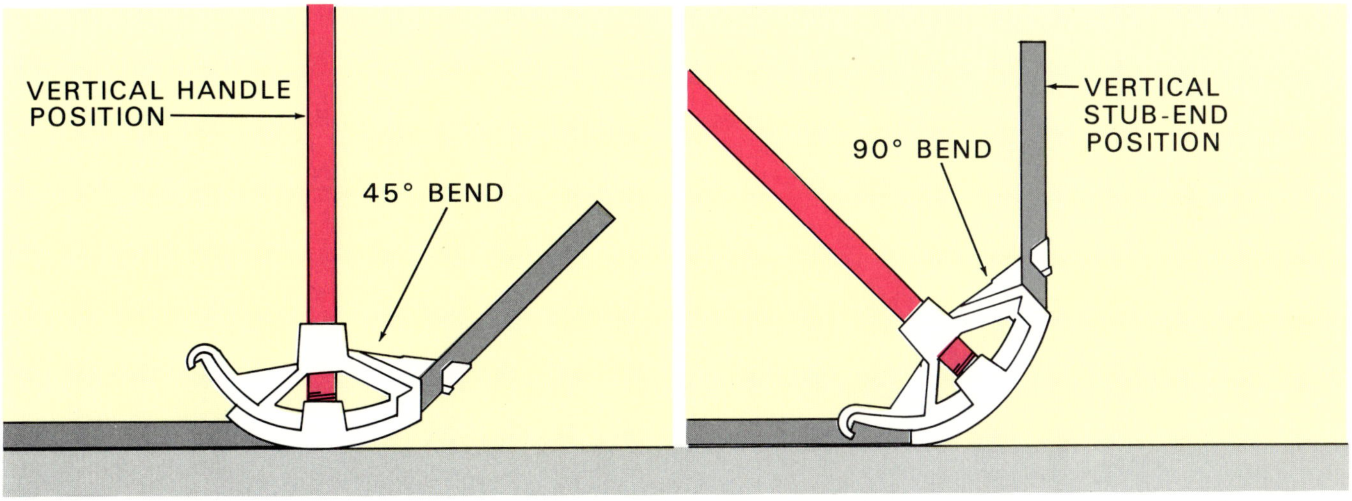

FIGURE 318. Vertical handle position indicates 45 degree bend. Vertical stub-end position indicates a 90 degree bend.

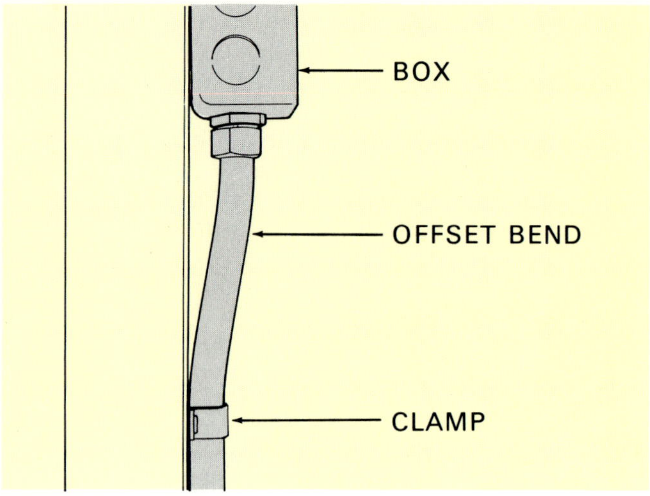

FIGURE 319. Offset bend to make straight-in connection with knockout in box.

4. *An offset bend is often required to make the conduit go straight into a box (Figure 319). Proceed as follows to make an offset bend (Figure 320):*

   Draw a chalk line on the conduit when making multiple bends to guide tool placements. (Several manufacturers now provide a line on the conduit for this purpose).

   Place the conduit on the floor and hook the bender close to one end. Pull on the bender until the conduit is at a **45° angle** to make the first bend. Then turn the conduit over and hook the bender under the center of the first bend. Bend again until the short end is parallel to the main piece of conduit.

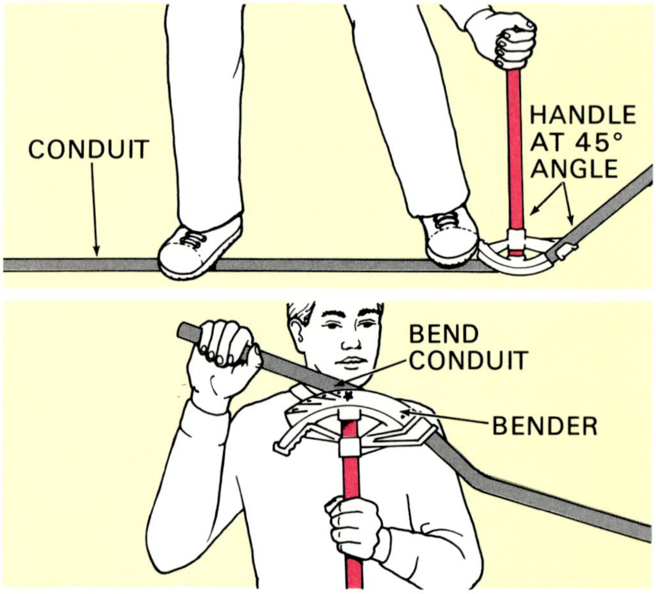

FIGURE 320. Two 45 degree bends on opposite sides produce an offset bend.

3. INSTALLING ELECTRICAL METALLIC TUBING

The conduit must be installed first and individual wires are pulled into it after it is installed. EMT must be supported within 0.92 m (3 feet) of every outlet or box and at least every 3.1 m (10 feet) between outlets or boxes.

When EMT is installed correctly, **the conduit serves as the equipment ground.** The conduit takes the place of a bare grounding wire found in NM cable, so only two wires are required inside the conduit for 120-volt or for 240-volt circuits. Circuits supplying both 120 volts and 240 volts, such as the range and dryer require 3 wires. All connections must be made very tight so they are electrically and mechanically connected at every point in the system. Only metal boxes may be used. If there is a loose connection anywhere, the conduit, boxes and equipment frames will not be correctly grounded.

EMT cannot be threaded. However, the code states that where integral couplings are used, such couplings may be factory threaded. Where coupled together by threads, the connector must be of a type to prevent bending the tubing at any part of the threads.

Two methods are commonly used to install EMT for residential wiring as follows:

1. *Drill holes through center of framing and couple short lengths of conduit for each run (Figure 321).*

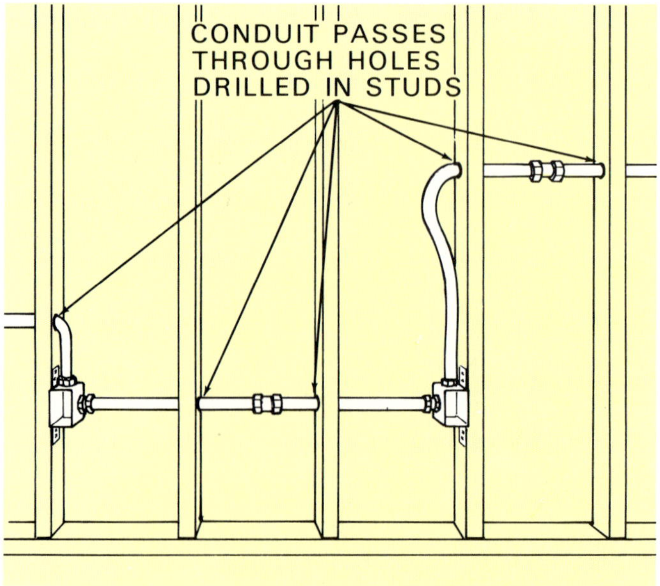

FIGURE 321. Couple short lengths of EMT through holes drilled in center of framing.

2. *Notch framing and cover notches with steel plates.*

   Use this method only where necessary. Notches large enough for EMT weaken the framing to some extent (Figure 322). Use compression connectors and couplings to make box connections and to join sections of EMT during assembly.

3. *Mount the boxes loosely on the framing until the conduit connections are complete.*

   If your conduit measurements are off by a fraction of an inch, you can adjust the box slightly as required. Make every connection mechanically and electrically solid to insure a good equipment ground with the conduit.

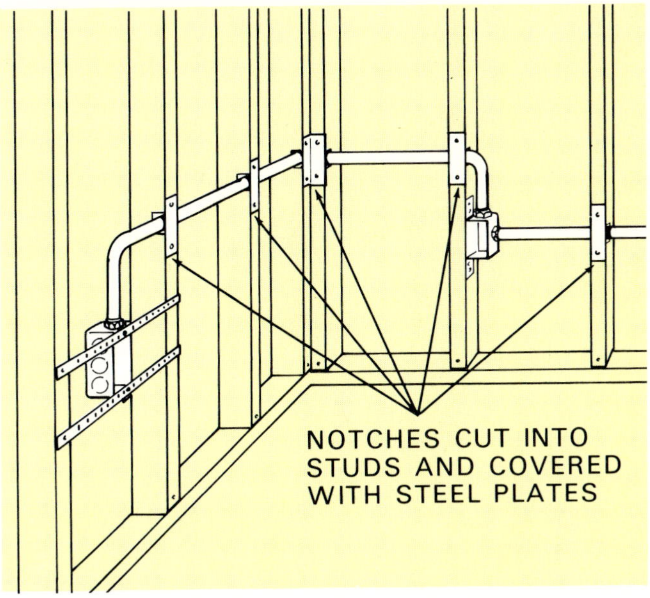

FIGURE 322. Notches may be used for some runs.

B. Installing Wiring in Metallic Conduit

♣ After all conduit is installed, you are ready to pull the wiring into place. Remember that only two wires per 120-volt or 240-volt circuit are required for EMT, IMC or rigid conduit. No splices are permitted inside the conduit and no substitutions for color coding can be made. Wires must be continuous from box to box.

For short runs, you can probably push the wires from one box to another. On longer runs, or those with several bends, you need a tool called a fish-tape. The fish-tape is a flexible steel tape, usually 3.2 mm (⅛-inch) wide and 15–30 m (50- or 100-feet) long with a loop on the end (Figure 323). When pushed into the conduit, it moves more easily than insulated wires. Wires are fastened to the loop (called a hook) and pulled through the conduit. Leave 15 cm (6 inches) of wire extended for connections at each outlet.

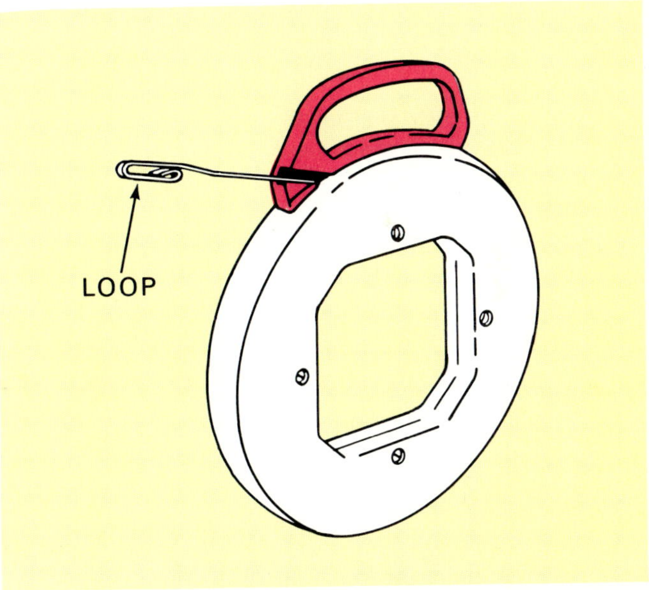

FIGURE 323. Fish-tape helps pull wires through conduit more easily.

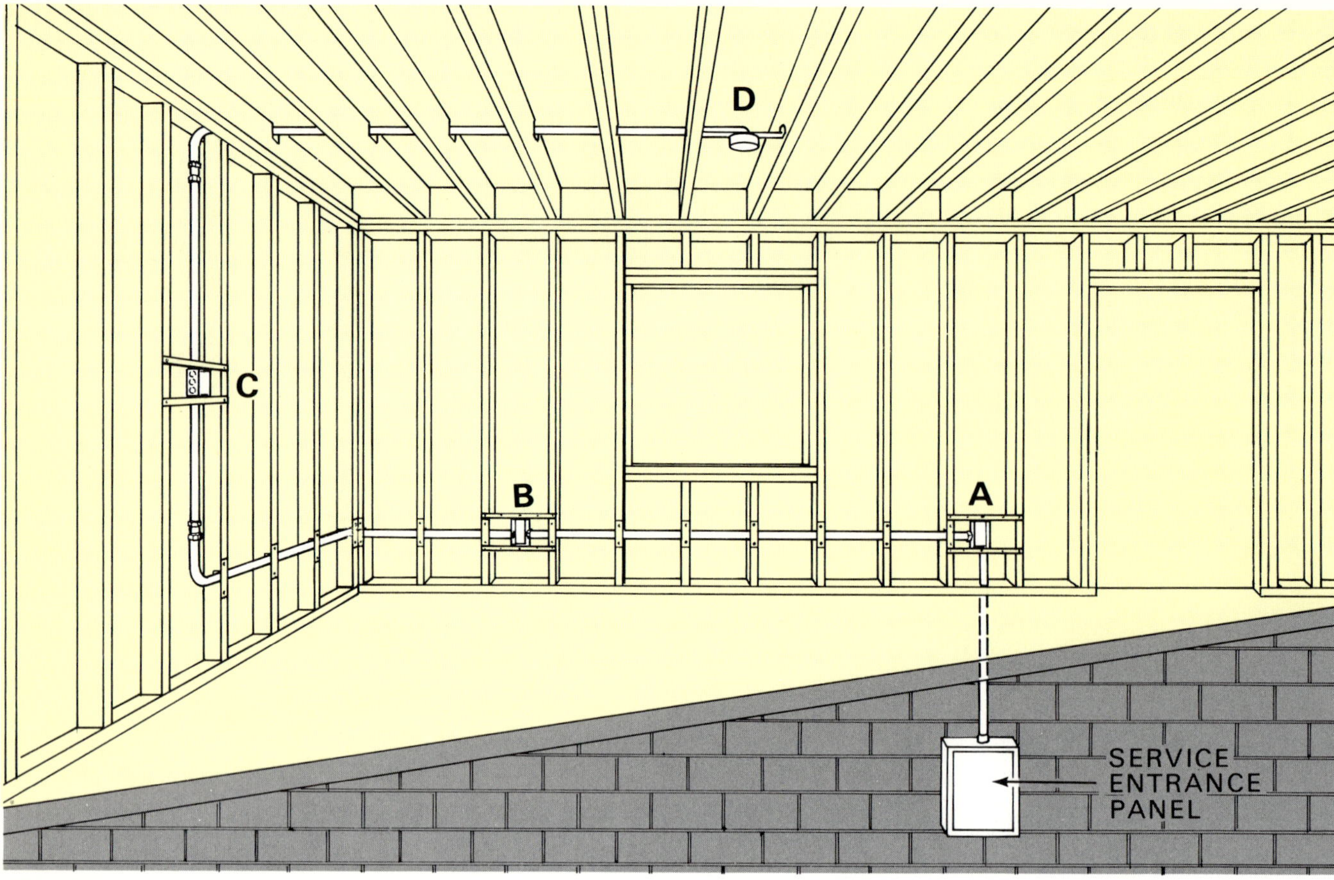

FIGURE 324. Insert wires into conduit and push or pull from point "A" to service entrance panel, from "A" to "B," from "B" to "C" and between "C" and "D."

Assume you are to pull wire into the conduit in Figure 324. Proceed as follows:

1. *Pull wires from outlet "A" to Service Entrance Panel.*

   Insert the fish-tape at the SEP in the basement and push to outlet "A" above entrance panel in the basement. Attach black and white wires to the hook. Wrap wires tightly so they won't pull loose, and make a smooth attachment so they won't catch on joints. Pull tape and wires through to the SEP. Leave enough length extended for connections at the SEP. Cut the wire at outlet "A", leaving correct length for connections.

2. *Push wires from "A" to "B" (Figure 324).*

   For a level, straight section such as this one, the tape is not needed. Push the black and white wires from "A" to "B" and cut.

3. *Pull wires between "B" and "C" (Figure 324).*

   Although short, this section contains two right angle bends, so attach black and white wires to fish tape for easier pulling.

   Insert fish tape at "C" and push to "B". Pull wires through and cut.

4. *Pull wires between "C" and "D" (Figure 324).*

   This is a switch connection and you must **use only black wires** or the color being used for hot conductors. No substitutions are allowed as in nonmetallic sheathed cable switch wiring.

   Use the fish-tape again for pushing through bends. Insert tape at "D" and push through to "C." Attach two black wires, pull through to "D" and cut.

5. *Connect wiring as for nonmetallic sheathed cable except, as noted earlier, a grounding wire is not needed.*

   The conduit itself serves as a grounding conductor.

VI. Estimating Wiring Costs

You will no doubt have had some experience as an electrician before you are called upon to estimate costs of a wiring job. Cost estimation is an important responsibility of the electrical contractor. He must be able to estimate material and labor costs accurately to stay in business.

Upon completion of this section, you will be able to follow procedures for estimating wiring costs for house wiring.

There are two ways of approaching an estimate, as follows:

A. Making a Detailed Analysis of Plans, Specifications and Labor Requirements.

B. Making a Quick Estimate.

A. Making a Detailed Analysis of Plans, Specifications and Labor Requirements

The most accurate method is to take the house plan and specifications for the job and list every detail of electrical wiring to be done. This would include a measure of the wire, conduit, or cable required. Count the number of boxes, outlets, switches and other devices needed. Include service entrance costs, plus a detailed estimate of the labor required to install all of the devices, the service entrance and wiring.

A chart similar to Tables I and II will be helpful in computing material costs.

Specifications for the wiring installation usually include the following provisions:

- —The **installation** shall **conform to** NEC and local *codes.*
- —**Only UL**-listed **materials** shall be used.
- —**Minimum wire size** allowed in branch circuits.
- —**Owner shall approve** or **select fixtures.**
- —**Warranty** of materials and workmanship, usually **one year.**

After all material and labor costs have been computed, the contractor must add **overhead** costs and a **profit margin.** Overhead and profit margins may vary greatly from one section of the nation to another. Fifteen percent for each, or a total of thirty percent may be common if housing starts are high in a fast growing area. Home building may be slow in another area and the total for overhead and profit combined may be as low as 15 or 20 percent.

B. Making a Quick Estimate

There are several variations of a quick way to estimate installation costs for house wiring. Two of them are discussed as follows:

1. Averaging Outlet Costs.
2. Estimating Separate Costs.

1. AVERAGING OUTLET COSTS

In its simplest form, the contractor arrives at an average cost for installing each outlet. He then multiples the average cost by the total number of outlets in the house to arrive at an estimate of wiring costs. The total includes all receptacles, lighting outlets and switch outlets. For example, suppose a house has a total of 70 outlets. Based on past experience, the contractor figures an average cost of $15.00 per outlet for all installation costs including overhead and profit. The estimate then becomes 70 × $15.00 = $1050.00

Fixture costs for the above arrangement are arrived at by agreement on a fixed amount, usually called a fixture allowance. The amount is added to the estimate. The fixtures are selected or approved by the owner. If the cost exceeds the figure agreed on, the difference is paid by the owner.

2. ESTIMATING SEPARATE COSTS

A more accurate method is to determine separately the cost of installing 120-volt, 240-volt and 120/240-volt outlets plus the separate cost of the service entrance. Using this procedure to estimate costs for the standard house plan, proceed as follows:

NOTE: These are examples only. True costs may vary greatly, depending on local conditions.

1. *Estimate the cost for 120-volt outlets.*
   73 outlets × cost per outlet
   73 × $12 = $876.00
2. *Estimate cost for 240-volt outlets.*
   5 outlets × cost per outlet
   5 × $80 = $400.00
3. *Estimate cost for 120/240-volt outlets.*
   3 outlets × cost per outlet
   3 × $85 = $255.00
4. *Estimate cost for service entrance.*
   $3 per ampere of service entrance × No. of amperes
   $3 × 150 = $450.00
5. *Total 1 through 4.*
   $876 + $400 + $255 + $450 = $1981.
6. *Add 25% for overhead and profit margin.*
   25% × $1981 = $495
   $495 + $1981 = $2476
   Total estimated cost = $2,476.00

When using averages, good judgment must still be applied to each cost estimate. Look for items that might cost more than usual and include the extra costs in your estimate. For example, a 240-volt basement circuit that can run parallel with the joists costs less to install than one of the same length that must run across the joists, requiring the holes be drilled in each joist. Also, a service entrance run that is longer than usual will add considerably to installation costs.

VII. Wiring Applications for Rural Areas

If you live on a farm or in a rural area, and plan to wire one or more buildings in addition to the residence, you need to plan carefully before you begin wiring. Garages, service areas, utility buildings and barns must be included in wiring plans to provide for electrical service needs in each one.

Upon successful completion of this section, you will be able to determine the size and type of wiring needed for the rural area, to install and wire a meter pole and to install wiring in agricultural buildings.

You should consult your local power supplier for help in planning and to obtain their specific requirements. If you only have one building or service area other than the residence, you may be able to wire a circuit for it from the service entrance panel in the house, depending on the load. However, if you have two or more buildings other than the residence, you should consider installing a meter pole for use as a central distribution point to the residence and other buildings (Figure 325).

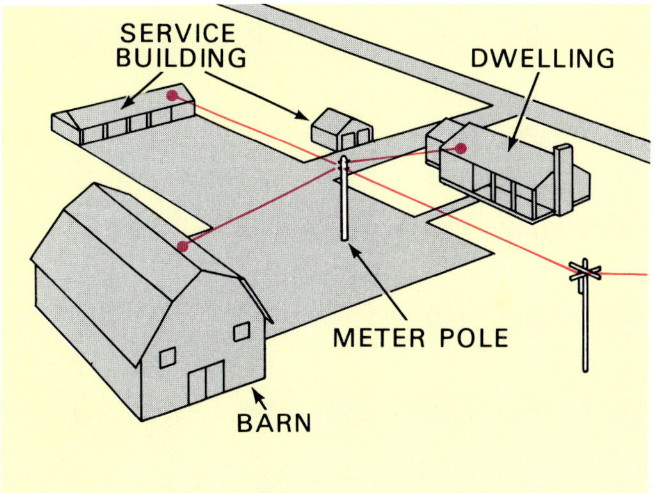

FIGURE 325. Meter pole serves as central distribution point to all buildings.

🍁 Each building must have a disconnecting means (usually called just a disconnect) and a service entrance panel. The disconnecting means may be a switch or a pullout block.

NOTE: If you have a disconnect used only for a water pump, it should be on a separate circuit ahead of the main disconnect, if it is used to provide fire protection.

To be sure of adequate power in each building, start with plans for a 3-wire service to each. The cables from the meter pole to buildings are called service drops or underground service laterals.

WARNING! Never attempt to disconnect power at a power supplier transformer. High voltage lines must be handled only with special equipment and by trained personnel. Many serious accidents occur annually when lay persons attempt to disconnect power at a high voltage transformer. Call your power supplier to perform this task.

Wiring applications for agricultural buildings are discussed as follows:

A. Determining Size of Wire and Service Equipment to Install.
B. Locating and Wiring the Meter Pole.
C. Wiring Service Drops to Buildings and Service Areas.
D. Wiring Agricultural and Other Utility Buildings.
E. Installing Remote-Control Wiring.
F. Installing New Cable and Boxes in Old Buildings.

A. Determining Size of Wire and Service Equipment to Install

In previous discussions of wire size as used in residential wiring, the size of wire was determined by the **ampacity** required in the circuit. In farm wiring, ampacity is also very important but two other factors must be considered. They are **mechanical strength** and **voltage drop.**

Mechnical strength is important in overhead wiring because of the possibility that wind or ice may cause the wires to break if they are not large enough. The Code requires overhead spans to be at least No. 10 wires for unsupported spans up to 15 m (50 feet) and No. 8 for longer spans.

Voltage drop is the reduction of voltage between the power source and the load, such as a motor. Voltage drop is not a problem in most residential circuits because the distances are usually not long. On farms, however, some circuits may be 92 m (300 feet) or more in length. The size of the wire must be larger on such long circuits to keep voltage drop within the normal 2 percent maximum for the load being served.

Voltage drop in wires is similar to the loss of water pressure in a long pipe. A 1.3 cm (½ inch) water pipe at 276 kPa (40 psi) would carry enough water to an irrigation outlet if it is only 4.6 to 6.1 m (15 to 20 feet) from the tank. However, if it is 153 or 183 m (500 or 600 feet) away, the pipe would be too small to maintain the desired pressure. Voltage drop in wire occurs because of total resistance on long runs if the wires are not large enough to carry the load.

Other factors shown in Table XIII affecting wire size are:

- —The **type of wire** used.
- —**Whether wire is run in conduit, cable, buried in the earth or overhead** in the air.
- —**Whether the overhead wires are single or in triplex,** which is a 3 wire cable assembly.

Table XIII shows sizes of wire required for various amperages at distances of 15 to 92 m (50 to 300 feet), based on 2 percent voltage drop. To the left of the double line are the smallest sizes of wire allowed for the amperages shown. To the right of the double line are sizes required for various distances. For example, for a 15-ampere load, the smallest wire allowed for any length circuit is No. 12, in cable, conduit or buried in the earth, for the types of wire listed. If the circuit is overhead in the air, No. 10 wire is the smallest allowed. If the circuit were 31 m (100 feet) long, No. 8 wire would be required to carry the 15-ampere load. For a 92 m (300-feet) 15-ampere circuit, use No. 4 wire. Note that this is for aluminum conductors and for 240 volts.

Determining the size of wire and service equipment to install is discussed as follows:

1. Computing Service Load for Residences.
2. Computing Service Load for Each Building Supplied by Two or More Branch Circuits.
3. Computing the Total Load for the Entire Service.

1. COMPUTING SERVICE LOAD FOR RESIDENCES

The load for the rural residence is computed in the same way as a residence in a city. This procedure is outlined in detail in Section IV., "Installing Service Entrance Equipment."

2. COMPUTING SERVICE LOAD FOR EACH BUILDING SUPPLIED BY TWO OR MORE BRANCH CIRCUITS

If a building has two or more circuits, the load must be computed in keeping with Code Section 220–40.

Computing the service load for each building is discussed as follows:

a. Computing Lighting Load.
b. Computing Receptacle Load.
c. Computing Motor Load(s).

a. Computing Lighting Load

The lighting load should be computed at the rate of 1.5 amperes per outlet at 120 volts. Assume 10 lighting outlets in a building, for example. The total lighting load is 10 × 1.5 or 15 amperes. If a larger or smaller load is to be installed, use the actual figure, such as flood lights that might average 250 watts each. To find amperes, use the formula: amperes = watts ÷ volts.

TABLE XIII. MINIMUM WIRE SIZES FOR AMPERAGE LOADS AND DISTANCES (Aluminum up to 400 Amperes, 230–240 Volts, Single Phase, Based on 2% Voltage Drop)

Minimum Allowable Size of Conductor	In Cable Conduit, Earth		Overhead in Air*		Length of Run in Feet — Compare size shown below with size shown to left of double line. Use the larger size.											
Load in Amps	Types R,T,TW	Types RH, RHW, THW	Bare Covered Conductors Single	Triplex	50	60	75	100	125	150	175	200	225	250	275	300
5	12	12	10		12	12	12	12	12	12	12	10	10	10	10	8
7	12	12	10		12	12	12	12	12	10	10	10	8	8	8	8
10	12	12	10		12	12	12	10	10	8	8	8	8	6	6	6
15	12	12	10		12	12	10	8	8	8	6	6	6	4	4	4
20	10	10	10		10	10	8	8	6	6	6	4	4	4	4	3
25	10	10	10		10	8	8	6	6	4	4	4	4	3	3	2
30	8	8	10		8	8	8	6	4	4	4	3	3	2	2	2
35	6	8	10		8	8	6	6	4	4	3	3	2	2	1	1
40	6	8	10		8	6	6	4	4	3	3	2	2	1	1	0
45	4	6	10		8	6	6	4	4	3	2	2	1	1	0	0
50	4	6	8		6	6	4	4	3	2	2	1	1	0	0	00
60	2	4	6	6	6	6	4	3	2	2	1	0	0	00	00	00
70	2	2	6	4	6	4	4	3	2	1	0	0	00	00	000	000
80	1	2	6	4	4	4	3	2	1	0	0	00	00	000	000	4/0
90	0	2	4	2	4	4	3	2	1	0	00	00	000	000	4/0	4/0
100	0	1	4	2	4	3	2	1	0	00	00	000	000	4/0	4/0	250
115	00	0	2	1	4	3	2	1	0	00	000	000	4/0	4/0	250	300
130	000	00	2	0	3	2	1	0	00	000	000	4/0	250	250	300	300
150	4/0	000	1	00	2	2	1	00	000	000	4/0	250	250	300	300	350
175	300	4/0	0	000	2	1	0	00	000	4/0	250	300	300	350	400	400
200	350	250	00	4/0	1	0	00	000	4/0	250	300	300	350	400	400	500
225	400	300	000		1	0	00	000	4/0	250	300	350	400	500	500	500
250	500	350	000		0	00	000	4/0	250	300	350	400	500	500	500	600
275	600	500	4/0		0	00	000	4/0	250	300	400	400	500	500	600	600
300	700	500	250		00	00	000	250	300	350	400	500	500	600	600	700
325	800	600	300		00	000	4/0	250	300	400	500	500	600	600	700	750
350	900	700	300		00	000	4/0	300	350	400	500	600	600	700	750	800
375	1M	700	350		000	000	4/0	300	350	500	500	600	700	700	800	900
400		900	350		000	4/0	250	300	400	500	600	600	700	750	900	900

*Conductors in overhead spans must be at least No. 10 copper or No. 8 aluminum for spans up to 50 feet and No. 8 copper or No. 6 aluminum for longer spans.

Courtesy of *Agricultural Wiring Handbook*

b. Computing Receptacle Load

The receptacle load should also be computed at 1.5 amperes per outlet at 120 volts. These are receptacles that are normally used for portable tools and equipment, such as drills, saws and clippers. Eight outlets, for example, at 1.5 amperes = 12 amperes total receptacle load.

c. Computing Motor Load(s)

This section must be computed in keeping with Code requirements outlined in Sections 430–22, 430–24 and Table 430–148, as follows:

—For a **single** motor, the rating must be 125 percent of the motor full-load current rating. For example, if you have only one motor in the building and its rating is 5 amperes, multiply 5 × 125% = 6.25 or 6 amperes.

—For more than one motor in a building, rate the largest motor at 125 percent and its rated full-load amperage. For additional motors and other large permanently connected equipment items, list the full amperage load.

Table XIV shows the full-load current for electric motors to help you determine ratings on motors that show only horsepower ratings.

After computing all lighting, receptacle and equipment loads for the building, add the total for all loads in the building to include lighting, receptacle and equipment loads.

NOTE: Another factor which must be considered is loads that operate without diversity, meaning, the equipment that will be operating at the same time to produce the heaviest load. The loads in each building must be considered carefully to arrive at a correct total for the load operating at the same time.

Proceed as follows to determine the load for each building supplied by two or more circuits:

1. *Add up the ampere load for all 120-volt circuits.* To arrive at the equivalent for a 240-volt circuit, divide the total amperage by two.
2. *Add up the amperage load for all 240-volt circuits.*
3. *Add 1 and 2 together.*
4. *As an example of a service center (shop building), list the lighting, receptacle and motor load and other large equipment as follows:*

8 lighting outlets at 1.5 a =	12 a
8 receptacle outlets at 1.5 a =	12 a
Total 120 v load =	24 a
Divide 120 v total by 2 for 240 equivalent, 24 ÷ 2 =	12 a, 240 v
Heater, 260 W =	11 a, 240 v
Tool Grinder, 373 W (½ hp) =	5 a, 240 v
Table Saw, 560 W (¾ hp) =	7 a, 240 v
Drill Press, 746 W (1 hp); 8 a × 1.25% =	10 a, 240 v
Welder =	40 a, 240 v

Total Service Center = 85 amperes, 240 volts.

NOTE: Code Table 430-148 uses headings of 115v and 230v for columns 3 and 4 - Table XIV is changed to headings of 120v and 240v to be consistent with Table XV.

TABLE XIV. FULL LOAD CURRENTS IN AMPERES FOR SINGLE PHASE AC MOTORS (From Code Table 430–148)

(Second, third & fourth columns From Code Table 430–148)

Power*			
SI (kW)	US (HP)	120v amperes	240v amperes
.124	1/6	4.4	2.2
.186	¼	5.8	2.9
.248	⅓	7.2	3.6
.373	½	9.8	4.9
.559	¾	13.8	6.9
.746	1	16	8
1.119	1½	20	10
1.491	2	24	12
2.237	3	34	17
3.729	5	56	28
5.593	7½	80	40
7.457	10	100	50

*Table XIV provides the full load current in amperes for commercially-manufactured motors with the size designation in both SI and US units.

Add all the equipment items that are permanently connected to arrive at the load.

The loads most likely to be operating at one time to produce the heaviest loads are: the welder at 40 amperes, the heater at 11 amperes and half of the lighting at 6 amperes for a total of 57 amperes without diversity.

5. *Apply factors in Table XV to the building as follows:*

TABLE XV. METHOD FOR COMPUTING INDIVIDUAL FARM BUILDING LOADS FOR OTHER THAN DWELLINGS (Code Table 220–40)

Ampere Load at 240 Volts	Demand Factor Percent
Loads expected to operate without diversity, but not less than 125 percent full-load current of the largest motor and not less than the first 60 amperes of load	100
Next 60 amperes of all other loads	50
Remainder of other load	25

a. You found that the load without diversity is 57 amperes.
b. The largest motor is one hp, drawing 8 amperes, 125% × 8 amperes = 10 amperes.
c. Not less than the first 60 amperes of all other load.
   (1) *From items a, b & c above, determine which is the larger of the three.*
   After this determination is made, apply the 100 percent demand to this amount.
   (2) *Subtract the largest of the above values, 60 amperes from the total connected load, 85 amperes, (85 – 60 = 25) amperes.*
   The next 60 amperes of the remainder has a demand factor of 50 percent. Subtract the 60 amperes from the above remainder. Since the remainder is less than 60, you show the actual amount, 25 amperes at 50 percent, or approximately 13 amperes. If the load were large enough to have a remainder after the first two steps, you would apply a demand factor of 25 percent.
   In the service center, however, the demand is 60 + 13 for a total of 73 amperes, 240 volts.

First 60 amperes at 100 percent demand factor =	60 amperes
Remaining 25 amperes at 50 percent demand factor =	13 amperes
Demand =	73 amperes 240 volts

Size of service entrance equipment should be 100 amperes to allow for future needs in the service center.

The wire size required for the conductors from the meter-pole to the service center is No. 1 aluminum cable in conduit or buried in the earth for distances up to 100 feet.

3. COMPUTING THE TOTAL LOAD FOR THE ENTIRE SERVICE

Repeat the process in Section 2 for each building, other than the residence.

As stated earlier, each building must have a service entrance with a disconnecting means. Size of the SEP is based on the demand load.

TABLE XVI. METHOD FOR COMPUTING TOTAL FARM LOAD
(Code Table 220-41)

Reprinted by permission from NFPA 70, National Electrical Code®
Copyright 1984, National Fire Protection Association, Boston, MA

Individual Loads Computed in Accordance with Table XV	Demand Factor (Percent)
Largest load	100
Second largest load	75
Third largest load	65
Remaining loads	50

To this total load, add the load of the dwelling.

To illustrate the application of Table XVI, assume you are planning a system for a farm with the following buildings and load demands, for buildings other than the residence:

Swine farrowing house	115 amperes
Service Center	73 amperes
Poultry house	36 amperes
Barn	26 amperes
Pumphouse	8 amperes

Apply demand factors from Table XVI to determine the total load. List each building with its demand factor percent.

Largest load (building with largest ampere load)	
Swine farrowing house, 115 a 100 % =	115 a
Second largest load	
Service center, 73 a at 75 % =	55 a
Third largest load	
Poultry house, 36 a at 65 % =	23 a
Remaining loads at 50 percent	
Barn, 26 a =	13 a
Pumphouse, 8 a =	4 a
Residence, 150 a at 100% =	150 a
Total Farm Load =	360 a

To determine sizes of wire required for meter pole and for service drops to each building above, refer to Table XIII which is based on 2 percent voltage drop.

Suggested size of service entrance for each building is as follows:

Swine farrowing house - 115 amperes. One hundred fifty amperes recommended. This allows for future expansion if needed.

Service center - 73 amperes. One hundred amperes is the minimum size service entrance. However, the installation of one additional item of shop equipment could exceed the limit. It would be preferable to install a **150-ampere service entrance** to allow for additional load requirements.

Poultry house - 36 amperes. Sixty-ampere service entrance is recommended.

Barn - 26 amperes. The minimum size of 30 amperes is barely above present requirements. **Sixty-ampere service** entrance is recommended to allow for future needs.

Pumphouse - 8 amperes. Power requirements for the pumphouse are not likely to increase much above present needs. **Thirty-ampere service** entrance is sufficient for the single circuit.

Meter pole wires must be large enough to carry the total farm load. Based on the 360-ampere above, the meter pole wires must be 350 MCM (thousand circular mils) for cable or conduit, Types R, T and TW at 38.1 m (125 feet). Instructions for wiring meter poles are given in the following section (B).

Wire sizes for each building above are as follows, as indicated in Table XIII for Types R, T and TW at various distances.

Swine farrowing house - 150 amperes, No. 2/0 for distances to 30.5 m (100 feet) and No. 3/0 45.8 m (to 150 feet).

Service center - 150 amperes, No. 2/0 for distances to 30.5 m (100 feet) and 3/0 to 45.8 m (150 feet).

Poultry house - 50 amperes, No. 3 wire for distances to 30.5 m (100 feet) or No. 2 to 45.8 m (150 feet).

Barn, 60 amperes, No. 3 wire for distances to 30.5 m (100 feet) or No. 2 to 45.8 m (150 feet).

Pumphouse - 30 amperes, No. 6 wire for distances to 30.5 m (100 feet) or No. 4 for 45.8 m (150 feet).

Residence - 150 ampere, 2/0 wire for distances to 30.5 m (100 feet) and 3/0 for 45.8 m (150 feet).

B. Locating and Wiring the Meter Pole

Locating and wiring the meter pole is discussed as follows:

1. Locating the Meter Pole.
2. Wiring the Meter Pole.

1. LOCATING THE METER POLE

One reason for installing the meter on a pole is to reduce the length of wire or cable required for two or more buildings. Other advantages are that repairs or changes can be made to wiring in one building without affecting service to the others and, usually, smaller conductors can be installed to service each building because of shorter distance.

Normally, the meter pole is located near a point in the center of the electrical load to be served. An exception might be if load demand to one building is much greater than the others. In such a case, it may cost less to locate the meter pole somewhat nearer the heavy demand to reduce the amount of larger wire required. Select the location that is most convenient and least costly (Figure 326).

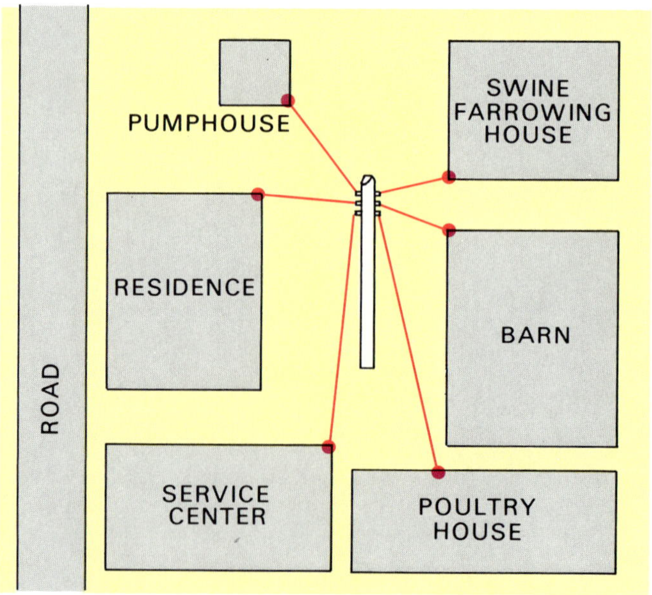

FIGURE 326. Locate the meter pole near the center of the electrical load to be served.

2. WIRING THE METER POLE

In most localities, the power supplier installs the meter pole. The power supplier also brings the wiring to meet the owner's wiring at the top of the pole. It is usually the owner's responsibility to install the wiring from that point. In every case, the power supplier should be contacted for specific requirements for the installation. In many instances, **current transformer** metering is installed on the meter pole, especially where the service size exceeds 200 amperes. Where current transformer metering is installed, large conductors are not needed up and down the pole. You must know the requirements of your local code and power supplier before you begin. In all cases, the neutral is grounded at the meter pole, using a different procedure than for grounding house wiring, as explained in the next section.

Wiring the meter pole is discussed as follows:

a. Wiring the meter pole with one conduit.
b. Wiring the meter pole with two conduits.
c. Wiring the meter pole with service entrance cable.
d. Wiring the meter pole when current transformer (CT) metering is used.

a. Wiring the Meter Pole with One Conduit

NOTE: The conduit with wires installed, the meter base, the disconnect switch and ground wire should be mounted on the pole before the pole is erected.

In this installation, you will run the two hot wires and the neutral wire for the power line connection at the top of the pole, down to the meter socket. Only the two hot wires run back to the top of the pole.

Proceed as follows:

1. *Install insulator racks or eyebolts and connectors near top of pole.*

   Mount one rack or connector for conductors from the power supplier, and one for each building that will be served from the pole (Figure 327). Fasten with heavy screws, or with bolts through the pole if heavy icing conditions are likely.

2. *Install the conduit and entrance head.*

   Select conduit type and size as in Section V and install on pole. Mount the top of the conduit higher than the insulators to provide for a drip loop. Fasten conduit in place with conduit straps. Attach entrance head with required number of holes (Figure 328).

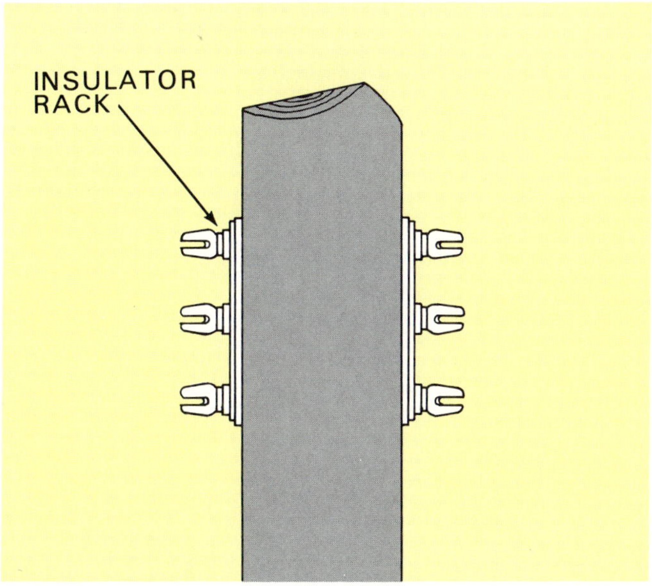

FIGURE 327. Mount one rack or eyebolt for power supplier conductors and one for each building to be served.

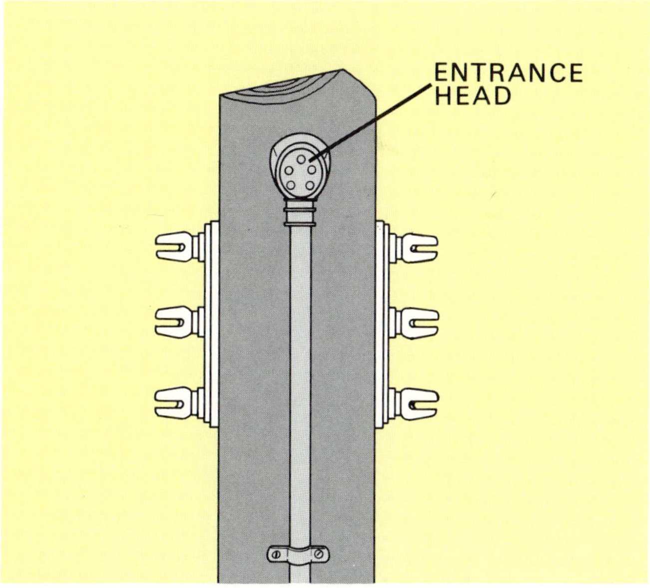

FIGURE 328. Select an entrance head with the required number of holes and mount high enough to form drip loops.

3. *Install meter base.*

Follow procedures as in Section D. Mount the meter about 1.4–1.7 m (4½–5 feet) above ground (Figure 329).

4. *Install wiring.*

Push the neutral wire (white) and two hot wires (usually black) down through the conduit to the meter base.

Leave about 1.2 m (48 inches) at the top of the pole to be attached to the power supplier lines. Connect the neutral and two hot wires to the meter base.

NOTE: The Code does not allow a disconnect to be installed where only one conduit is installed for meter pole wiring.

5. *Install hot wires from the meter base to the distribution wires.*

Inside the same conduit, install two more hot wires. The neutral connections are made at the top of the pole by splicing each one to the power line neutral, so another neutral is not required in the conduit.

Connect the bottom end of the two hot wires to the lower (output) terminals on the meter base. At the entrance head, you will connect the two hot wires to the outgoing wires to other buildings when installed (Figure 330).

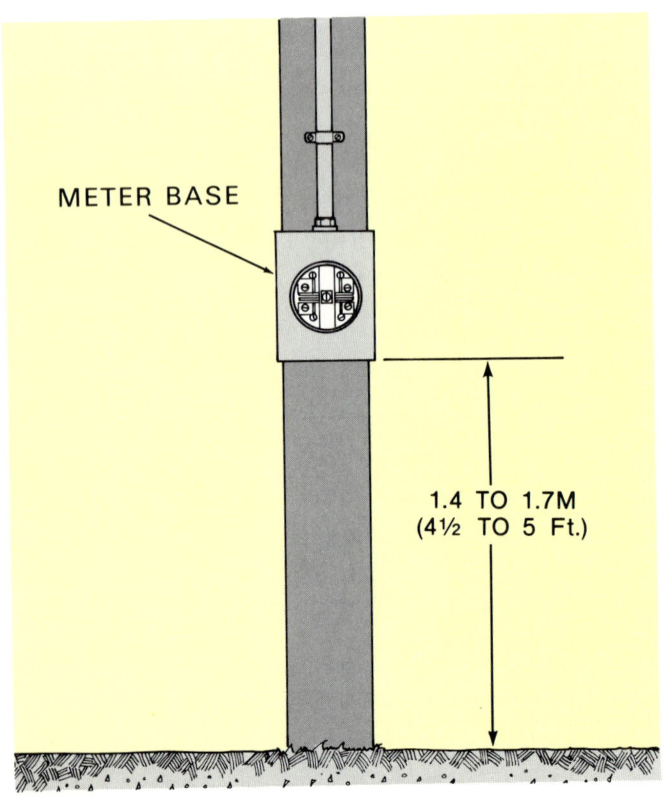

FIGURE 329. Mount the meter base 1.4 to 1.7 m (4½ to 5 feet) above ground.

6. *Connect neutral from power line.*

Splice a wire between the power line neutral and the outgoing circuit neutrals, using a heavy split-bolt connector (Figure 331).

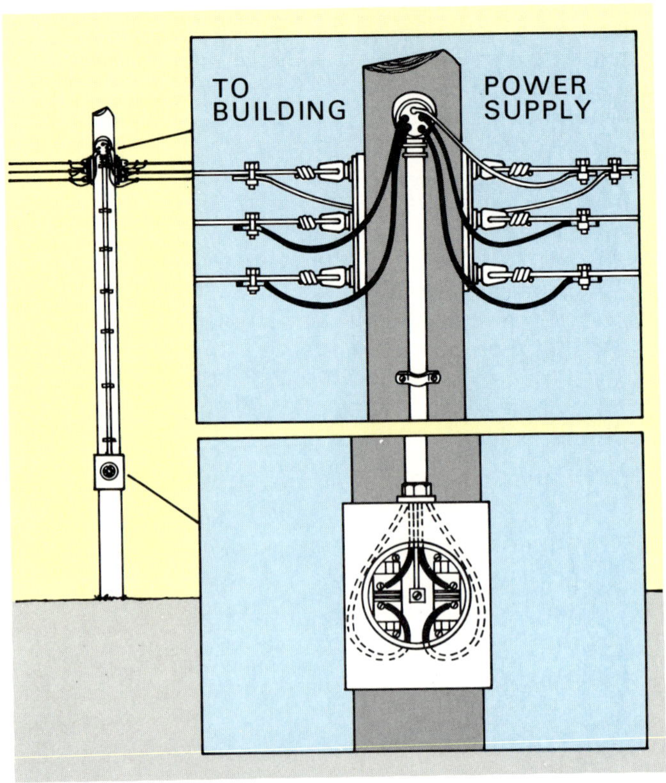

FIGURE 330. Wiring in meter base and at top of pole.

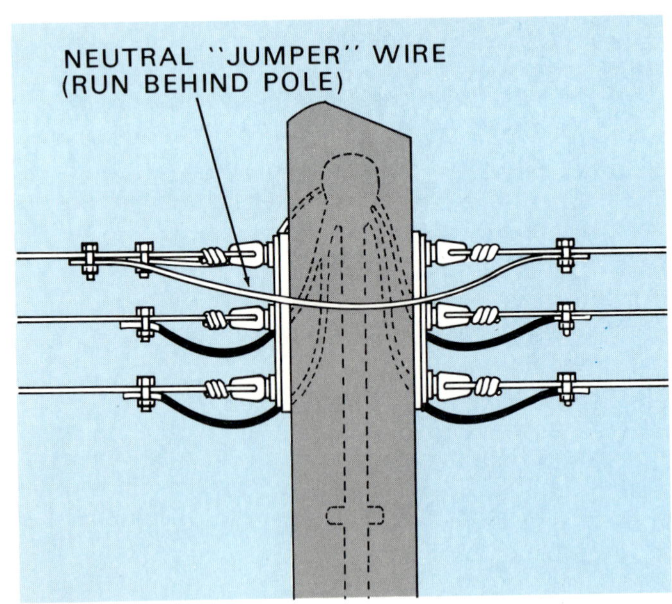

FIGURE 331. Wire between power supply neutral and outgoing neutrals.

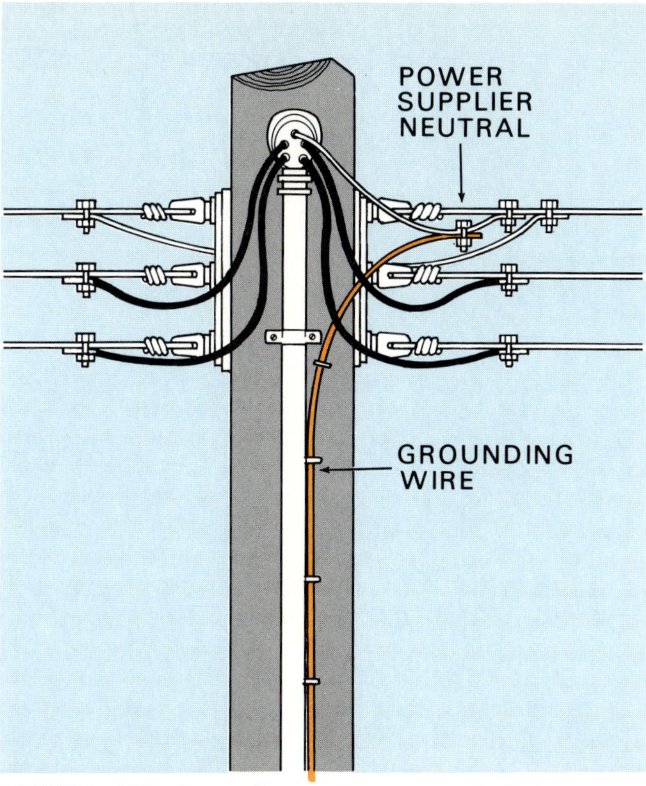

FIGURE 332. Grounding wire connected to power supplier neutral and stapled to pole.

7. *Connect the grounding wire.*

   On previous ground connections, you have connected the grounding wire to the service entrance and from there to the cold water pipe and/or a "made" ground. Because of the increased danger from lightning on pole installations, most power suppliers recommend connecting the grounding wire to the neutral at the top of the pole.

   Splice the grounding wire to the neutral and run it down the pole direct to the ground rod or

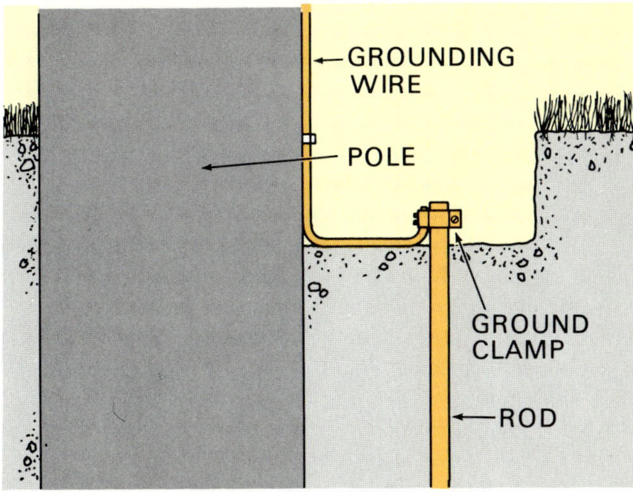

FIGURE 333. Grounding wire attached to grounding rod with approved clamp.

water pipe (Figure 332). No. 6 wire is usually large enough, but some power suppliers require No. 4. Staple the grounding wire to the pole alongside the conduit for protection, or to the opposite side of the pole, if your Code requires it. To protect the grounding wire when installed with a ground rod, dig a hole 15 or 20 cm (6 or 8 inches) deep next to the pole. Select and install the rod as in Section IV, "Installing Service Entrance Equipment." A 1.3 cm (one half inch) copper rod 2.4 m (8 feet) long is a common size. Attach the ground wire with an approved clamp and cover with soil after inspection (Figure 333). If the ground wire attached to the pole is subject to damage by vehicles driving past or other hazards, nail a narrow strip of wood to cover it to a height of 2.4 to 3.1 m (8 to 10 feet), for protection.

b. Wiring the Meter Pole with Two Conduits

Proceed as follows:

1. *Run the neutral and two hot wires in one conduit for connecting the power supplier lines to the meter base (and to a disconnect switch, if installed).*
2. *Mount a second conduit with neutral and two hot wires to run from the meter base or disconnect switch to the top of the pole (Figure 334).*

   Since circuits in both conduits have neutral wires, it is not necessary to connect the neutral from the power line to neutrals of outgoing circuits. Instead, connect the neutral from the meter base or disconnect switch to the outgoing circuits at the top of the pole.

   Check with the power supplier if a **disconnect switch** is desired for the meter pole. **Where a disconnect switch is installed, it should be a double-pole, double-throw type, with terminals for connecting a standby generator for emergency power. The switch must be in the "off" position before the generator is started** to prevent current from feeding in reverse to the power supply lines. This can cause serious injury or death to repairmen working on the lines.

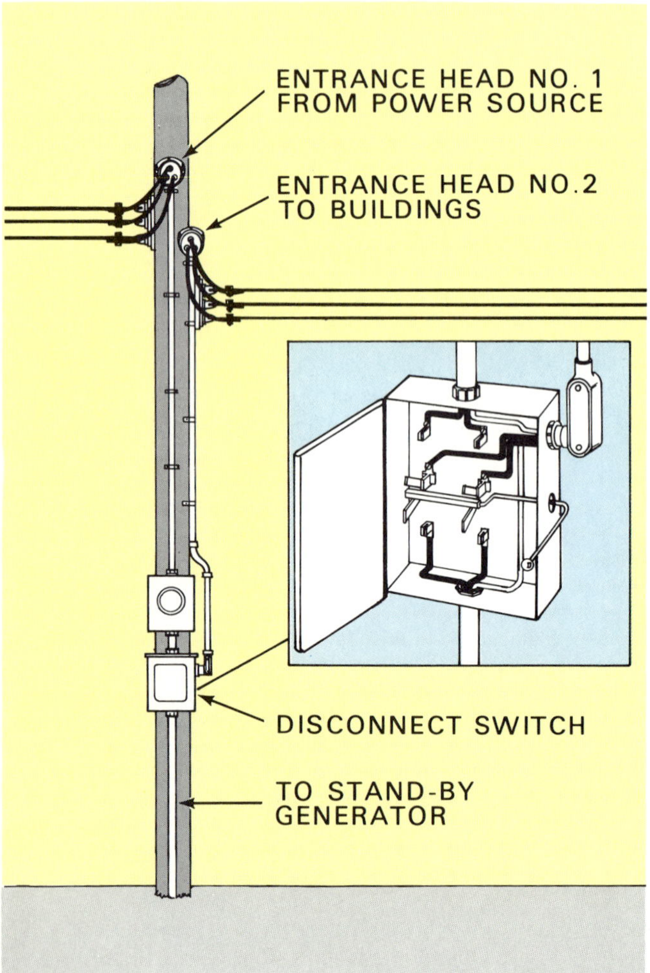

FIGURE 334. Two conduits each contain neutral and two hot wires. They connect power supplier lines to meter base and meter base to outgoing wires at the top of the pole.

c. Wiring the Meter Pole with Service Entrance Cable

Install three-wire service entrance cable following the same procedures outlined above for two conduits. Fasten the cable with straps and protect it from damage where required.

d. Wiring the Meter Pole When Current Transformer (CT) Metering Is Used

Current transformer metering is now frequently installed on the meter pole, especially where total loads approach or exceed 200 amperes. In these installations, large conductors are not needed down and up the meter pole (Figure 335).

The current transformer is normally installed on the meter pole by the power supplier. Operating at a maximum of 5 amperes to the meter, small gage wires are installed between the current transformer and meter. Service drops to buildings are spliced directly to the service drop conductor from the power supplier unless there is a pole-top switch installed.

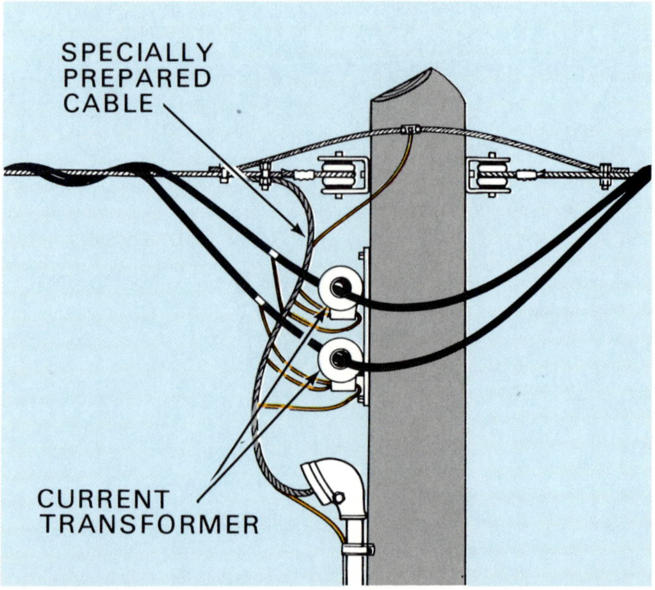

FIGURE 335. Current transformer (CT) metering permits the use of smaller gage wiring from power supplier wires to the meter.

C. Wiring Service Drops to Buildings and Service Areas

Refer to Table XIII to determine the size of wire needed for each service drop or underground service lateral.

Wiring service drops to buildings is discussed as follows:

1. Wiring Overhead Service Drops to Buildings.
2. Wiring Underground Service Laterals to Homes and Buildings.

1. WIRING OVERHEAD SERVICE DROPS TO BUILDINGS

1. *Cut conductors to length.*

   Unroll and cut the conductors (triplex or individual wires) to required length on the ground. Allow about 1.2 m (four feet) at each end for attaching to insulators.

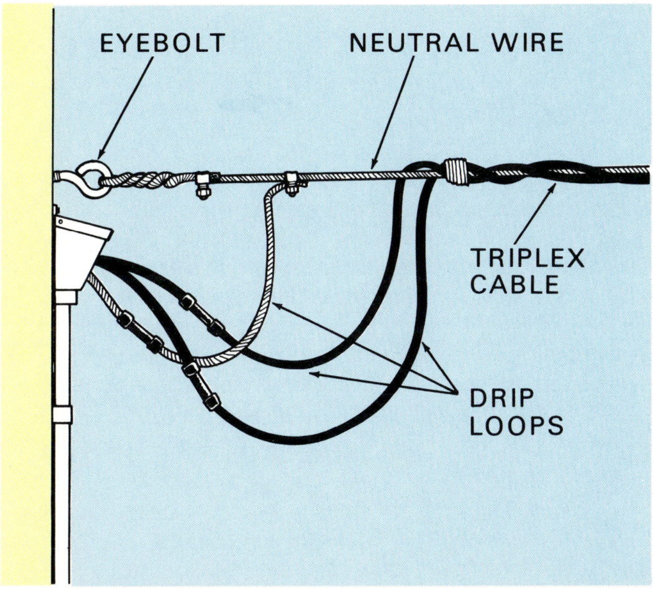

FIGURE 336. Triplex cable is usually fastened to the building by means of an eyebolt or insulated connector.

2. *Attach conductors to insulator(s) on pole and to building.*

   Triplex cable is generally favored for service drops. It is made up of a bare neutral wire with two insulated cables wrapped around it for support. Triplex cable is usually fastened to the meter pole and to the building by means of an eyebolt or insulated connector (Figure 336). Fasten the eyebolts securely to the pole and buildings. If mounting on a wood wall, screw the eyebolt directly into a stud or other framing member in back of the siding or sheathing. Loop the neutral wire through the eyebolt and fasten with a crimp fastener or split-bolt connector.

3. *Splice wires for feeder connections.*

   Splice the heavy outdoor wires with split-bolt solderless connectors. They may be used to tap a wire, as with drip loop connections (Figure 337), or to splice two ends together. Proceed as follows:

   Remove the insulation from the wires and place together. Remove the nut from the split bolt. Place the bolt over the wires. Replace the nut and tighten with a wrench. Cover the splice with insulating tape. Wrap the tape around the splice to a thickness equal to conductor insulation or greater.

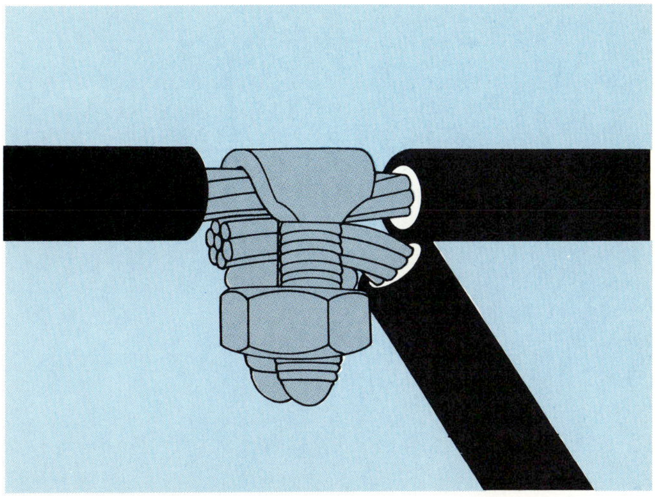

FIGURE 337. Split-bolt solderless connectors are used to splice the heavy outdoor wires. Splice must be wrapped with insulating tape.

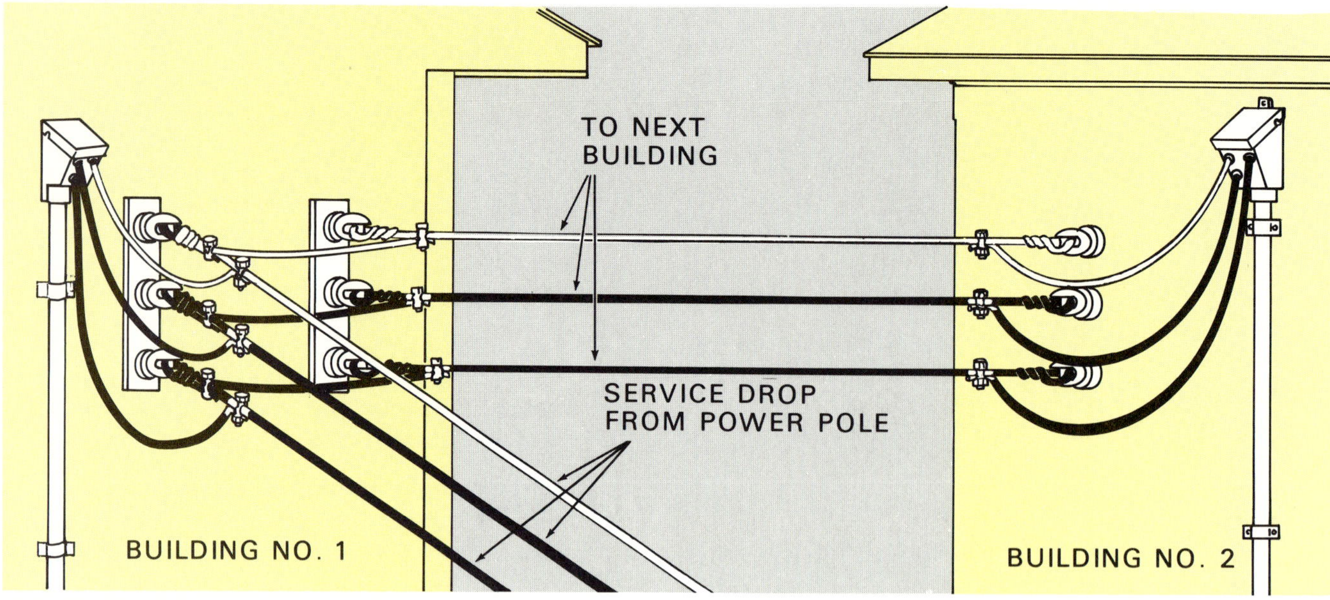

FIGURE 338. Wiring for two buildings fairly close together. Tap the service drop from the pole and extend to the second building.

4. *Attach wires to house or other buildings.*
   Install the insulator or insulator rack as in Section IV. C., "Locating and Mounting the Service Drop." Procedures are the same. Remember to observe the height requirements for the overhead wires as in Section IV. C.
5. *Install service entrance.*
   Follow procedures in Section IV. D., "Installing Service Entrance." Procedures are the same except that you do not install a meter.

   A 30-ampere capacity service entrance may be installed if the building does not have over 2 circuits. For all others, install service equipment rated according to directions in Table XV. Remember that each service entrance must have a disconnect, overload protection and must be grounded.

   If you have two buildings close together and both contain fairly small loads, you may serve both with one circuit from the pole. Use wires large enough to carry both loads. Run the three wires from the pole to one building. Then extend wires to the next building as shown in Figure 338. If only 120 volts are needed in the second building, tap only the neutral and one hot wire for a 2-wire circuit. Install complete service entrance at each building.

2. WIRING UNDERGROUND SERVICE LATERALS TO HOMES AND BUILDINGS

Underground wiring is becoming popular for use in homes, farms, and many other places. It is simple to install with trenching equipment. Newer types of wire are low in cost and there are fewer maintenance problems. With underground wiring, there are no service interruptions caused by ice storms, falling branches and high winds. Tall equipment can be moved freely without the hazard of striking overhead wires.

Two types of cable are approved for underground wiring. They are Type UF (underground feeder) and Type USE (underground service entrance). Both types can be used to run feeder circuits buried directly in the earth from the pole to homes and buildings.

Proceed as follows:

1. *Install conduit with wire or service entrance cable from top of pole to meter base and to the disconnect switch.*
   Leave 1.2 or 1.5 meters (four or five feet) at top of pole for connections to power suppliers service drop.
2. *Install conduit from disconnect switch to trench for underground cable.*
   Conduit end at trench must have bushing installed to prevent damage to cable from rough

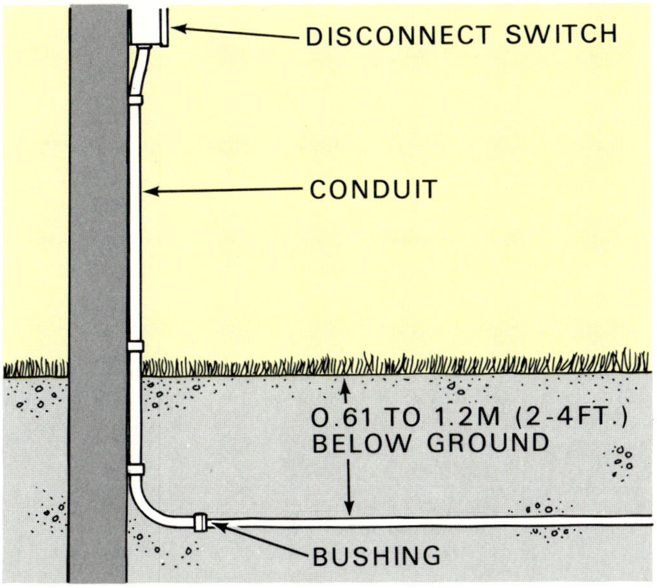

FIGURE 339. Conduit should extend 0.61 to 1.2 m (2 to 4 feet) below ground with bushing attached to protect cable.

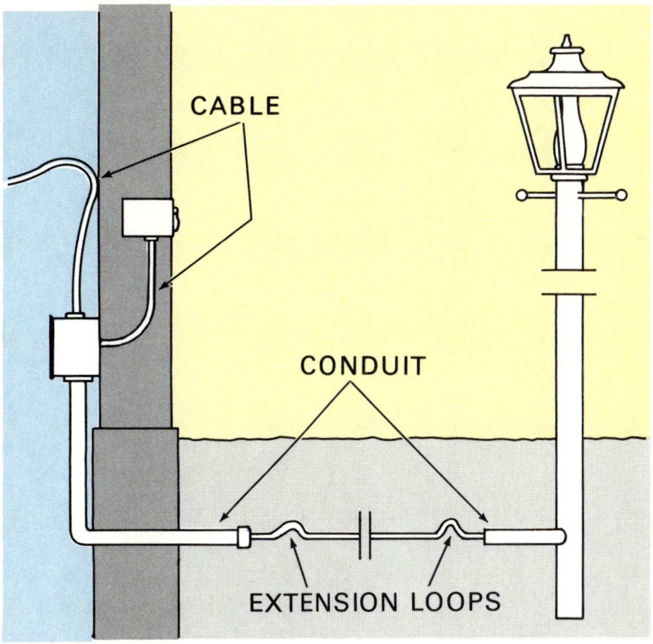

FIGURE 341A. Circuit extension for post lantern.

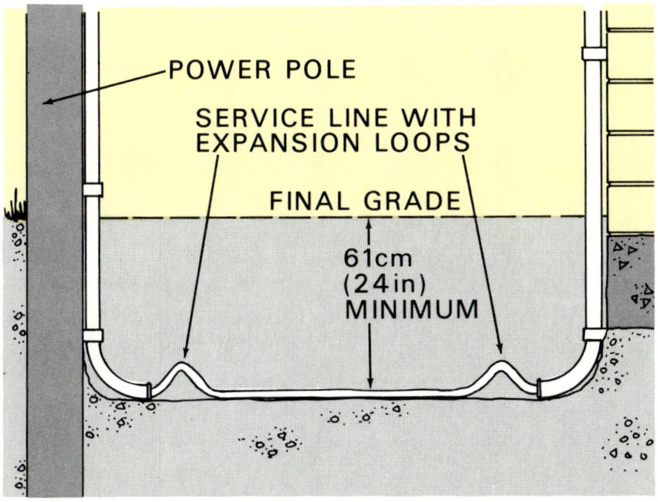

FIGURE 340. Lay cable in a trench at least 61 cm (24 inches) deep. Leave expansion loop at each end.

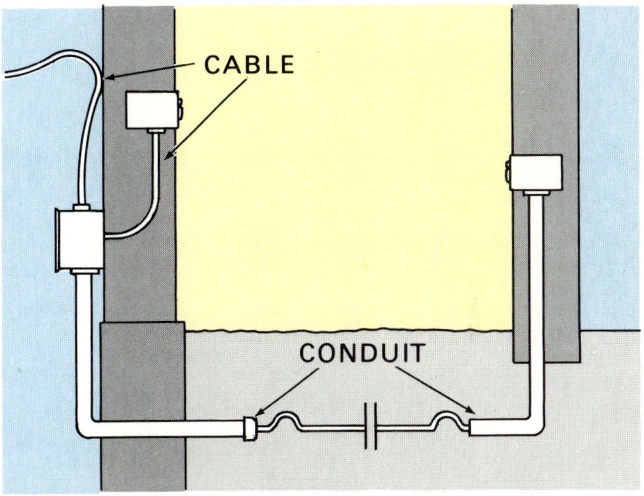

FIGURE 341B. Circuit extension to another building.

edges (Figure 339). Conduit should protect cable from the switch to a point 0.61 to 1.2 m (2 to 4 feet) below grade level.

3. *Install cable in trench.*

   Trench must be at least 61 cm (24 inches) deep. Local codes may require up to 1.2 m (48 inches). Lay cable with expansion loop at each end (Figure 340). When cable is located beneath driveways or other heavily loaded surfaces or if sharp rocks may damage the cable, protection should be provided with conduit. Underground conductors run under a building must be in conduit. The conduit must extend beyond the outside walls of the building. The Code doesn't say how far.

   Similar procedures may be followed to extend a circuit from inside the house to a post lantern (Figure 341A) or to another building (Figure 341B).

4. Follow procedures in Section IV. D4., "Installing Underground Service Entrance Cable."

D. Wiring Agricultural and Other Utility Buildings

Wiring agricultural buildings is discussed under the following headings:

1. Types of Wire.
2. Types of Wiring Devices, Boxes and Fittings.
3. Procedures for Wiring Buildings.
4. Stray Voltage.

1. TYPES OF WIRE

Type NM nonmetallic sheathed cable can be used only in normally dry locations. For wiring most agricultural buildings, either Type NMC, which may not be available or Type UF (underground feeder) cable is recommended. Both have insulation and covering that permits their use in dry, moist, damp or corrosive locations and in outside or inside walls of masonry, block or tile. An added advantage for Type UF is that it may be used underground. Also, it may be exposed to sunlight on the exterior of a building if not subject to physical damage.

All cables should be surface mounted to permit proper wiring inspection and maintenance, and to minimize damage from rodents. The cables should be secured by non-metallic cable straps with corrosion resistant nails. The cables should be secured at three-foot intervals on vertical surfaces and two-foot intervals on horizontal surfaces and within eight inches of each box or device.

Portable tools and equipment are likely to be used at times in most agricultural buildings. To insure equipment grounding at every outlet, install cable and receptacles with ground in all agricultural buildings.

Metal conduit and armored cable should not be installed in agricultural buildings where moist and/or corrosive conditions exist. Rust and corrosion may gradually cause separation of the joints. This action exposes conductors inside the conduit and may cause a short circuit. Also, it leaves the circuit without effective equipment grounding since conduit or armor serves as the equipment grounding conductor.

If more than 3 conductors are needed, as for a 3 phase fan motor or control, then it is convenient to use **non-metallic conduit**. There are several types of conduit available from different manufacturers. Some insurance companies are concerned about toxic fumes from this material (in instances of fire) so it is advisable to check with your insurance company before installing.

In agricultural buildings, conduit may be subject to severe physical abuse so heavy walled conduit should be used. Type 80 heavy-walled rigid PVC conduit (Figure 342) is designed for both above ground and underground applications. It should not extend from a warm, moist part of a building to a cold part because of condensation collecting in the conduit. It should be installed using the correct fittings and glued to the boxes or to the threaded connectors. **Non-metallic flexible tubing** has been developed to replace electrical metallic tubing (EMT) (thin-wall conduit). One type has concentric corrugations that give it high strength and bendability without heavy weight (Figure 342). Special couplings and quick connect terminations fit into the concentric corrugations to make it easy to attach to special single-gang or two-gang non-metallic boxes. Another style of this material has spiral configuration and can be made watertight by using silicone calking where the conduit enters the threaded connectors. This type of watertight non-metallic conduit is used in such wet locations as milking parlors, and fruit and vegetable processing areas.

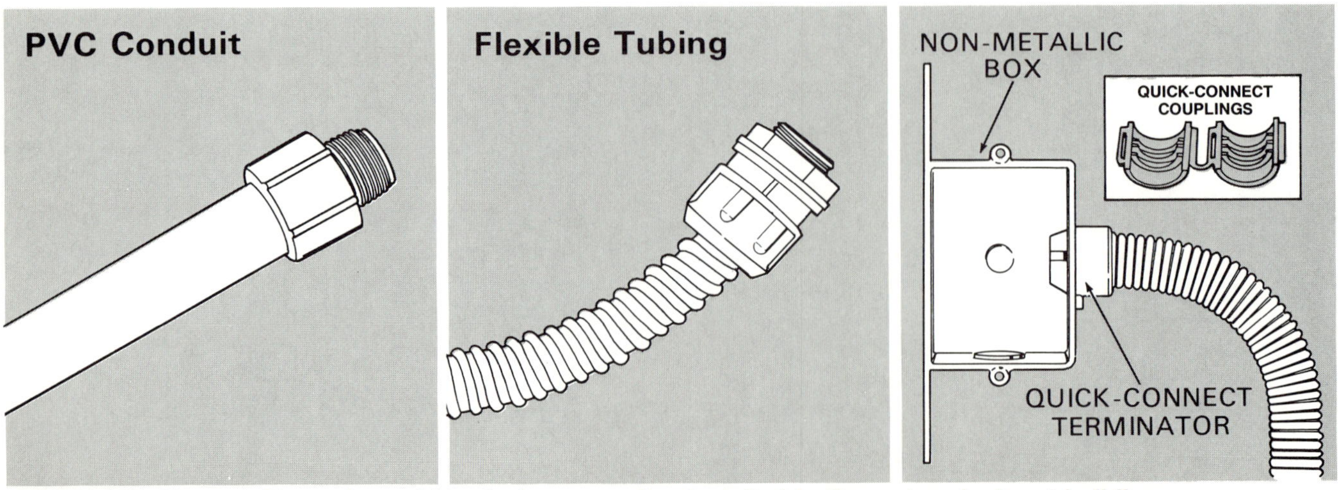

FIGURE 342. Nonmetallic surface-wiring devices are often installed in Group I agricultural buildings.

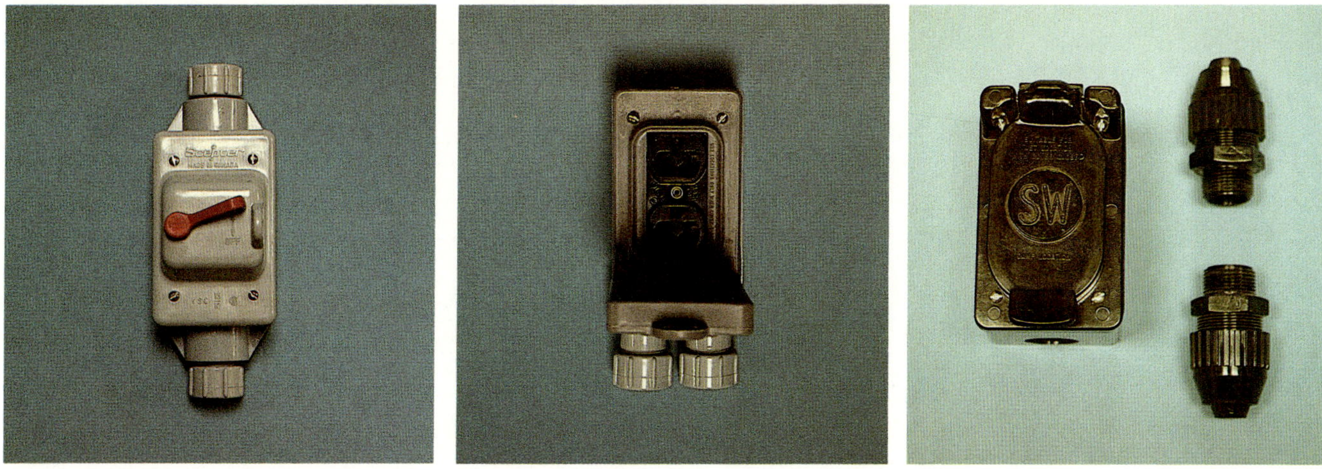

FIGURE 343. Group II agricultural buildings require devices, boxes and fittings that are both dust-tight and watertight and made of corrosion-resistant materials.

2. TYPES OF WIRING DEVICES, BOXES AND FITTINGS

The Code restrictions for agricultural buildings, Article 547, make it desirable to divide farm buildings into two types for selection of wiring devices, boxes and fittings. For purposes of this discussion, they will be divided into Group I and Group II.

Included in Group I agricultural buildings are those that do not require special materials for wiring. Examples are open barns, service centers and machinery and equipment storage buildings.

Group I buildings may be wired with the same type of materials used for residential wiring in most cases. **Number 12 AWG should be the minimum size wire used in agricultural buildings.** Surface wiring is often installed in farm buildings to save on materials and labor. You do not drill holes through framing members for surface wiring. Switches, duplex receptacles and lighting outlets may be self-contained.

Group II buildings are those that house livestock or poultry in confinement and are totally enclosed and environmentally controlled or in buildings or areas of buildings where one or more of the following conditions exist:

—Where **excessive dust** or **dust with water** may accumulate from litter dust or feed dust.

—Where **poultry and animal excrement** may cause corrosion vapors in the confinement area.

—Where **corrosive particles** may combine with water.

—Where the **area is damp and wet** by reason of periodic washing with water and sanitizing agents.

Group II buildings are mostly confined to northern areas with extremely cold winters.

Where required, insurance rates may be reduced for those who install approved wiring and materials.

Such buildings must have materials installed that meet strict Code requirements, as follows:

—Cables shall be Types UF, NMC, SNM or other cables or raceways suitable for the location.

—All wiring devices, boxes and fittings must be dust-tight and watertight and made of corrosion-resistant materials (Figure 343).

Lighting fixtures in agricultural buildings must comply with the following:

—A lighting fixture that may be exposed to moisture from condensation or building cleaning water must be watertight.

—Any lighting fixture that may be exposed to physical damage must be protected by a suitable guard (Figure 344). Also, a fixture that may be exposed to water from condensation and/or building cleaning water or solution must be watertight.

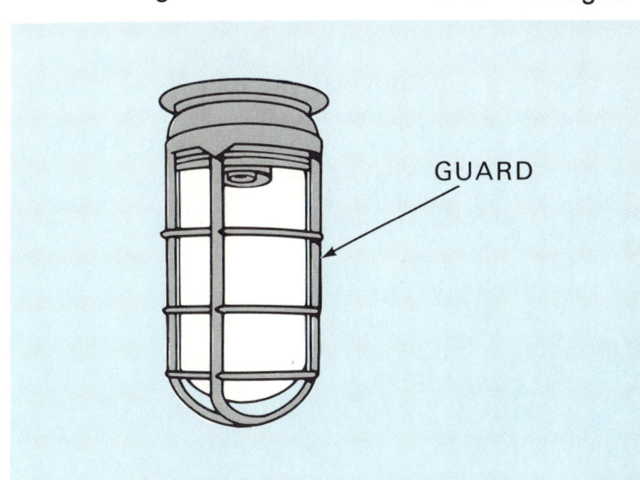

FIGURE 344. Any lighting fixtures in Group II buildings that may be exposed to physical damage must be protected by a suitable guard.

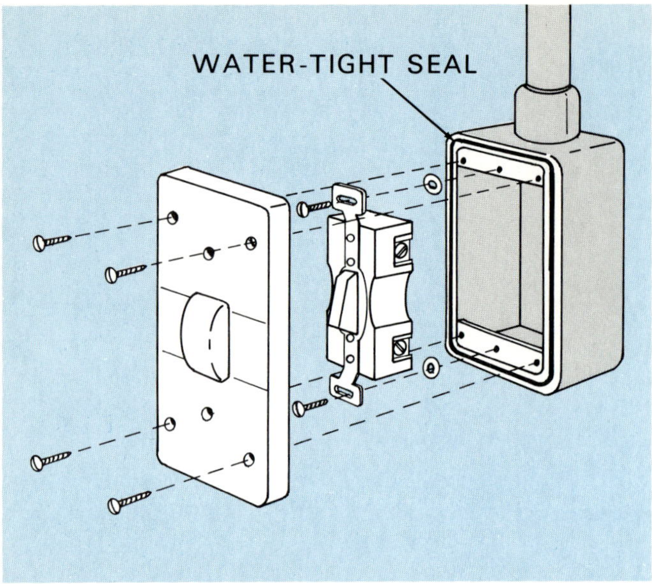

FIGURE 345. Nonmetallic boxes and corrosion-resistant, dust-tight and watertight devices are required for Group II buildings.

Switches, circuit breakers, motor controls and **fuses** must have a weatherproof, corrosion-resistant enclosure with a close-fitting cover. Nonmetallic boxes are available that are corrosion resistant and very durable. Such boxes may be fitted with dust-tight and watertight devices. Also, complete units are available that will meet the requirements (Figure 345).

Following are other items insurance companies recommend producers give consideration when building Group II livestock or poultry buildings:

—Install as much as possible of the electrical system, such as the SEP, time clocks, and switches outside the animal area in an office or workroom. If the SEP is located in the livestock area, it should be mounted in a moisture resistive enclosure, on fire-resistive material such as asbestos board.

—The SEP should be located on an interior division wall. However, if mounted on an outside wall, there should be a minimum clearance of 25.4 mm (1-inch) air space between the SEP and the wall.

—In those areas where the wiring system is subject to physical damage (such as inside a pen of livestock), the electrical wire should have physical protection from damage such as with the use of metal conduit. A separate grounding wire should be installed in the metal conduit in such areas. This would insure grounding continuity in case the conduit corrodes and separates at some point.

—If the service entrance head is located outside of the building, the service entrance conduit should be connected to either the side or bottom of the SEP.

—Electrically-heated livestock waterers or water lines heated with an electric heat tape should be grounded utilizing a two wire w/g system for protection from electrocution.

—A lightning surge arrestor should be installed on the building's incoming electrical service entrance lines or at the SEP.

—The electrical power supplier should be requested to review the electrical blueprints for the building and/or to visually inspect the electrical apparatus for an auxiliary powered generator should one be utilized.

—The electrical circuitry for the water supply system of the building should be wired ahead of the main electrical disconnect at the building. Then if the electrical system at the building is shut off, there would still be a supply of water to the structure.

3. PROCEDURES FOR WIRING BUILDINGS

Wire the agricultural buildings following the same procedures as for residential wiring except that the cable is usually mounted on the surface. The special surface-wiring devices that may be used include duplex receptacles, lampholders, single-pole switches and three- and four-way switches. They are shock-proof and corrosion-resistant. Some types provide extra terminals, allowing additional circuits to be added later as needed.

Proceed as follows to wire surface-mounted circuits and devices.

1. *Run cable on surface from SEP to all switches and outlets on circuit.*

   Leave 30 cm (12 inches) of slack for each device to be mounted (Figure 346). Terminals in surface mounted devices eliminate the need for splices and solderless connectors. Therefore, only short lengths of wire are needed for terminal connections at each outlet. Run cable in protected areas of joists, timbers and studs

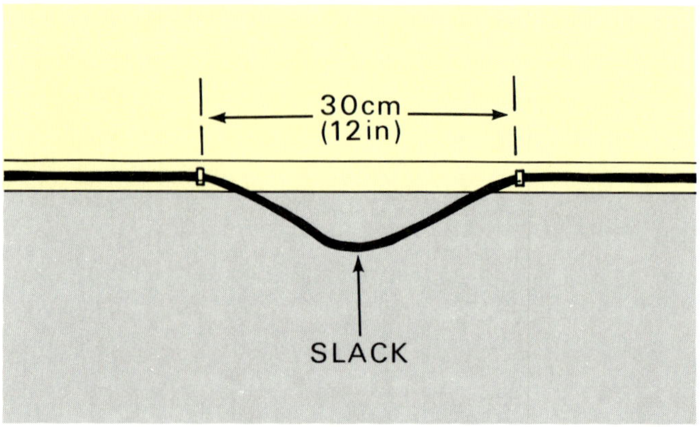

FIGURE 346. Surface-mounted devices require only short lengths of wire for terminal connections. Leave 30 cm (12 inches) of slack at each outlet.

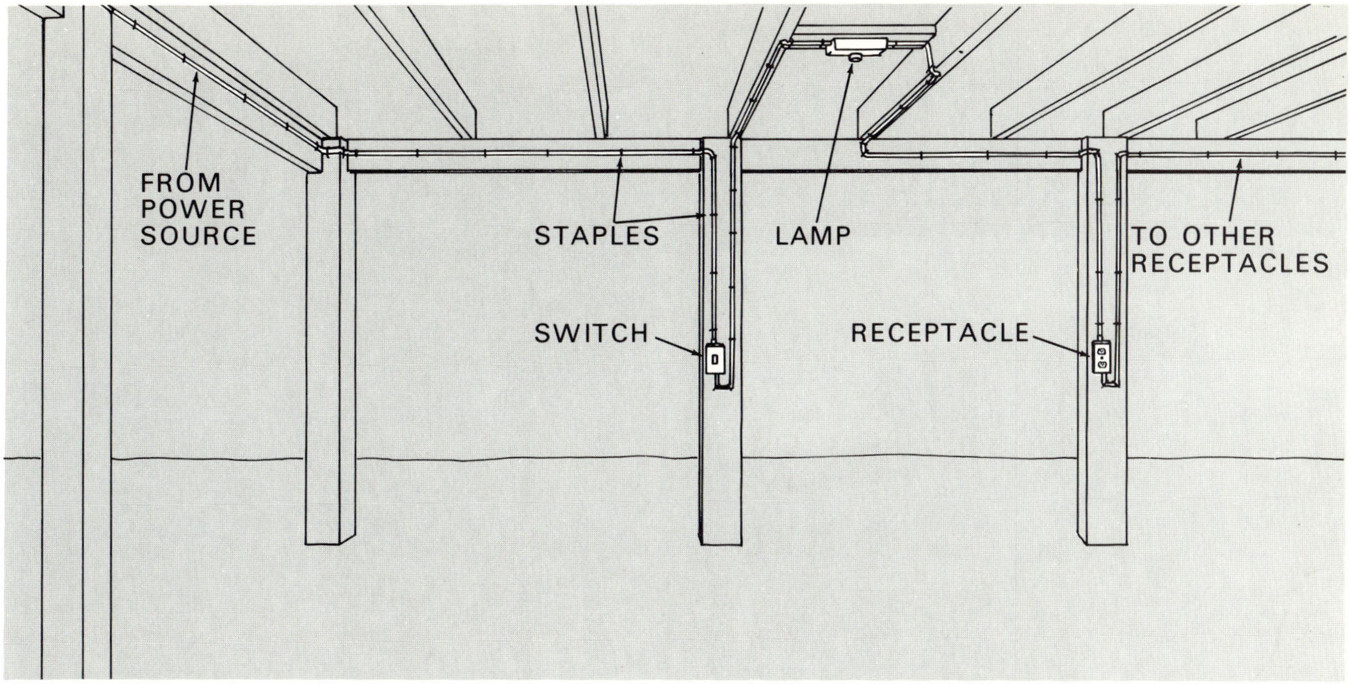

FIGURE 347. Follow closely the contour of the building and keep cable and lampholders in protected areas of joists, timbers and studs.

where possible, following closely the contour of the building (Figure 347).

2. *Mount surface-type switches, lampholders and receptacles.*

Mount overhead lampholders in recessed positions between joists for protection. A 2.5 cm × 10.2 cm (1″ × 4″) or 5.1 cm × 10.2 cm (2″ × 4″) nailed between or across top edges of joists permits lampholders to be mounted directly on the surface (Figure 347).

To mount an individual device, proceed as follows:

(1) *Remove cover.*

(2) *Mount base to surface (Figure 348).*

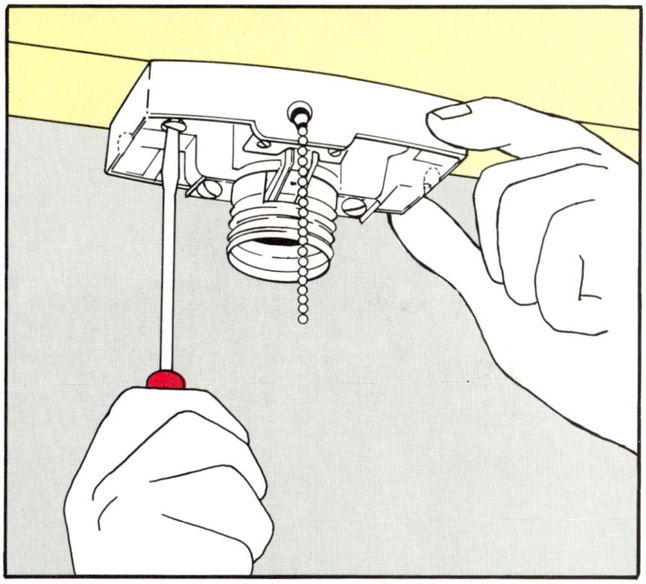

FIGURE 348. Mount the base of the surface-mounted device first.

(3) *Remove knockouts as required (Figure 349).*

(4) *Remove outer sheath of cable and connect wires to screw terminals, white to silver, black to brass (Figure 350).*

(5) *Connect grounding wire.*

(6) *Attach cover (Figure 351).*

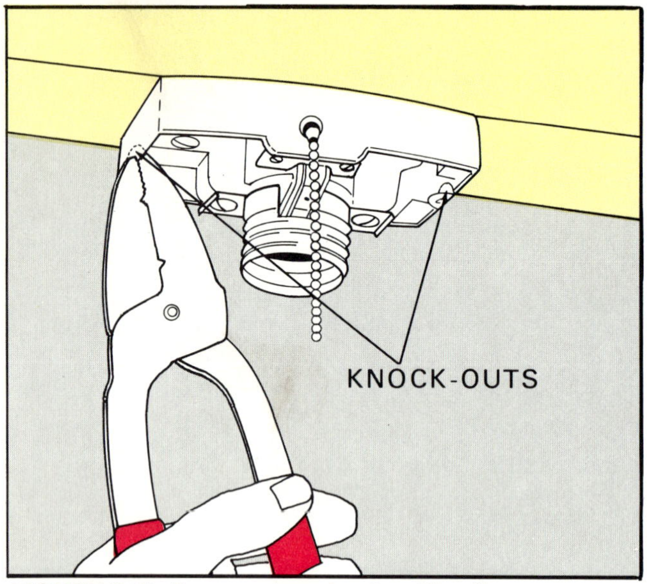

FIGURE 349. Remove knockouts with pliers.

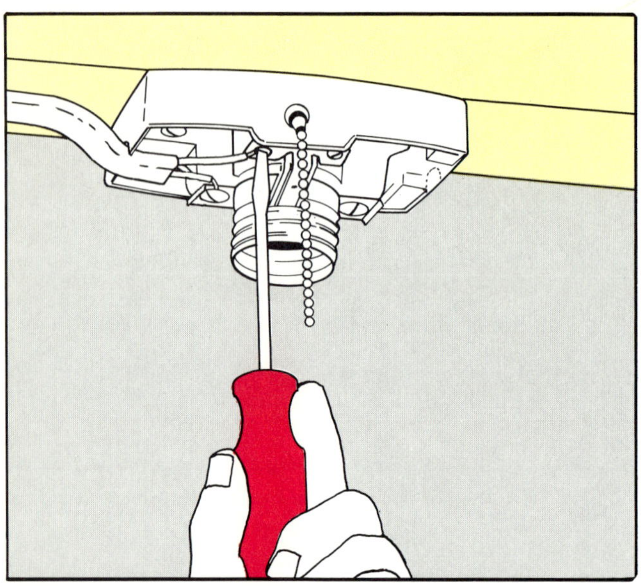

FIGURE 350. Connect wires to terminals, white to silver, black to brass and grounding wire to ground terminal.

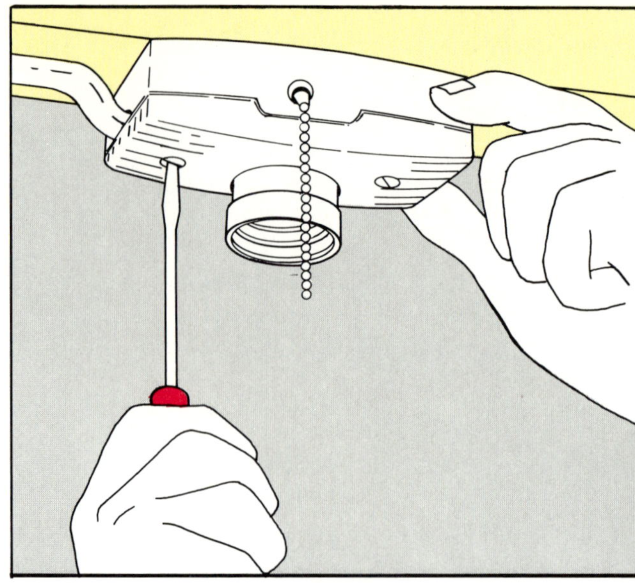

FIGURE 351. Attach cover plate.

4. STRAY VOLTAGE

Small electrical currents passing through livestock in confinement situations can cause severe production losses and poor animal health. These small electrical currents are usually caused by low alternating current (AC) voltages on the grounded neutral wires of the farm wiring system and are ofen referred to as **stray voltage**. The voltage and resulting currents may originate off the farm and be brought to the farm through the grounded neutral wiring network.

Authorities estimate that 20 percent or more of the nations dairies have stray voltage in the milking parlor exceeding ½-volt (AC). Voltages of alternating current exceeding ½-volt between feed troughs or other metal to wet concrete floors can be detected by livestock. The resulting "shock" and aggravation can lead to substantial monetary losses. One case study documented that an 80-cow dairy herd suffered a 3,000 pound reduction in milk production per cow over a period of two years, and over $15,000 was spent in an attempt to solve the problem.

Reduction of stray voltage in livestock confinement areas generally can be accomplished if a careful analysis of the problem is made. Proper measurement techniques, identification of the stray voltage sources and application of recommended techniques in construction and bonding provide the most satisfactory approach for eliminating stray voltage. To obtain more information on the symptoms, causes, measurement and reduction of stray voltage, contact your state land-grant university or your local cooperative extension service office.

E. Installing Remote-Control Wiring

Remote-control low-voltage (24-volt) systems offer some advantages for use in agricultural wiring to control lights from several locations. Also, some people prefer to install remote-control switching systems in residences.

The outlets for remote-control systems, such as lights, are wired in the usual way with 120-volt circuits. The advantage lies in the low-voltage, inexpensive cable used for the control circuits to the switches. Number 18 insulated cable may be used. If you have a central yard light, for example, that you want to control from two or three locations that are a considerable distance apart, low-voltage wiring to each switch is less expensive. And because it is low-voltage, it is safer.

The remote-control system consists of the following materials other than those usually installed (Figure 352): **one low voltage transformer 24 volts, No. 18 three conductor cable for outdoor use, a low-voltage remote-control relay at each outlet, and push-button switches.** Also available is a weather-resistant metal outdoor unit containing a built-in 24-volt transformer and control relay (Figure 353). The unit has two compartments, the upper one for low-voltage connections and the lower for 125 volts. The unit is mounted with two machine screws. A circuit wiring diagram is included

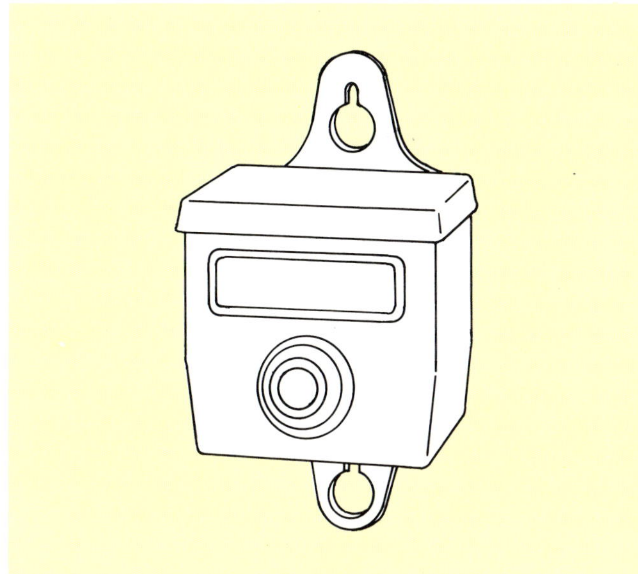

FIGURE 353. Weather-resistant outdoor unit.

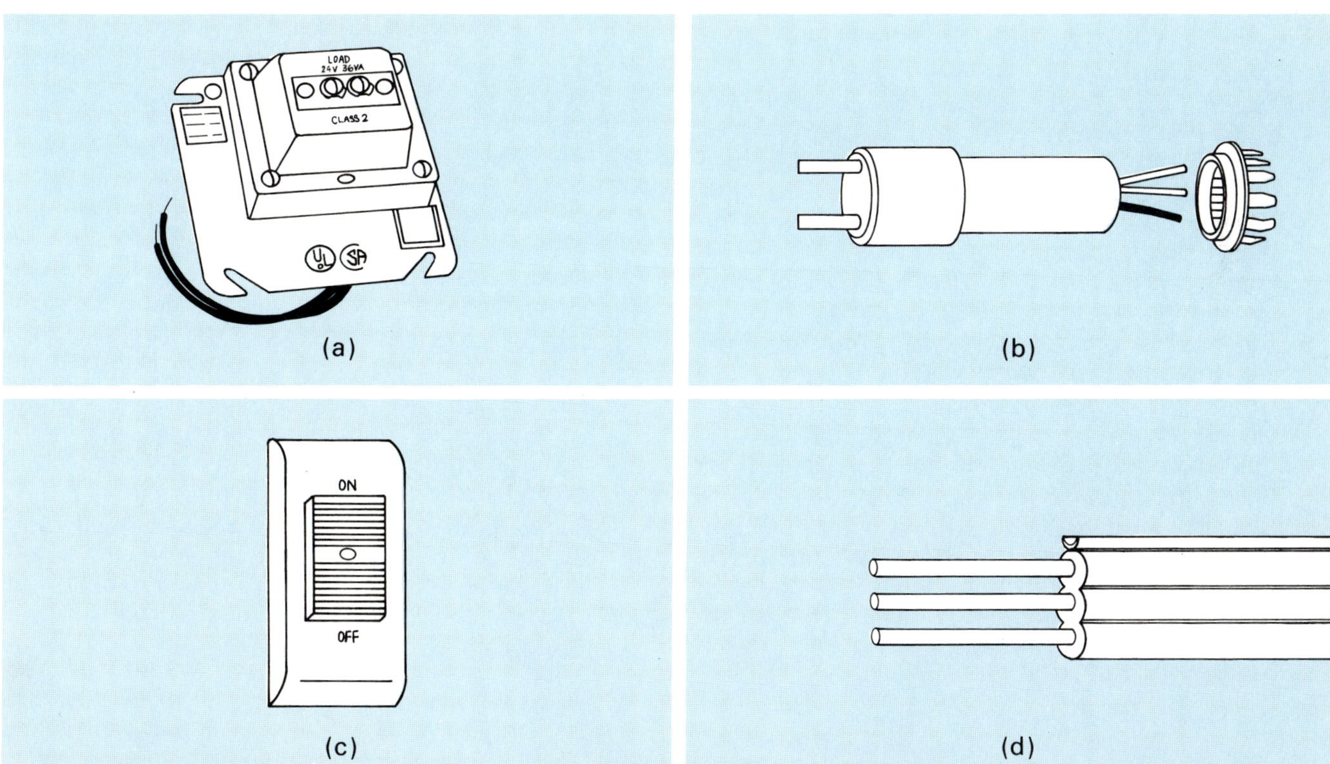

FIGURE 352. Components of the remote-control system are (a) Low-voltage transformer, (b) Low-voltage remote-control relay, (c) Low-voltage switch and (d) Three-conductor low-voltage cable.

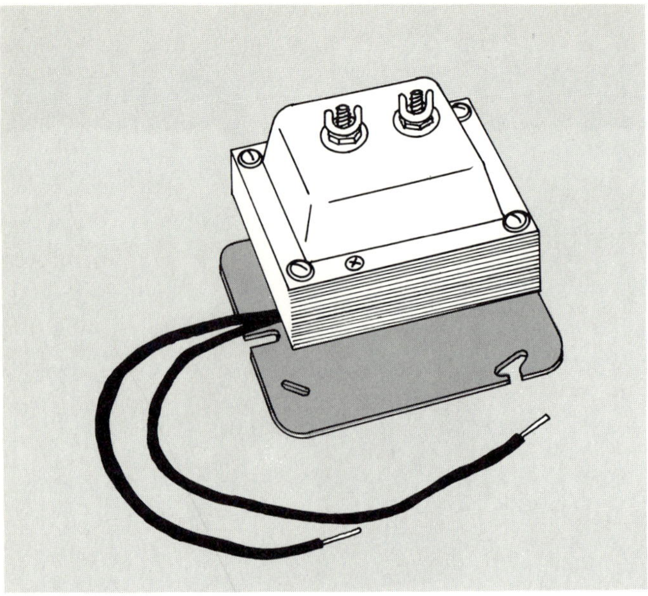

FIGURE 354. Twenty four-volt transformer.

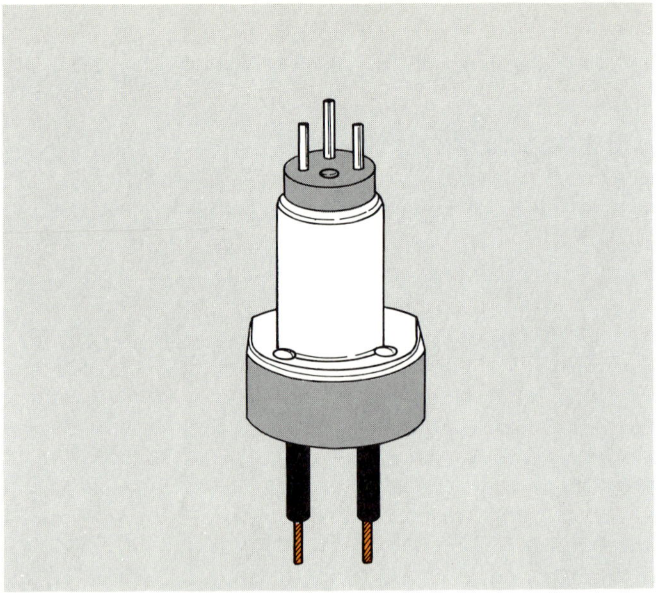

FIGURE 355. Low-voltage relay operates on 24-volts.

with each unit. Follow the manufacturer's instructions for installing the outdoor type.

Proceed as follows to install remote-control wiring:

1. *Install a 24-volt transformer (Figure 354).*

   Mount the transformer in a central location, since it will serve all the remote-control circuits. The transformer reduces voltage from 120 volts to 24 volts as required for the low-voltage system.

2. *Install a low-voltage relay at each lighting outlet box (Figure 355).*

   Mount a low-voltage relay in a knockout of each outlet box where a light is to be controlled. Connect the white wire from the source to the load (light). Connect the black wire from the source to the relay. Connect the other conductor from the relay to the load (light) inside the box. Three low-voltage wires extend outside the box for connecting to each switch.

3. *Run the low-voltage cable from the transformer to the relay and then to each switch position for that light.*

   On long spans, a messenger (support) cable may be required to support low-voltage cables from pole to buildings.

   Most of the cables used in low-voltage wiring consist of three parallel wires. Along one side of the cable is a ridge or a rib to identify the common conductor. **The common conductor (on the rib side) must be in the same position at each switch throughout the circuit** (Figure 356). Some switches have numbered terminals making it easier to connect the common conductor to the correct terminal at each switch.

   Allow about 15 cm (6 in.) of extra cable at each relay. The low-voltage cable is lightly insulated and does not require any further protection. It can be stapled on the surface or run through interior walls.

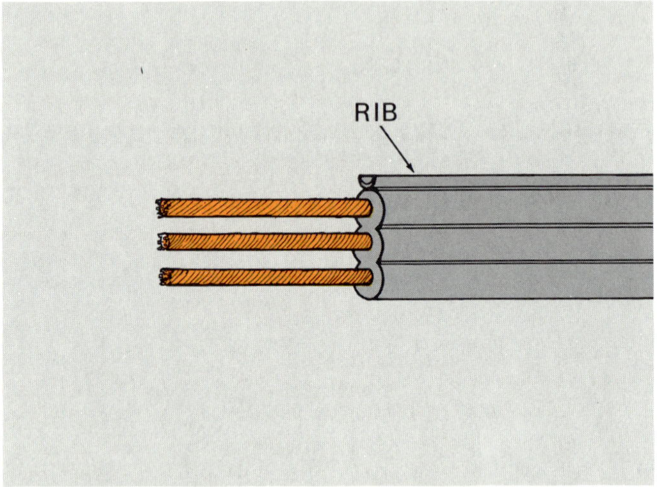

FIGURE 356. Common conductor identified by a rib on the top conductor of the cable must be connected to the same terminal position on every switch.

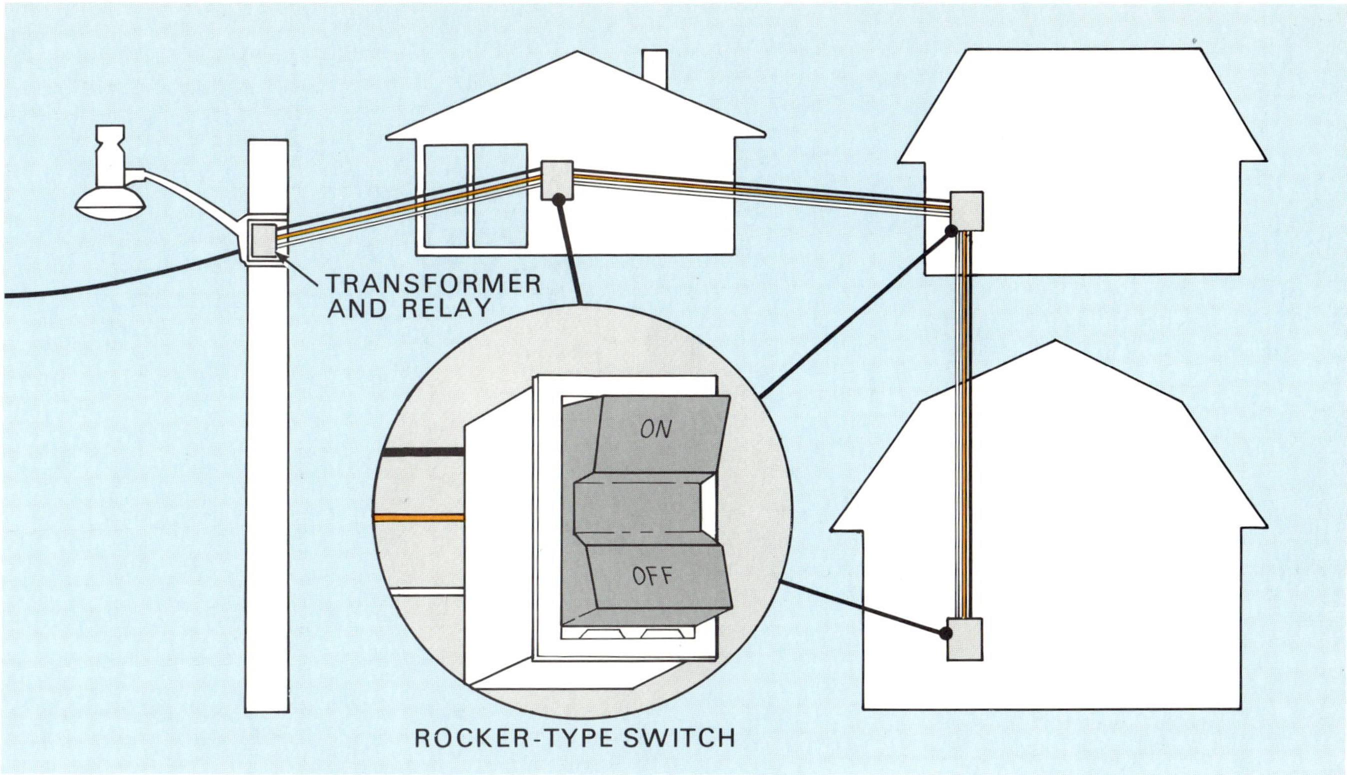

FIGURE 357 Typical yard-light installation of remote-control system.

4. *Install the switches (Figure 357).*

   You may install as many switches as desired on any circuit. One type has a rocker-type switch which turns the light on when pressed at the top and off at the bottom. Others are the momentary-contact type that push to turn on and push again to turn off. These control a ratchet-type relay.

5. *Connect the switches.*

   Connect wiring from the transformer to the relay and to each switch. Follow the manufacturer's instructions. Figure 357 illustrates a typical yard light installation. More switches could be added if needed.

F. Installing New Cable and Boxes in Old Buildings

In old buildings or homes, you may have problems extending circuits to new receptacles or fixtures. The following procedures will make the job easier. Turn off power to circuits at SEP before you start. For step-by-step procedures for connecting circuits, see pp. 78–91.

NOTE: In the following illustrations, grounding wires have been omitted for the sake of clarity. Insert grounding wires as in pages 78-91.

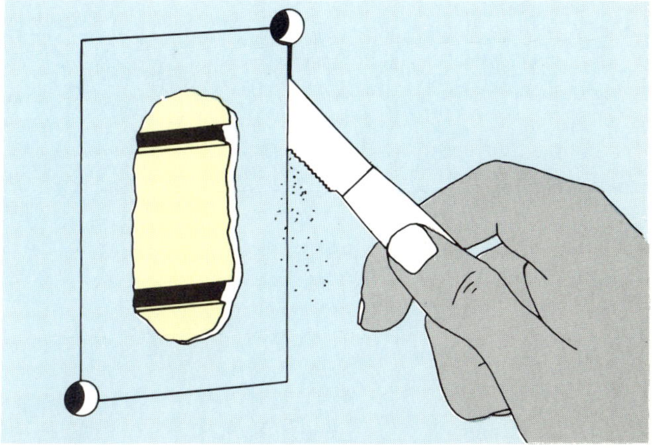

FIGURE 358. Use a template or outline of your box where the box is to be mounted. Drill holes at opposite corners and cut opening with a keyhole saw or hacksaw blade.

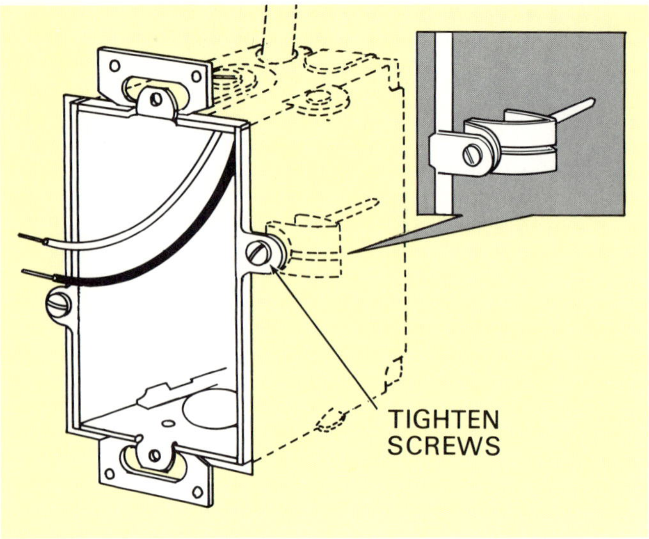

FIGURE 359. Install cable in box and fasten box in place with special expansion brackets and screws on each side.

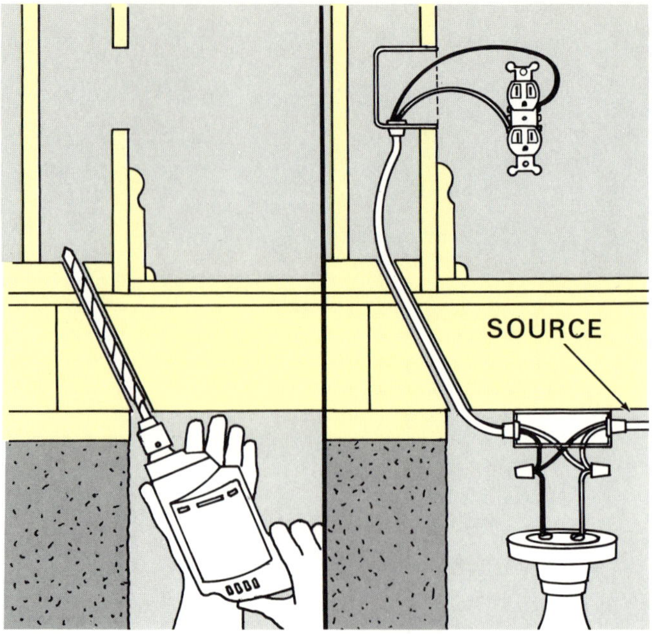

FIGURE 360. In basement, crawl space or first-floor closet in two-story home, use a long-shank bit to bore a diagonal hole through floor to new outlet opening. Use fish-tape to pull cable from closet fixture to new receptacle above.

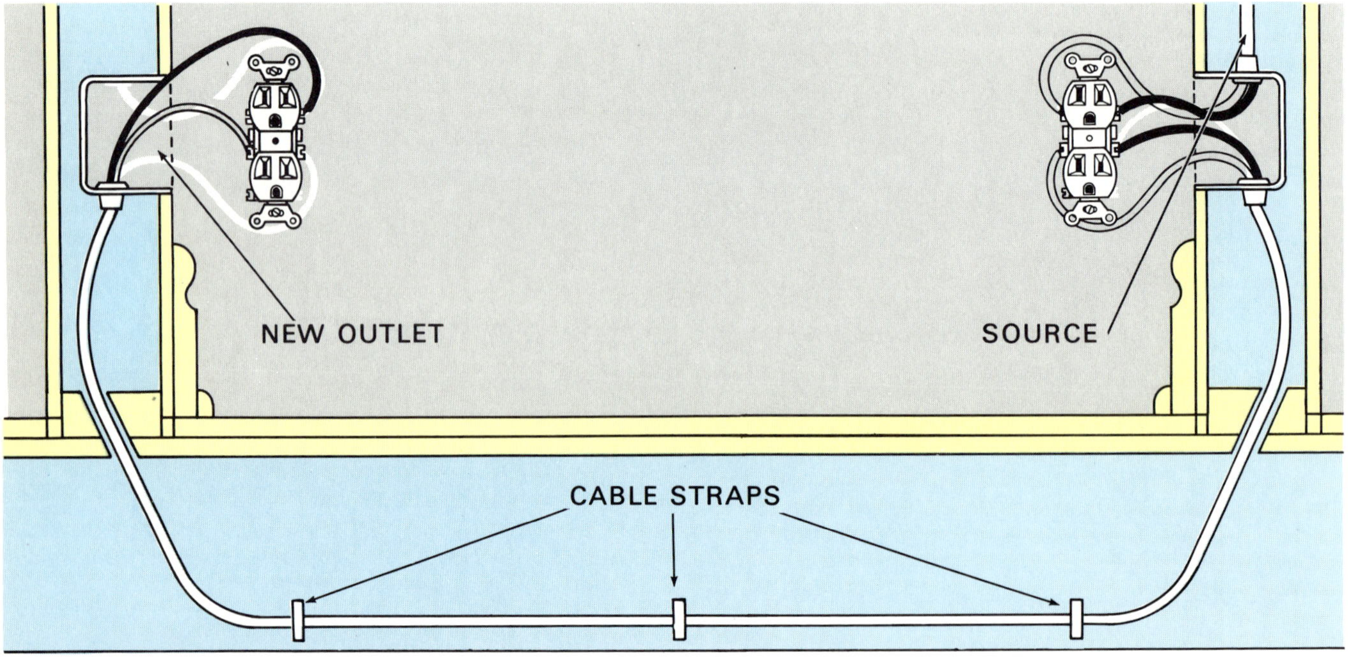

FIGURE 361. In basement or crawl space, run cable from old receptacle to new by boring two diagonal holes (or straight up on interior wall) to reach new opening. Staple cable to joist every three feet.

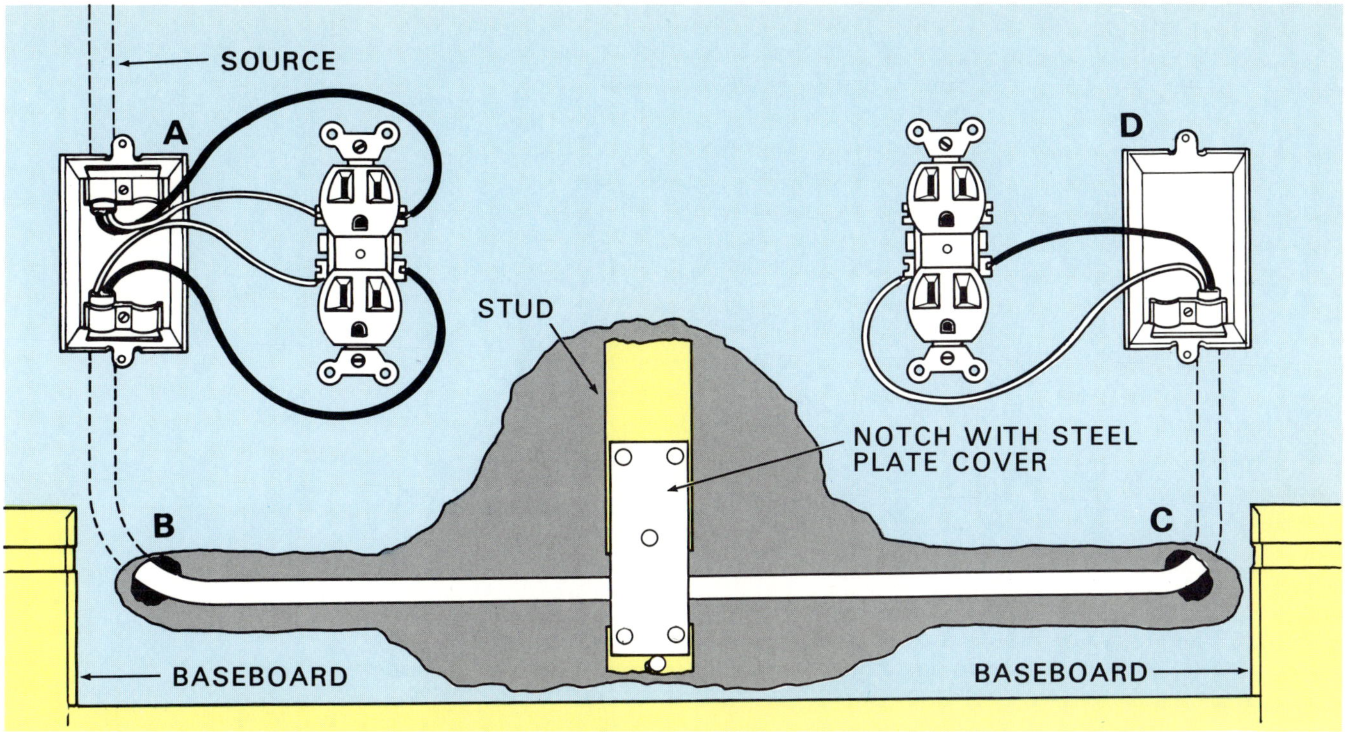

FIGURE 362. To run cable from old receptacle to new in the same room, cut opening D for new box. Remove the baseboard. Cut holes B and C and notch channel in plaster and stud. Fish cable from A to B to C and D. Install cable in new box and fasten in place. Cover notch in stud with steel plate and replace baseboard.

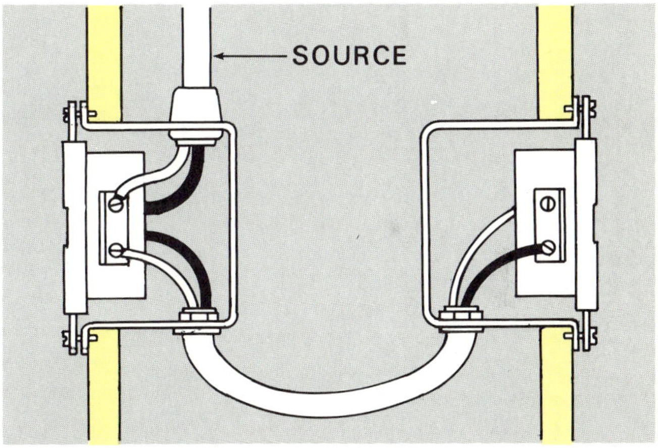

FIGURE 363. For back-to-back outlets, cut hole for new box opposite the old one in the next room, but offset slightly to allow room for the box. Run cable from old receptacle to new box. Clamp cable to new box and install the box in the opening.

NOTE: If you are replacing a receptacle in a box that does not have a means of grounding, you must install either a nongrounding type or a ground fault circuit interrupter. However the GFCI must not be connected to any other receptacles downstream.

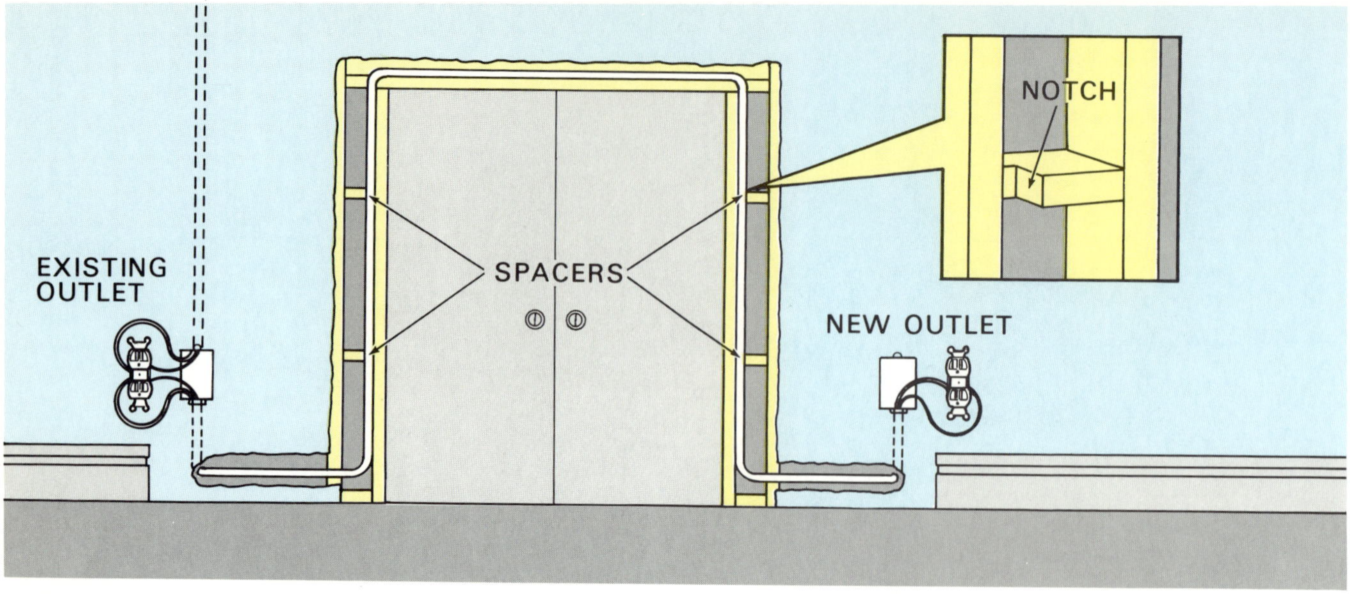

FIGURE 364. To run cable around a door frame, remove baseboards and door trim. Notch the wall plaster and the wood spacers between the frame and jamb. Run cable from old receptacle around door frame and to new box. Clamp cable to new box, install the box in the opening and fasten in place. Cover notches with steel plate.

Acknowledgments

Recognition is given to the following for photographs and information.

INDIVIDUALS

Anderson, Paul M., Agricultural Engineering Department, Pennsylvania State University

Barr, Robert C., National Fire Protection Association, Boston, MA

Bear, W. Forrest, Agricultural Engineering Department, University of Minnesota

Bennett, Wayland, Agricultural Engineering Department, Texas Tech University

Blanton, Lloyd H., Agricultural Education Department, Clemson University

Braker, Clifton R., Agricultural Engineering Department, University of Arkansas

Brown, Robert H., Chairman, Agricultural Engineering Division, University of Georgia

Bruwelheide, Ken, Agricultural and Industral Education, Montana State University

Bryan, C. E., Reynolds Aluminum, Richmond, VA

Bullard, Ben B., Texas Farm Bureau, Waco, TX

Carlson, Axel R., Extension Engineer, University of Alaska

Carter, James K., Jr., Agricultural Mechanics Department, Reedley College

Clouse, James P., Agricultural Education Department, VPI & SU

Collins, Edmond B., Cooperative Extension Service, West Virginia University

Cranford, William, Central Power and Light Co., Corpus Christi, TX

Crawford, Fred M., Extension Agricultural Engineer, University of Missouri

Crawford, Ronald, Washington State Department of Education

Cross, Ward H., Texas Farm Bureau Insurance Companies, Waco, TX

Daum, Donald R., Agricultural Engineering Extension, Pennsylvania State University

Davies, Roger, Raco, Inc., South Bend, IN

Davis, John R., Agricultural Engineering Department, Oregon State University

Deming, D. R., Raco, Inc., South Bend, IN

Doering, Floyd J., Wisconsin State Department of Education

Downey, Gary, Grinell Mutual Reinsurance Co., Grinell, IA

Durham, Dwight H., Southwire Co., Carrollton, GA

Durkee, James, Department of Vocational Education, University of Wyoming

Espenschied, Roland F., Agricultural Engineering Department, University of Illinois

Farmer, Gary, Georgia State Department of Education

Faulkner, T. L., Alabama State Department of Education

Fiester, Fred, Claverack R. E. C., Towanda, PA

Froehlich, Nord, Square D Company, Lexington, KY

Gerrish, John, Agricultural Engineering Department, Michigan State University

Giguere, John F., Sault Ste. Marie, Ontario

Gilbertson, Ozzie, Agricultural Education Department, University of Nebraska

Gilden, Robert O., Agricultural Engineer, ES-USDA, Washington, DC

Gilman, Francis E., Extension Agricultural Engineer, University of New Hampshire

Ginn, Olin W., Georgia Power Company, Atlanta, GA

Grant, Lee, Agricultural Engineering Department, University of Maryland

Griffith, Lowell, Illinois Power Company, Decatur, IL

Hansen, Herbert, Agricultural Education Department, Oregon State University

Hansen, Hugh J., Western Regional Agricultural Engineering Service, Oregon State University

Harvey, L. M., Electrical/Electronic Department, The Sault College of Applied Arts and Technology

Hayles, Jasper A., Agricultural Education Department, Arkansas State University

Hinkle, Charles N., Agricultural Engineering Department, Purdue University

Hoerner, Thomas A., Agricultural Engineering Department, Iowa State University

Hogsett, Ordie L., Extension Safety Specialist, University of Illinois

Holler, Ivan, Wayne-White Cooperative, Fairfield, IL

Huff, Edward R., Agricultural Engineering Department, University of Maine

Jacobs, Clinton O., Agricultural Education Department, University of Arizona

Killough, Frank B., Alabama State Department of Education

Kimmons, Marion L., Agricultural Mechanics Department, West Virginia University

Kindschy, Dwight, University of Idaho

Kingsley, Curtis, Georgia State Department of Education

Kruesi, William R., General Electric Company, Fairfield, CT

Land, Bill, Georgia Power Company, Atlanta, GA

Law, S. Edward, Agricultural Engineering Department, University of Georgia

Lawrence, John A., Agricultural Education Department, University of Idaho

Lively, Robert M., Gifford-Hill & Company, Inc., Lubbock, TX

Matt, Stephen R., Industrial Arts Department, University of Georgia

McBride, Robert, Vo-Ag Teacher, Kenton, OH

McClure, W. D., Agricultural Engineering Department, Texas A&M University

McFate, Kenneth L., Food and Energy Council, Inc., Columbia, MO

McLendon, Derrell, Agricultural Engineering Department, University of Georgia

Meeks, Leon, Electrical Construction, North Georgia Technical School

Miller, Bruce W., Central Power and Light Company, Corpus Christi, TX

Mitchell, Kenneth K., Tennessee State Department of Education

Mitchell, Martin L., New Hampshire State Department of Education

Moore, Carlos, Arizona State Department of Education

Moore, James, Electrical Inspector, Decatur, IL

Moore, Milo J., Extension Engineer Specialist, University of Vermont

Moss, Carter, Georgia Power Company, Atlanta, GA

Mumford, Michael, Agricultural Engineering Department, University of Nebraska

Munson, Maynard, Iowa Power Company, Des Moines, IA

Nelson, Robert, Mohawk College of Applied Arts and Technology, Hamilton, Ontario

O'Brien, Michael, Agricultural Engineering Department, University of California-Davis

Page, Foy, Vocational Instructional Services, Texas A&M University

Parrish, Bob, Residential Wiring, DeKalb Occupational Education Center

Patton, Bob, Oklahoma State Department of Vocational Technical Education

Pearson, A. J., Asst. Dir., National Joint Apprenticeship and Training Committee, Lanham, MD

Phillips, Andrew, Director, National Joint Apprenticeship and Training Committee, Lanham, MD

Plummer, Robert K., Maine State Department of Education and Cultural Affairs

Priebe, Don, Agricultural Education Department, North Dakota State University

Reed, Joe, Agricultural Engineering Department, University of Georgia

Rice, Charles, Agricultural Engineering Department, University of Georgia

Ridenour, Harlan E., Ohio Agricultural Education Curriculum Materials Service

Searls, Dean, Adams Electric Cooperative, Camp Point, IL

Sheets, J. T., Electrical Inspector, Decatur, GA

Slyter, Damon E., Kansas State Department of Education

Smith, Kenneth L., A. O. Smith Harvestore Products, Inc., Eureka, IL

Smith, Norman, Agricultural Engineering Department, University of Maine

Stetson, LaVerne E., Agriculture Engineering Farm Electric Research, University of Nebraska

Surbrook, Truman, Agricultural Engineering Department, Michigan State University

Tart, C. V., North Carolina State Department of Public Instruction

Todd, John D., Agricultural Education Department, University of Tennessee

Trollope, D. A., Apprentice Department, Mohawk College

Warnock, William K., Agricultural Engineering Department, University of Arkansas

Watts, Thomas, Vermont State Department of Education

West, William, West Virginia State Department of Education

Wills, William D., Agricultural Engineering Department, California Polytechnic State University

Wilson, Donald E., California State Department of Education

Winsett, Ivan, Ronk Industries, Nokomis, IL

Wood, Jay, Washington State Department of Education

INDUSTRY

A.O. Smith Harvestore Products, Inc., Eureka, IL

Black and Decker, Towson, MD

Blackhawk Industries, Dubuque, IA

Bush Grip

Bussman Manufacturing, Earth City, MO

Carlson Electrical Services, Inc., Cleveland, OH

Channellock, Inc., Meadville, PA

Crouse-Hinds, Syracuse, NY

Farm Electric, Inc., Athens, GA

Georgia Power Co., Athens, GA

General Electric Company, Fairfield, CT

Gifford-Hill & Company, Inc., Lubbock, TX

Gould ITE, Springhouse, PA

Honeywell, Minneapolis, MN

Hubbell, Harvey, Inc., Christianburg, VA

Ideal Industries, Inc., Sycamore, IL

I.T.T. Holub, Sycamore, IL

Klien Tolls, Inc., Chicago, IL

Larkin, G.E. Company, Hickory, NC

Levitron, Little Neck, NY

National Fire Protection Association, Boston, MA

Raco, South Bend, IN

Reynolds Aluminum, Richmond, VA

Slater Electric, Inc., Glen Cave, NY

Southwire Company, Carrollton, GA

Square D Company, Lexington, KY

Texas Farm Bureau Insurance Company, Waco, TX

Thomas and Betts, Inc.

3M Company, St. Paul, MN

Wiremold Company, West Hartford, CT

Part 2

Maintaining the Lighting and Wiring System

Preface

The information in this book is something everyone should know. It deals with the maintenance and repair of the common lighting and wiring problems found in every home, shop and on the farm.

The skills to be learned from the step-by-step procedures given here will prepare you to do these jobs. They are also necessary if you expect to engage in a course of learning pursuant to becoming an electric wiring and maintenance technician.

The general objectives of this manual are to help you to become proficient in the following jobs:

1. Replacing attachment plugs.
2. Replacing lamp sockets.
3. Replacing heater plugs.
4. Replacing service cords.
5. Replacing fuses, and resetting or changing circuit breakers.
6. Replacing single-pole switches
7. Replacing 3-way switches.
8. Replacing 4-way switches.
9. Replacing convenience, dryer, and range outlets.
10. Replacing incandescent lighting fixtures.
11. Replacing and maintaining fluorescent lighting fixtures.
12. Replacing ground fault circuit interrupters.

Specific learning outcomes will result in you being able to perform as follows:

1. Select the proper tools for doing each job.
2. Take the proper safety precautions when working with electrical wiring.
3. Apply the recommendations given by the National Electric Code.
4. Recognize faulty electrical wiring, lighting fixtures and switches.
5. Remove worn or broken parts.
6. Replace worn or broken parts.

Introduction

Some people are inclined to put off doing anything about replacing or repairing service cords, or seeing that proper size fuses are used, or that broken switches are replaced. Such neglect can be costly.

If your service cords or service outlets are not kept in good condition, they can cause serious inconvenience. Worse yet, your property or even your life, may be in danger with certain types of neglect such as shown in Figures 1 and 2.

In a study made by the National Fire Protection Association, 163,800 fires were reported in one year in buildings that started from some electrical cause. In another study, the International Association of Electrical Inspectors found the causes of fires in 8,132 homes were as follows:

Service cords	7.8%
Christmas tree wiring	0.6%
Heating appliances and lamps	9.4%
Motor-winding burnouts	21.4%
Transformer burnouts	3.3%
Radios	0.8%
Televisions	4.2%
Miscellaneous wires & devices	40.1%
Electrical origin (no details)	12.4%

Figure 1. This service cord should have been replaced when it started to show wear. Probably the fuse that protects the circuit is too large.

The maintenance jobs discussed here are ones you can soon learn to do for yourself. Little time is required to do them and very little skill is needed. Once you have learned, you save money by not having to call an electrician for each repair job. You are more likely to fix worn or damaged wiring units before they become a hazard or a serious inconvenience.

The repair and maintenance jobs described here are the more common ones found in the home, on a farm, or around a business.

They are grouped under the following headings:

VIII. Repairing Service Cords
IX. Maintaining the Permanent Wiring System

Figure 2. A damp or wet floor plus a defective metal socket can be a serious hazard. This metal socket should be replaced with a plastic type that provides electrical insulation.

VIII Repairing Service Cords

"Service cord" refers to all types of flexible cords that connect an appliance to a convenience outlet. They are more likely to cause trouble than any other part of the wiring system. Some service cords have one end permanently connected to the appliance. Others can be disconnected from the appliance as well as from the convenience outlet. The latter type may be equipped with a lamp socket, heater plug or plug-in block for attaching other service cords.

After you have used service cords for a year or more, you can expect some of the less durable ones to become worn, frayed or damaged. These ought to be repaired. You can repair most of them yourself. This will save the inconvenience and the danger of using worn-out cords in your house. Do not try to "get by" with worn cords or defective plugs.

In Section VIII you will learn the following service-cord maintenance jobs:

A. Replacing Attachment Plugs.

B. Replacing Lamp Sockets.

C. Replacing Heater Plugs.

D. Replacing Service Cords.

A. Replacing Attachment Plugs

The attachment plug may also be called an "attachment-plug cap," a "plug cap," or "plug." In this publication it is referred to as an "attachment plug" and as a "plug." Upon completion of this section, you will be able to **replace attachment plugs** on various types of service cords. Replacing attachment plugs is discussed under the following headings:

1. Tools and Materials Needed.
2. Disconnecting the Cord.
3. Removing Attachment Plug from Cord.
4. Preparing Cord for Standard or Core Plugs.
5. Tying Holding (Underwriters') Knot.
6. Connecting Cord to Attachment Plug.

TOOLS AND MATERIALS NEEDED

Tools:

Small screwdriver
Pocket knife
Small side-cutting pliers
Scissors

Materials:

New attachment plug

There is a wide variety of attachment plugs, some of which are shown in Figure 3. In selecting one for your needs, keep in mind how yours will be used.

Most table and floor lamps and other devices used in the home are equipped with plugs that have only two prongs. Most power tools and equipment have a third prong for grounding. Keep this in mind when repairing cords and replacing plugs.

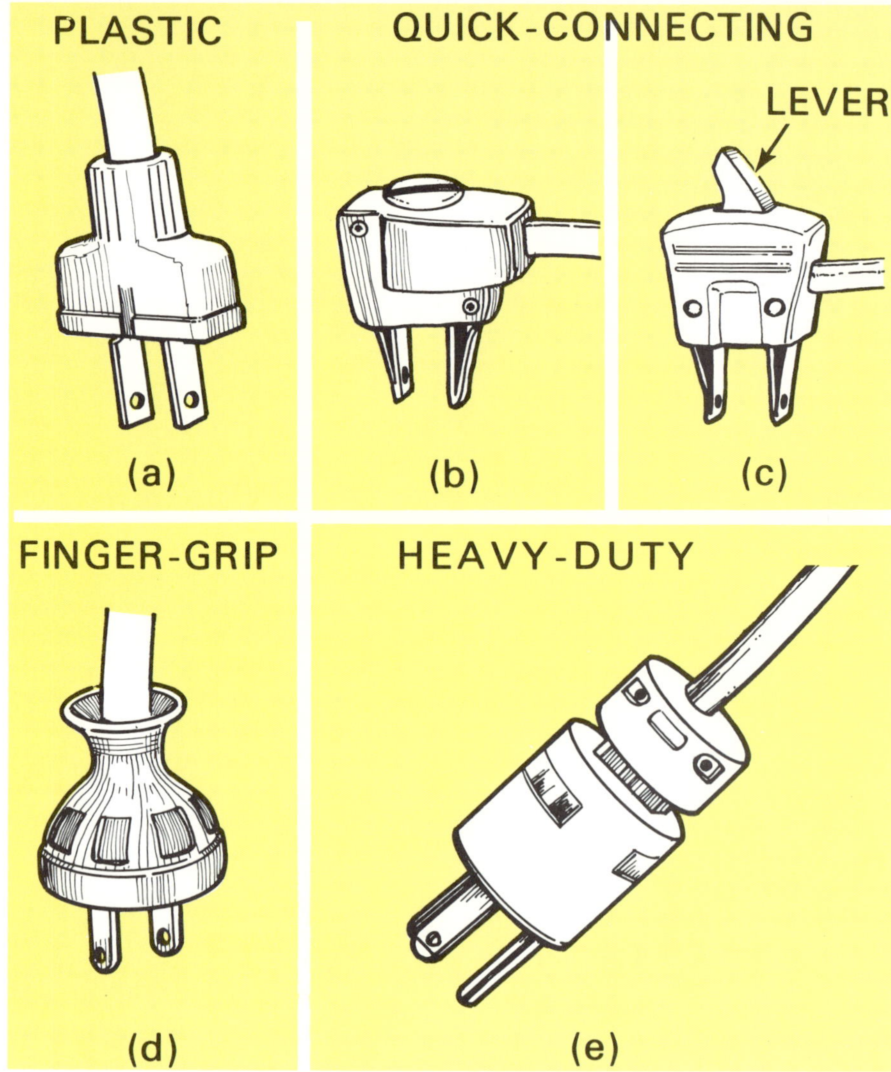

Figure 3. Kinds of attachment plugs. (a) Plastic — inexpensive, difficult to remove from convenience outlet, easily broken or crushed. (b) and (c) Quick-connecting plugs — plastic, fairly durable, easily attached to cord, satisfactory for light electrical loads, used on parallel cords only (Figure 9d). (d) Rubber or plastic finger-grip type — for removal from convenience outlet. Rubber type very durable. (e) Heavy duty plugs should be the grounding type if used in damp basement areas or for outside connections. New Code requirements call for 15 and 20 ampere attachment plugs to be of "dead-front construction." This means that the cord entrance at the back of the plug and the front of the plug where the prongs extend must both be sealed so no wiring can be reached without taking the plug apart.

DISCONNECTING THE CORD

To disconnect the cord, proceed as follows:

1. *Grasp attachment plug between your thumb and finger to remove it from receptacle (Figure 4).*

   Repeated jerking on the cord to remove the attachment plug (Figure 5) will damage it. The wires and insulation stretch and break, or the wires break loose in the attachment plug.

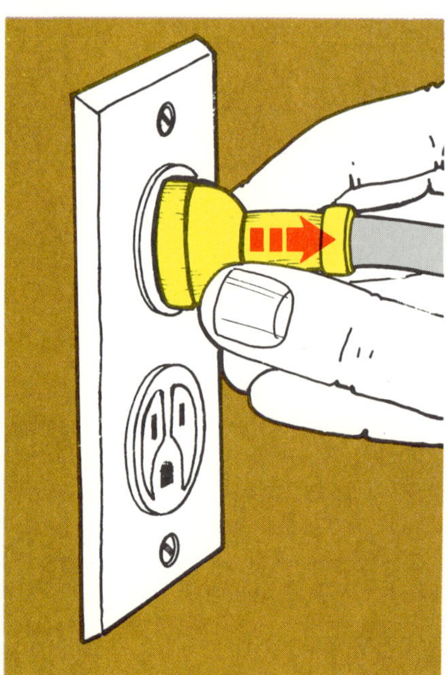

5304

Figure 4. The correct way to remove plug from receptacle.

Figure 5. Do not remove plug by pulling on cord.

5305

REMOVING ATTACHMENT PLUG FROM CORD

Procedures for removing the attachment plug from the cord are as follows:

1. *Cut off plug and frayed end of cord (Figure 6).*

   Use knife, scissors or cutting pliers.

2. *If plug is broken (Figure 7a), or is of the molded type (Figure 7b), throw it away.*

   The type plug shown in Figure 7b has the prongs and electrical connections molded into the attachment plug.

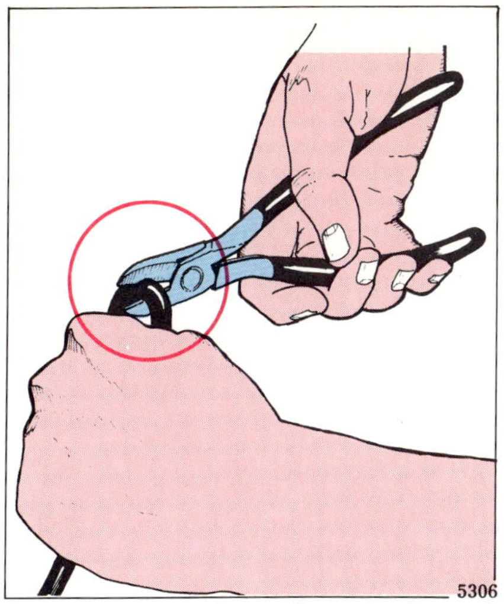

Figure 6. Cut frayed end of cord with pliers, knife or scissors.

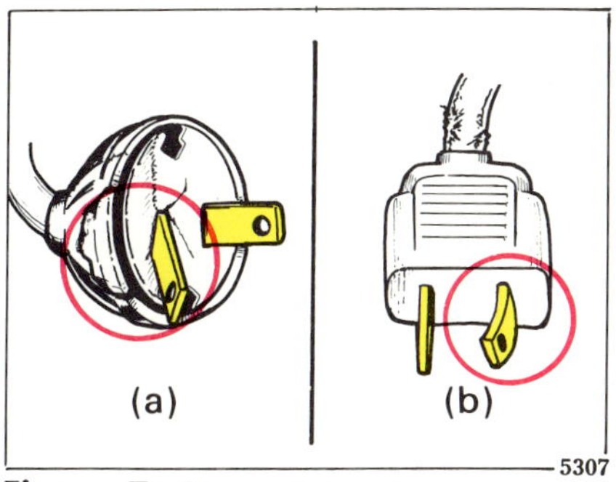

Figure 7. Do not try to re-use (a) broken plugs or (b) molded plugs—ones with wires molded into the plug body.

There is no way to reach the wire connections without destroying the plug.

3. *If attachment plug is still in good condition, and is of the replaceable type, remove cord wires and throw them away.*

There are three general types of attachment plugs: (a) standard, (b) dead-front, and (c) quick-connecting. The way to remove the wires will depend on the kind of plug you have. Figure 8 shows the different types of plugs and procedures to use in removing the wires. If you are replacing a plug with a grounding connection (Figure 3), you will have 3 wires.

a. If it is a **standard plug** (Figure 8a), proceed as follows:

(1) *Remove insulating disk.*

(2) *Loosen both screws. Do not remove them.*

If it is a *grounding plug,* it has 3 screws to loosen.

(3) *Remove wires.*

b. If it is a dead-front plug, proceed as follows for one type (Figure 8b):

(1) Unscrew threaded cap and slide back.

(2) Press side-release buttons and pull dead-front forward.

(3) Turn terminal screws to release wires.

c. If it is a **quick-connecting plug** (Figure 8c), proceed as follows:

(1) *Press prongs together.*

Some quick-connecting plugs have a lever on the end (Figure 3c). Pulling the lever releases the cord.

(2) *Pull core from plug body.*

(3) *Spread prongs to release cord.*

PREPARING CORD FOR STANDARD OR CORE PLUGS

If you are using a **quick-connecting, attachment plug (Figure 8c) and parallel cord,** leave out this series of steps. Proceed to the section, "Connecting Cord to Attachment Plug," page 194.

If you are using a **standard, dead-front** or **core plug,** proceed as follows:

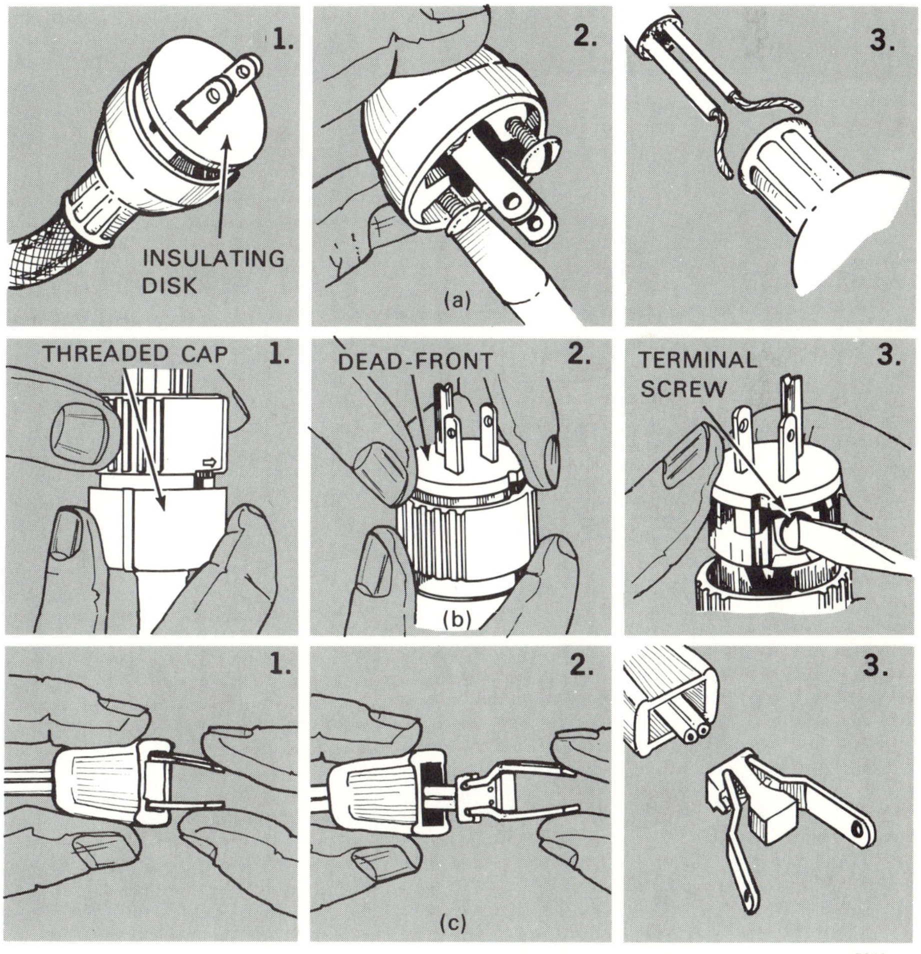

Figure 8. Common types of plugs: (a) standard, (b) dead-front and (c) quick-connecting.

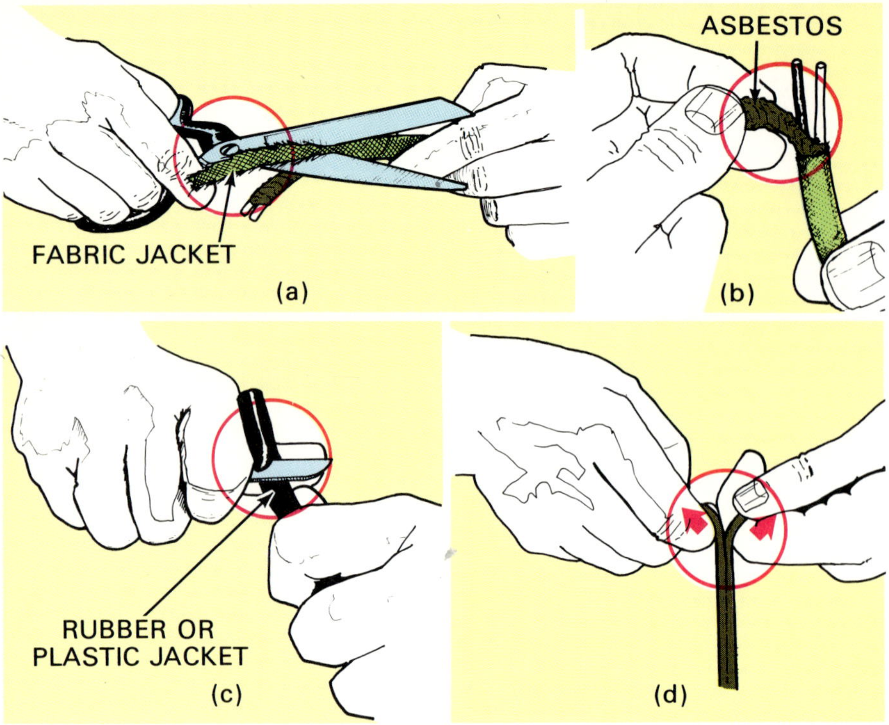

Figure 9. (a) Heater cord with fabric jacket. (b) Heater-cord wires are wrapped with heat-insulating material. (c) Hard-service cord has a tough rubber or plastic jacket. (d) Parallel cord has no jacket.

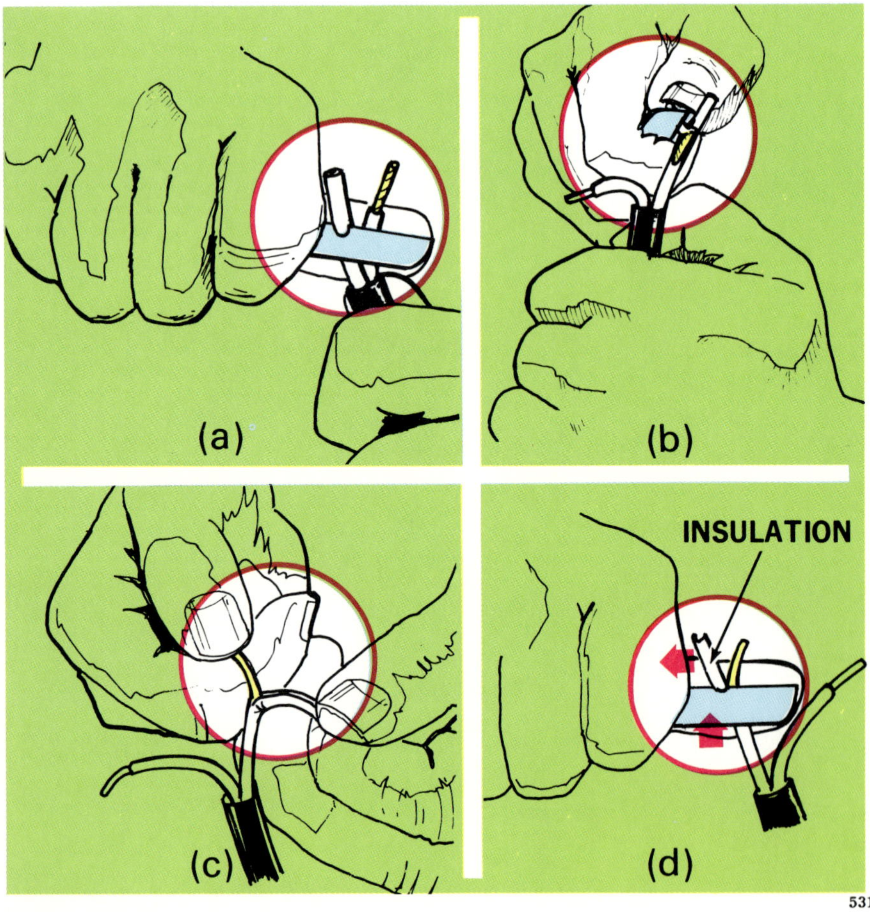

Figure 10. Steps for removing insulation from wires.

1. *Remove about 2 inches of (fabric or rubber) outer jacket.*

   The **jacket** is the outer covering that binds the insulated wires together. **Parallel cord** (Figure 9d) has no separate jacket; the insulated wires are pulled apart for a distance of about 2 inches. **Heater cords** have a layer of asbestos under the jacket (Figure 9b). Remove all of exposed asbestos.

2. *With knife remove ½ inch of insulation from each wire (Figure 10).*

   To avoid cutting the wire strands: (1) slope knife until it is almost parallel with wire, (2) cut through rubber insulation to form lip (Figure 10a), (3) pull lip toward cut end of wire (Figure 10b), (4) lay back remaining insulation (Figure 10c) and (5) cut off (Figure 10d).

3. *Unwrap fine thread from both wires and cut it off (Figure 11).*

4. *Scrape each wire clean with*

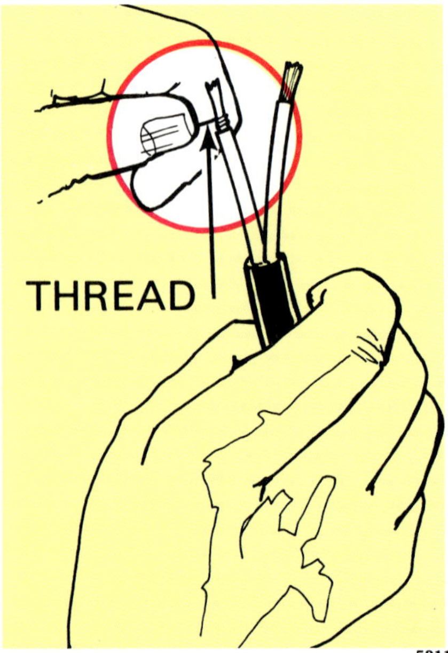

Figure 11. Unwrap thread from around both wires and remove by cutting thread.

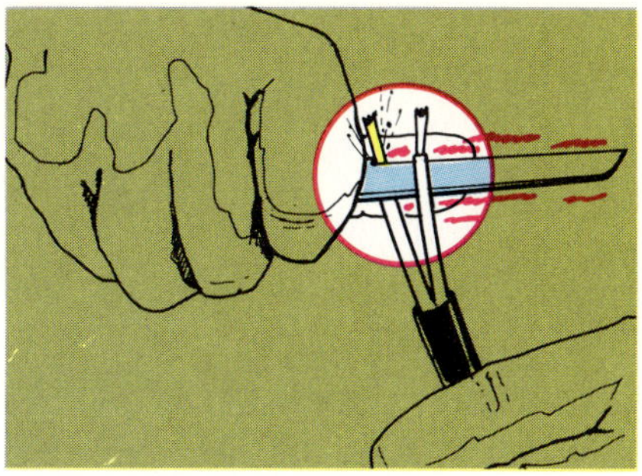

Figure 12. Scraping wires to improve electrical connection.

dull part of knife blade (Figure 12).

Scrape wires until they are bright. This makes a good electrical connection.

5. *Twist strands of each wire together tightly (Figure 13).*

Figure 13. Twisting strands of each wire.

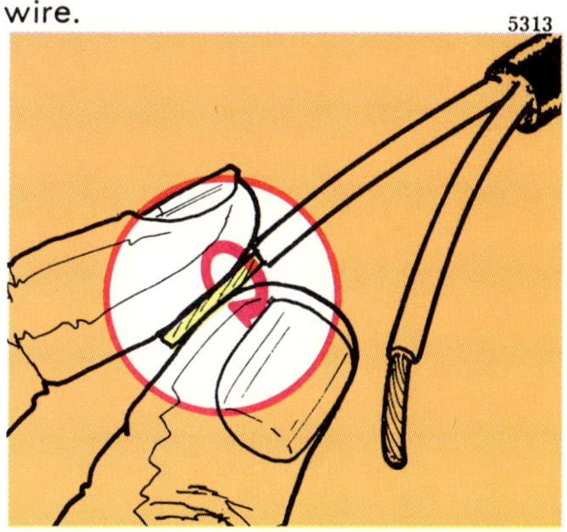

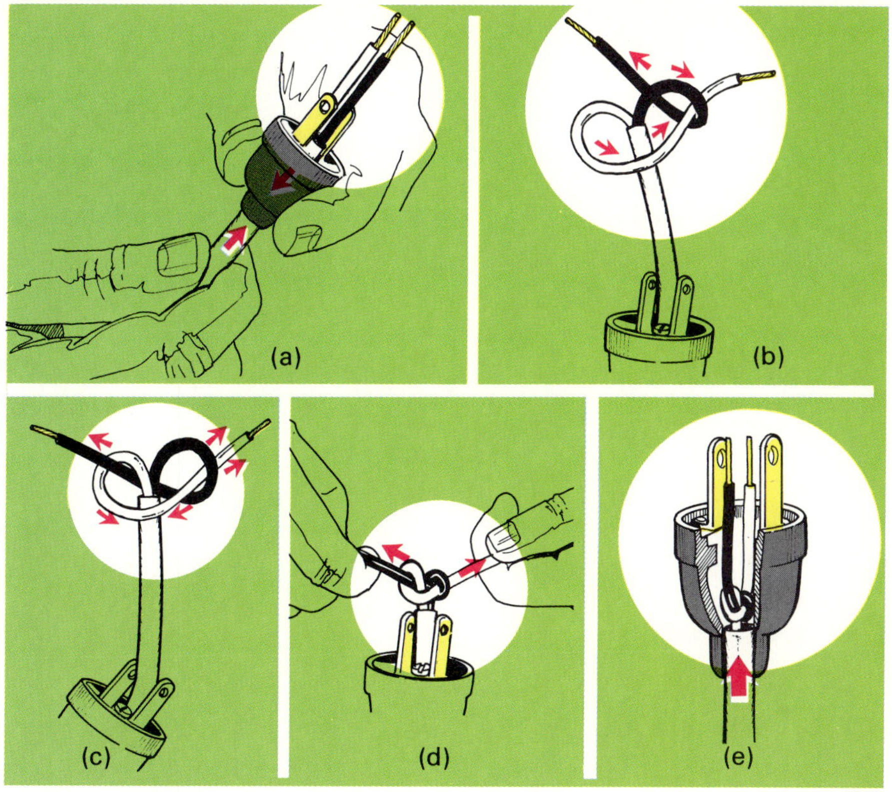

Figure 15. Tying a holding (Underwriters') knot.

Twisting binds the wire strands together so they will not fan out and cause a short circuit when fastened under the screw terminal.

6. *If cord is fabric covered, wrap end of fabric with electrical tape to prevent unraveling (Figure 14).*

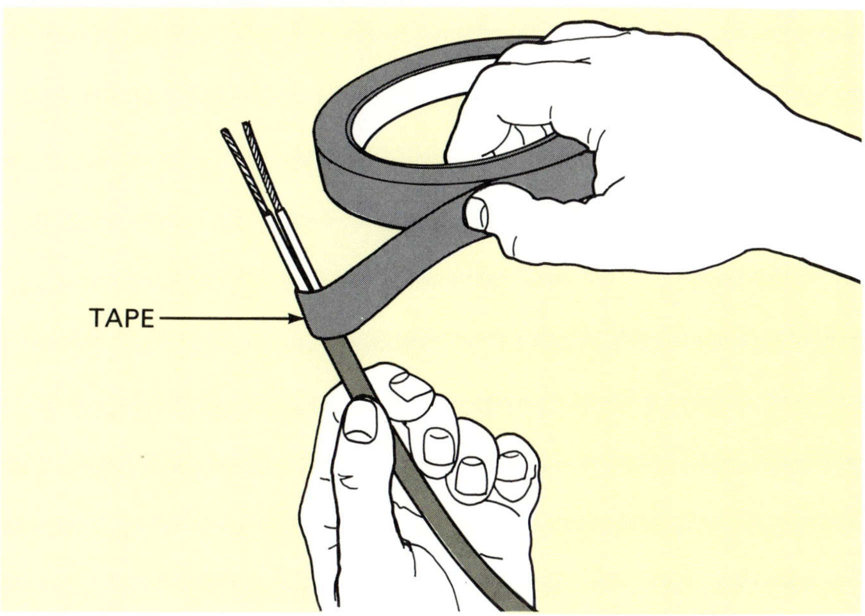

Figure 14. If cord cover is fabric, wrap end of fabric with tape to prevent unraveling.

TYING HOLDING (UNDERWRITERS') KNOT

The Underwriters' or holding knot (Figure 15) is used with standard and core-type plugs only. When the knot is tied, it is larger than the hole in the neck of the plug. When there is strain on the cord, the holding knot takes the strain and protects the wire connections inside the plug.

Proceed as follows:

1. *If new attachment plug is used, take plug apart as shown in Figure 8a.*
2. *Insert prepared wires into plug (Figure 15a).*
3. *Loop right-hand wire as shown in Figure 15b.*
4. *Loop left-hand wire as in Figure 15c.*
5. *Tighten knot (Figure 15d).*

   Tighten so free wire ends are of equal length.

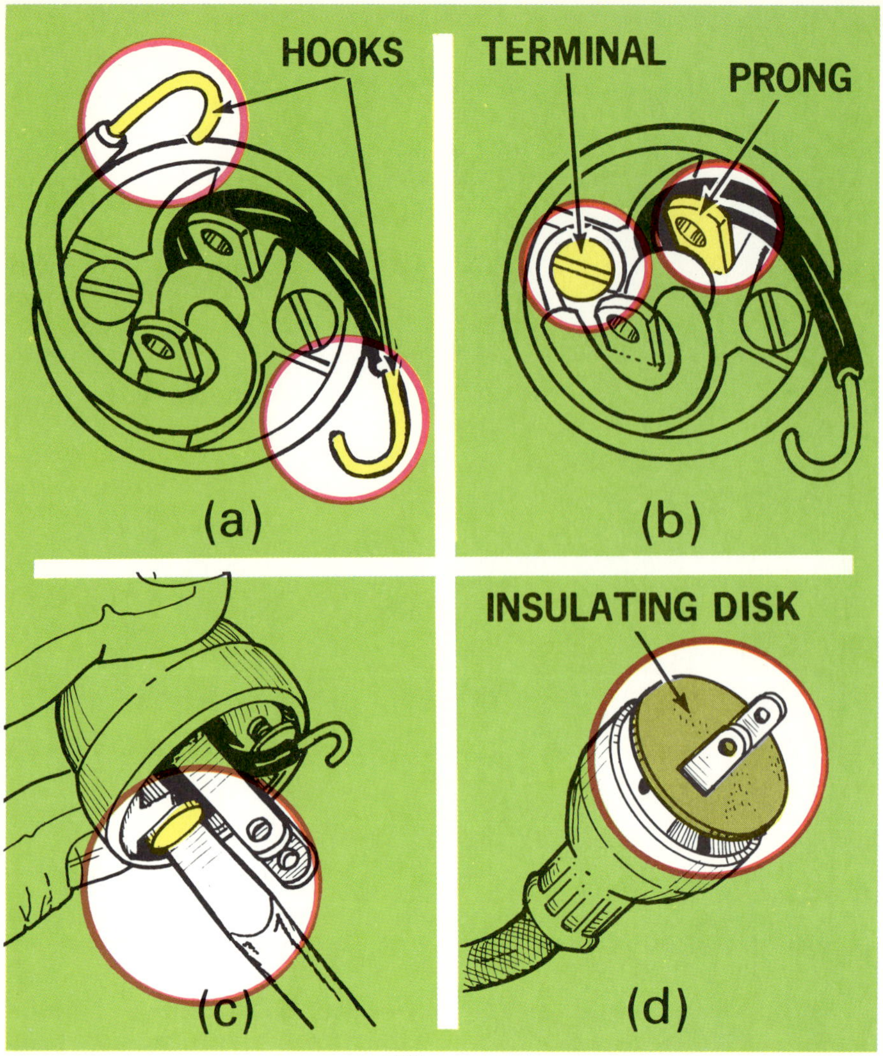

5316

Figure 16. Steps for connecting a service cord to a standard plug.

6. *Pull knot firmly against hole in plug (Figure 15e).*

CONNECTING CORD TO ATTACHMENT PLUG

The kind of plug you have will determine how you will connect the cord: **standard, dead-front,** and **quick-connecting.** Look at the plugs in Figures 16, 17 and 18. Follow the directions for the kind of plug you have.

a. If you have a **standard plug** (Figure 16), proceed as follows:

1. *Bend bare ends of each wire into a hook (Figure 16a).*
2. *Bend one wire around a prong and put wire hook under screw (Figure 16b).*

   Be sure that hook points in the same direction that screw turns. If pointed the opposite direction, wire tends to spread out from under the screw head.

   If you are connecting a **3-wire polarized plug** (Figure 3c), connect *green* wire to terminal of round or U-shaped prong first.

3. *Tighten screw. Be sure there are no fine wires projecting out from the screw terminal (Figure 16c).*
4. *Repeat steps (2) and (3) with second wire.*
5. *Replace insulating disk (Figure 16d).*

b. If you have a **dead-front plug** (Figure 17), proceed as follows:

1. *Insert wires through plug body to terminals and tighten screws.*
2. *Slide plug body toward terminals until spring lock engages.*
3. *Turn threaded cap until hand-tight.*

Figure 17. Steps for connecting a service cord to a dead-front plug.

5317

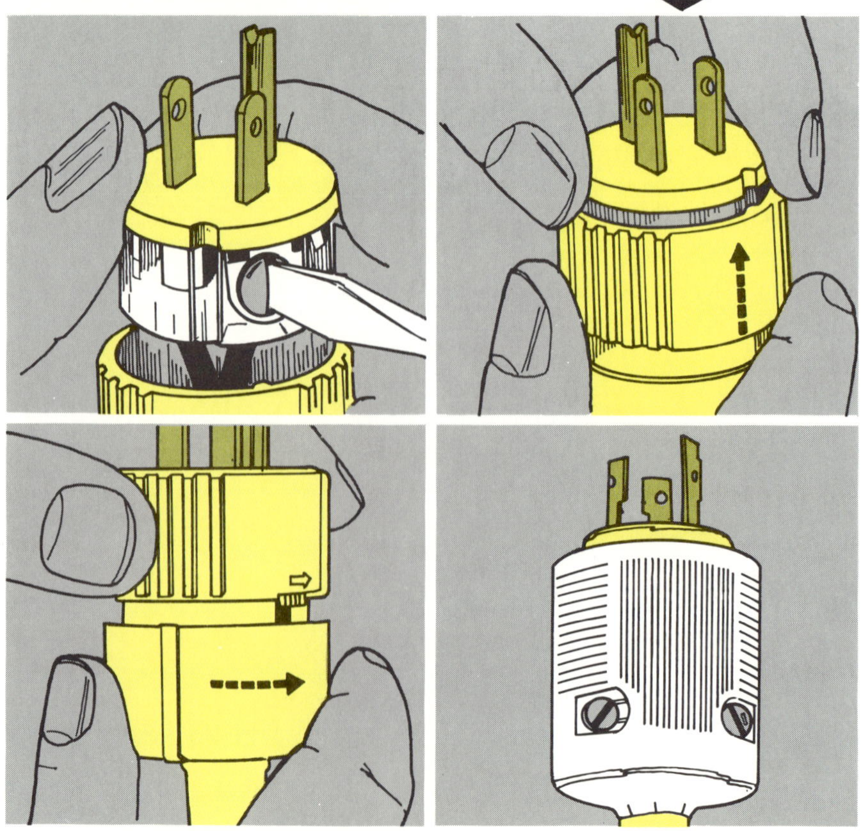

4. *Another type assembles body halves using two assembly screws.*

c. If you have a **quick-connecting plug** (Figure 18), proceed as follows:

1.. *Insert wires through outer shell. Do not remove insulation (Figure 18a).*

2. *Insert wires into core (Figure 18b).*

3. *Press prongs together (Figure 18c).*

   This action causes a small metal tooth on each prong to pierce the insulation and extend into the wire. One tooth contacts one wire and the other tooth the other wire.

4. *Push core back into shell (Figure 18d).*

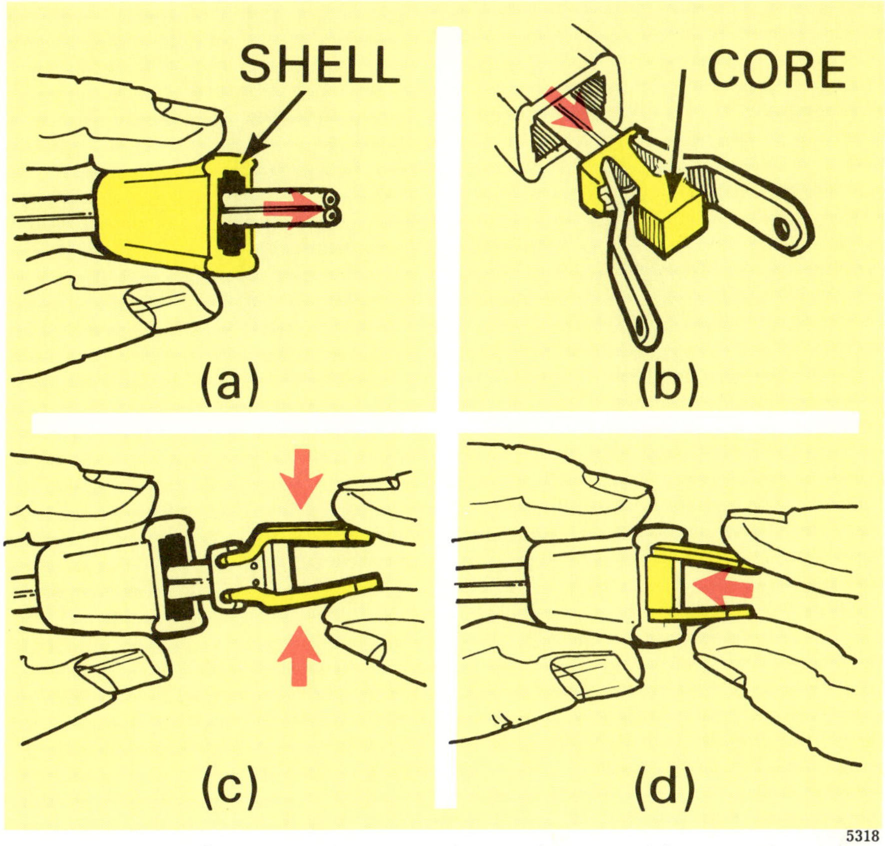

Figure 18. Steps for connecting a service cord to a quick-connecting plug.

B. Replacing Lamp Sockets

As shown in Figure 19, there are four different types of lamp sockets. The first two — the **candelabra** and **intermediate** — are for decorative purposes such as Christmas tree lights. There is little you can do about replacing these.

The **medium-base** lamp socket is most commonly used for general lighting. It is standard for home, farm and office lighting, and will take lamp sizes up to 300 watts.

The **mogul-base** lamp socket is used for lamps of 300 watts or larger.

The medium-base and mogul-base lamp sockets are the lamp sockets you often need to replace.

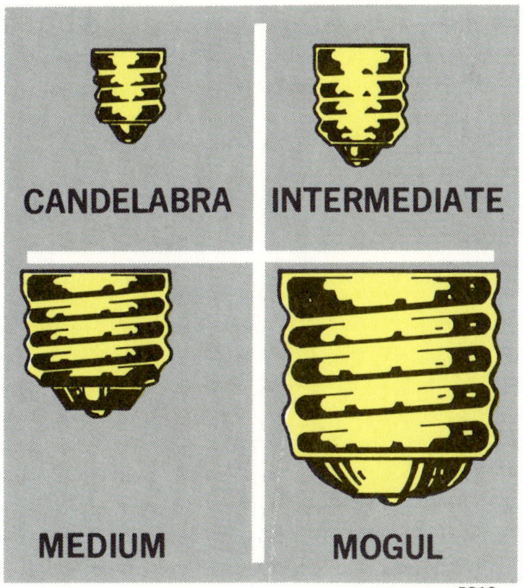

Figure 19. Lamp sockets in general use in 'homes. The candelabra and intermediate types are for decorative lighting. The medium-base type is most common. The mogul-base type is for lamps of 300 watts or more.

The medium-base and mogul-base lamps are alike except for size. In their simplest form — without a switch — the socket consists of a screw shell, a center contact, an insulating block (to which the other two parts are fastened) and two terminals (Figure 20, inset).

Figure 20 shows the path followed by the electric current from the cord through the socket

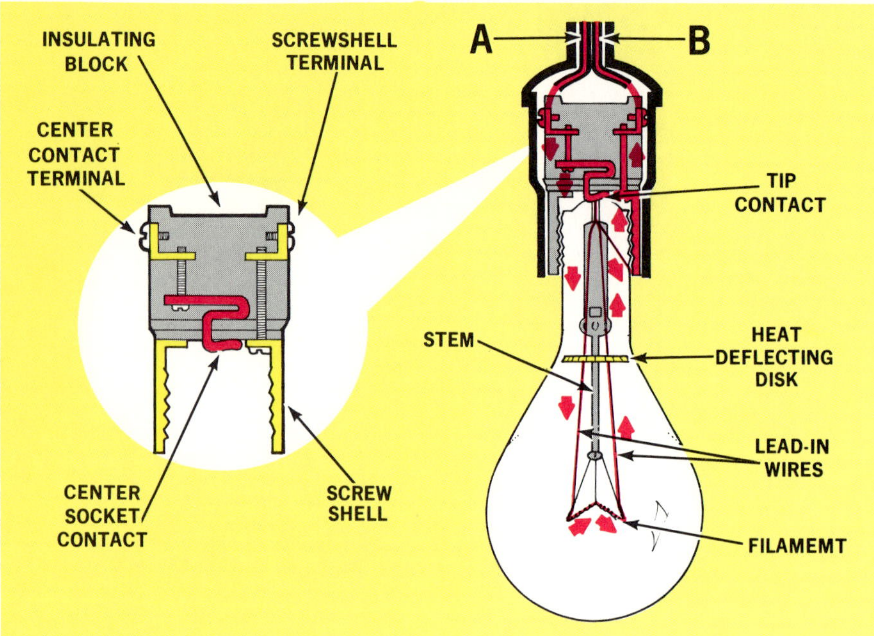

Figure 20. Sectional view of a socket and lamp. Red lines show path of current flow through the socket and lamp.

and lamp and back to the other wire in the cord. Assume the electric current is entering on wire A. The red lines indicate the path. It passes through the center contact terminal to the center socket contact then to the tip contact of the lamp bulb. From there it passes along the lead-in wire conductor to the filament in the lamp. As the current passes through the filament, the filament gets hot and gives off light.

The current then returns on the other lead-in wire up to the screw shell of the bulb. Since the bulb screw shell fits snugly inside the socket screw shell, the current passes easily from one to the other. From the socket screw shell the current passes out through the screw-shell terminal and wire B.

Light created by the above method is called **incandescent** light. This is the type of light discovered by Thomas Edison and is the most common type of light used in residences. **Fluorescent** light is used extensively by commercial and industrial facilities and requires equipment different from that used by incandescent lights. Fluorescent lighting is discussed later in this publication.

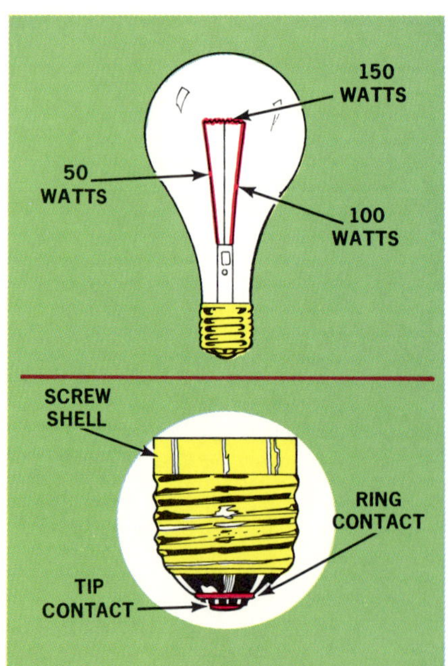

Figure 21. A 3-light bulb has two separate filaments and three contact points on the base: the tip contact, the ring contact and the screw shell.

Figure 21 shows how a 3-light lamp works — the kind often found in floor and table lamps. When the switch is turned to the first position, usually the lower wattage filament is energized to give low intensity light. In the second position, only the higher wattage filament is energized to give higher intensity light. In the third position, both filaments are energized to give its full lighting capacity. Some lamps work in reverse order — high, medium, then low. Both medium and mogul bases are used for these lamps, but the sockets have three electrical contacts instead of two. The lamp base also has three contacts as shown in Figure 21, inset. However, when replacing either of these sockets, the procedures are the same as for standard sockets. Only two cord wires need to be attached to the socket screw terminals.

When purchasing a socket for replacement, you should consider how it will be used. **Plastic, porcelain** and **rubber** (Figure 22) are safest where the floor and surroundings are moist or wet. Their use avoids a chance of shock or burn as shown in Figure 2.

Metal sockets are used where appearance is important and where surroundings are dry, such as on living-room floor lamps and table lamps. For ceiling drop lights, **porcelain** or **plastic** is desirable. For light-duty extension cords about the home, **plastic sockets** are light weight and safe. Extension cords for use outside or in farm buildings and shops should have **rubber sockets.** They are both safe and durable.

The instructions that follow apply to replacing metal sockets. Procedures for other sockets are similar.

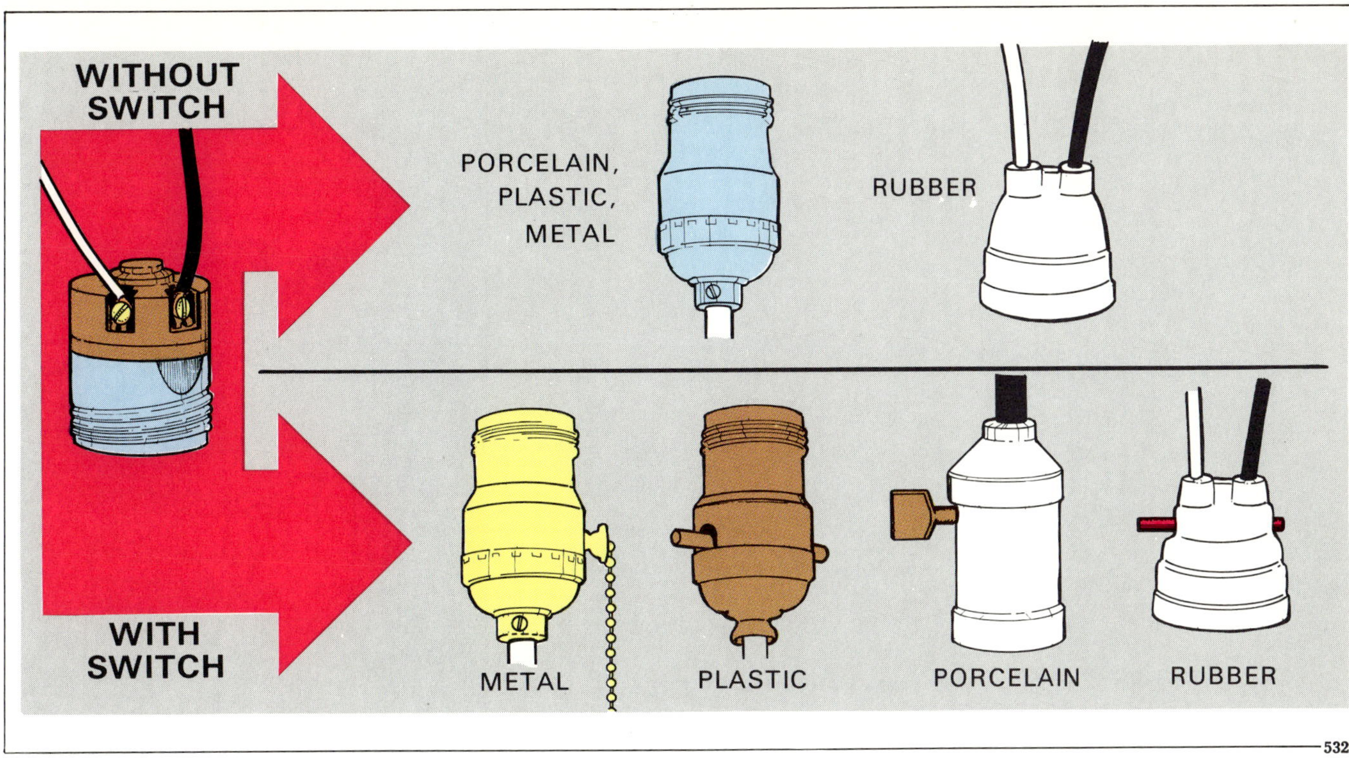

Figure 22. Lamp sockets are made of different materials and either with or without switches.

Upon completion of this section, you will be able to **replace common types of lamp sockets.**

Replacing lamp sockets is discussed under the following headings:

1. Tools and Materials Needed.
2. Removing the Socket.
3. Replacing the Socket.

TOOLS AND MATERIALS NEEDED

Tools:

Small screwdriver
Pocket knife or scissors
Small side-cutting pliers

Materials:

1 lamp socket

REMOVING THE SOCKET

To remove the socket, proceed as follows:

1. *Always pull attachment plug from outlet (Figure 23a). If socket is permanently wired to house current, disconnect main switch (Figure 23b), remove fuse or turn off circuit breaker.*
2. *Hold socket in your hand and press shell at point marked "press" (Figure 24).*

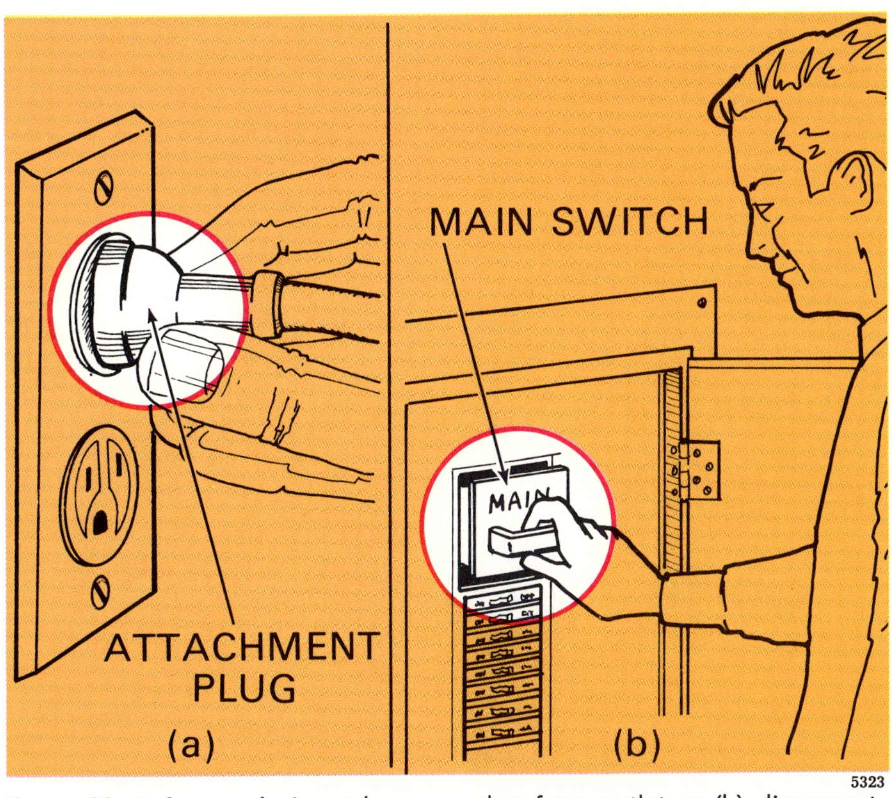

Figure 23. Before replacing a lamp socket, always disconnect the electric power by (a) pulling attachment plug from outlet or (b) disconnecting main switch, removing fuse or turning off circuit breaker.

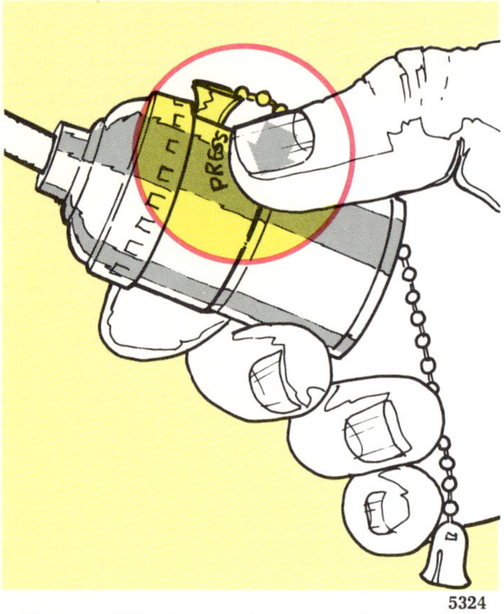

Figure 24. Removing cap from shell by pressing with thumb.

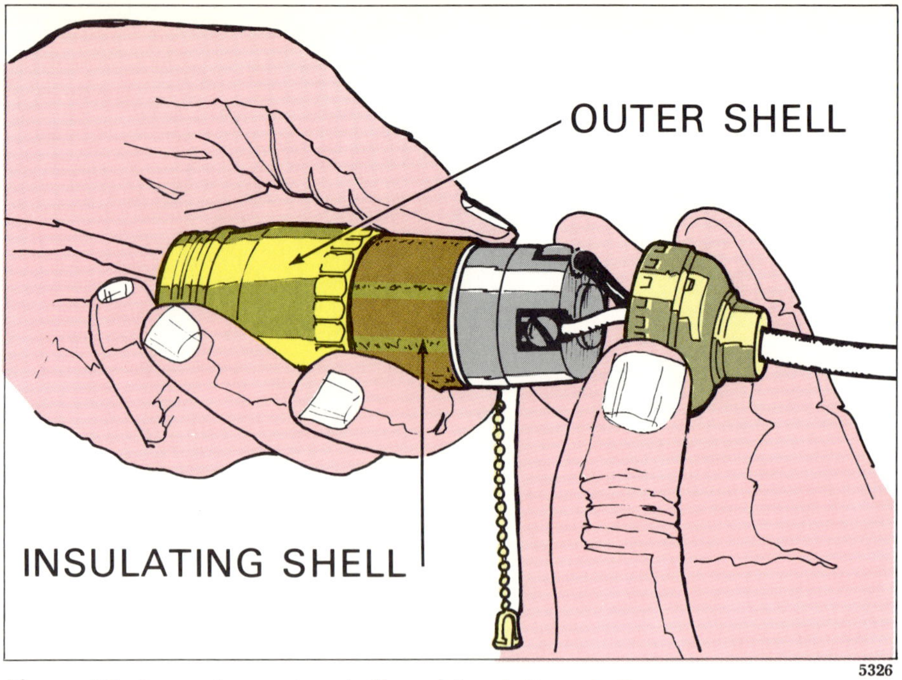

Figure 26. Removing outer shell and insulating shell.

You will usually find the marking to the left of the switch opening.

With plastic or porcelain sockets the top may be held in place with a threaded ring or may snap into place. Some porcelain sockets are held together by two bolts that are removed from inside the screw shell.

3. *Hold cap steady while you apply pressure and a gentle twisting action on shell (Figure 25).*

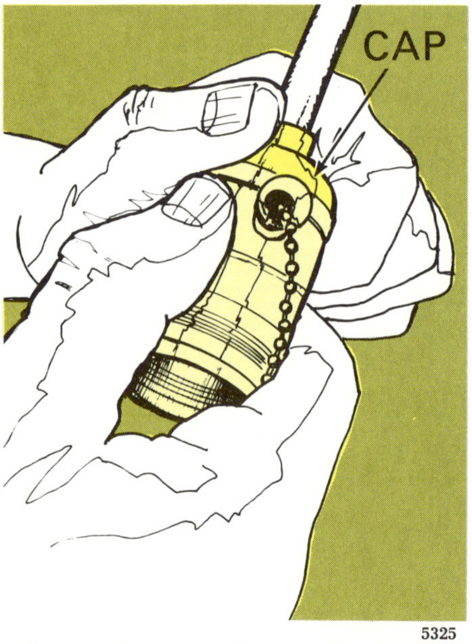

Figure 25. A gentle twisting action on shell helps in removing cap.

4. *Remove outer shell and insulating shell (Figure 26).*

Note the condition of the insulating shell. If it is black and charred, the socket has probably been heavily overloaded, causing it to get too hot. It should be replaced with a new socket. This often happens when an electric iron or other heavy wattage appliance is connected to a lamp socket.

Most sockets with switches are designed for 250 watts; without switches they will handle 660 watts. Most irons are 1000 watts or more.

If the socket is in use in a basement, milk room or other damp place, the insulating shell may be damp. When it is damp, and the power is on, it will conduct electric current. It may shock you when you touch it unless your shoes are dry. It should be replaced with a plastic or porcelain socket.

5. *Loosen terminal screws and remove wires (Figure 27). Do not remove screws.*

6. *Untie holding knot if there is one.*

7. *Remove cap from cord.*

8. *Examine cord and wire ends.*

If insulation is charred or cord is frayed, cut off damaged section, and follow procedures for "Preparing Cord for Standard or Core Plugs," page 191.

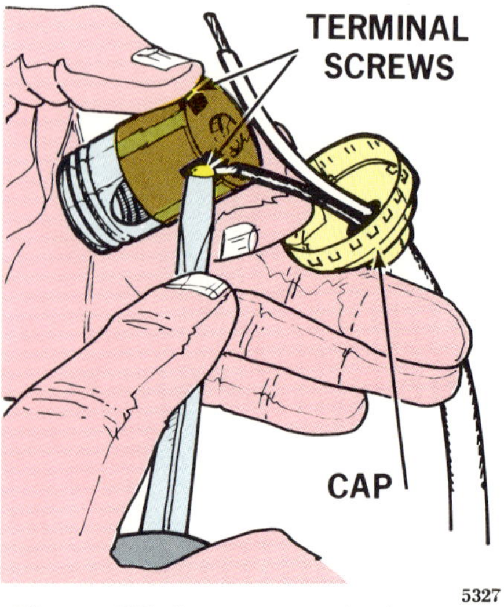

Figure 27. Loosen terminal screws and remove wires.

REPLACING THE SOCKET

Procedures for replacing the socket are as follows:

1. *Insert cord through hole in cap of new socket (Figure 28).*

   Be sure cardboard insulating shell is in cap if it is a metal socket.

2. *If cord is fabric covered, wrap end of fabric with plastic tape to prevent unraveling (Figure 14).*

3. *Tie the two wires into a holding (Underwriters') knot (Figure 15).*

4. *Fasten black or marked wire to brass terminal on socket body (Figure 29).*

   Be sure each wire end is wrapped around the screw terminal in a clockwise direction. As you tighten the screw, the wire tends to draw closely around the screw terminal.

5. *Fasten white wire to aluminum screw terminal.*

6. *Put socket back together.*

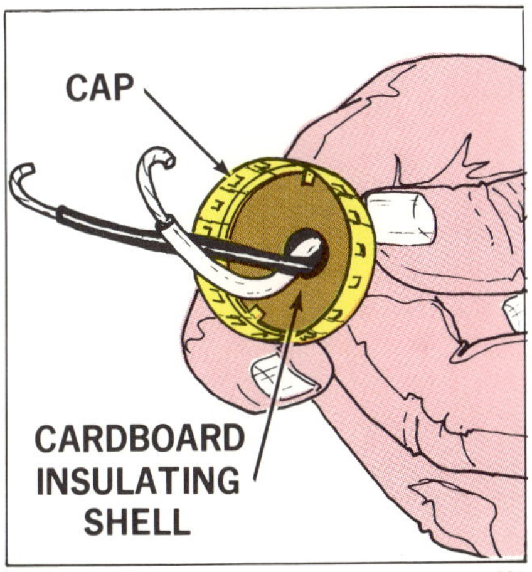

5328

Figure 28. Inserting cord in hole in cap.

5329

Figure 29. Fasten black or marked wire to brass screw terminal. Fasten white wire to aluminum screw terminal.

C. Replacing Heater Plugs

Heater cords tend to get out of condition quickly. This is because they are affected by the intense heat of the appliance and because of their frequent use. Some people ruin a heater cord by pulling on the cord rather than on the appliance plug, when they disconnect the cord from the appliance (Figure 30).

Quite often a short circuit develops in or near the heater plug and "blows" a fuse. If this has happened, check the plug to see if it is in good enough condition to use again. If it is loose, broken or discolored from heat, replace it with a new heater plug.

When an appliance does not heat, it is probably caused by an open circuit. Possibly one of the cord wires has broken and

5330

Figure 30. A quick way to ruin an appliance cord.

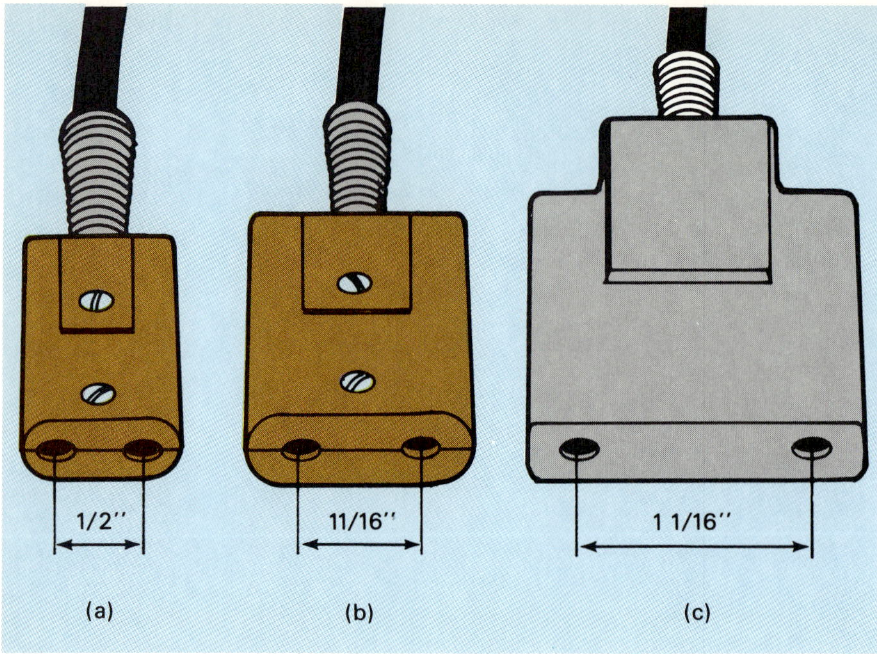

5331

Figure 31. Heater plugs. (a) Appliance plug. (b) Iron plug. (c) Plastic plug with cord molded into top. This type cannot be repaired.

the ends separated. If so, it is most likely to be broken next to the heater plug.

When you are buying a new heater plug, keep in mind that there are several sizes; but only two are commonly used. One is an **appliance plug** which is small in size. The holes for the appliance prongs are only ½ inch apart from center to center (Figure 31a). The other is larger and is called an **iron plug** (Figure 31b). The holes on the iron plug are 11/16ths inch from center to center.

Upon completion of this section, you will be able to **repair or replace appliance plugs and iron plugs.**

Replacing heater plugs is discussed under the following headings:

1. Tools and Materials Needed.
2. Removing the Heater Plug.
3. Repairing the Cord.
4. Replacing the Heater Plug.

TOOLS AND MATERIALS NEEDED

Tools:

Small screwdriver
Pocket knife
Small side-cutting pliers

Materials:

New heater plug

REMOVING HEATER PLUG

To remove the heater plug, proceed as follows:

1. *Disconnect heater cord from convenience outlet and from appliance.*
2. *Remove small bolts (Figure 32).*
3. *Open plug carefully. Note position of wires and terminal clips.*

   You will need to replace them in the same positions.
4. *Slide cord protector back on cord (Figure 33).*
5. *Loosen screws on clips and remove wires (Figure 34).*
6. *Examine plug.*

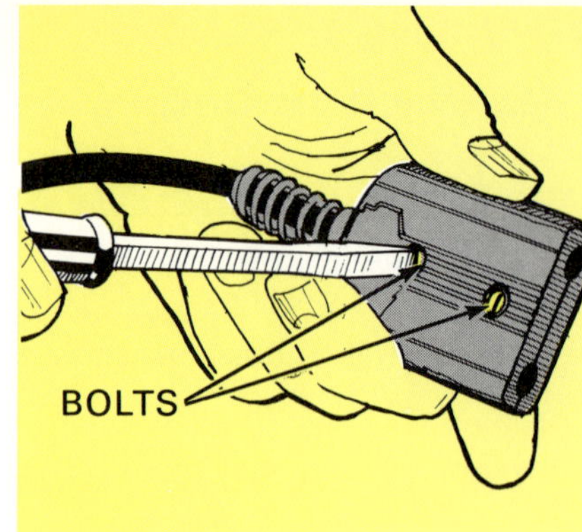

5332

Figure 32. To disassemble heater plug, (a) remove small bolts.

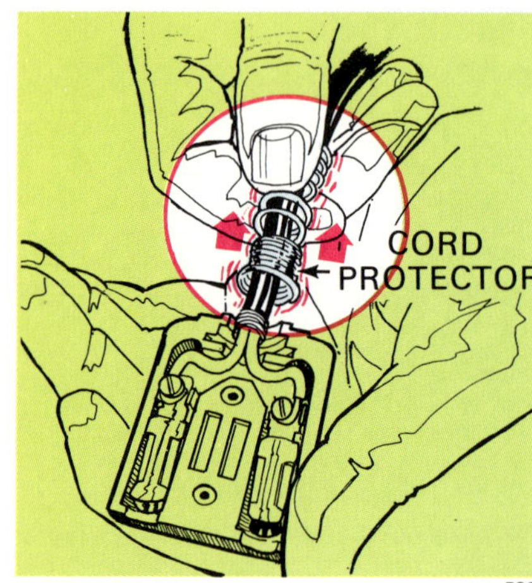

5333

Figure 33. Sliding cord protector back on cord.

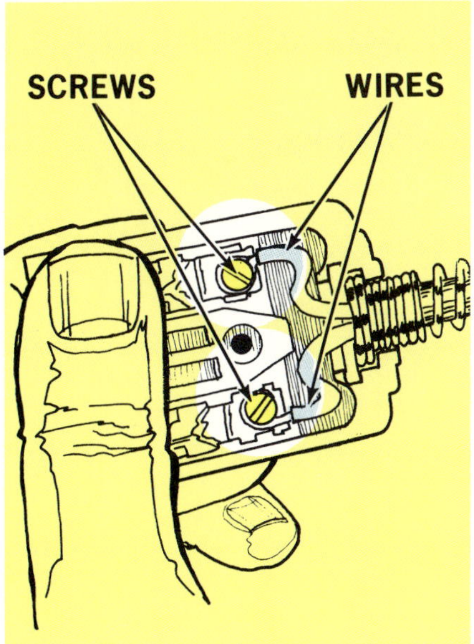

5334

Figure 34. To remove wires, loosen screws on clips.

If the plug is **burned, cracked, broken or discolored with heat,** throw it away. If the plug body appears to be in good condition, examine the terminal clips. Heat causes them to lose their spring action thus making poor contact. If they fit loosely on appliance prongs, throw plug away.

REPAIRING THE CORD

For repairing the cord proceed with the steps that follow:

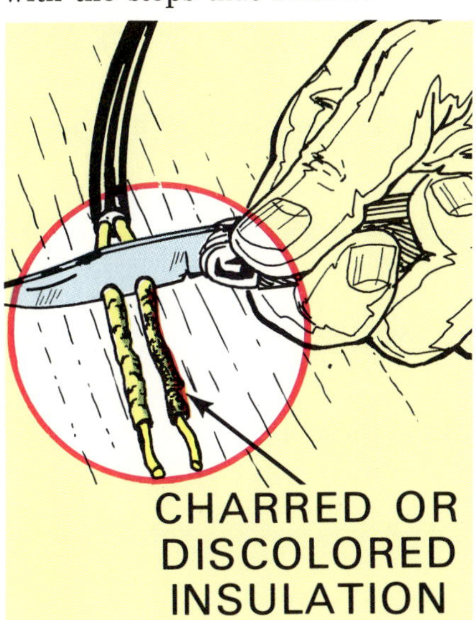

5335

Figure 35. Removing charred or discolored insulation.

1. *Remove all wire that has charred or discolored insulation (Figure 35).*

   If the cord has been in use for some time, the rubber insulation around the wires in the plug is usually brittle or discolored. Unless these wires are cut back to where the insulation is good, a short circuit may soon develop after you repair the plug.

2. *Determine amount of cord jacket to be removed (Figure 36).*

   Lay free ends of both wires in plug channels to find out how much cord jacket, if any, must be removed. Be sure both wires are the same length.

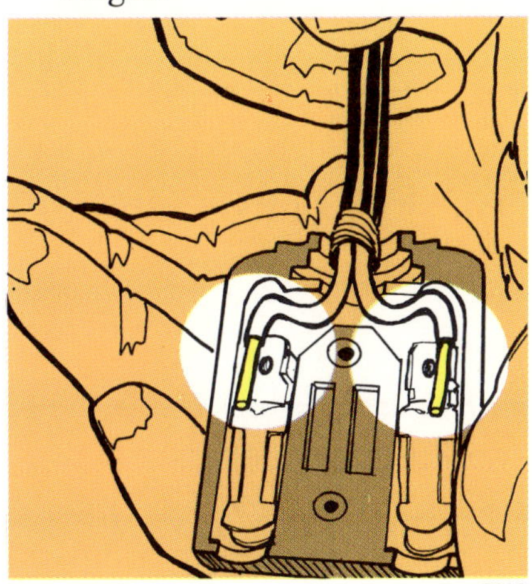

5336

Figure 36. Check to determine how much cord jacket to be removed.

3. *Prepare cord for connecting to plug.*

   (Follow instructions on page 191, "Preparing Cord for Standard or Core Plugs.")

   Heater cords have an asbestos wrapping around the insulated wires and under the cord jacket. Remove the asbestos back as far as the jacket has been removed. Some prefer to wrap a portion of it around each insulated wire that extends to the terminal clip. This is satisfactory if there is enough space for the wire and wrapping in the heater plug channels.

REPLACING THE HEATER PLUG

Procedures for replacing the heater plug are given as follows:

1. *Attach each wire to a terminal clip (Figure 37).*

   Wrap bare wire around terminal post in same direction that screw tightens.

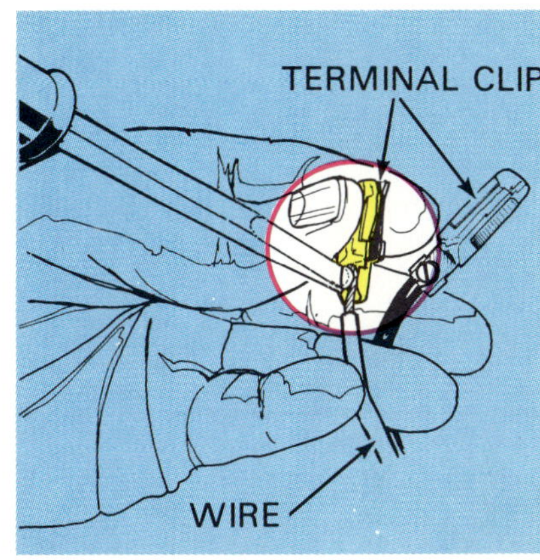

5337

Figure 37. Attaching wire to a terminal clip.

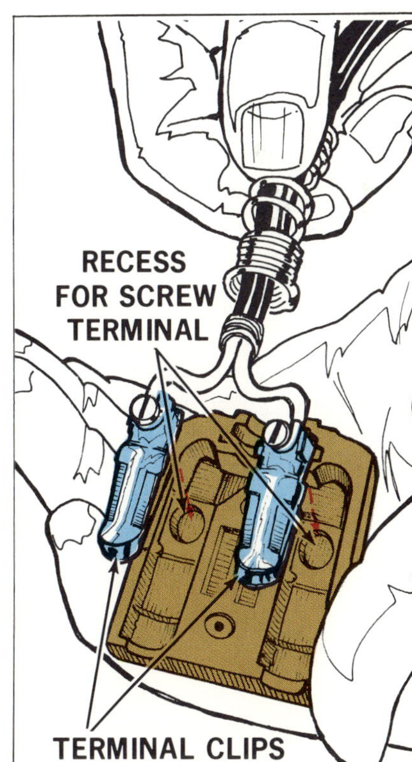

5338

Figure 38. Replacing terminal clips in plug.

2. *Place terminal clips in position in plug (Figure 38).*

Note the small hole in each channel that holds the clips. This is to receive the end of the screw terminal.

3. *Hold terminal clips in position with your thumb while replacing cord protector (Figure 39).*

4. *Press wires to terminal clips into channels in plug.*

5. *Place other half of plug into position (Figure 40).*

Do not force into position. If the two halves do not fit well, remove and check for cause.

6. *Replace bolts or clips that hold plug together.*

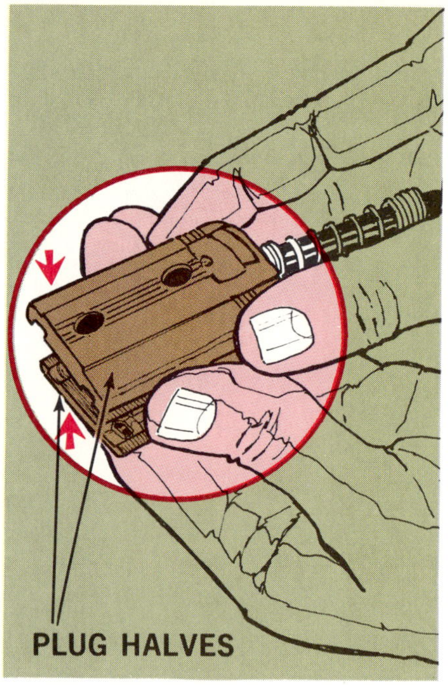

5340

Figure 40. Replacing plug halves into position.

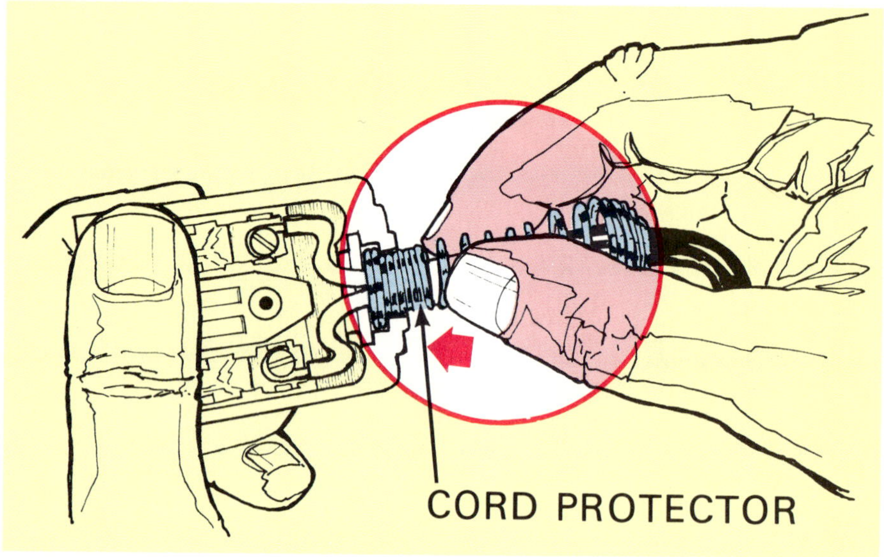

5339

Figure 39. Replacing cord protector.

D. Replacing Service Cords

The cord itself frequently becomes worn. You may notice that the jacket is frayed (Figure 41a) or that the insulation is nicked or slivered (Figure 41b).

Another type of wear, that is not always evident, occurs where there are frequent and sharp bends at one place in the cord (Figure 41c). This may cause the wires to break and the ends to separate without the jacket showing undue wear.

The conditions in Figure 41 (a) or (b) can be dangerous. This is particularly true if a person touches the bare wire in the cord and at the same time touches a water pipe or furnace register. People have been shocked badly, or even killed, in this way. A cord in this condition,

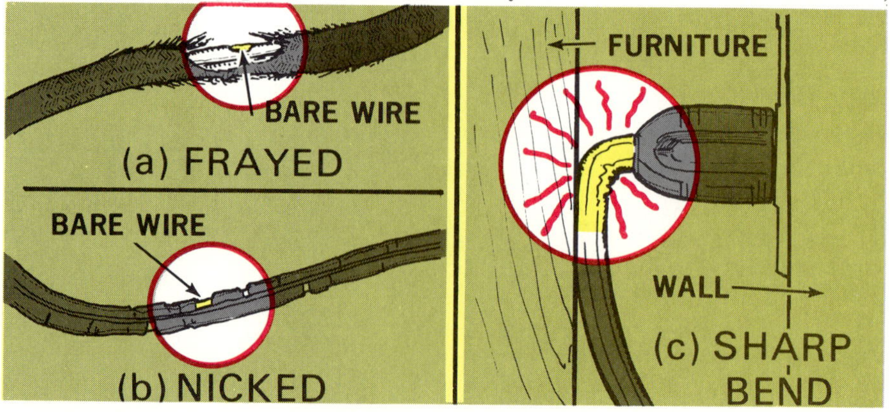

5341

Figure 41. Service cords need to be replaced when (a) frayed, (b) nicked, or (c) damaged by sharp bends.

FOR GENERAL USE IN HOME

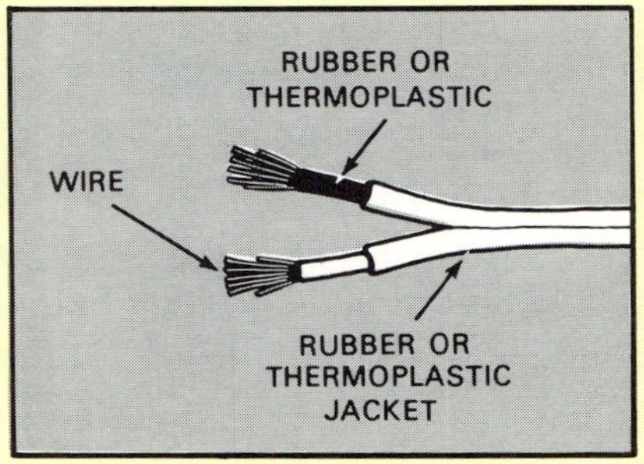

RUBBER PARALLEL CORD—(Type SP for rubber, type SPT for plastic) A durable cord for small household appliances such as radios, fans, floor lamps, and table lamps. It is not suitable for heating appliances. The covering is grooved so the wires can be separated by simply pulling them apart. Available in sizes 18 and 16.

FOR HEATING APPLIANCES

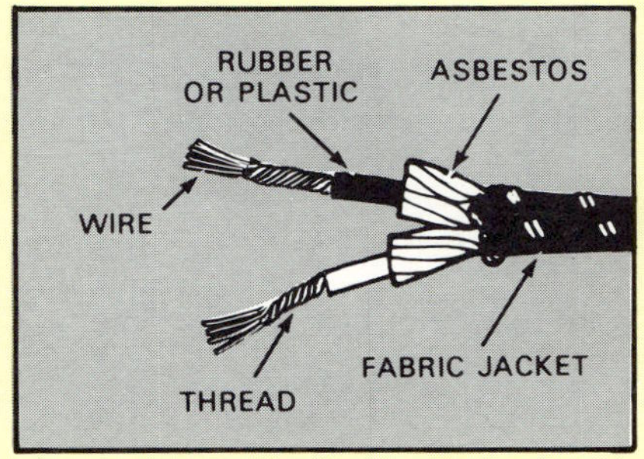

HEATER CORD—(Type HPD) For use with heating appliances. Each wire has an asbestos covering. The jacket is braided. It must be kept dry. Available in sizes 18 to 12. (If cord is to be used in damp locations, such as in a milk room, use a rubber- or plastic-jacketed heater cord such as types HSJ, HS, HPN, ASFJ or AFS. Most dealers will have to order these. Available in sizes 18 to 12.

FOR USE IN UTILITY BUILDINGS, BARNS, & DAMP LOCATIONS

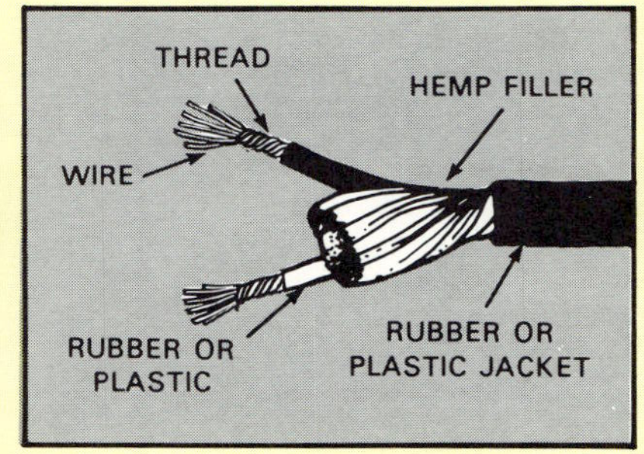

JUNIOR HARD SERVICE CORD (Type SJ for rubber covering, type SJT for plastic covering) Will stand hard usage. Can be used in damp places. Desirable for use on portable equipment, in barns, milk houses and basements. Desirable for general-use extension cords. Thermoplastic type is SJT. Available in sizes 18–12.

FOR USE IN SEVERE CONDITIONS

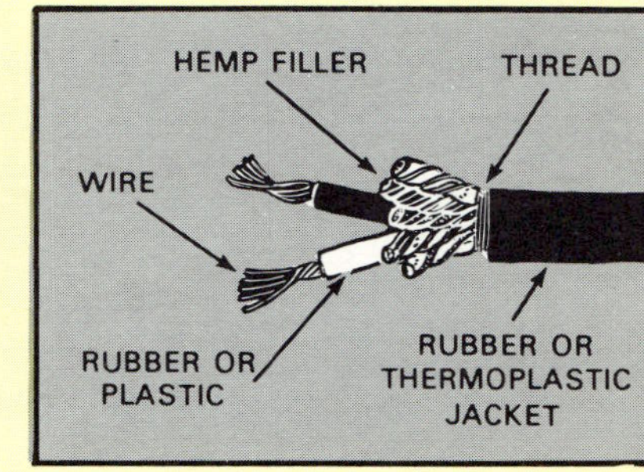

HARD-SERVICE CORD—(Type S, SO and ST) Same as hard-service cord above except rubber or thermoplastic jacket is thicker and very tough. Will stand more mechanical abuse than other cords. Desirable for portable equipment used outdoors. Available in sizes 18 to 10.

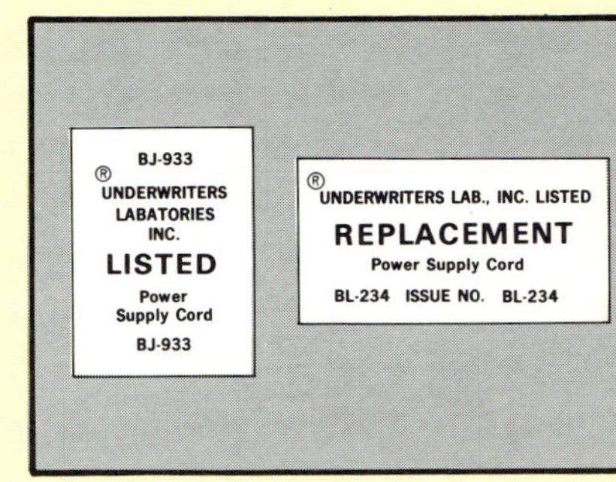

(Left) Typical Underwriters' label, used on appliance supply cords for many years, has been replaced with a new label (right) which appears on all replacement cords. Underwriters' approval tags should be on any new cords you buy; or on the reel on which the cord is wound.

5342

Figure 42. Kinds of service cords.

used in a damp place or outdoors, is especially dangerous. Under these conditions, your body provides a direct path to ground through which the electric current will travel.

Upon completion of this section, you will be able to **replace service cords on appliances.**

Replacing service cords is discussed under the following headings:

1. What Kind of Cord to Use.
2. Replacing Detachable Service Cords.
3. Tools and Materials Needed.
4. Disconnecting Cord from Equipment.
5. Reconnecting Cord to Equipment.

WHAT KIND OF CORD TO USE

When you replace a cord, it is important to select the right one. There are many different kinds of cords — such as for heating equipment, for hard service, for light service, for moist conditions, for grease, etc. Some cords are made to meet two or more of these conditions. For example, a cord serving a portable motor needs to be resistant to moisture and grease and designed for hard service. The type of cord called the "hard-service type" meets these needs. Figure 42 shows the different cords and how they are used.

REPLACING DETACHABLE SERVICE CORDS

When you have selected the right cord, decide what length is needed. Be sure the wire size in the new cord is no smaller than the one you are replacing.

When you buy, be sure the cord is labeled with an Underwriters' inspection tag*, as shown at the bottom of Figure 42; or that their name appears on the spool on which the cord is wound.

To remove or reconnect the attachment plug, socket or heater plug, follow the instructions in A, B, and C of Section VIII.

TOOLS AND MATERIALS NEEDED

Tools:

Small screwdriver
Pocket knife
Small side-cutting pliers
Small adjustable wrench
Soldering iron (if there are solder joints)

Materials:

Penetrating oil

For **solder connections:**

6 inches resin-core solder

1 roll plastic electrical tape

For **solderless connections:**
2 solderless connectors

DISCONNECTING CORD FROM EQUIPMENT

To disconnect the cord, proceed as follows:

1. *Disconnect attachment plug from outlet.*
2. *Follow cord to see where and how it connects to equipment (Figure 43).*

   With most **floor and table lamps** the cord feeds through the base up through a pipe to the socket (Figure 43a). The socket is threaded to the pipe. With **other equipment** the cord may be attached to a fixed terminal-post block (Figure 43b) or be soldered to other wires (Figure 43c). (If the job looks too complicated to you, take it to a repair shop.)
3. *Disconnect cord from appliance.*

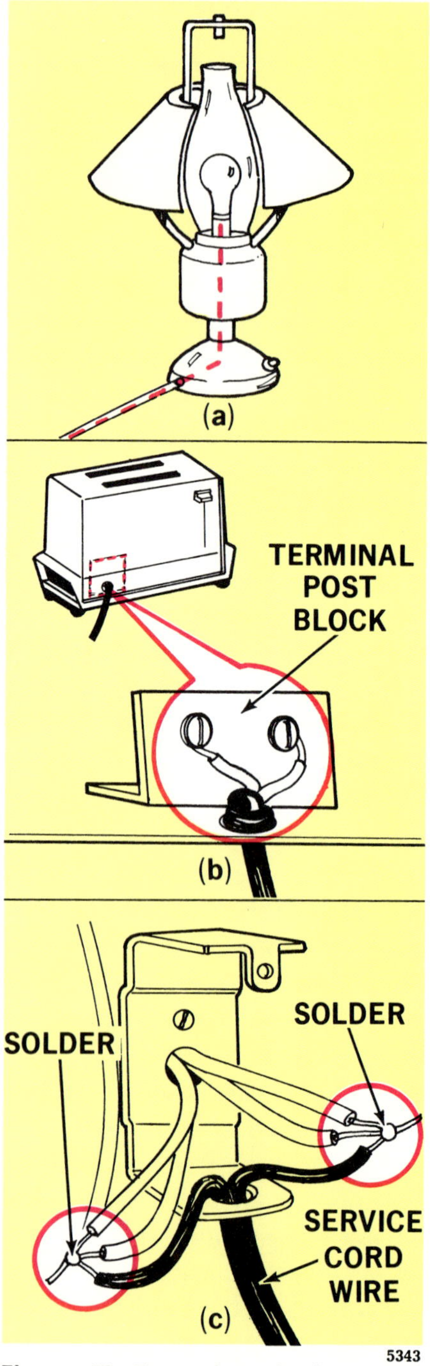

Figure 43. Examples of where service cords connect to equipment.

*Underwriters Laboratories, Inc. is a nonprofit organization which has established tests to determine if supplies and equipment meet certain safety standards. It offers its independent judgment to manufacturers submitting equipment for test. Comprehensive tests are made to insure the service and safety or cords. These tests are based on use and abuse to which cords will be subjected. They include crushing and impact tests, accelerated aging tests at higher temperatures, abrasion or wear-resistance tests, and tests on insulation thickness, conductor size and quality of insulation. Only a cord like the ones which pass these tests may use the label as shown in Figure 42.

With **floor lamps, table lamps and bridge lamps,** remove the shade, lamp bulb and diffusing bowl, if any (Figure 43a). The socket wire connections will be similar to the ones shown in Figures 26 and 27.

If your appliance has a **terminal-post block,** turn screws or nuts to left to loosen (Figure 43b). If "frozen" in position, use one or two drops of penetrating oil on screw or nut.

For **soldered connections,** remove tape and melt solder from connections (Figure 44a). With motors there are often two (or more) wires that fasten to one service-cord wire. It is important that the same two wires be reconnected in the same way when the new cord is added.

If wire ends are fastened with a solderless connector, twist connector to left. Remove when loose (Figure 44b).

4. *Remove knot from cord.*

A simple knot is often tied in the cord between the terminal connections and the appliance casing. The knot takes the stress off the electrical connections in case the cord is jerked.

5. *Remove cord from appliance.*

When removing a cord from a **floor or table lamp,** tie the end of the new cord to the end of the old one. As the old cord is withdrawn, the new one is pulled into place.

6. *Remove attachment plug from cord.*

Follow procedures on page 190, "Removing Attachment Plug from Cord."

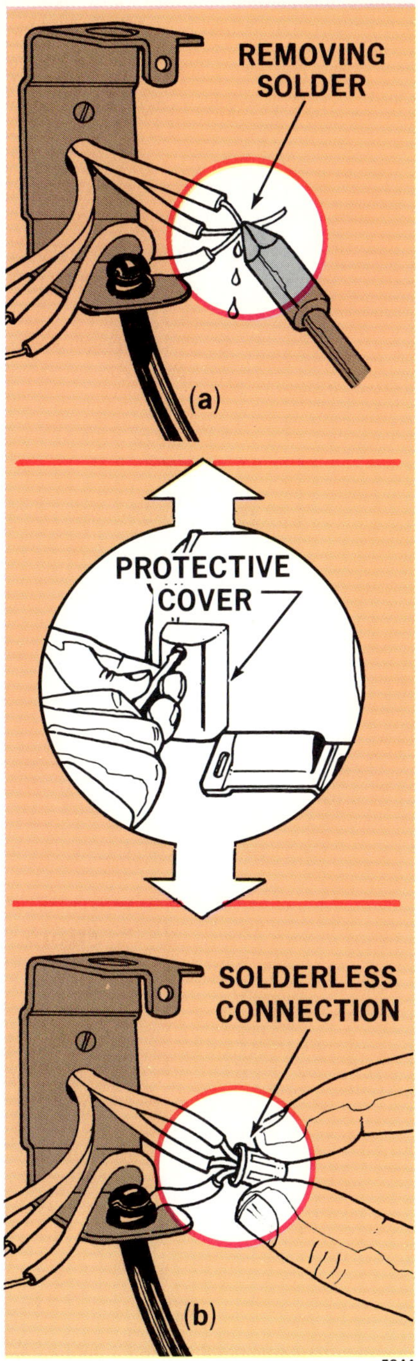

Figure 44. To replace faulty service cord on electric motor, remove protective cover (inset). Wire connections may be (a) soldered or (b) solderless type.

RECONNECTING CORD TO EQUIPMENT

Procedures for reconnecting the cord are as follows:

1. *Connect attachment plug to new cord.*

Follow instructions on page 189 under the section, "Replacing Attachment plugs."

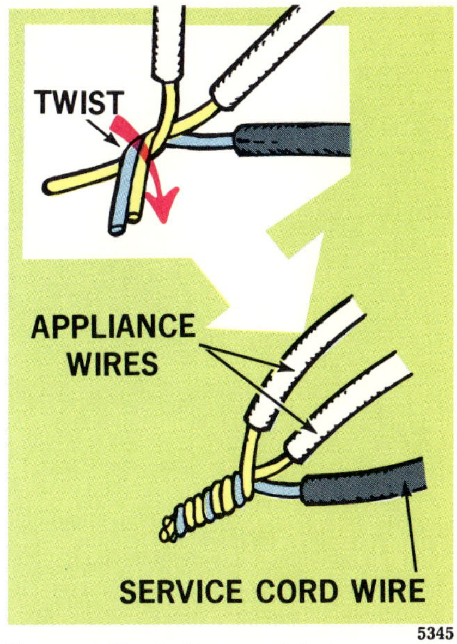

Figure 45. Before soldering, twist wires together.

2. *Prepare cord for attachment to appliance following steps on page 191* "Preparing Cord for Standard or Core Plugs."
3. *Thread new cord into appliance.*
4. *Tie simple knot in cord (if one was used on old cord to help take strain off of the connections.*
5. *Attach wires to appliance connections.*

   a. For **lamp-socket connections,** follow procedures on page 199, "Replacing the Socket."

   b. **If wires are to be soldered:**

   (1) *Scrape appliance wire connection clean with knife.*

   (2) *Twist one wire of cord and appliance wire together.*

   Motors may have two (or more) wires to connect to one cord wire (Figure 45).

   (3) *Connect remaining wires in same manner.*

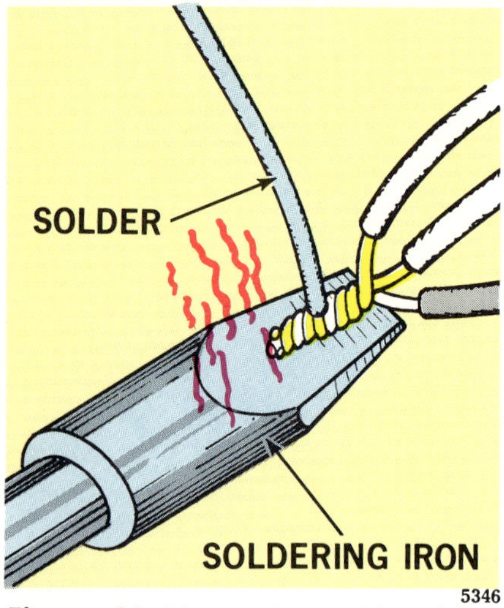

Figure 46. Heat wires with soldering iron to melt solder.

(4) *Heat splice with soldering iron until wire is hot enough to melt solder (Figure 46).*

Use resin-core solder. (Acid-core solder tends to "eat up" the wire.)

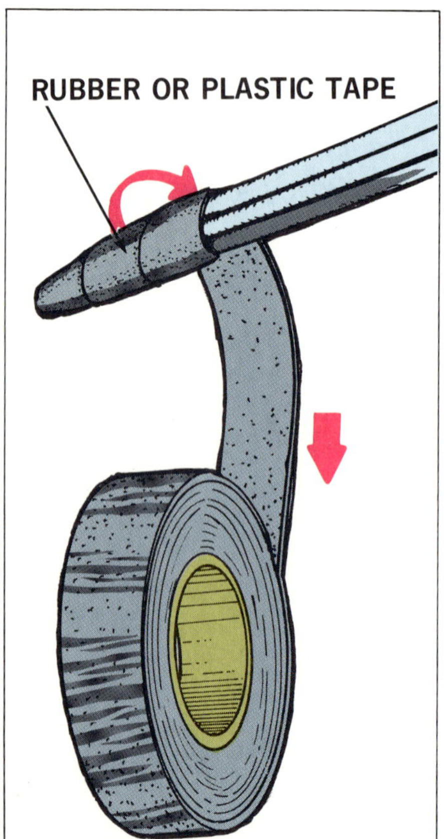

Figure 47. Cover soldered connection with plastic tape.

(5) *Solder second splice.*

(6) *Apply plastic tape over each splice (Figure 47).*

Extend the tape past the bare wire on to a portion of the old wire insulation. The tape gradually bonds together for permanent insulation.

c. If **solderless connector** is used:

(1) *Hold wires parallel with each other (Figure 48).*

If one wire is smaller, let it extend a little past the larger one.

(2) *Place connector over the wire ends and turn it to the right.*

The connector cuts its own threads and binds the wires together securely.

If any bare wire is still exposed, unscrew the connector and cut off enough wire end to allow the connector to completely cover the exposed area when it is tightened into place.

6. *Replace remaining parts on appliance.*

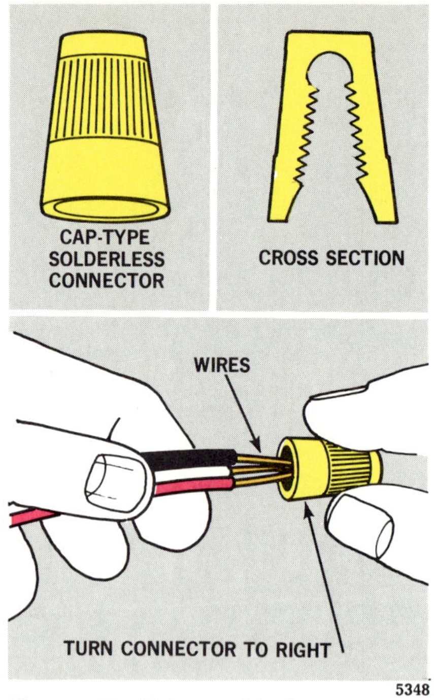

Figure 48. Using solderless connector.

IX. Maintaining the Permanent Wiring System

The permanent wiring system consists of the wires in the house walls and ceiling, the wall switches, the convenience outlets and the service-entrance equipment. (The latter includes the main switch and fuse equipment where power is supplied from the main line.) Permanent wiring also includes the overhead wiring between buildings and the surface wiring in shops, garages, and other buildings.

Except for the replacement of lamps and blown fuses, the permanent wiring system needs much less maintenance than service cords. However, when it does need repair, it may cause you more inconvenience than a service cord needing repair.

Discussed in this section are the maintenance and repair jobs that occur most frequently. They are included under the following headings:

A. Replacing Fuses, or Resetting or Changing Circuit Breakers.
B. Replacing Single-Pole Switches.
C. Replacing 3-Way Switches.
D. Replacing 4-Way Switches.
E. Replacing Convenience, Dryer or Range Outlets.
F. Replacing Incandescent Lighting Fixtures.
G. Replacing and Maintaining Fluorescent Lighting Units.
H. Replacing Ground Fault Circuit Interrupters.

A. Replacing Fuses, or Resetting or Changing Circuit Breakers

You have seen the electric current go off because a circuit was overloaded causing a fuse to blow or a circuit breaker to trip (Figure 49a); or possibly a short circuit in some appliance or extension cord may have caused it (Figure 49b).

Whatever the cause, it is fortunate that your wiring system provides some kind of protection.

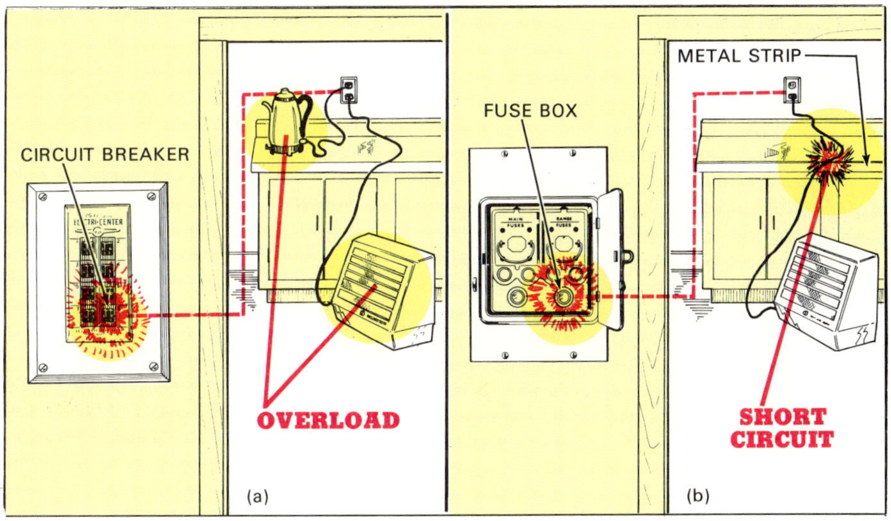

5349
Figure 49. Whether a circuit breaker (a) is used, or a fuse (b), its job is to protect your wiring against overloads and short circuits.

Without it, your home or other buildings might be set on fire.

It is important that you understand what this protection is and how it works. It is part of maintaining your wiring system.

HOW WIRING PROTECTION IS PROVIDED

Wiring-system protection can be provided by either of two means: (1) **fuses** or (2) **circuit breakers.** Both devices are intended to be the "weak link" in your wiring system.

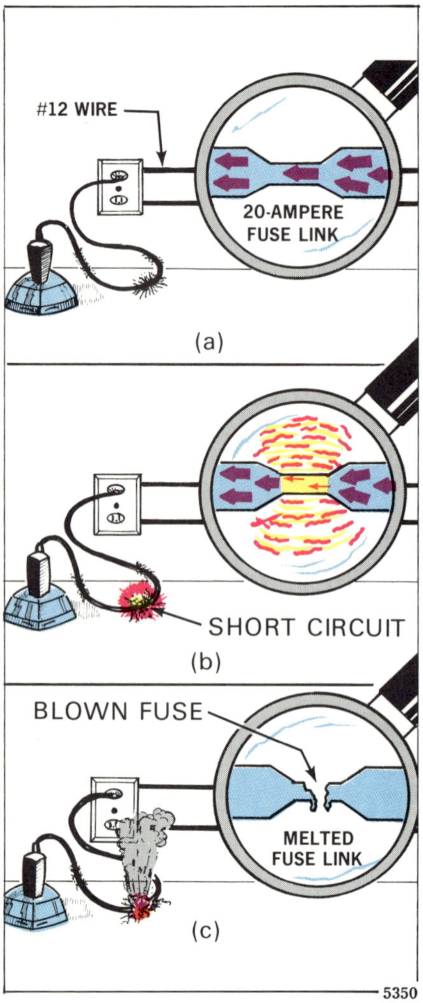

Figure 50. How a **fuse** works. (a) Animated movement of electrons (current) through a fuse when operating within its rated capacity. The narrow portion of the fuse link is designed to handle a definite amount of current flow. (b) When the current flow becomes greater than the capacity of the fuse link, the link heats. (c) Heat melts the fuse link (blows the fuse) which opens the circuit so no further current can flow.

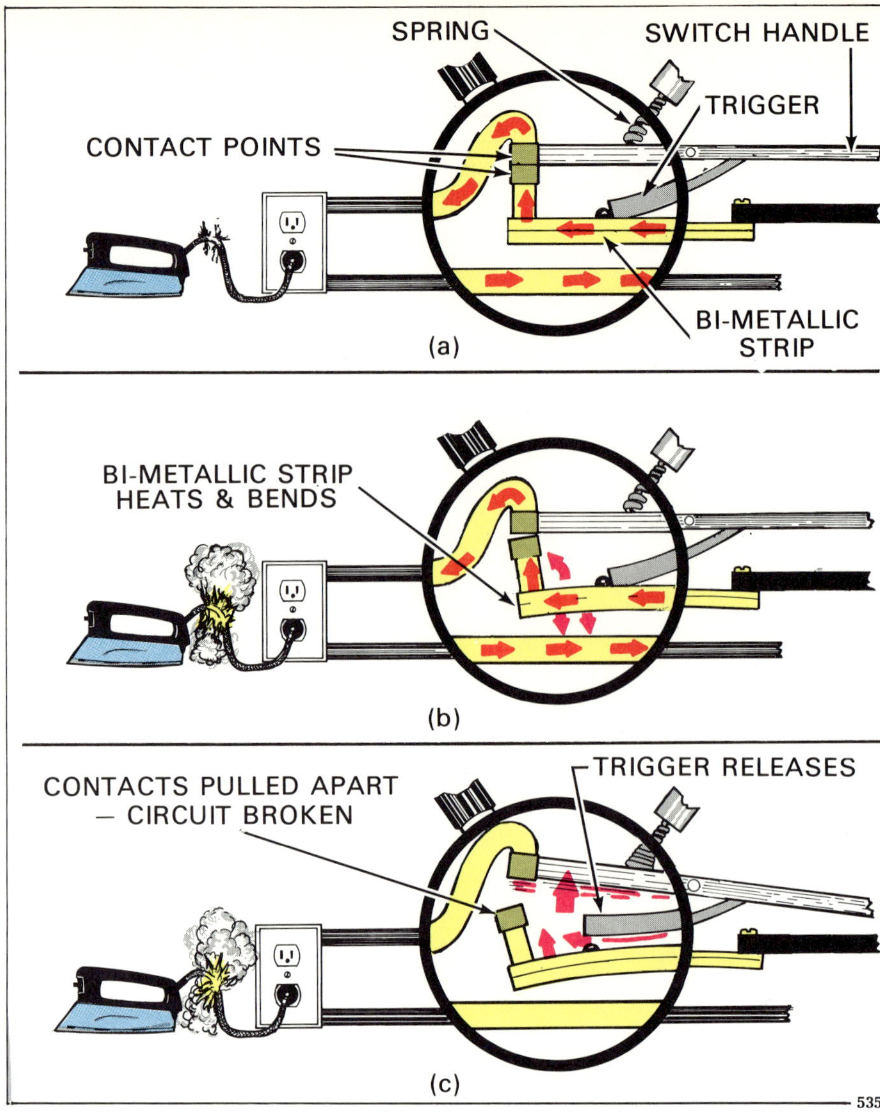

Figure 51. How a **circuit breaker** works. (a) Animated movement of electrons (current) through a circuit breaker when operating within its rated capacity. (b) With an overload, the bi-metallic strip heats and bends. (c) When heated sufficiently, the bi-metallic strip bends enough to release the trigger. The spring pulls the contacts apart, thus opening the circuit. After the bi-metallic strip has cooled, it straightens to its original position as in (a). Then, you can raise the switch handle and reset the breaker which turns on the current.

If you are using **fuses** and have installed the proper size, the fuse will heat and "blow" before there is over-heating of any part of the wiring system (Figure 50). You will have to insert a new fuse before service can be restored on a fuse-protected circuit.

The **circuit breaker** also heats when the load gets too heavy. The sensitive part is a bi-metallic strip. When the strip becomes hot enough, it bends (Figure 51) and releases the trigger on the switch. This automatically opens the circuit until the metal strip cools and you reset the breaker by hand.

Both types of wiring protection provide ample safety for your wiring system unless you tamper with them. It is difficult to tamper with circuit breakers. But, until recently, it has been rather easy to tamper with fuses.

(Tampering reduces the protection provided by the fuses.) New fuse-panel installations, that meet the present requirements of the National Electrical Code, are tamper resistant.

Figure 52. Common type of plug fuse.

In order to meet the problem of fuse tampering, manufacturers of fuses have developed **tamper-resistant fuses.** One type is shown in Figure 53. It replaces the common plug fuse (Figure 52). The tamper-resistant fuse has an adapter that locks into position in the fuse socket. The design of the fuse base along with the adapter arrangement makes it almost impossible to tamper with this type of fuse.

There are three different sizes of adapters. They are for 15-ampere circuits, for 20-ampere circuits and for 30-ampere cir-

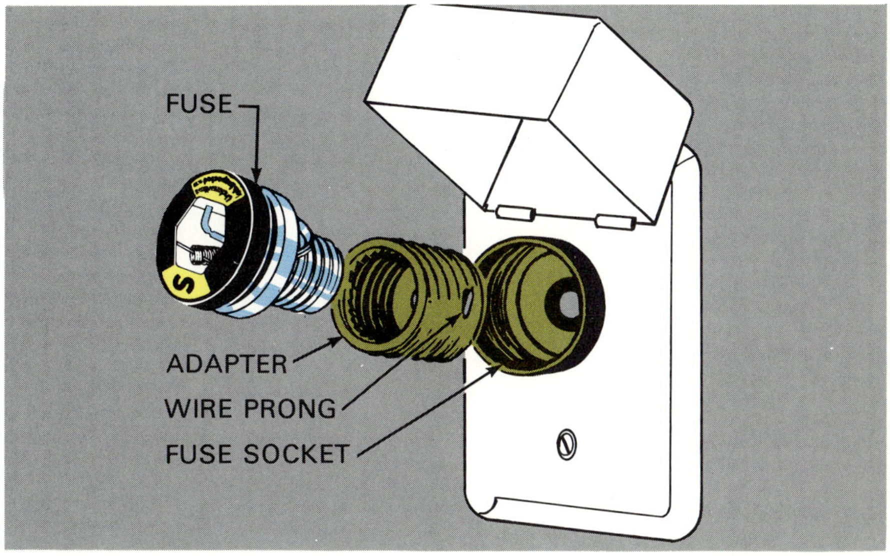

Figure 53. Tamper-resistant fuse with special adapter that fits in screw-socket base. Wire prong that extends from side locks adapter in place.

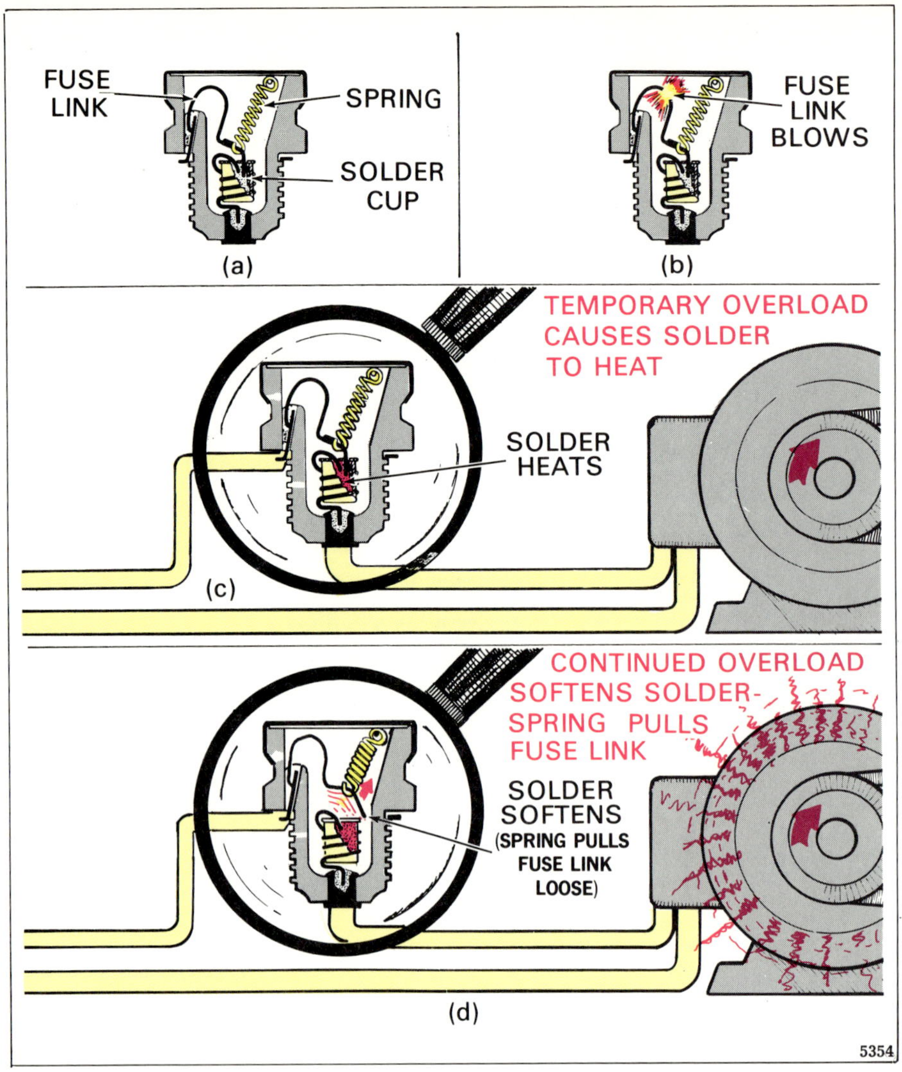

Figure 54. Operating principle of a time-delay fuse. (a) Principal parts consist of a fuse link, solder cup and spring. (b) In case of a short circuit, the fuse link blows as with a common fuse. (c) When there is a temporary overload, such as when a motor starts, the solder in the solder cup heats, but cools again when the overload stops. (d) If there is a continuous overload, the solder softens enough for the spring to pull the fuse link from the solder cup.

cuits. Each of these adapters differs in length and in thread type. Consequently, once an adapter is locked into place for a 15-ampere fuse, a fuse of higher amperage can not be inserted.

When a 15- or 20-ampere **fuse blows rather frequently,** some people replace it with a 30-ampere fuse. This is a form of tampering called **overfusing.** Often the cause of frequent fuse blowing is a motor on the circuit that starts a heavy load. The motor may pull from three to five times more current while starting than after it reaches full speed. This is often enough current to cause a common fuse to blow. No doubt this has irritated many people enough to cause them to overfuse.

Since this heavy current for starting the motor lasts for a very short time, there is no problem of overheating the wiring circuit. To meet this need, a "time delay" fuse — sometimes called a "dual-element" or "delayed-action" fuse — was developed. Figure 54 shows how this type of fuse works.

Note in Figure 54a that the three main parts consist of a fuse link, a solder cup and a spring. When a **short circuit develops,** the fuse link blows (Figure 54b) the same as in a regular fuse. But for **moderate overloads,** such as when a motor is starting, instead of the fuse link blowing, the solder cup starts to heat (Figure 54c). If there is a continuous overload, the solder in the solder cup softens, and the spring pulls the fuse link out of the solder cup (Figure 54d) thus opening the circuit. This is the arrangement that provides the delayed action needed for starting motors.

Note in Figure 55a that **the tamper-resistant fuse is a time-delay type.** You can also get cartridge fuses of the time-delay type (Figure 55b).

Circuit breakers provide the same time-delay action for motor starting that is provided by time-delay fuses. With the types of electrical loads found in most residences and businesses at the present time, it is best that all circuits be protected with time-

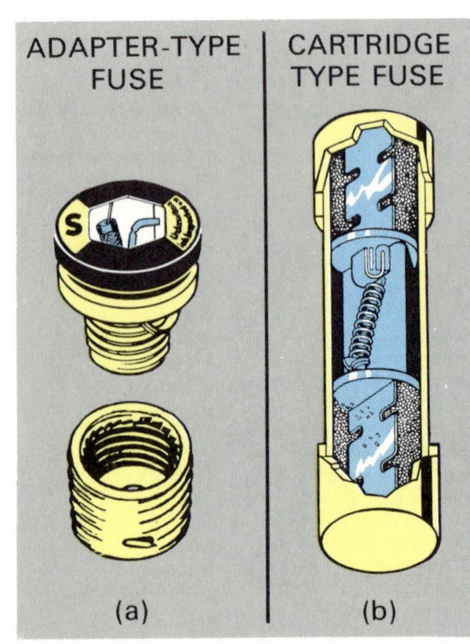

Figure 55. Time-delay fuses. (a) Adapter-type fuse for tamper-resistant use. (b) Cartridge-type fuse with the same type of delayed action as provided in the plug fuse.

delay fuses or circuit breakers. This is the protection the **National Electrical Code now specifies for all electrical circuits, with 30-ampere capacity or less, in residences.**

Upon completion of this section, you will be able to **replace fuses and circuit breakers** in service entrances and those protecting individual circuits.

How to replace fuses, or reset or change a circuit breaker, is discussed under the following headings:

1. Tools and Materials Needeed.
2. Locating Cause of Trouble.
3. Re-establishing Service on Circuit Breaker.
4. Re-establishing Service on Plug-Fuse Panel.
5. Re-establishing Service on Cartridge Fuse Panel.
6. Removing Old Circuit Breaker.
7. Installing New Circuit Breaker.

TOOLS AND MATERIALS NEEDED

Tools:

Medium screwdriver
Slip-joint pliers

Materials:

New fuse(s) or circuit breaker(s)

LOCATING CAUSE OF TROUBLE

It is usually easy to see the cause of the trouble, such as a short in the equipment cord you are using (Figure 49b), or an overload due to too much equipment (Figure 49a).

If you know the cause of the trouble, disconnect the appliances, omit the next three steps and proceed with step 4.

If you do not know the cause of the trouble, proceed as follows:

1. *Check lights fed by other circuits to make certain the power is on.*
2. *Check for equipment that is hot or has a burned odor.*

   If found, disconnect it from outlet and proceed with step 4. If not, proceed with step 3.
3. *Disconnect any equipment that has been added to the circuit or turned on within last 20 or 30 minutes.*
4. *Locate the circuit breaker, or fuse box, that protects this circuit (Figure 56).* It is probably located in the service entrance panel (Figure 56a and b) where the service entrance cable enters the building.
5. *Locate which fuse has blown or which circuit breaker has opened.*

   (a) If your circuit is **protected with a circuit breaker** (Figures 56b, and 58):

   (1) *Note which breaker switch is in the "OFF" or "TRIPPED" position (Figure 58).*

   (2) *Proceed with steps outlined under the section, "Re-establishing Service on Circuit Breaker," page 212.*

   (b) If your circuit is protected with **plug fuses** (Figure 56a):

   (1) *Inspect each fuse to see which one is blown.*

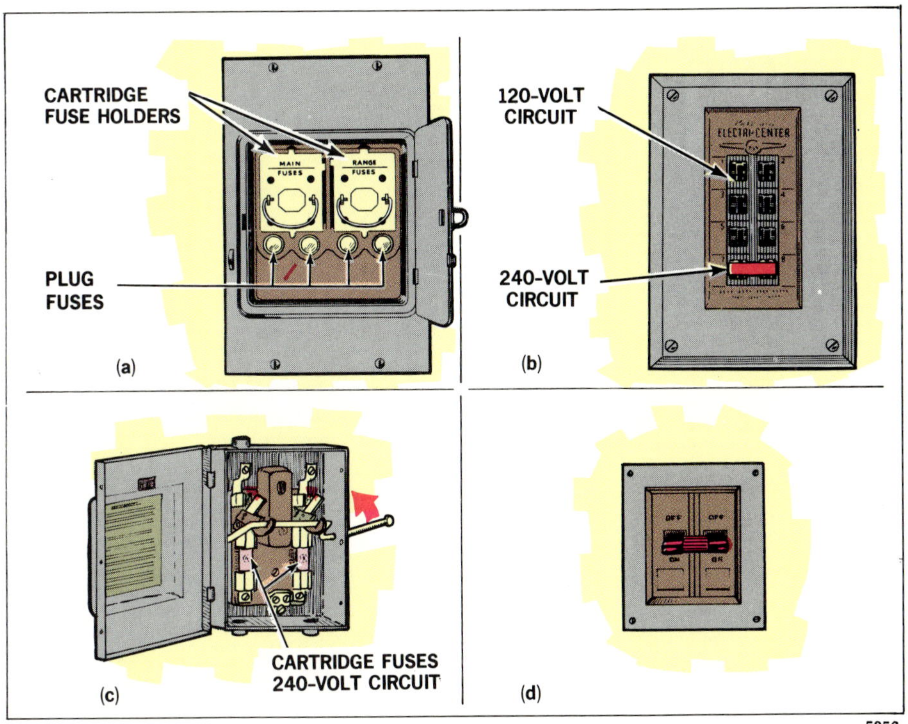

Figure 56. (a) Service-entrance **fuse panel** commonly used in homes to protect most or all of the circuits. (b) Service-entrance **circuit-breaker** panel commonly used in homes. (c) 240-volt fused switch for individual equipment circuit such as for a water heater. The type shown is for cartridge fuses. (d) 240-volt circuit-breaker switch for an individual equipment circuit.

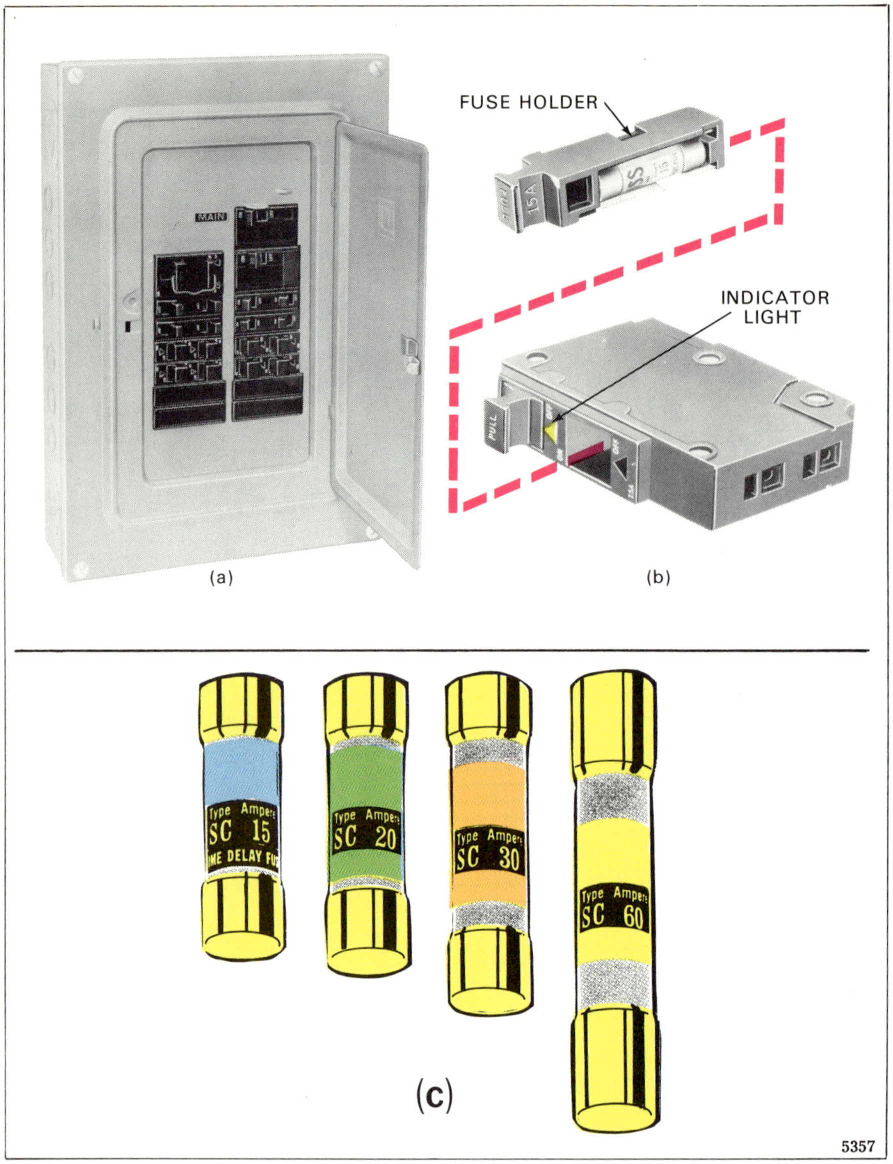

Figure 57. (a) A pull-out fuse-holder panel is similar in appearance to a circuit-breaker panel. Cartridge fuses are used in place of plug fuses. (b) When fuse is blown, the indicator light glows. Pull-out fuse holder is removed when cartridge fuse is to be replaced. (c) Fuse sizes (Type SC) have been approved by the Underwriters' Laboratories for use with these panels so it is impossible to overfuse. All of the fuses are of the time-delay type.

If none of the fuses appear to be blown, it may be due to a defective fuse. The procedures that follow may have to be repeated with each fuse until the defective one is found.

(2) *Proceed with steps outlined under the section, "Re-establishing Service on Plug-Fuse Panel," page 213.*

(c) If your circuit is protected with **cartridge fuses** (Figures 56c and 57):

Proceed with steps outlined under the section, "Re-establishing Service on Cartridge Fuse Panel," page 214.

RE-ESTABLISHING SERVICE ON CIRCUIT BREAKER

To re-establish service on a circuit breaker, proceed as follows:

1. *Check instructions on front of breaker panel (Figure 58).*

   The procedures given will reset the trigger mechanism.

2. *Reset circuit breaker and move to "ON" position (Figure 58).*

   If the breaker will not remain in the "ON" position, wait for a minute or two and reset again. The heater strip requires a short time to cool before the breaker can be reset.

   If, after a cooling period, the breaker still will not stay in the "ON" position, the short circuit or overload condition is still present or the circuit breaker is faulty. If you are sure there is no overload or short on the circuit, proceed to the section, "Removing Old Circuit Breaker," page 216.

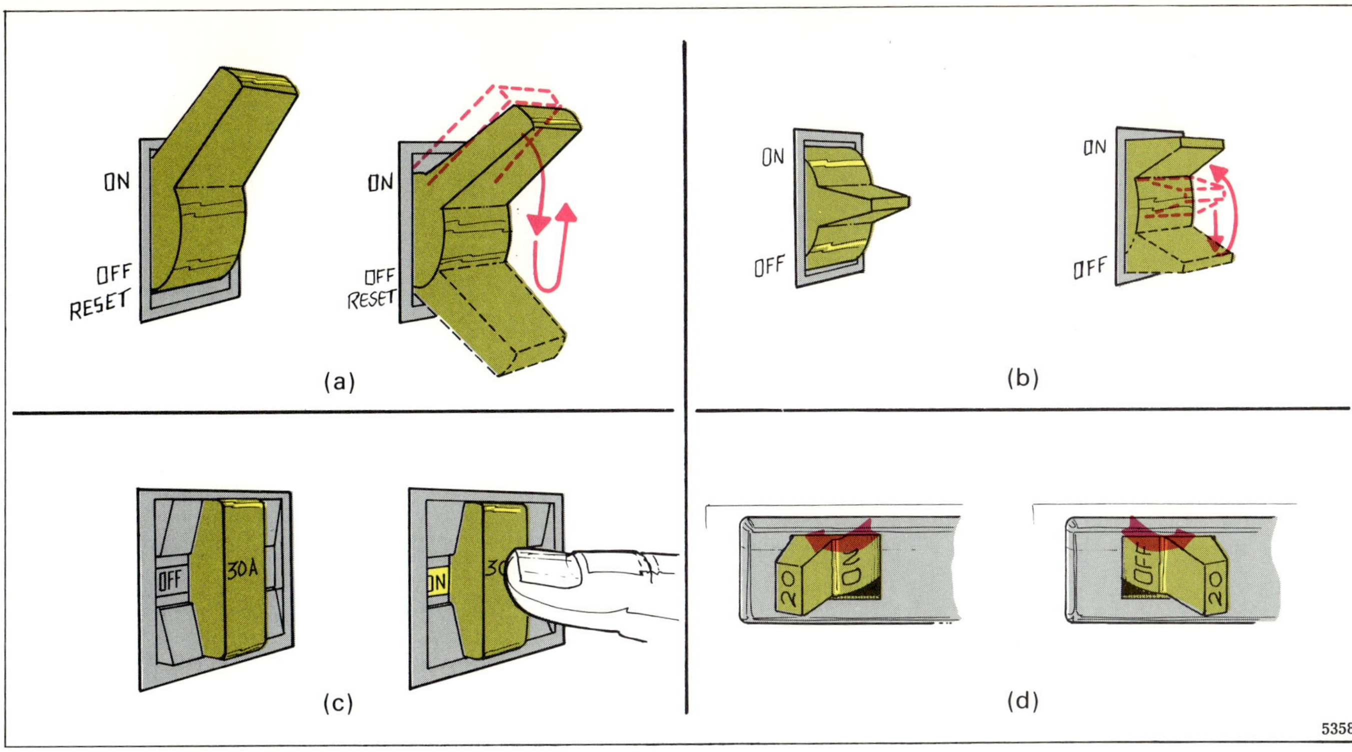

Figure 58. Procedures for setting different kinds of circuit breakers. (a) When "tripped," lever is past the "ON" position. To reset, handle is moved to "RESET" and back to "ON." (b) When tripped, handle is in center position between "ON" and "OFF." To reset, move handle to "OFF" position and back to "ON." (c) Push-type breaker. When tripped, "OFF" shows in window of switch. Press in on switch and release. If reset, "ON" will show in window when switch returns to normal position. (d) Circuit breaker.

RE-ESTABLISHING SERVICE ON PLUG-FUSE PANEL

To re-establish service on a plug-fuse panel, proceed as follows:

1. *Pull main switch (Figure 59a or b).*

Pulling the main fuse block of the **service entrance switch** (Figure 59a) will discontinue service all over the building. If this causes inconvenience to other operations in the building, leave the current on and proceed directly to the next step.

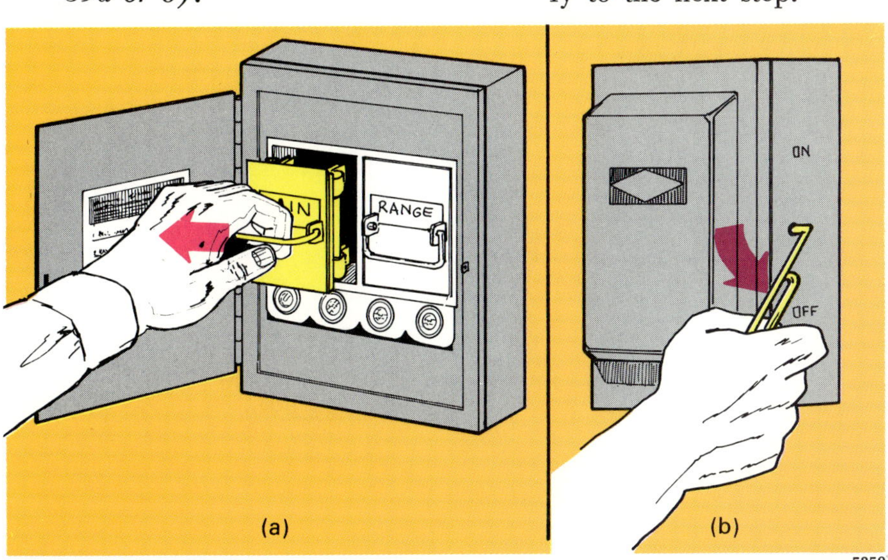

Figure 59. Disconnect power at service entrance switch before removing blown fuses.

2. *Screw out blown fuse (Figure 60a).*

   If the main switch is "ON," be sure you are standing on a dry floor. If the floor is wet, stand on a dry board or rubber mat (Figure 60b). This protects you against shock in case you happen to touch a live part while screwing out the fuse. Do not touch metal parts on box or any pipes.

3. *Examine fuse to determine the type (Figure 61).*

4. *Note number on top or bottom of fuse plug that shows amperage (Figure 61).*

5. *Replace with fuse of correct size.*

   If the fuse you removed was an old-style common fuse, replace it with the same size tamper-resistant type if possible. It costs very little more to modernize your wir-

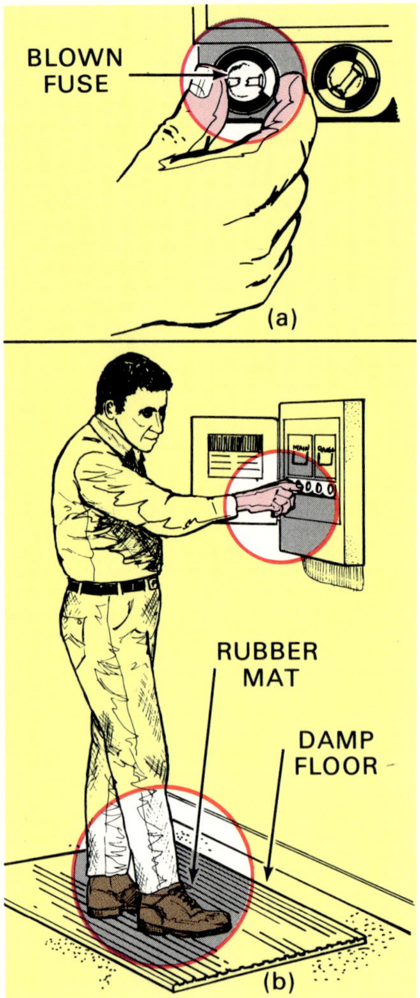

Figure 60. (a) Identifying blown fuse. (b) Use safety precautions in replacing fuses.

ing-protection system by installing tamper-resistant fuses.

6. *Throw main switch to "ON" if it was turned off.*

Figure 61. Examine fuse to be replaced to determine type and amperage.

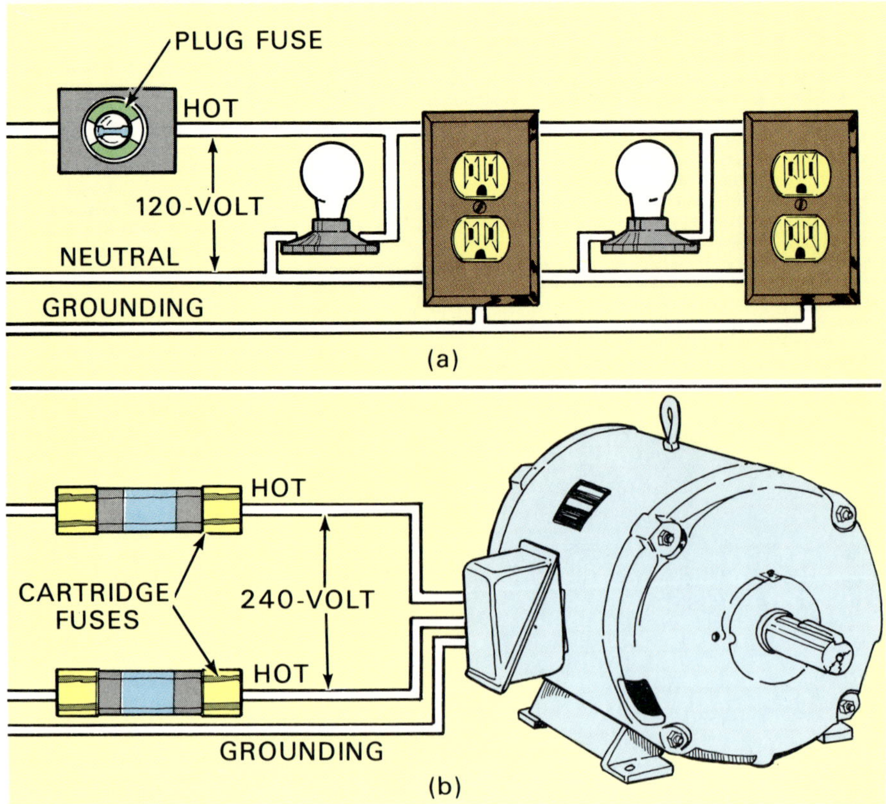

Figure 62. (a) A single fuse protects the 120-volt system, but (b) two fuses are used in the 240-volt system.

RE-ESTABLISHING SERVICE ON CARTRIDGE-FUSE PANEL

Before re-establishing service on a cartridge fuse panel, you must first determine if it is of the SC fuse type or the standard cartridge fuse type.

To re-establish service on a cartridge fuse panel, proceed as follows:

a. If your fuse panel is designed for the **type SC fuses** (Figure 57), proceed as follows:

1. *Check to see which indicator light is glowing.*
2. *Remove pull-out fuse holder next to it.*
3. *Replace fuse with type SC cartridge fuse of the same size.*

Type SC fuses are the ones shown in Figure 57. They are of different design and size than the standard cartridge fuse shown in Figure 65. If you are replacing a standard 20-ampere fuse you can use either a 15-ampere size or 20-ampere size. But if you are replacing a 15-ampere SC fuse, you will have to use another 15-ampere fuse. A 20-ampere fuse will not fit the holder.

4. *Insert fuseholder into original socket.*

The fuseholders are also designed so you can not make the mistake of putting a 20-ampere fuseholder into a 15-ampere socket.

b. If your fuse panel, or fused switch, is similar to those shown in Figures 59 and 63, it uses

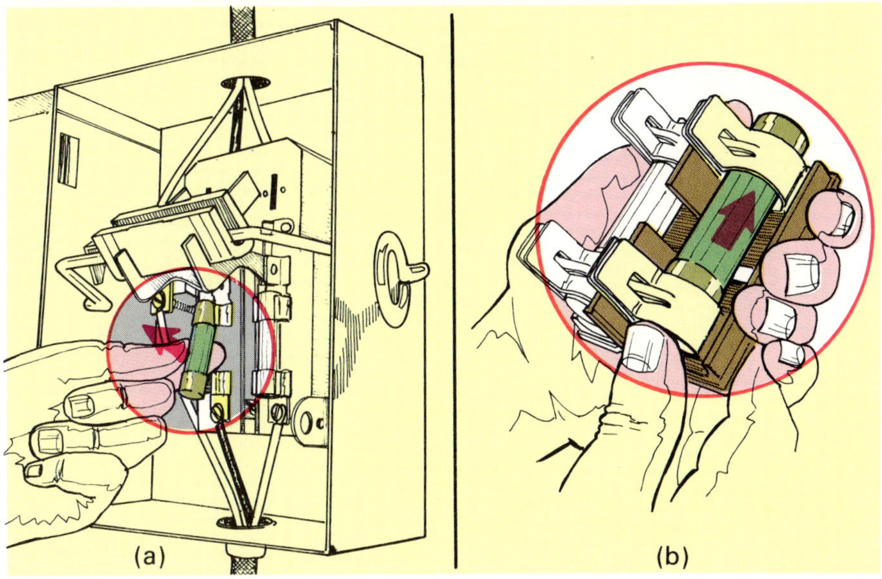

Figure 63. Check for faulty fuse by removing and replacing with new fuse.

Standard cartridge fuses like those shown in Figure 64. Keep in mind that cartridge fuses of this type are most commonly used on 240-volt circuits. In that case, the circuit has two fuses to be checked (Figure 62b) instead of one as with a 120-volt lighting circuit (Figure 62a). Proceed as follows:

1. *Remove main fuse holder (Figure 59a) or throw main switch to "OFF" (Figure 59b).*
2. *Remove one fuse from fuse clips (Figure 63a or b.)*

   Sometimes the outer shell is browned from heat. If so, it is the one most likely to be blown. If both look the same, it makes no difference which one is checked first.
3. *Examine old fuse to determine which kind it is (Figure 64).*
4. *Replace old fuse with new one of same amperage rating.*
5. *Throw main switch to "ON."*
6. *See if service is restored. If not, proceed with 7.*
7. *Throw main switch to "OFF."*
8. *Remove new fuse and put old one back in place.*
9. *Remove second old fuse and replace with new one (Figure 65a or b).*
10. *Throw main switch to "ON."*
11. *See if service is restored on circuit. If not, it is probable both fuses have blown. Proceed with 12.*
12. *Throw main switch to "OFF."*
13. *Replace first fuse checked with new one.*

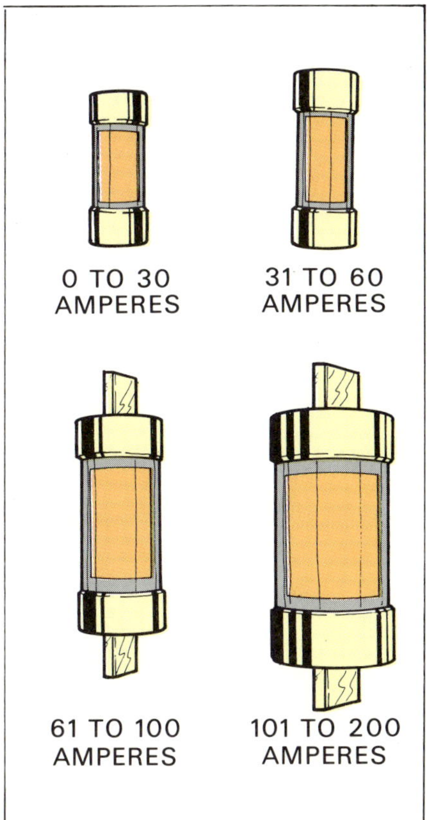

Figure 64. Kinds of standard cartridge fuses.

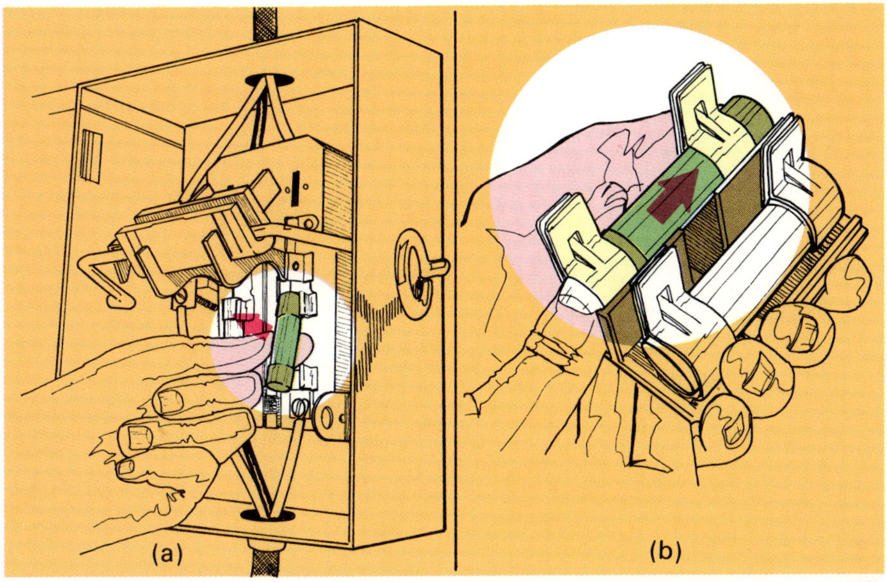

Figure 65. If first fuse replaced is not faulty, replace second fuse.

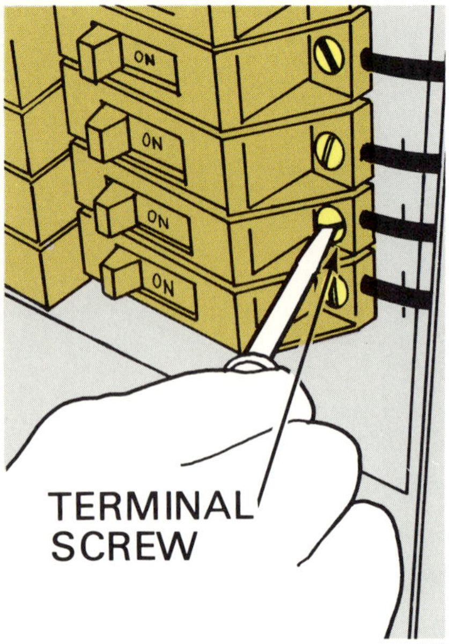

5366

Figure 66. Loosen terminal screw and remove wire to remove circuit breaker.

14. *Throw main switch to "ON."*

REMOVING OLD CIRCUIT BREAKER

To remove an old circuit breaker, proceed as follows:

1. *Throw main switch to "OFF."*
2. *Remove cover to switch box.*
3. *Loosen terminal screw and remove the wire on the circuit breaker giving trouble (Figure 66).*
4. *Pull circuit breaker toward you.*

Usually the end of the circuit breaker nearest the center of the box is held in place by the electrical contacts (Figure 67). The opposite end is secured by a metal projection on the circuit breaker into a slot or under a screw. It may be removed by continued pressure on the circuit breaker after the electrical contacts have released (Figure 68).

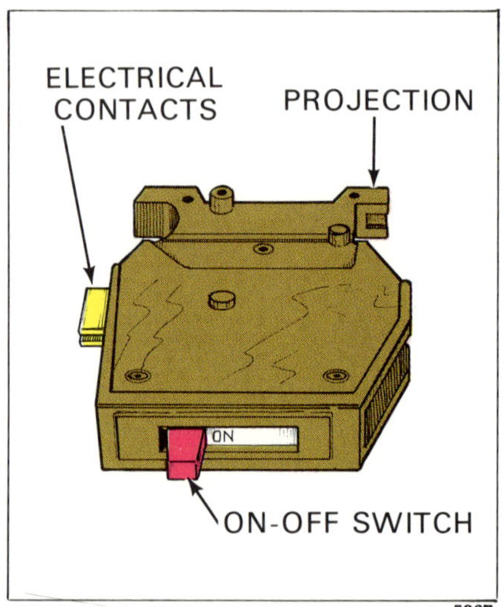

5367

Figure 67. Circuit breaker.

INSTALLING NEW CIRCUIT BREAKER

To install new circuit breaker, proceed as follows:

1. *Insert metal projection on back of circuit breaker into the slot or under the screw (Figure 68).*
2. *Press electrical contacts onto prong of electrical bus bar (Figure 67).*
3. *Loosen the terminal screw.*
4. *Insert the circuit wire under the terminal screw.*
5. *Tighten the terminal screw.*
6. *Replace switch box cover.*
7. *Throw main switch to "ON."*
8. *Turn circuit breaker to "ON."*

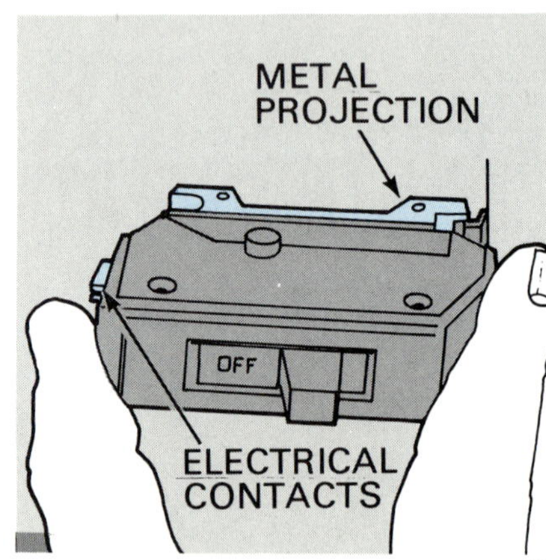

5368

Figure 68. Provisions for attaching circuit breaker to the service panel.

B. Replacing Single-Pole Switches

Single-pole switches (and dimmers) are used when a light is turned on, or off, from only one location. Figure 69 shows the use of this type of switch in a bathroom. Switches will wear with use until they finally fail completely. Note in the simplified sketch (Figure 69b) that there are only two wires connected to the switch. This makes it very simple to replace. Light dimmers are wired like switches. They are widely used in rooms where varying levels of light are desired. The dimmer controls the lighting level and also turns the light on or off (Figure 70).

When purchasing a new switch or dimmer it is often difficult to decide which one to purchase. There are a wide variety of trade names and a wide range of prices. Your first check should be to see that it has the UL label attached or molded on it. This is assurance that the switch or dimmer has met certain safety standards established by Underwriters' Laboratories.

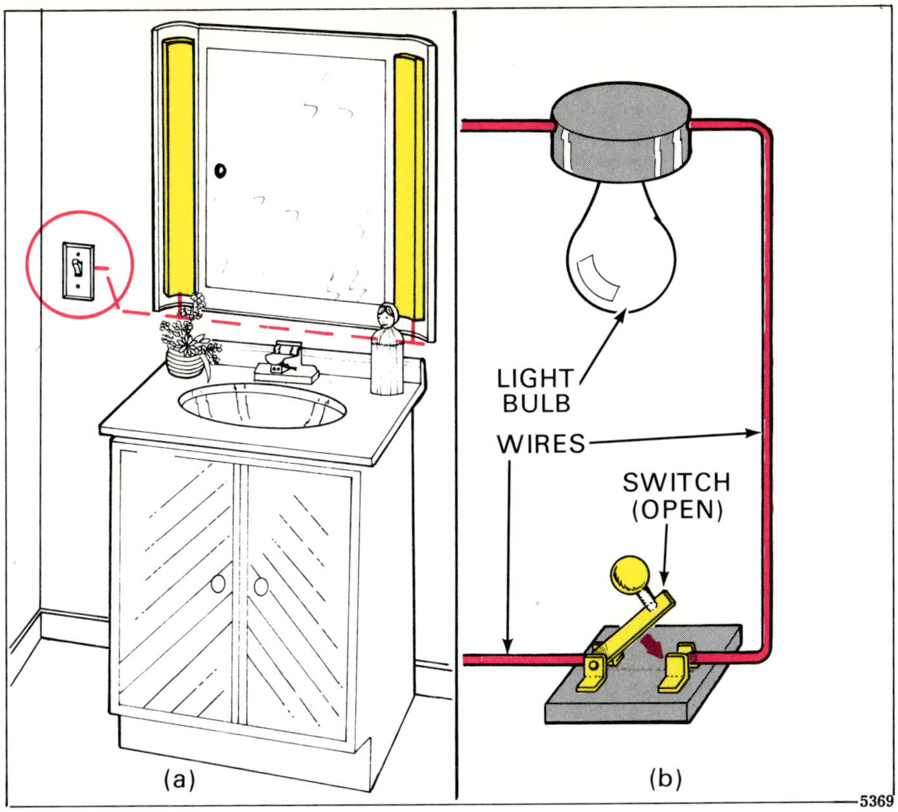

Figure 69. (a) A single-pole switch controls lights from only one location. (b) Simplified sketch of how a single-pole switch works. (An exposed switch such as this one is a hazard and should never be used.)

If your house is wired with aluminum wire be sure the switch is marked "CO/ALR" on the mounting strap.

Upon completion of this section, you will be able to **replace single-pole switches** in the home, the shop or other buildings.

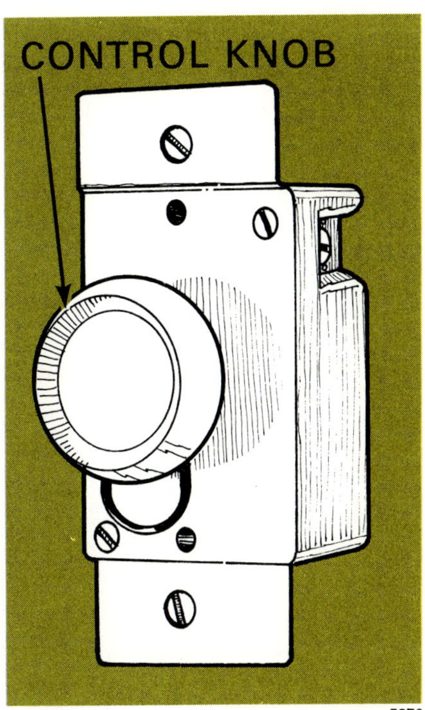

Figure 70. Incandescent dimmer.

Figure 71. Fluorescent dimmer.

Replacing single-pole switches is discussed under the following headings:

1. Tools and Materials Needed.
2. Removing Old Switch.
3. Installing New Switch.
4. Installing Light Dimmer.

TOOLS AND MATERIALS NEEDED

Tools:

Screwdriver
Pocket knife
Electrician's pliers
or thin-nose pliers
Neon test lamp*

Materials:

New single-pole switch, or dimmer.

REMOVING OLD SWITCH

To remove an old switch, proceed as follows:

1. *Pull main switch or remove fuse (or trip circuit breaker) on circuit supplying power to switch.*

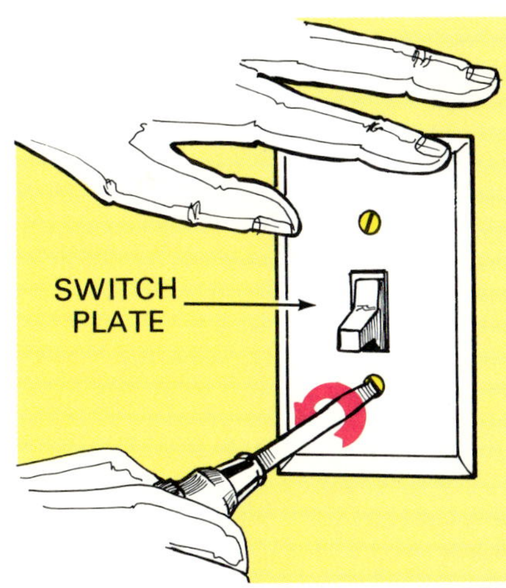

Figure 72. Removing switch plate.

*A neon test lamp is inexpensive and available almost any place where electrical supplies are sold. It will glow on an extremely small amount of current and a wide range of voltages. For this reason it is more accurate, and can be used under a wider range of conditions and is safer than a light-bulb test lamp. A regular light-bulb test lamp will not work satisfactorily on this kind of test unless it is of low wattage as compared to the lamp controlled by the switch.

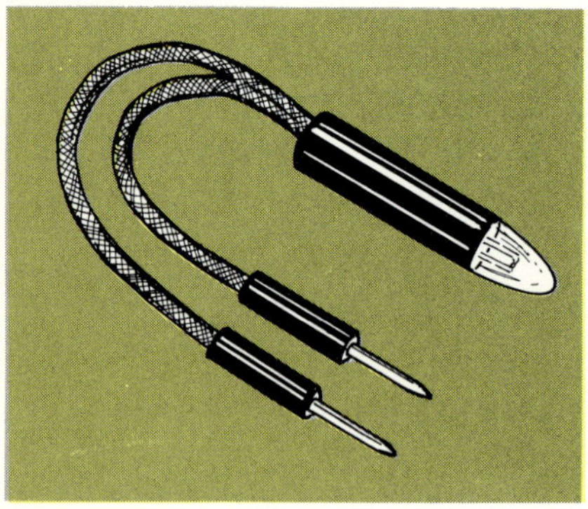

5373

Figure 73. Neon test lamp.

2. *Remove screws holding switch plate (Figure 72).*

3. *Remove switch plate.*

If there is any question about the circuit being

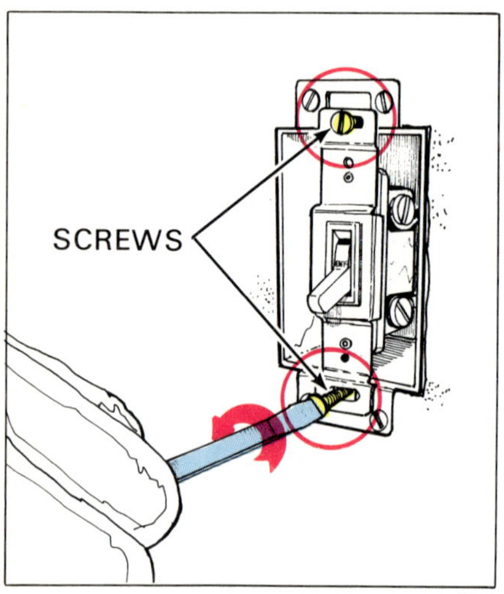

5374

Figure 74. Removing screws holding switch in outlet box.

"dead," check it with a neon test lamp (Figure 73). If lamp glows, the circuit is still "hot." Check further for main switch that controls the circuit.

4. *Remove screws holding switch in place (Figure 74).*

5. *Pull switch from outlet box (Figure 75).*

There should be plenty of slack wire in the outlet box back of the switch.

6. *Loosen screw terminals and*

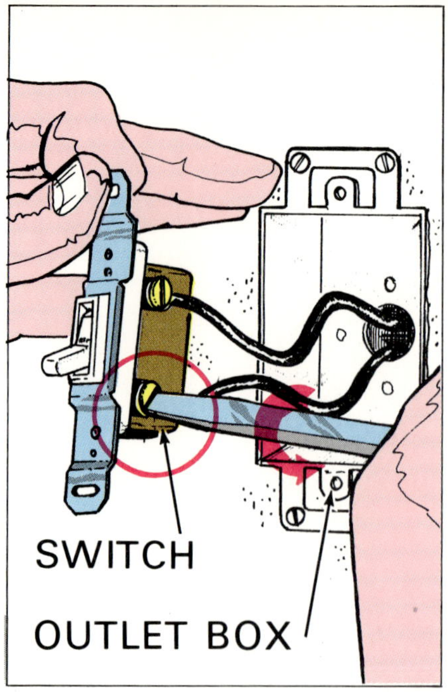

5375

Figure 75. Removing switch.

unhook wires from switch (Figure 75).

If the switch is of the back-wired type, wires have been pushed into holes on the back of the switch. These wires may be removed by applying pressure with a small screwdriver to the metal clip in the hole nearest the wire. The wire may then be pulled out (Figure 76). Some back-wire devices require loosening the

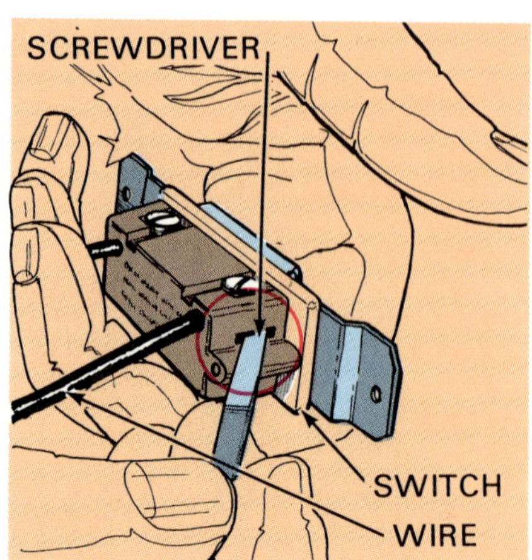

5376

Figure 76. Removing wire held by pressure clamp from back wired switch.

screw nearest the wire to release the wire (Figure 77).

INSTALLING NEW SWITCH

To install a new switch, proceed as follows:

1. *Scrape bare ends of wire to remove dirt and corrosion.*

2. *Loosen screw terminals on new switch.*

Do not remove them. Some back wired switches require loosening the screw (Figure 77) while others (Figure 76) do not.

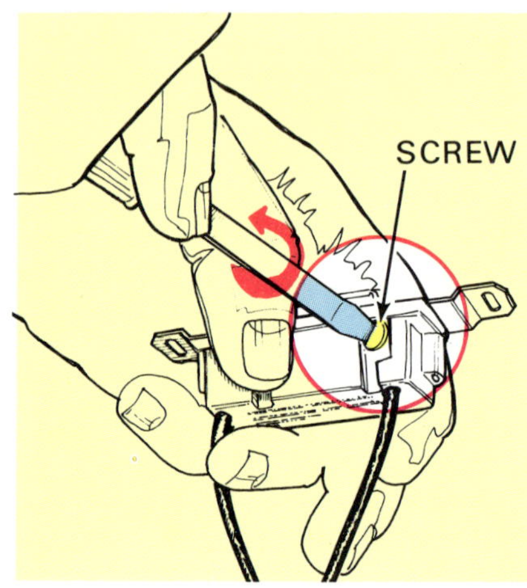

5377

Figure 77. Loosening screw to remove wire from back-wired switch.

3. *Examine switch to see which end should be up.*

The "OFF" and "ON" markings in the switch handle should read right-side up.

4. *Connect wires to terminal posts (Figure 78) or insert wires into back of switch (Figures 76 and 77).*

Either wire can be connected to either post. If the switch is equipped with 4 screws or 4 insert openings, be sure to connect wires on opposite sides (Figure 79).

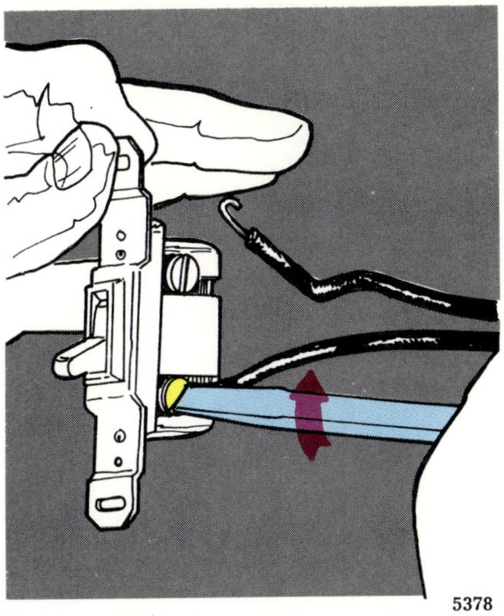

Figure 78. Connecting wires to terminal posts.

Be sure the hook on each wire extends around the terminal post in the same direction that the screw tightens. If aluminum wires are used, wrap the wire at least ⅔ of the way around the terminal screw, tighten the screw until snug, and then tighten an additional half turn.

Switches vary in design. Many of them are front-connected. Some have the terminal screws on the sides while others have them on the ends of the switch. Others use back wiring. If back wiring is used, be sure the wire is locked in place. Some back-wire devices tighten the electrical wire with a screw (Figure 77). Others are secured with a pressure clip (Figure 76). Switches having no conductor marking or having back-wired terminals or screwless terminals of the push-in type may not be used with aluminum conductors. Switches marked "CO/ALR" on the mounting strap are approved for direct connection to aluminum conductors.

5. *Push switch into outlet box.*

If you have trouble getting the slack wire back into the box, bend both wires so they will "accordian" back into the outlet box (Figure 80).

6. *Replace screws that hold switch in place (Figure 81).*

Note that the holes in the switch mounting straps are slotted. This is so the switch may be adjusted to a straight up-and-down position even though the box is set crooked.

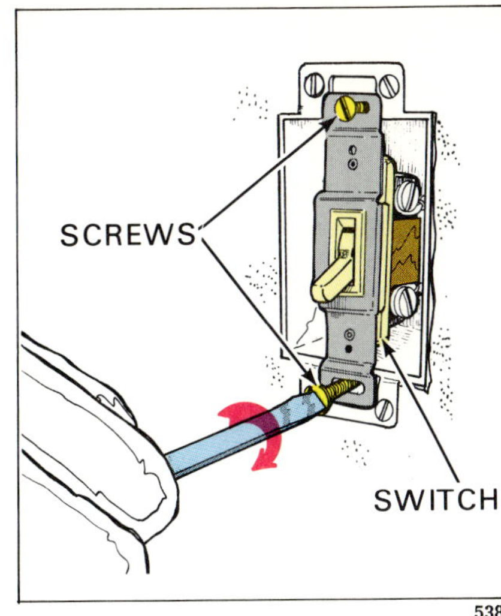

Figure 81. Replace screws to hold switch in place.

7. *Replace switch plate and screws.*

If the switch plate is of plastic material, do not tighten screws too much. It may crack the plate.

8. *Re-establish service to circuit.*

9. *Test switch to see if it works.*

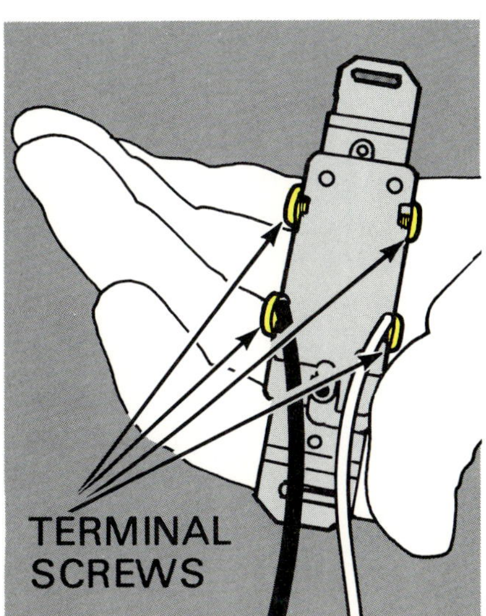

Figure 79. Switch equipped with four terminal screws.

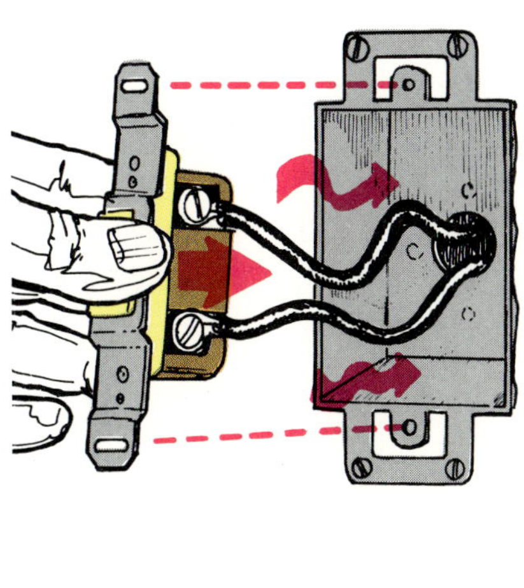

Figure 80. Bend wires and push back into outlet box.

INSTALLING LIGHT DIMMER

To install a light dimmer, proceed as follows:

Follow the same steps observed in steps 1 through 7. in the preceding section. Push dimmer knob onto control rod (Figure 82). Complete steps 8 and 9 in the preceding section.

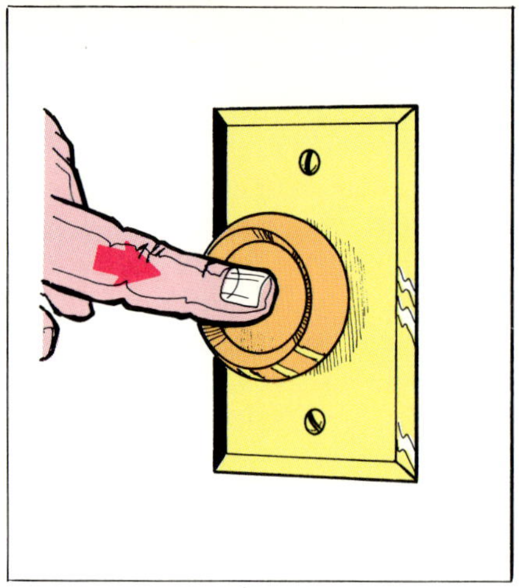

Figure 82. Push to install control knob on dimmer.

C. Replacing 3-Way Switches

Three-way switches are used so that a light (or lights) may be turned on or off at either of two locations (Figure 83). Some people call them "2-way" switches but that is not correct. The term "3-way" refers to the number of wire connections, not to the number of switches. They are widely used because of their convenience in saving steps when passing from one room to another. A 3-way switch is used for a yard light when you wish to turn the light on or off at either of two locations, such as either at the house or at another building.

The quality of 3-way switches can be determined in the same manner as with single-pole switches.

Figure 84 shows the operating principle of 3-way switches Figure 84e shows one type of commerical switch and how it is connected. Other types are shown in Figure 85.

It is more difficult to replace a 3-way switch than a single-pole switch. In fact, you will have trouble reconnecting a new 3-way switch unless you know how the switches work together, or unless you follow very carefully the procedure given here. If no one has tampered with the switch—changed the wire connections or completely removed the switch—it will not be necessary to use all of the following procedures.

If it has *four wires connected to it*, it is a 4-way switch. Follow procedures in job D, page 226.

It is usually difficult to know which of the two switches is giving trouble until you test them.

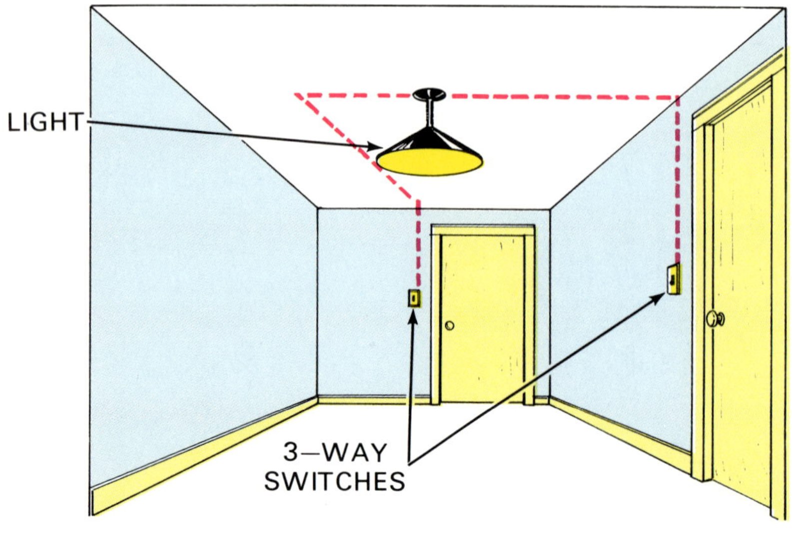

Figure 83. Three-way switches turn a light on or off at two different locations.

*These procedures are for the standard snap-type switches in most common use. The same testing procedures will not work with mercury switches or low-voltage remote-control switches.

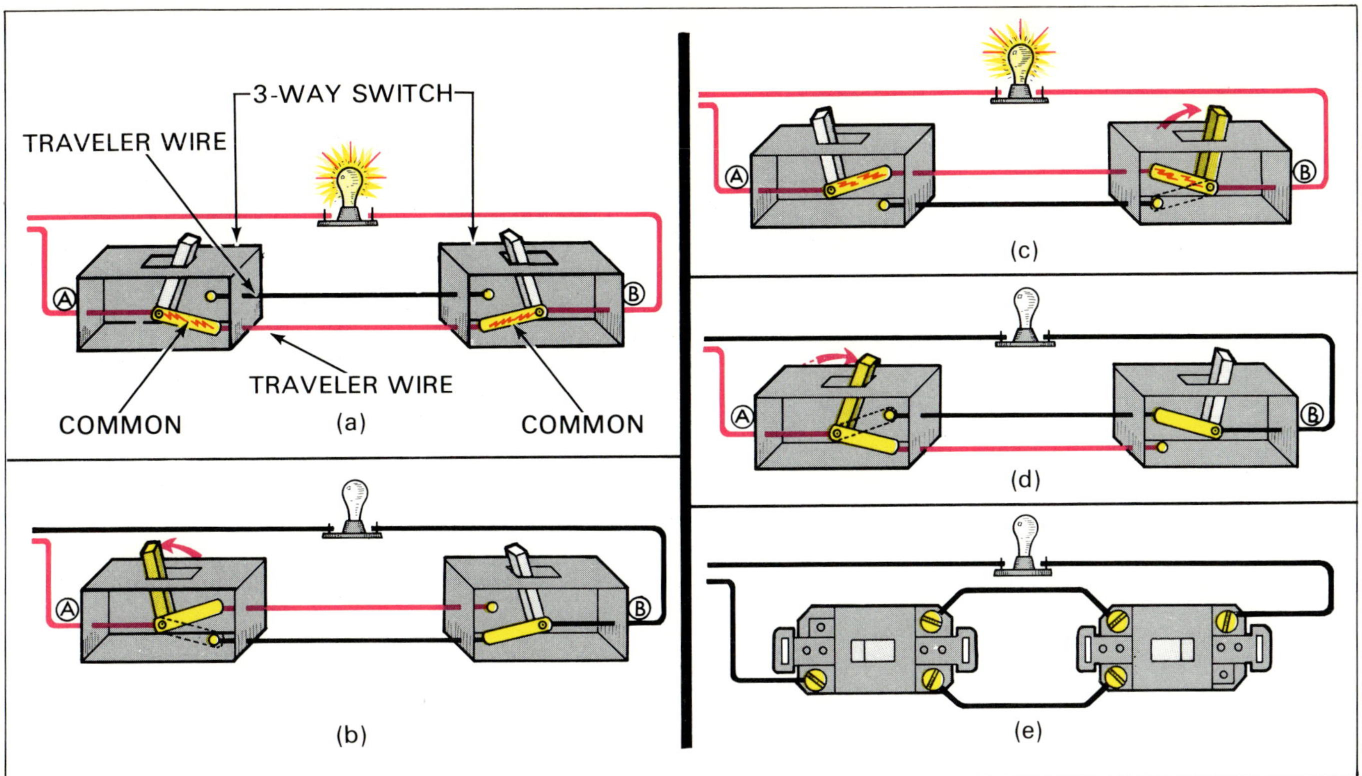

Figure 84. How 3-way switches work: (a) Current entering common wire "A" flows through the left-hand switch and lower traveler wire, then through the right-hand switch-leg "B" to the light. The light is **on.** The path being followed by the current is shown in red. (b) Light is turned **off** when left-hand switch is moved to opposite position which connects it to the upper traveler wire. Right-hand switch is still connected to lower traveler wire. (c) Light is turned **on** when right-hand switch is moved so it connects to the upper traveler wire to complete the circuit. (d) With left-hand switch changed to connect to the lower traveler wire, the circuit is broken. The light is **off.** (e) Wire connections as they appear on standard commercial-type switches.

Either switch may turn the light on at times, while at other times both switches have to be worked to get any light. This may lead you to think that both switches are faulty. The trouble is usually in only one switch.

Upon completion of this section, you will be able to **replace 3-way switches correctly.**

How to replace 3-way switches is discussed under the following headings:

1. Tools and Materials Needed.
2. Testing for a Faulty Switch.
3. Locating Common Wire and Terminal.
4. Disconnecting the Switch.
5. Installing the Switch.

TOOLS AND MATERIALS NEEDED

Tools:

Screwdriver with insulated handle
Pocket knife
Pliers (preferably thin-nosed pliers)
Neon test lamp
Pencil

Materials:

New 3-way switch

TESTING FOR A FAULTY SWITCH

To test for a faulty switch proceed as follows:

1. *Work both switches until light burns (Figure 86).*
2. *Snap each switch to opposite position.*
3. *Remove screws holding switch plate on one of the switches (Figure 72).*
4. *Remove switch plate.*

CAUTION — The wires are still "hot" so the switch may be tested. There is no danger of shock if you are standing on a dry floor and if you keep away from any plumbing or other metal objects while testing the switch. If you must stand on the ground or on a damp or wet floor, provide electrical insulation by using a dry board or box.

5. *Label the terminal screws on switch: "A", "B", and "C" (Figure 87).*

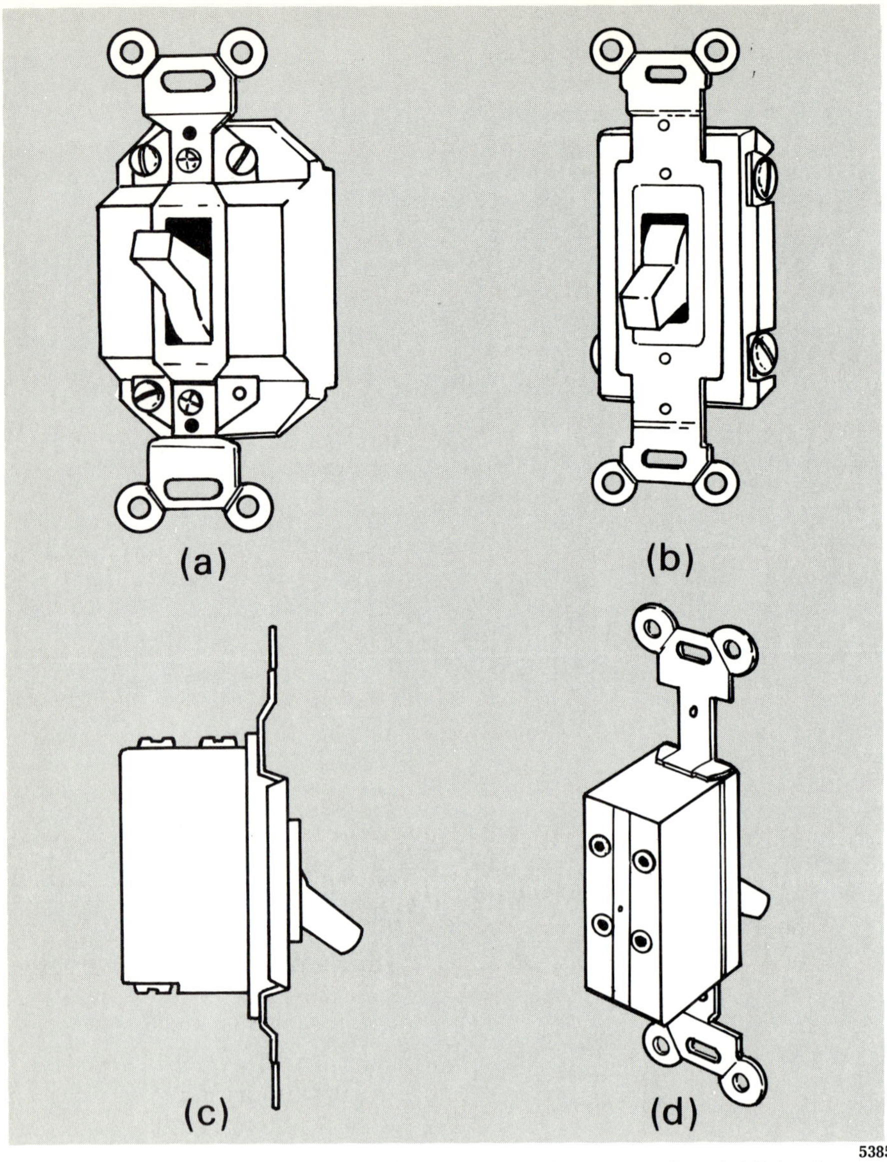

Figure 85. Types of 3-way switches in relation to wire connections: (a) front-connected, (b) side-connected, (c) end-connected and (d) back-connected.

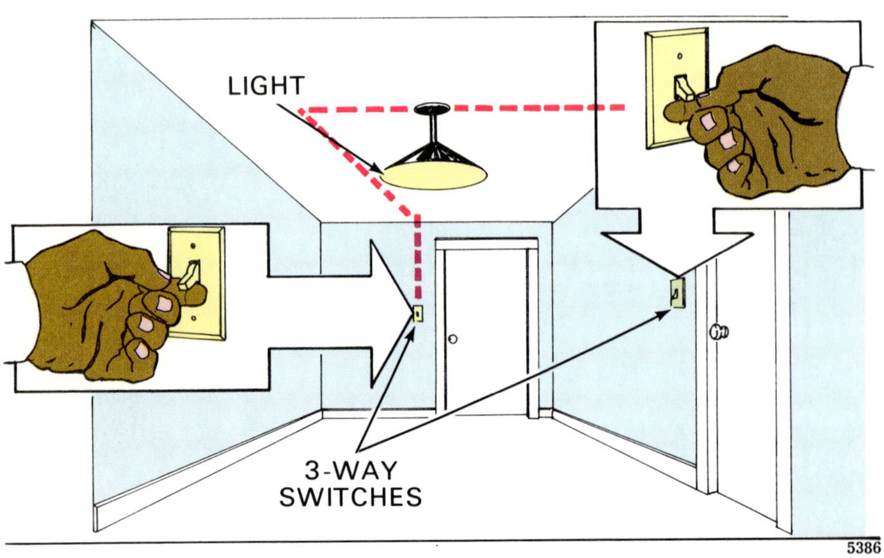

Figure 86. Work both switches until the light burns.

Pencil the letters on a paper next to the screw terminals. To use a paper, cut a hole the size of the switch box in a blank piece of paper and fasten it to the wall over the switch box.

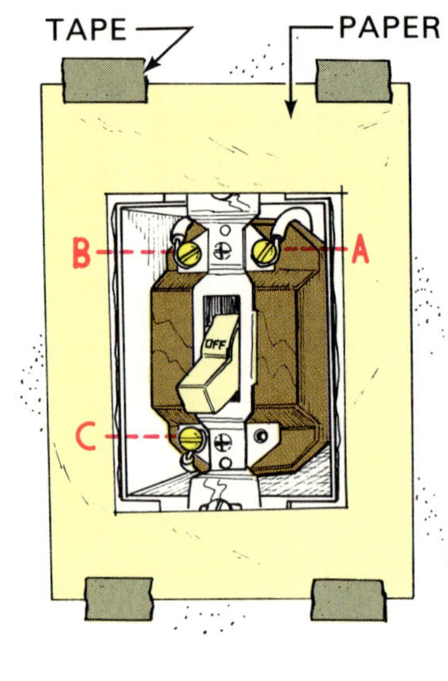

Figure 87. Label the terminals on the switch for identification while testing.

On **back-connected** switches, use the terminal screws for testing purposes. They are connected to the appropriate wires.

If your switch is a **front-connected** type (Figure 89a), you can test it without removal. Proceed to step 6.

If your switch **terminals are not accessible** from the front (Figure 89b, c and d), you must remove it from the receptacle. Remove the screws that hold the switch in place and pull it out (Figure 74). Then label each terminal.

6. *Touch test lamp wires to terminals A and B, then to terminals A and C, then*

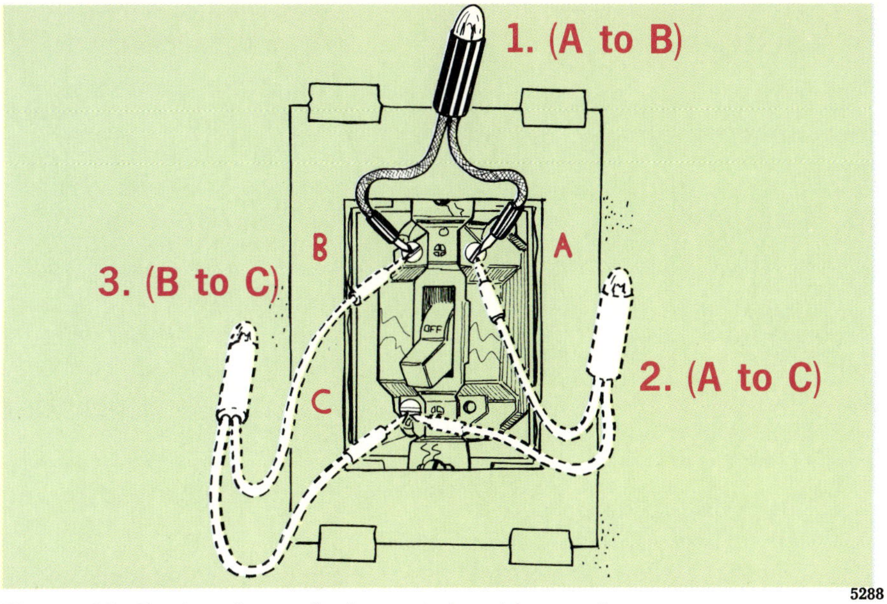

Figure 88. Testing for a faulty switch with test lamp.

to terminals B and C (Figure 88).

If the **light burns with two of the combinations,** your switch is connected by the so-called "hot" system. The remainder of these procedures do not apply. It is wired differently than shown in Figure 84. Employ a competent electrician to change your switch since it is important that certain safety precautions be followed.

If the **test light burns** with one of these combinations, the switch you are testing is faulty. It should be replaced with a new one. Proceed to next heading.

If the **test light does not burn** with any of these combinations, the switch you are testing is good. It is the other switch that is faulty. Reassemble this switch and start over again with the other switch.

LOCATING COMMON WIRE AND TERMINAL

When replacing a 3-way switch, your most important job is to find which is the **common terminal** (Figure 89) and which is the **common wire** that connects to it.

On most **front-connected switches,** the common terminal is either black or copper colored (Figure 89a).

On **side-connected switches,** the common terminal may be by itself on one side of the switch (Figure 89b).

On **end-connected switches,** the common terminal may be by itself on one end of the switch (Figure 89c).

On **back-wired switches,** the common terminal may be by itself on one side of the switch (Figure 89d), or it may be labeled.

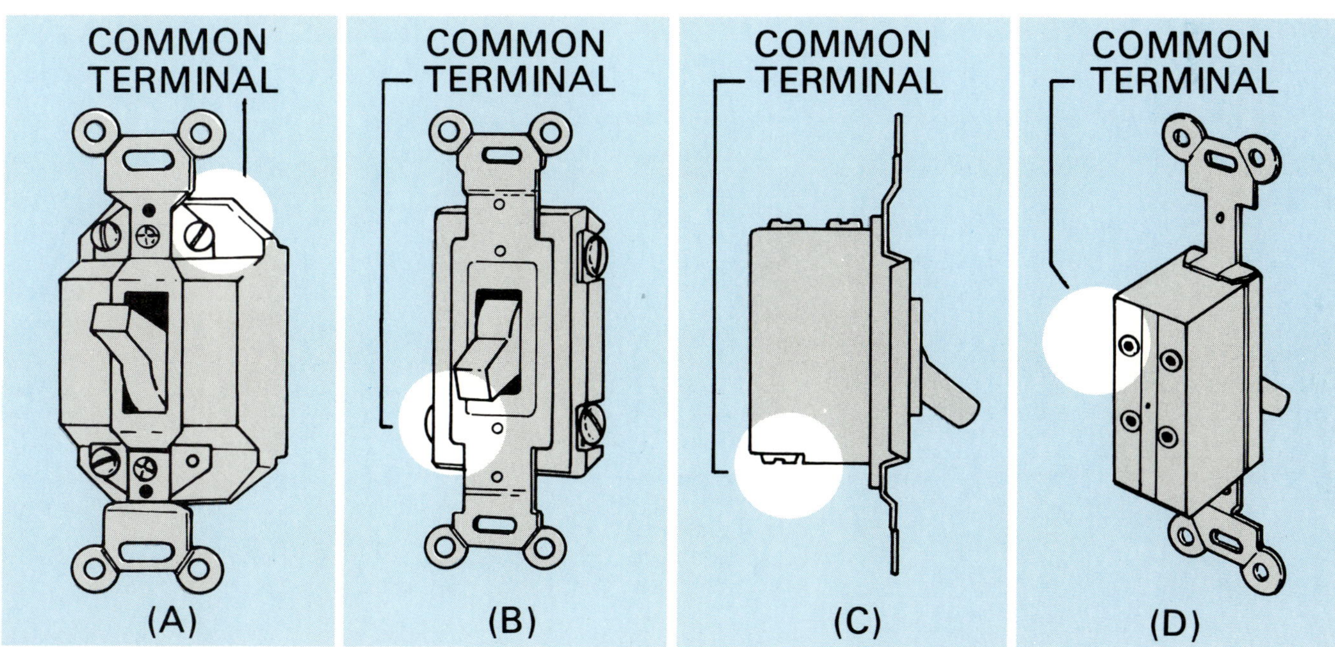

Figure 89. Identifying the common terminal.

If you are able to locate the **common terminal** on your switch, proceed to next heading.

If you **cannot tell which is the common terminal,** you probably have (a) a front-connected switch with all brass screws or (b) the switch has been disconnected.

a. If you have a **front-connected** switch with all brass screws, proceed as follows:

1. *Work both switches until light burns that is controlled by switches (Figure 86).*
2. *Turn light off at faulty switch.*
3. *Label the three terminal screws on faulty switch: "A", "B" and "C" (Figure 87).*

   Follow procedure outlined under the previous heading in step 5, page 221.
4. *Touch test-lamp wires to terminals A and B, then to terminals A and C, then to terminals B and C (Figure 88).*
5. *Draw a pencil line between the two letters of terminals that produced a light.*
6. *Snap the good 3-way switch to the opposite position.*
7. *Repeat step 4 on faulty switch.*

   The test lamp should again burn in one of the other positions.
8. *Draw pencil line between two letters of terminals that produced light.*

   The letter with two pencil lines to it is the common terminal. If you did not get a light in two positions with these procedures, you have probably made an error. Follow your steps through again.

b. If your **switch has been disconnected,** proceed as follows:

1. *Separate wires so ends are at least 2 or 3 inches apart.*
2. *Label each wire "A", "B" and "C" (Figure 87).*
3. *Throw main switch, or circuit breaker, to "ON" or replace fuse.*
4. *Touch neon test lamp to wires A and B, then to wires A and C, then .to wires B and C (Figure 88).*
5. *Draw a pencil line between the two letters representing the wires that produced a light (if you were able to get a light).*
6. *Snap other switch to opposite position.*
7. *Repeat step 4.*
8. *Draw a pencil line between the two letters representing the wires that produced a light (if you were able to get a light).*

   If you have **two pencil lines between lettered terminals,** the switch that was removed was the faulty one. The letter with two lines to it is the common terminal. Proceed to next reading.

   If you have only *one line between your lettered terminals*, the other switch is the faulty one. Proceed to step 9.
9. *Remove faulty switch from outlet box but do not remove wires from terminals.*
10. *See if you can locate the common terminal (Figure 89).*

    If you can locate the common terminal, proceed to next heading. If you cannot, proceed with the next step.
11. *Throw the main switch or circuit breaker to "OFF" or remove the fuse to the circuit on which you are working.*
12. *Twist bare ends of wires together at switch box where switch has already been removed.*
13. *Disconnect wires from faulty switch.*
14. *Repeat main steps 2 through 5.*

    You will get a light in two positions with this test. Be sure to draw a line between each of the lettered terminals that produces a light. The terminal letter with two lines to it is the common terminal.
15. *Proceed with next heading.*
16. *Repeat steps under a or b when you replace the other switch.*

DISCONNECTING THE SWITCH

To disconnect the switch, proceed as follows:

1. *Throw main switch, or circuit breaker, to "OFF" or remove fuse to circuit on which you are working.*
2. *Disconnect wire from common terminal and bend it away from switch so you can*

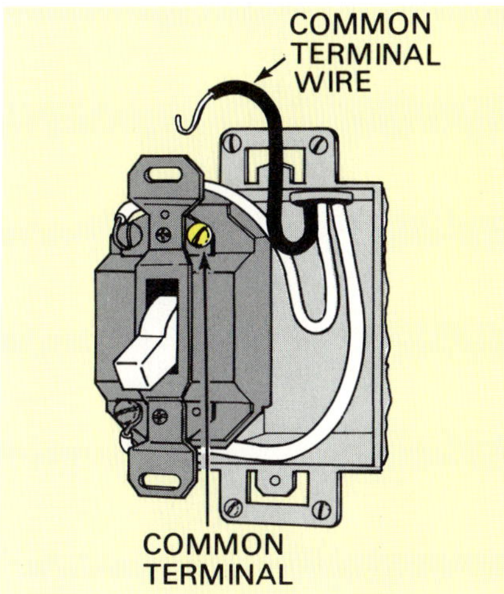

Figure 90. Identifying the common wire.

tell it from the other wires (Figure 90).

3. *Disconnect remaining two wires and bend them to another position (Figure 91).*

If you completely **disconnected the 3-way switch without identifying the common wire** or else forgot which wire it was, repeat procedures under "Locating the Common Wire and Terminal," page 223.

If you **have the common wire identified,** follow procedures under the next heading.

INSTALLING THE SWITCH

To install the switch, proceed as follows:

1. *Disconnect circuit if it is not already "dead."*
2. *Examine new switch to find common terminal.*

   It may be the "odd" screw. The common terminal may be like one of those in Figure 89.

   If all **terminal screws are the same color,** look for a wiring sketch on the box containing the switch. A sketch similar to Figure 92b may be on a slip of paper in the box. Or, the common terminal may be labeled.

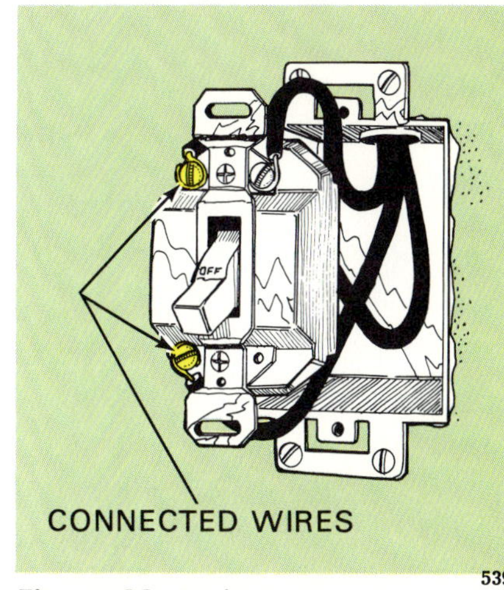

Figure 93. Tighten screws.

3. *Connect common wire to common terminal on new switch (Figure 92a).*
4. *Connect other two wires to remaining terminals (Figure 93).*

   If you have selected the **right wire** for the common, it makes no difference which one of the other two wires is connected to the other two terminals.

5. *Follow same procedures as for installing single-pole switch, page 218, steps 4 through 9.*

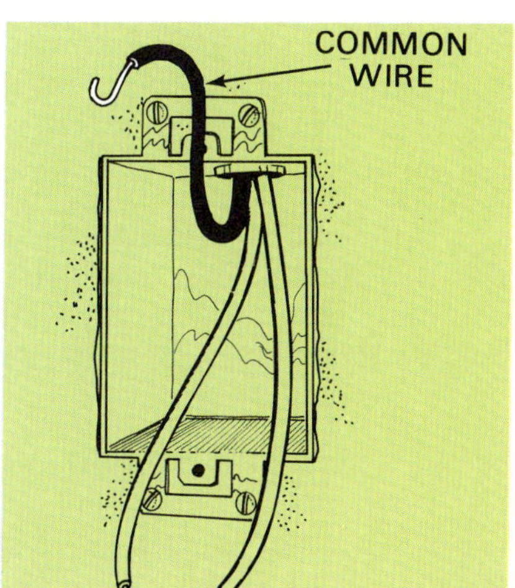

Figure 91. Separate common wire from other wires.

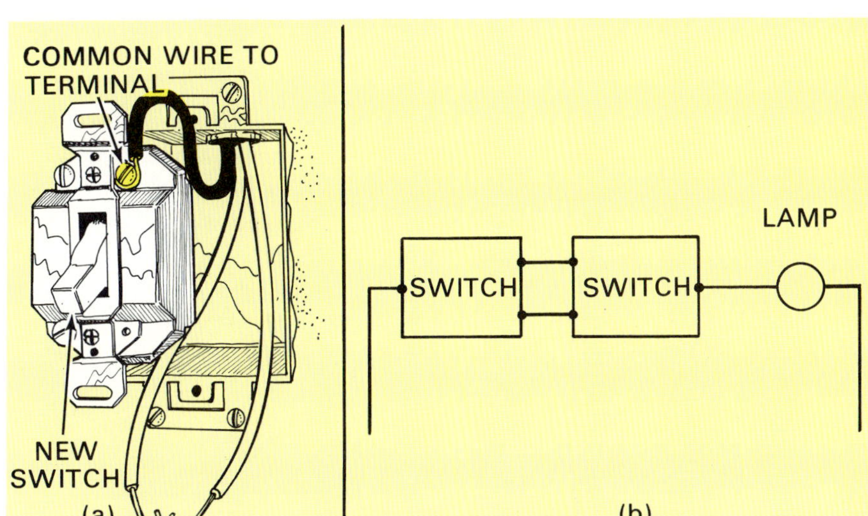

Figure 92. (a) Switch showing common wire connected. (b) Sketch showing how a switch should be connected.

D. Replacing 4-Way Switches

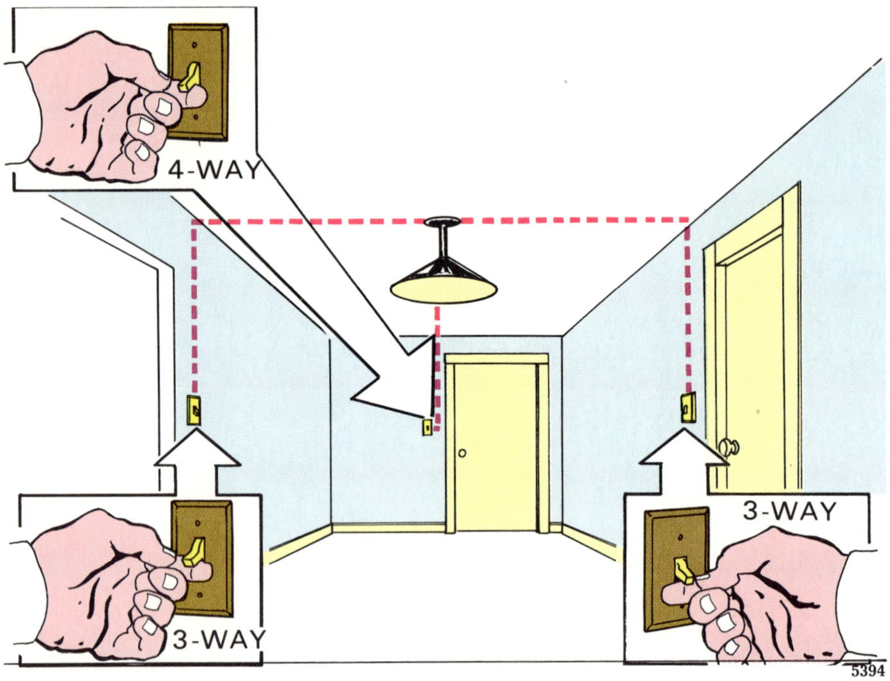

Figure 94. To operate a light from three locations, a 4-way switch is used with two 3-way switches. Additional switches would be of the 4-way type.

When a light is operated by switches at three different locations, two of the switches are 3-way switches and the other is a "4-way switch" (Figure 94). In fact, after two 3-way switches are installed, as many 4-way switches may be added as desired for controlling the same light (or lights) from other points.

Such switch combinations are often used in homes and about the farm and shop. One of their greatest advantages on farms is for controlling a centrally-located yard light from several different buildings.

Four-way switches, like 3-way switches, get their name from the number of wire connections on the switch. They are easier to replace than 3-way switches. However, the explanation of how a 4-way switch functions is a little more difficult to understand. Figure 95 shows how a 4-way switch works with 3-way switches.

The quality of 4-way switches can be determined in the same manner as with single-pole switches. Note page 216.

Upon completion of this section, you will be able to **replace 4-way switches in combination with 3-way switches.**

Procedures used in replacing 4-way switches are given under the following headings:

1. Tools and Materials Needed
2. Checking for Faulty Switch.
3. Installing New 4-Way Switch

TOOLS AND MATERIALS NEEDED

Tools:

Screw driver

Pocket knife

Pliers (preferably thin-nose pliers)

Materials:

New 4-way switch

CHECKING FOR FAULTY SWITCH

To check for a faulty switch, proceed as follows:

1. *Disconnect circuit at main switch, circuit breaker or fuse panel.*
2. *Remove old 4-way switch.*

   Follow procedures under "Removing Old Switch," page 217, steps 1 through 6.

   If there is any question as to whether or not the switch is a 4-way, check to see if it has four wires connected to it.
3. *Hook bare ends of wires from opposite cables together temporarily (Figure 96).*
4. *Reconnect circuit.*
5. *Operate remaining 3-way (and 4-way switches if any) to see if they operate properly.*

   If the **other switches turn the light on and off properly,** you have found the faulty switch. Proceed to next heading.

   If the **other switches work only in certain positions,** you have not found the faulty switch. In that case proceed with step 6.

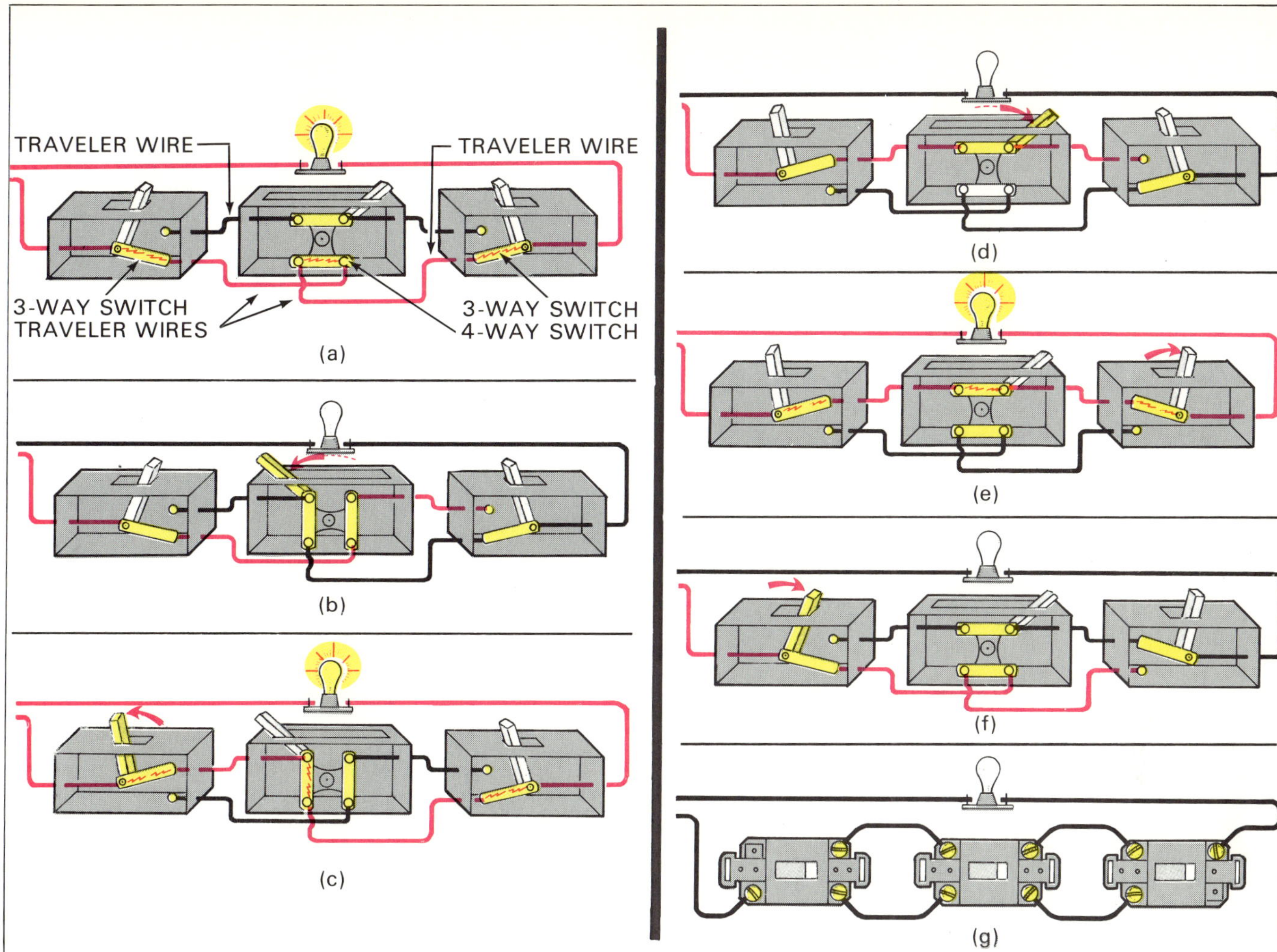

Figure 95. How a 4-way switch works: Note that a 4-way switch is used with two 3-way switches and that the design is different from that for 3-way switches. (a) One combination of switch positions will permit current to travel through the circuit and light the lamp. Red lines show path traveled by electric current. (b) Four-way switch is given ¼ turn. This combination of connections provides for an open circuit at the right-hand 3-way switch. Light is **off.** (c) Left-hand switch connection is changed from lower to upper traveler wire. Light is **on.** (d) Four-way switch is given ¼ turn. New combination of connections turn light **off.** Circuit is open at right-hand 3-way switch. (e) Right-hand 3-way switch is changed from lower to upper traveler wire. Light turns **on.** (f) Left-hand 3-way switch is changed from upper to lower traveler wire. Light is **off.** (g) Wire connections as they appear on standard commercial-type switches.

6. *Disconnect circuit.*
7. *Reinstall old 4-way switch following procedures given under C.*
8. *Check other 4-way switches, if there are any more, before checking the 3-way switches.*

   If you check the 3-way switches, follow procedures under Replacing 3-Way Switches, page 220. Do not operate any of the 4-way switches while testing the 3-way switches.

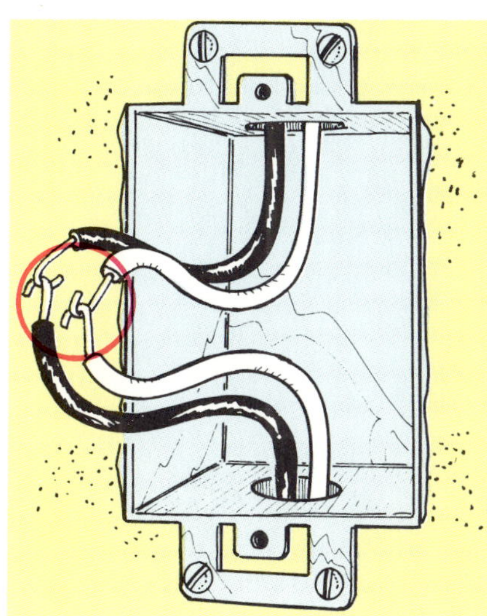

5396

Figure 96. To check for faulty 4-way switch, hook bare ends of wires from opposite cables together temporarily.

INSTALLING NEW 4-WAY SWITCH

Install the new 4-way switch according to the wiring diagram on the box. One type connects as follows:

1. *Disconnect service at main switch, circuit breaker or fuse panel.*
2. *Connect two wires from one cable to two lower terminals (Figure 97a).*

   Connect either wire to either terminal.
3. *Connect two remaining wires to two top terminals (Figure 97b).*

   Connect either wire to either terminal. Figure 95a through (f) shows the two lower traveler wires crossing over each other to connect at opposite screw terminals. This crossover is built into a commercial switch so connections can be made as shown in Figures 95a and 97b.
4. *Complete installation of switch. (Follow procedures in "Installing New Switch," steps 5 through 9, page 218).*

If your switch does not work it may be necessary to disconnect the two wires on one side of your switch and interchange them.

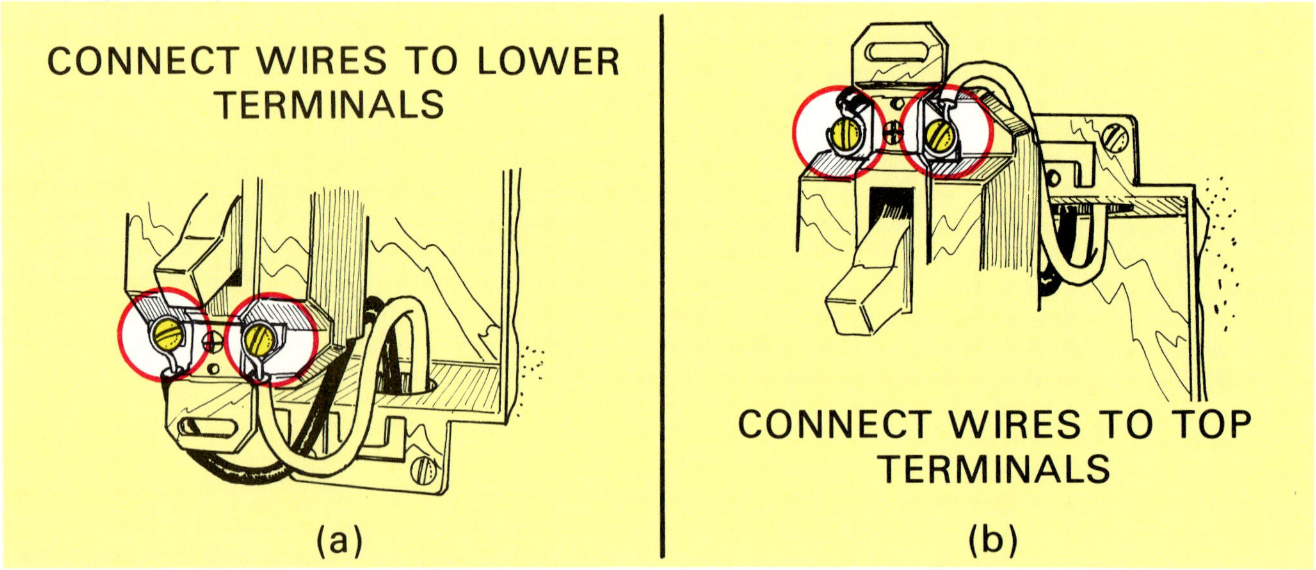

Figure 97. Installing new 4-way switch. (a) Two wires from one cable connected to lower terminals. (b) Wires from other cable connected to upper terminal.

E. Replacing Convenience, Dryer or Range Outlets

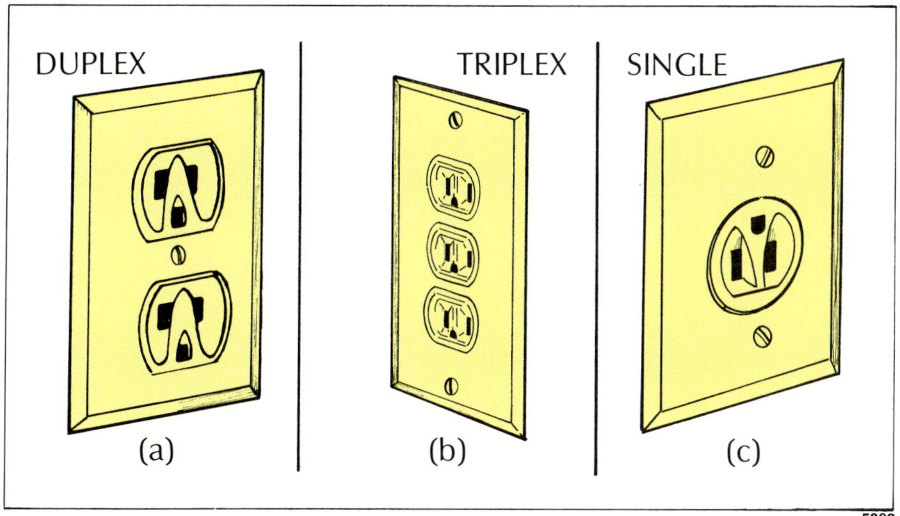

Figure 98. Types of convenience outlets: (a) duplex, (b) triplex and (c) single.

Convenience outlets — also called "receptacles"—have to be replaced much less often than switches. However, they eventually wear to a point where attachment plugs are not held securely.

By far the most common convenience outlet is the **duplex type** (Figure 98a). Some of the **triplex types are used** (Figure 98b), but very few of the **single types** (Figure 98c) are used.

New wiring installations have *grounding-type convenience outlets* (Figure 99). This type of receptacle can be used for connecting either a standard 2-prong plug or a 3-prong grounding plug. The grounding-type plug receptacle has provisions for attaching three wires to it. The third wire leads directly to the ground, or to a (metal) electrical conduit or sheathing which connects to a ground.

The **third prong** on the attachment plug connects to the third wire in the receptacle. Through a third wire in the service cord, it also connects to the frame or case on the equipment you are using. With this arrangement, when the insulation wears off of a "hot" wire and it comes in contact with the equipment frame, the electric current flows through the third wire to the ground connection. This is an important safety provision to keep you from getting shocked.

There are different **qualities of receptacles.** Some of the inexpensive ones, have only one contact strip in the receptacle for each prong on the attachment plug. These are not very durable. The better types have double contact strips — one to press against each side of each prong. These provide much better electrical connections. If you look into one of the slots of a receptacle, you can usually see if there is one contact or two per prong.

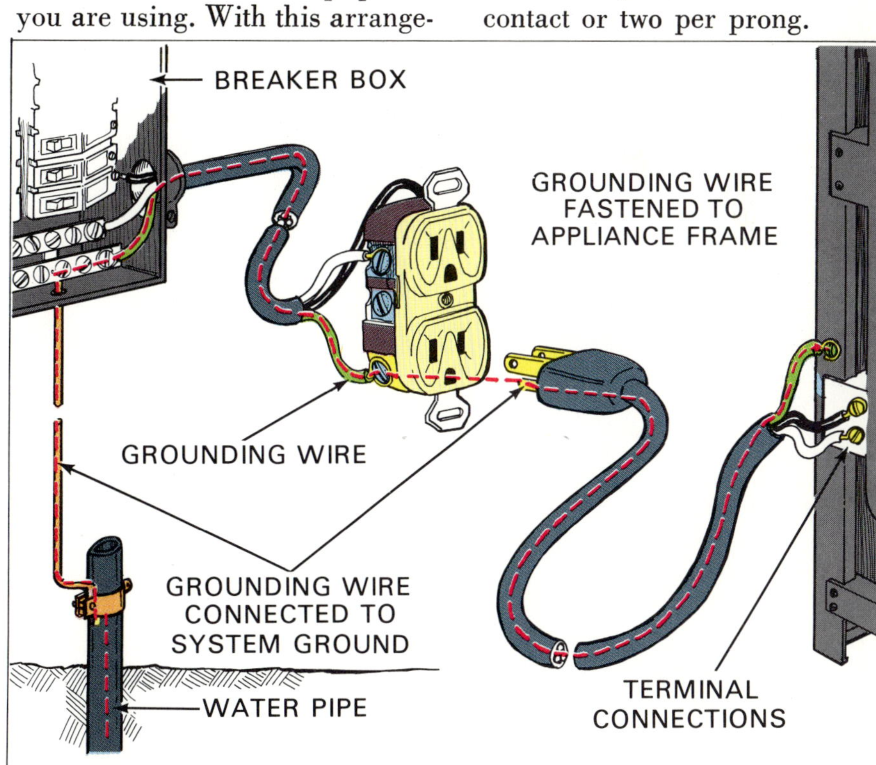

Figure 99. Grounding-type convenience outlet now required by the National Electrical Code.

NEMA CONFIGURATIONS FOR GENERAL-PURPOSE NONLOCKING PLUGS AND RECEPTACLES

			15 AMPERE		20 AMPERE		30 AMPERE		50 AMPERE		60 AMPERE	
			Receptacle	Plug	Receptacle	Plug	Receptacle	Plug	Receptacle	Plug	Receptacle	Plug
2-POLE 2-WIRE	1	125 V	1-15R	1-15P								
2-POLE 2-WIRE	2	250 V		2-15P	2-20R	2-20P	2-30R	2-30P				
2-POLE 3-WIRE GROUNDING	5	125 V	5-15R	5-15P	5-20R	5-20P	5-30R	5-30P	5-50R	5-50P		
2-POLE 3-WIRE GROUNDING	6	250 V	6-15R	6-15P	6-20R	6-20P	6-30R	6-30P	6-50R	6-50P		
2-POLE 3-WIRE GROUNDING	7	277 V AC	7-15R	7-15P	7-20R	7-20P	7-30R	7-30P	7-50R	7-50P		
3-POLE 3-WIRE	10	125/ 250 V			10-20R	10-20P	10-30R	10-30P	10-50R	10-50P		
3-POLE 3-WIRE	11	3φΔ 250 V	* 11-15R	* 11-15P	* 11-20R	* 11-20P	* 11-30R	* 11-30P	* 11-50R	* 11-50P		
3-POLE 4-WIRE GROUNDING	14	125/ 250 V	* 14-15R	* 14-15P	14-20R	14-20P	14-30R	14-30P	14-50R	14-50P	14-60R	14-60P
3-POLE 4-WIRE GROUNDING	15	3φΔ 250 V	* 15-15R	* 15-15P	15-20R	15-20P	15-30R	15-30P	15-50R	15-50P	15-60R	15-60P
4-POLE 4-WIRE	18	3φY 120/ 208 V	18-15R	18-15P	18-20R	18-20P	* 18-30R	* 18-30P	* 18-50R	* 18-50P	18-60R	18-60P

*These configurations were approved as suggested standard for future design 5-17-1965. All other configurations were approved as Nema Standard 5-17-1965.

Receptacles and plugs have different configurations (designs) for various voltages and amperages. Be sure the replacement receptacle is of the proper configuration for the electrical equipment it will serve. Table I shows the official configurations used by the National Electrical Manufacturers Association for all applications.

When buying a new convenience outlet, check to see that it has the UL (Underwriters' Laboratories) mark of approval stamped on it. This means it meets their established standards for safety. It does not necessarily mean it will be durable.

Electric dryers use a 30 ampere receptacle rated at 125/250 volts which uses the NEMA #10-30 configuration (Figure 100a). Electric ranges use a 50 ampere receptacle rated at 125/250 volts and uses the NEMA #10-50 configuration (Figure 100b). Surface mounted receptacles are most common.

Upon completion of this section, you will be able to **replace convenience outlets and range and dryer outlets.**

Procedures used in replacing convenience, dryer, or range outlets are given under the following headings:

1. Tools and Materials Needed.

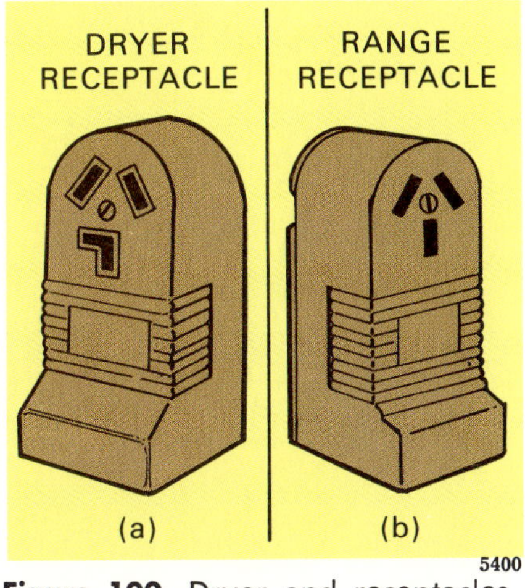

Figure 100. Dryer and receptacles.

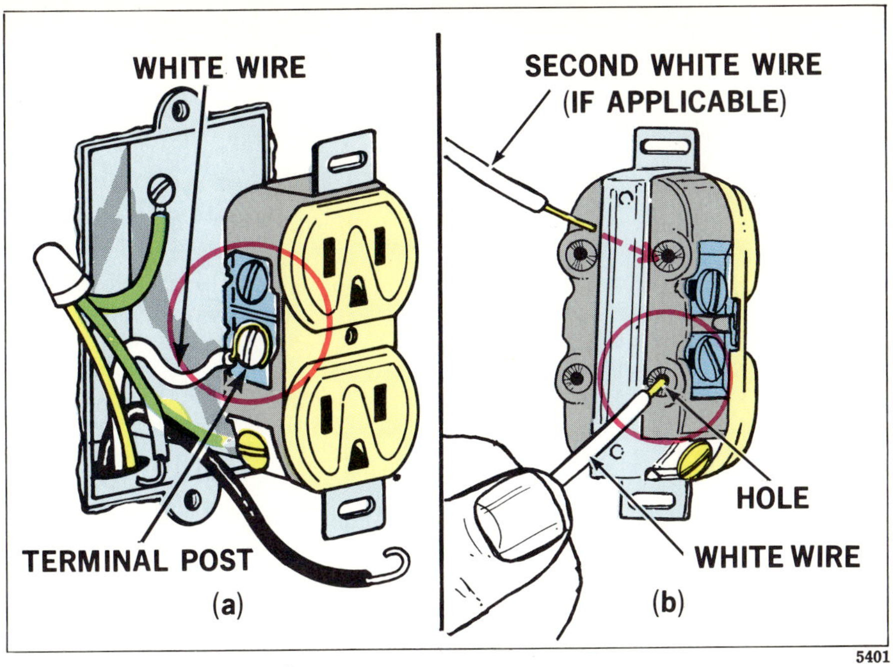

Figure 101. Attach white wire to aluminum-colored terminal post.

2. Removing Old Convenience Outlet.
3. Installing New Convenience Outlet.
4. Removing Old Dryer or Range Receptacle.
5. Installing a Dryer or Range Receptacle.

TOOLS AND MATERIALS NEEDED

Tools:

Screwdriver
Pocket knife
Pliers (preferably thin-nose pliers)

Materials:

New convenience-outlet receptacle

REMOVING OLD CONVENIENCE OUTLET

Follow procedures outlined under the section, "Removing Old Switch," page 217.

CAUTION — **When circuit has been disconnected be sure to plug in a floor or table lamp or check with a test lamp to make certain the circuit is "dead."**

INSTALLING NEW CONVENIENCE OUTLET

NOTE: *If aluminum conductors are used,* read section under "Installing New Switch", step 4, page 218.

1. *Connect white wire to aluminum-colored terminal post (Figure 101a), or push white wire into hole nearest the aluminum-colored terminal post (Figure 101b).*

Note the restrictions regarding aluminum conductors listed under the section "Installing New Switch," page 218.

If a second white wire was connected to the old convenience outlet, reconnect it to the second aluminum colored terminal or use the push-in connection nearest it (Figure 101b).

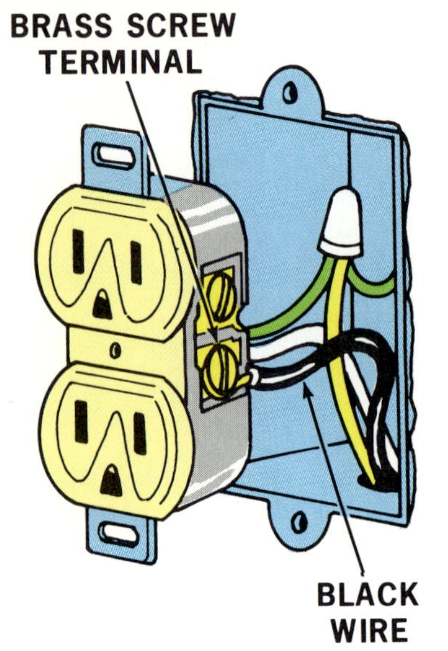

5402
Figure 102. Attach black wire to brass terminal post.

The extra white and black wires continue the circuit to the next outlet.

2. *Connect black wire to brass screw terminal on opposite side (Figure 102).*

   If a second black wire was connected to the old convenience outlet, reconnect it to the second brass terminal post or use the push-in connection nearest it.

3. *Connect grounding wire (if used) to grounding (green) terminal post (Figure 103).*

   If yours is a wiring installation that has been installed for a long time, there will probably be no grounding wire.

4. *Recheck connections.*

   An error in connecting convenience-outlet wiring will cause a short circuit.

5. *Complete installation. Follow procedures under the section, "Installing New Switch," steps 5 through 8, page 218.*

REMOVING OLD DRYER OR RANGE RECEPTACLE

Steps for removing one type of dryer or range receptacle are as follows:

1. *Turn the dryer or range main circuit breaker to "off," or pull the fuse plug serving it.*
2. *Locate and remove cover screw and remove cover (Figure 104).*

   Note that the white wire is usually attached to the center terminal.

3. *On one surface-mounted type, loosen, but do not remove, the terminal screws (Figure 105).*
4. *Lift the wires out of the terminal slots (Figure 106).*
5. *Remove the three mounting*

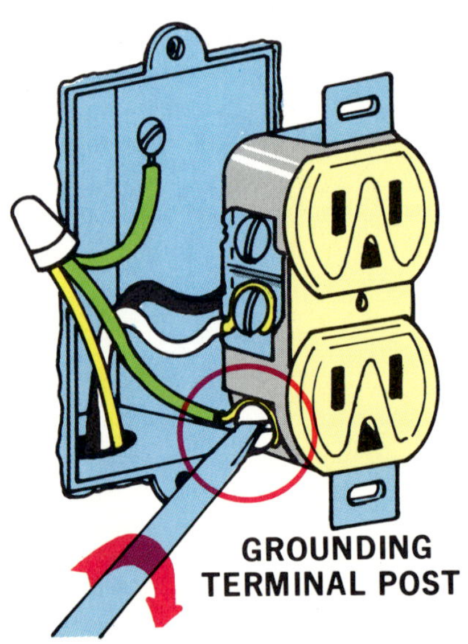

5403
Figure 103. Attach grounding wire (if used) to grounding terminal post.

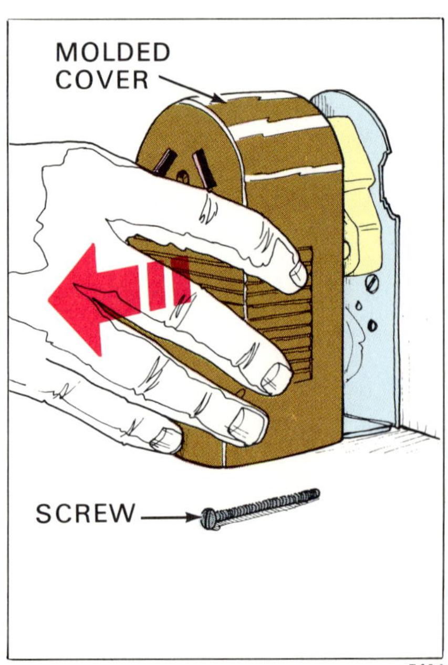

5404
Figure 104. Remove molded cover.

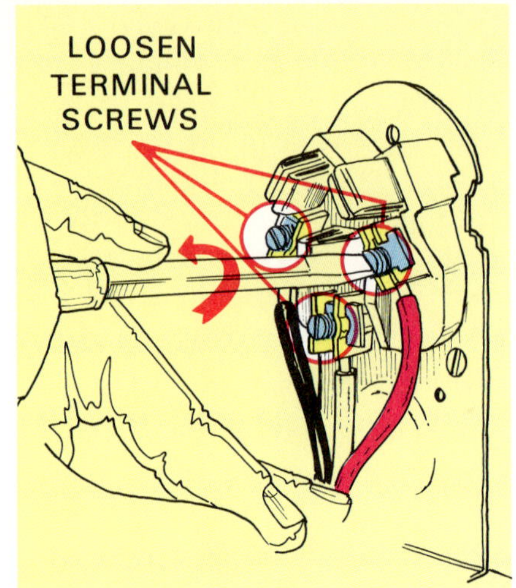

5405
Figure 105. Loosen terminal screws and remove plate.

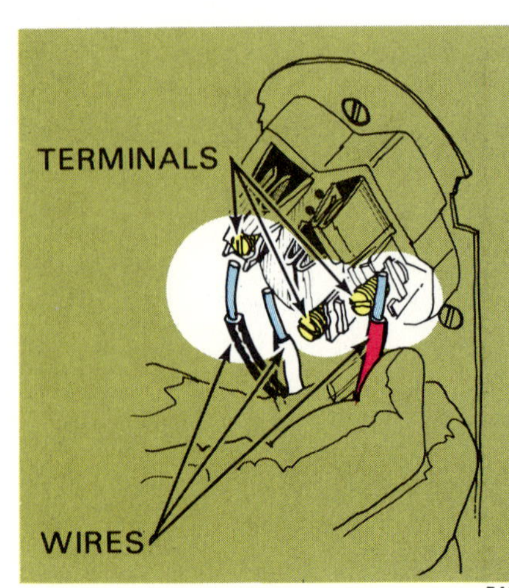

5406
Figure 106. Lift wires out of terminal slots.

screws holding the base plate to the wall.

6. *Loosen the two cable clamp screws and pull the cable from the receptacle base plate (Figure 107).*

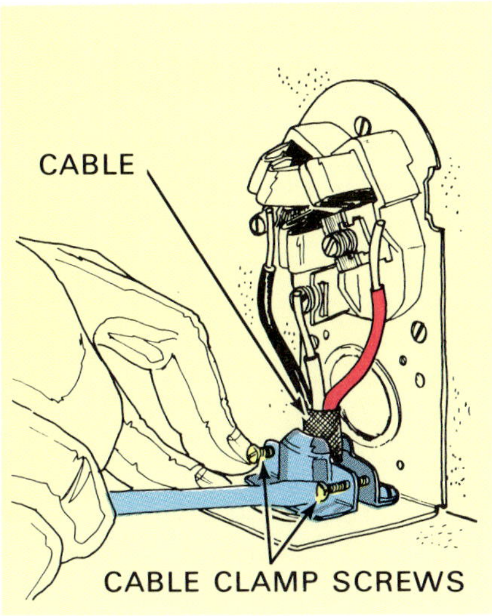

Figure 107. Loosen cable clamp screws and remove cable from receptacle base plate.

INSTALLING A DRYER OR RANGE RECEPTACLE

Proceed as follows to install a dryer or range receptacle:

1. *Follow instructions on carton or enclosure.*

   Check the receptacle rating to be sure it is NEMA #10-30 (AMP) for a dryer, or NEMA #10-50 (AMP) for a range.

2. *On one surface-mounted type, remove the receptacle cover (Figure 104).*

3. *Determine which knockout must be removed.*

   Using a hammer or screwdriver, remove it (Figure 108).

4. *Install the cable clamp by attaching one half of the cable clamp to the receptacle base with the two short screws provided with the receptacle (Figure 109a).*

5. *Attach the other half of the cable clamp using the two longer metal screws provided (109b).*

6. *Insert conductor cable through cable clamp while making sure the wire is long enough to reach the terminals. The exact length required is usually shown on the receptacle box.*

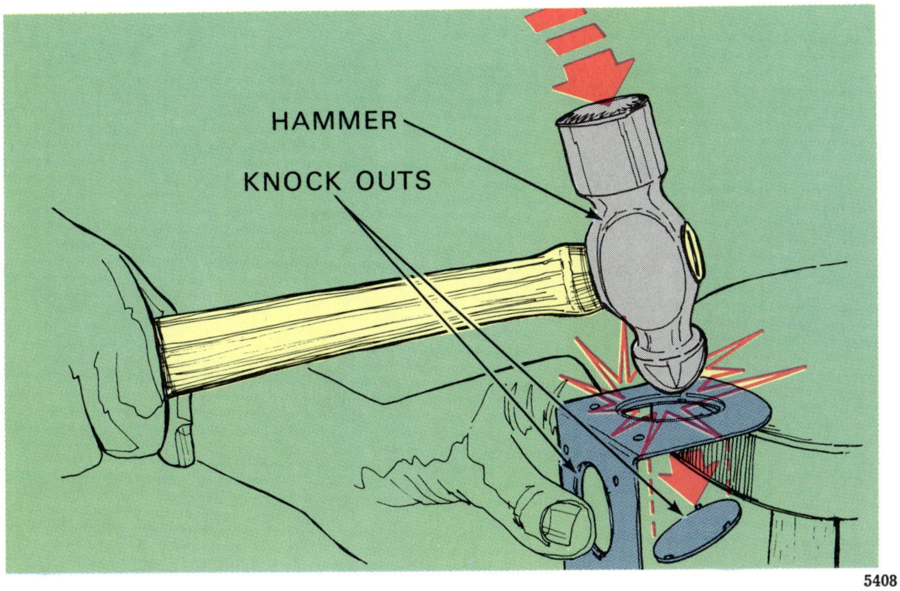

Figure 108. Remove knockout.

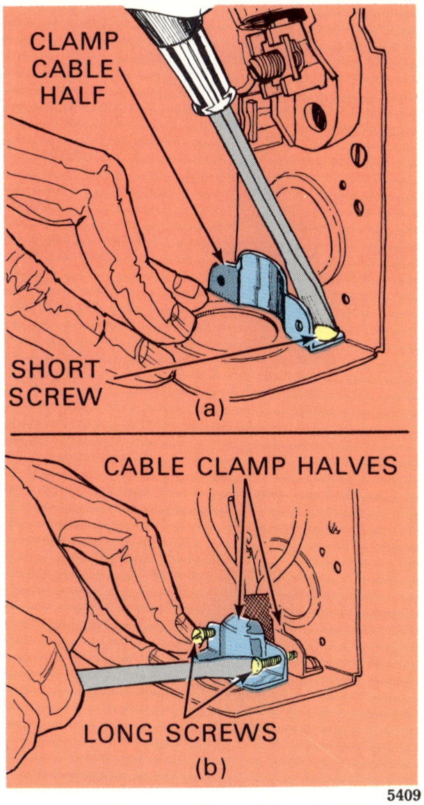

Figure 109. Install cable clamp.

7. *Tighten the cable clamp.*

8. *Install the base plate on the wall using the three wood screws provided (Figure 107).*

9. *Loosen the three terminal screws and slide them to the side (Figure 105).*

10. *If new conductors are used, check the length and note that the white wire may be about 1 inch shorter than the colored wires (Figure 110).*

11. *Strip insulation ¾ inch from the end of each wire.*

12. *Lay white wire into the middle terminal, then slide terminal screw plate over the wire and tighten the screw (Figure 111).*

13. *Repeat step 12 with the other two conductors and terminals.*

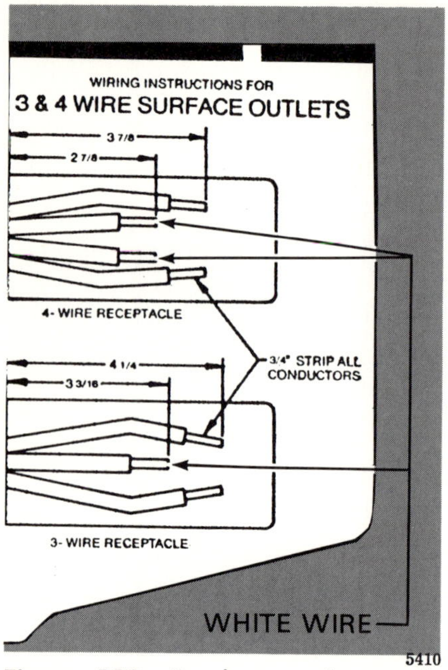

Figure 110. Conductor wires with new conductors.

14. *Install molded cover plate with long screw provided (Figure 104).*
15. *Insert range or dryer plug.*
16. Restore electrical service.

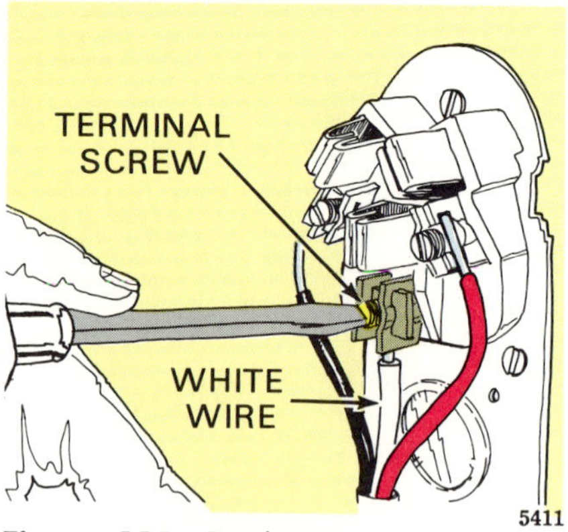

Figure 111. Replacing wires in receptacle.

F. Replacing Incandescent Lighting Fixtures

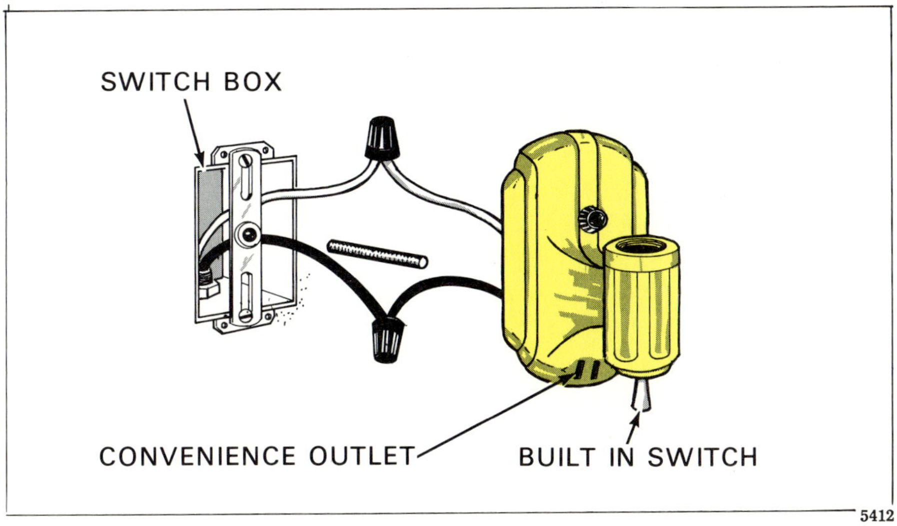

Figure 112. A side-wall lighting fixture is usually mounted on a switch box. Note that this lighting fixture is also equipped with a switch and convenience outlet.

Constant improvements are being made in the quality and quantity of light for use in the home, on the farm, and in business. To take advantage of some of these improvements you may need to change lighting fixtures (Figures 112 and 113).

Occasionally a lighting fixture becomes broken or damaged and has to be replaced. If it can be repaired without delay, you may prevent injury, especially if it is at some location where there is considerable risk if you can not see, such as over a stairway or near a ladder.

Replacing incandescent lighting fixtures is a rather simple operation. If you learn how to replace your own, you can often take advantage of improved lighting at relatively little cost.

In selecting a lighting fixture it is important from a safety standpoint that you check for the **Underwriter's label of approval.** Check with your local power distributor or electrical dealer if you are interested in getting information as to the kind and style of fixture to buy. They can give you recommendations for a particular room or for a particular use.

Fixtures are supplied either with or without **built-in switches.** If the fixture you are planning to replace is controlled by a wall switch, it is best to use one without the switch on the light.

Some fixtures that mount on the side wall are equipped with a **convenience outlet.** The convenience outlet is usually not as durable as the regular wall-type. However, if you are in need of a convenience outlet **for limited use** in the vicinity of the light, you will find the one in the fixture very helpful. Many people have this in mind when they select new lighting fixtures for the bathroom mirror. The built-in convenience outlet is then available for an electric razor.

Most ceiling lighting fixtures are installed on **outlet boxes** (Figure 113). If your old ceiling fixture is not mounted on an out-

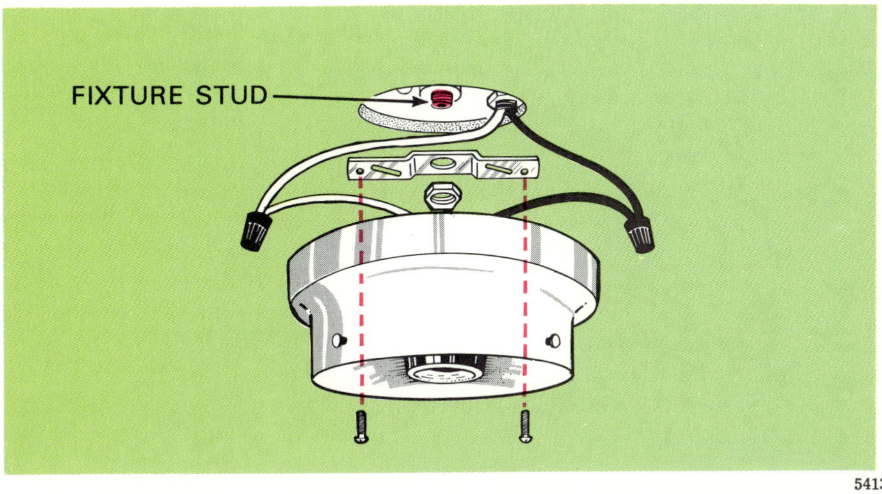

Figure 113. Ceiling fixtures are usually attached to a round outlet box in the ceiling.

let box, it is important from a safety standpoint that you install a shallow-type box similar to the one in Figure 116a. This type of box is available either with or without a fixture stud (Figure 113). With most fixtures a stud is not needed.

It is best to get an outlet box with the **cable clamps already installed** (Figure 118). If your ceiling is of material that will not hold a screw, you will also need a bar hanger (Figure 116c).

For sidewall fixtures a standard **switch box** is usually best. It will require two metal switch-box supports (Figure 116d) to hold it in position in sheet rock or plaster-and-lath walls.

Upon completion of this section, you will be able to **replace incandescent lighting fixtures.**

Procedures used in replacing incandescent lighting fixtures are given under the following headings:

1. Tools and Materials Needed.
2. Removing Old Fixture.
3. Installing New Outlet Box.
4. Mounting New Fixture.

TOOLS AND MATERIALS NEEDED

Tools:

Small screwdriver
Thin-nose pliers
Pocketknife

Materials:

2 solderless connectors (sometimes furnished with fixture)
2 No. 7 wood screws
1 outlet box for ceiling fixture, or
1 switch box for side-wall fixture

REMOVING OLD FIXTURE

To remove old fixtures, proceed as follows:

1. *Cut off current: pull main switch, remove circuit fuse or trip circuit breaker.*
2. *Remove light globe if there is one (Figure 114).*

   Most light globes are held in place with three thumbscrews that hold the edge of the globe (Figure 114a). Others are held in place with a cap nut which is screwed on to a central threaded stem (Figure 114a).
3. *Remove light bulb(s).*
4. *Disconnect fixture from wall or ceiling.*
5. *Disconnect feeder wires from fixture wires or fixture terminals.*

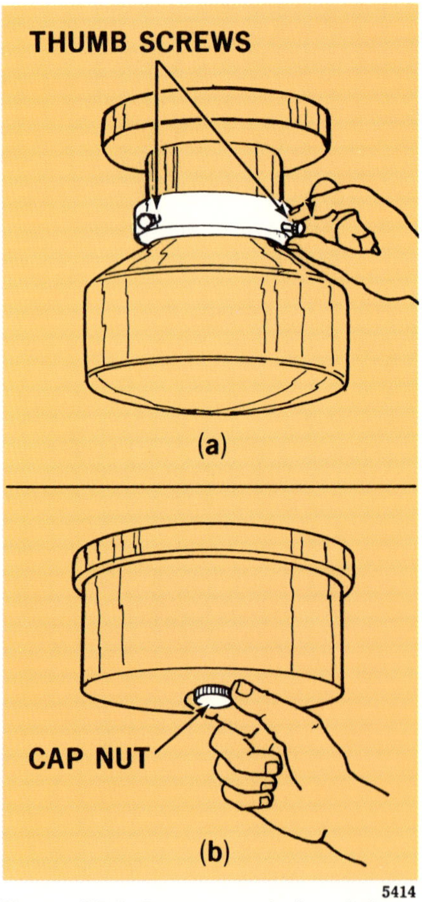

Figure 114. Removing light globes.

INSTALLING NEW OUTLET BOX

If your old fixture was attached to an outlet box or switch box, omit this operation and proceed with section on "Mounting New Fixture," page 238.

Figure 115. Some old fixtures were installed without outlet box.

If your old fixture was mounted directly to the ceiling (Figure 115) without an outlet box, as is sometimes the case in older installations, determine what kind of outlet box will be needed. Here are several ways of mounting outlet boxes for new fixtures:

If the **wall or ceiling is of wood,** the box can be attached directly to the wood and held in place with wood screws (Figure 116a).

If the **wall or ceiling is plaster on wood lath,** remove enough plaster to permit the box to be attached directly to the wood lath (Figure 116b).

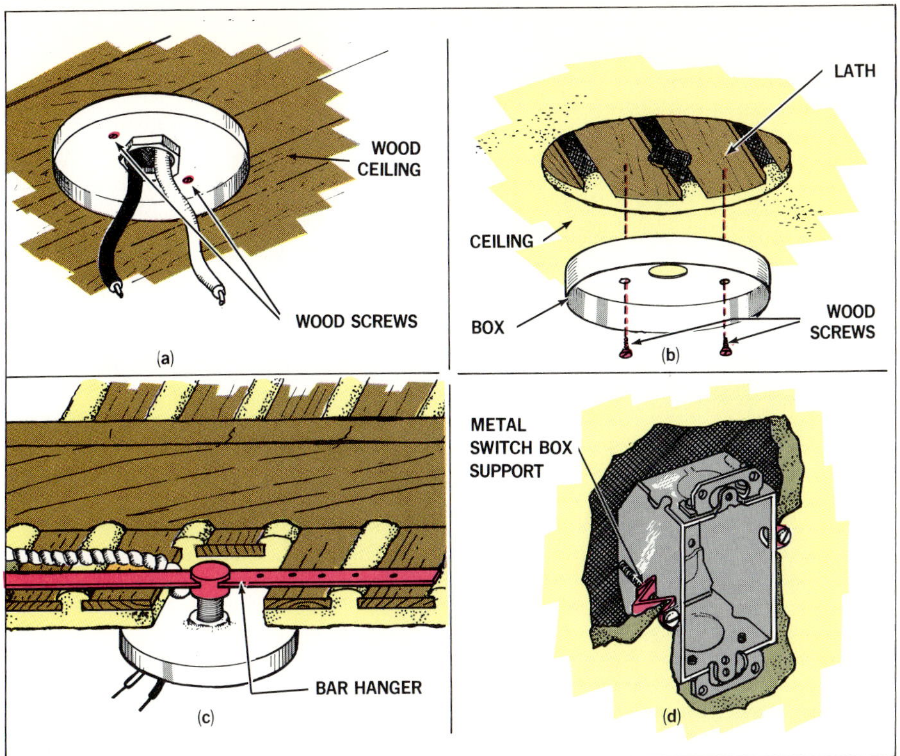

Figure 116. The kind of materials used for the ceiling or sidewall determine what box to use and how it shall be mounted. (a) With a wood ceiling, a shallow-type outlet box can be fastened directly to the ceiling with wood screws. (b) With a plaster-and-lath ceiling, the plaster may be chipped off to recess the box and the box fastened with screws, or (c) the box may be held with a bar hanger. (d) Side wall switch boxes may be attached to wood walls with screws, but in plaster or sheet-rock walls, metal switch-box supports are used.

If your **wall or ceiling is of plaster board,** or some other material that will not hold a screw satisfactorily, the type of support will depend on whether the fixture is to be mounted on the ceiling or sidewall.

For **ceiling boxes** a small hole is cut in the ceiling, a straight bar hanger inserted and the box attached to it (Figure 116c).

For **sidewall boxes** metal switch-box supports are used to hold the box in place (Figure 116d).

For any one of these methods proceed as follows:

1. *Place outlet box in mounting position temporarily to see which screen holes to use, which knockouts to remove in the box and how large to cut the hole for the box.*

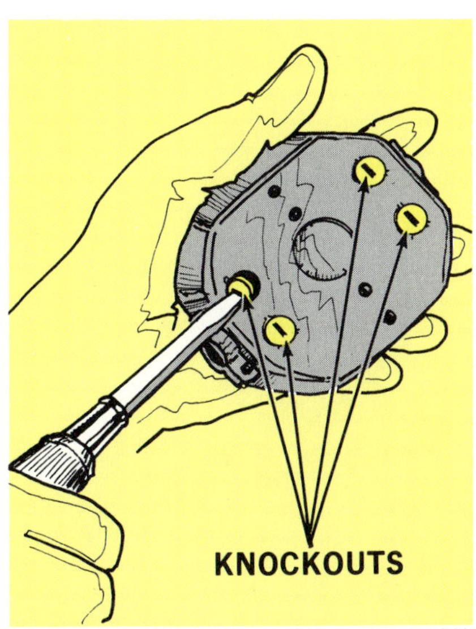

Figure 117. Removing necessary number of knockouts in outlet box.

2. *If box is to be recessed (Figure 116b and d), cut hole large enough to accept box.*
3. *Remove necessary number of knockouts (Figure 117).*

   Remove one knockout for each cable.

   If the fixture is fed by individual wires, remove one knockout for each wire.
4. *Loosen cable clamps in box (Figure 118).*

   Be sure the cable extends into the box far enough so that the clamp will grip the cable sheath.

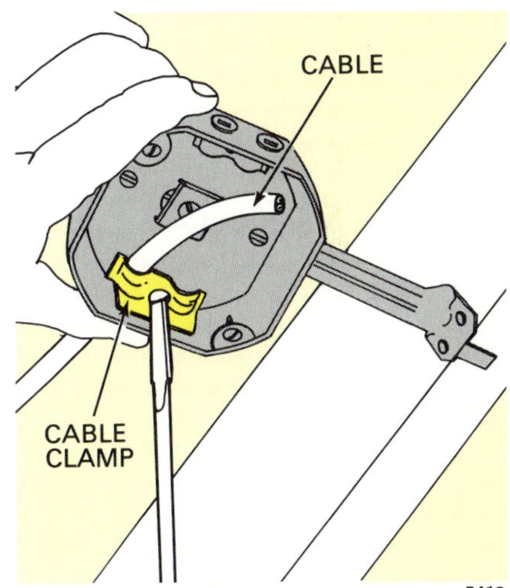

Figure 118. Loosening cable clamps and inserting cable in outlet box.

6. *Fasten box securely in place using one of the following methods:*
   - (a) For a **wood ceiling** attach box directly to ceiling with two No. 7, or larger, wood screws.
   - (b) For a **plaster-on-lath ceiling:**
     - (1) *Remove enough plaster to recess box against lath.*
     - (2) *Attach box to lath with two No. 7 or larger, wood screws.*

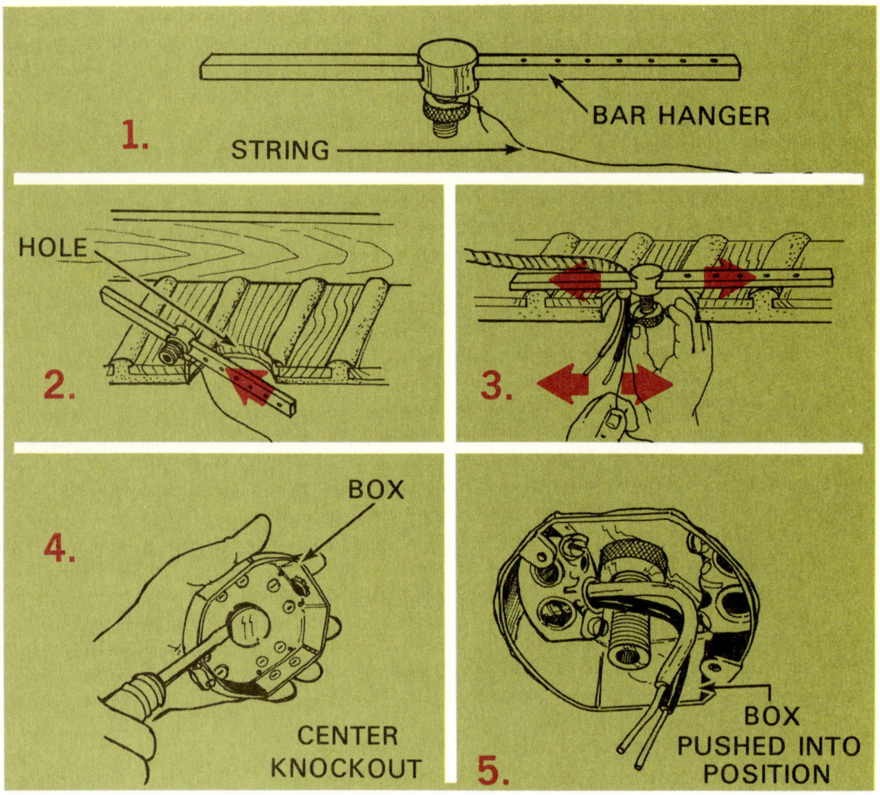

Figure 119. Mounting a ceiling outlet box to a bar hanger. (1) Tie string to center stud of bar hanger. (2) Insert hanger through hole in ceiling. (3) Use string to position bar so that fixture stud is in center of opening. If you can not center it, loosen the stud on the bar and slide stud to center of hole then tighten. (4) Remove center knockout from box. (5) With cable already attached to box, push it into position on center stud and fasten with nut.

(c) For a **plaster board or composition ceiling** where it is necessary to use a bar hanger, proceed as shown in Figure 119.

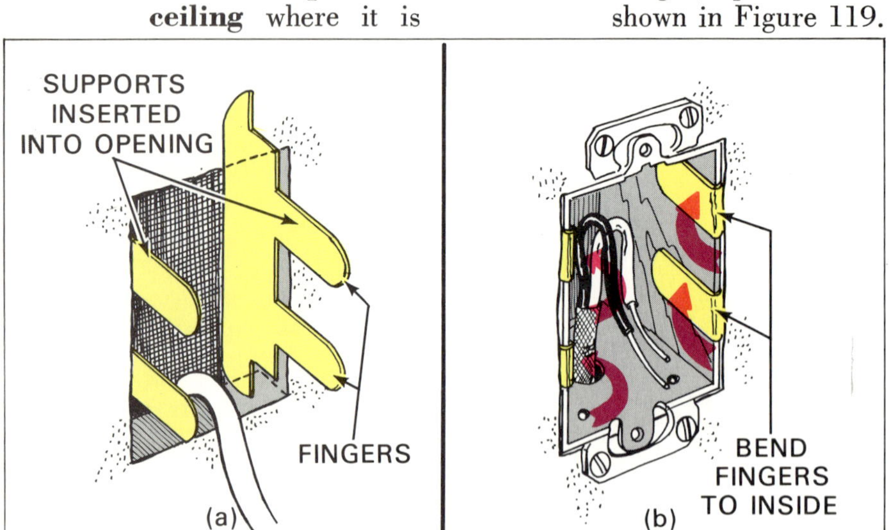

Figure 120. Mounting a switch box for a wall fixture. (1) After cutting hole in wall to receive box, insert one support and hold in position on one side of opening. Insert second support in same manner and hold in position on opposite side of opening. (2) Place outlet box (with cable already connected) between supports. Bend metal fingers over edge of box and press firmly against inside of box.

(d) For a **wall-mounted switch box,** using metal switch-box supports, proceed as shown in Figure 120.

MOUNTING NEW FIXTURE

Some lighting fixtures have mounting instructions included in the packing box. **If there are mounting instructions** with yours, read them over carefully and follow their procedures.

If there are no instructions with your fixture, proceed as follows:

1. *Remove insulation about 3/4 inch from end of each fixture wire if not already prepared.*

   Follow steps 2 through 5 starting on page 233.

   Some fixtures, like the first one in Figure 123a, have screw terminals mounted on the lamp socket instead of fixture wires. It is a **pull-chain fixture.** If so, omit this step. You may have to remove the screw ring from *around the socket and re*move the socket to get at the terminals.

2. *Remove insulation about 1/2 inch from each cable wire (unless it is already removed). Scrape wires clean.*

3. *Attach wires to fixture (unless your fixture mounts directly to a ceiling fixture stud).*

   If your fixture mounts directly to a ceiling fixture stud (Figure 126a), attach it to the box stud before connecting the wiring.

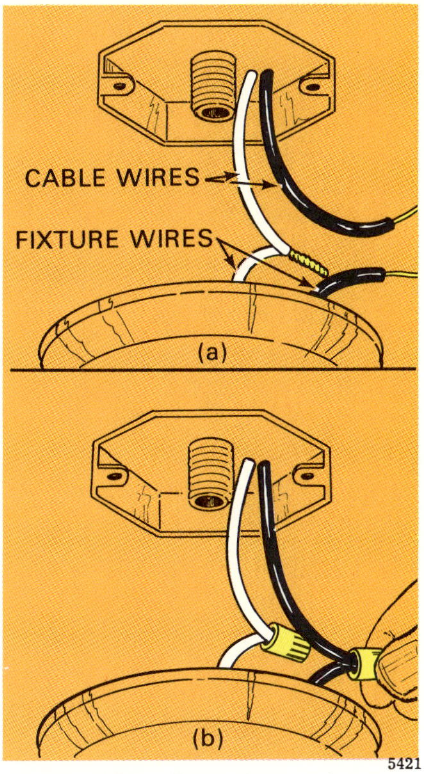

Figure 121. Fastening wires together using solderless connectors. (a) Matching and joining wires together. (b) Attaching solderless connectors.

(a) If your **fixture is equipped with screw terminals,** follow the procedures given on page 236.

(b) If your **fixture has lead wires:**

(1) *Select the white fixture wire(s) and the white wire extending from the outlet box.*

(2) *Hold wires parallel to each other (Figure 121a).*

A small wire should extend past the end of a larger wire.

(3) *Insert solderless connector over wire ends and turn connector to right (Figure 121b).*

The connector is threaded on the inside so it cuts its own threads on the wires at the same time it twists them together. Be sure no bare wire shows after the connector is in place. If bare wire is showing, remove the connector and cut off a short length of wire.

(4) *Select black wire(s) from fixture and black wire from outlet box. Repeat steps (2) and (3).*

If none of the fixture wires are black, you will find that one of each pair of fixture wires has a tracer in it. This is usually a group of colored threads woven into the

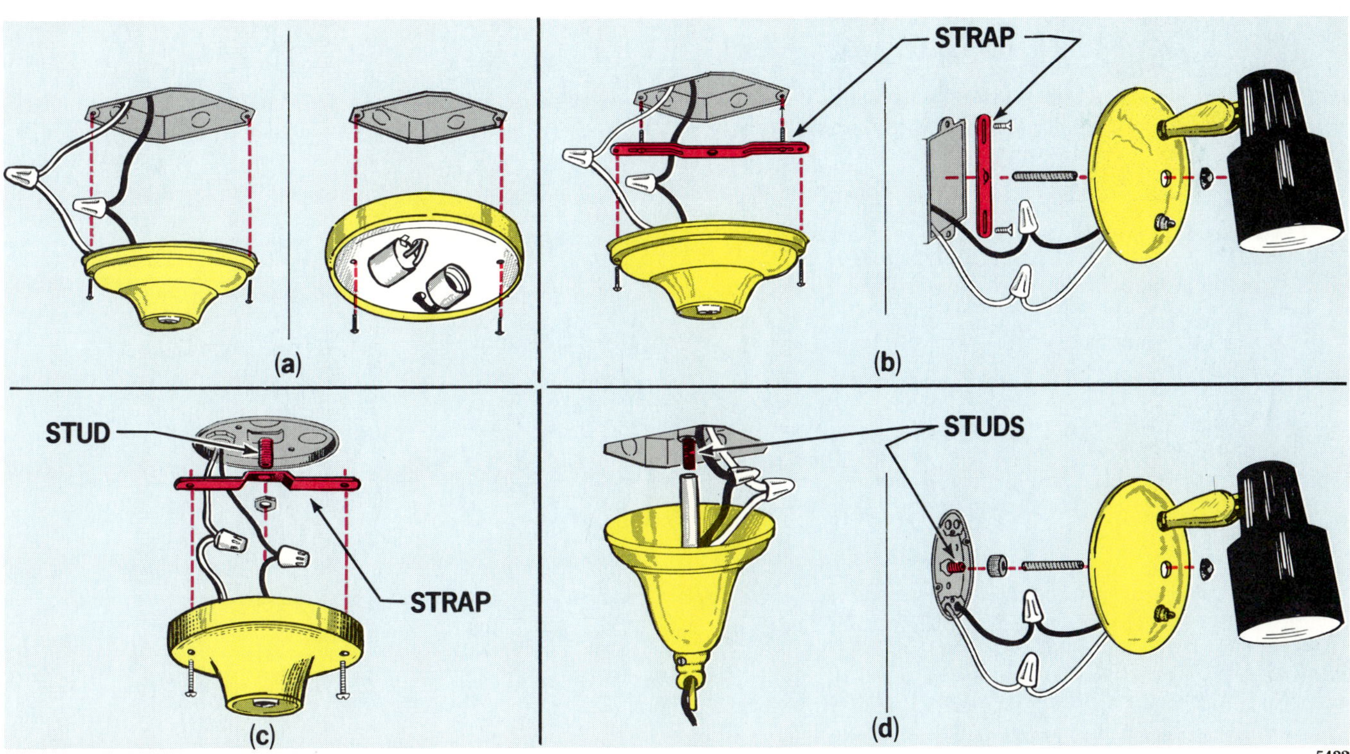

Figure 122. (a) Direct-to-box mount. (b) Strap-to-box mount. (c) Strap-to-stud mount. (d) Fixture-to-stud mount.

Figure 123. (a) DIRECT TO BOX MOUNT. (1) "Accordion" wires and push coverplate socket into position. (2) Insert bolts through plate into threaded ears on box and tighten gently. (b) KEYHOLE-SLOT FIXTURE. (1) Start bolts in ears of box (distribute heat insulating material in top of fixture if any was supplied). (2) Place fixture on box with bolts extending through keyhole slots. Rotate fixture so that bolts fit in narrow part of slot. (3) Tighten bolts.

5423

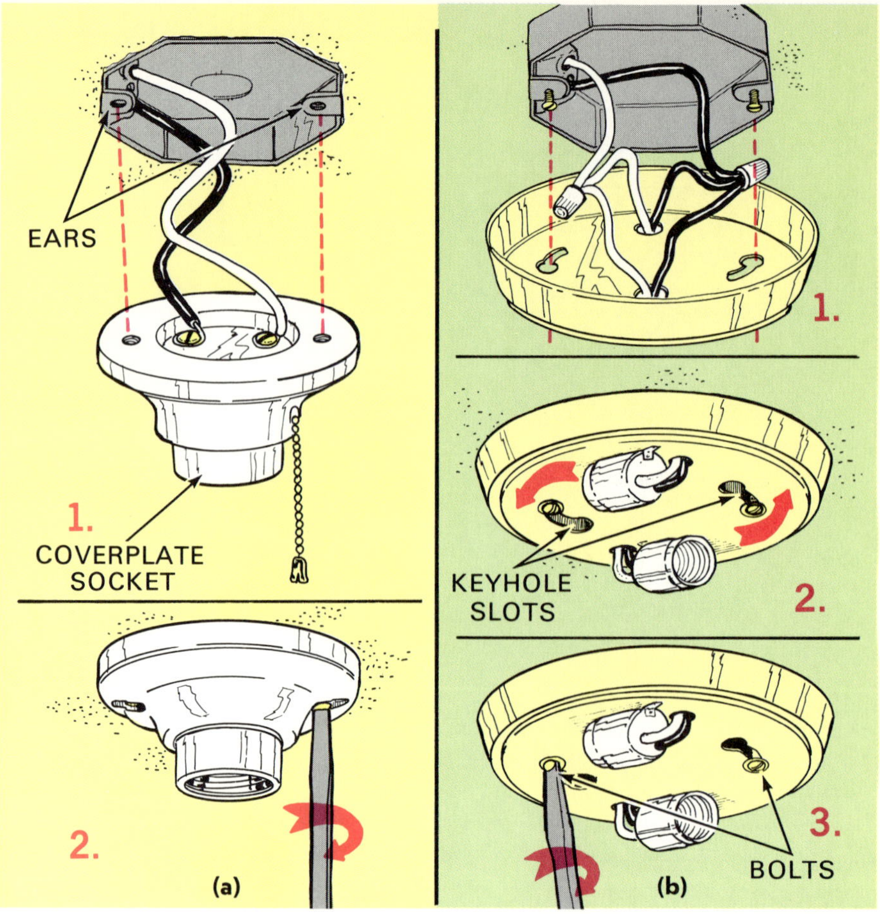

wire jacket. These are used in place of black wires.

4. *Determine how your fixture mounts on the box.*

There are four ways that fixtures commonly attach to an outlet box. The four methods are shown in Figure 122. Select the mounting that will work with your fixture.

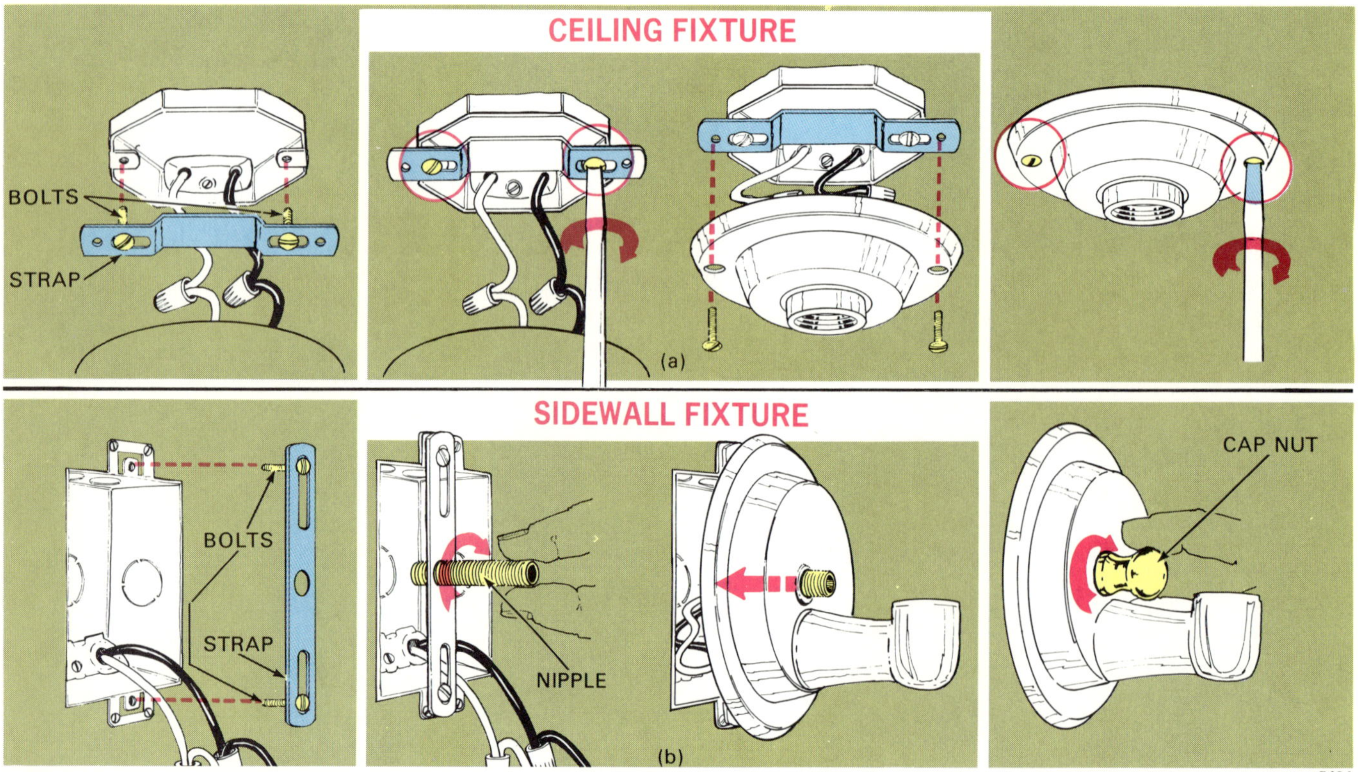

5424

Figure 124. (a) STRAP TO BOX MOUNT. (1) Insert bolts through slots in strap. (2) Attach to ears on box. (3) Position fixture so that holes are in line with mounting holes on strap. (4) Insert fixture bolts and tighten. (b) (1) Insert bolts through slots in strap. (2) Screw nipple into hole in strap. (3) Place fixture on nipple and at the same time "accordion" the wires into the box. (4) Tighten cap nut on nipple until fixture is held securely in place.

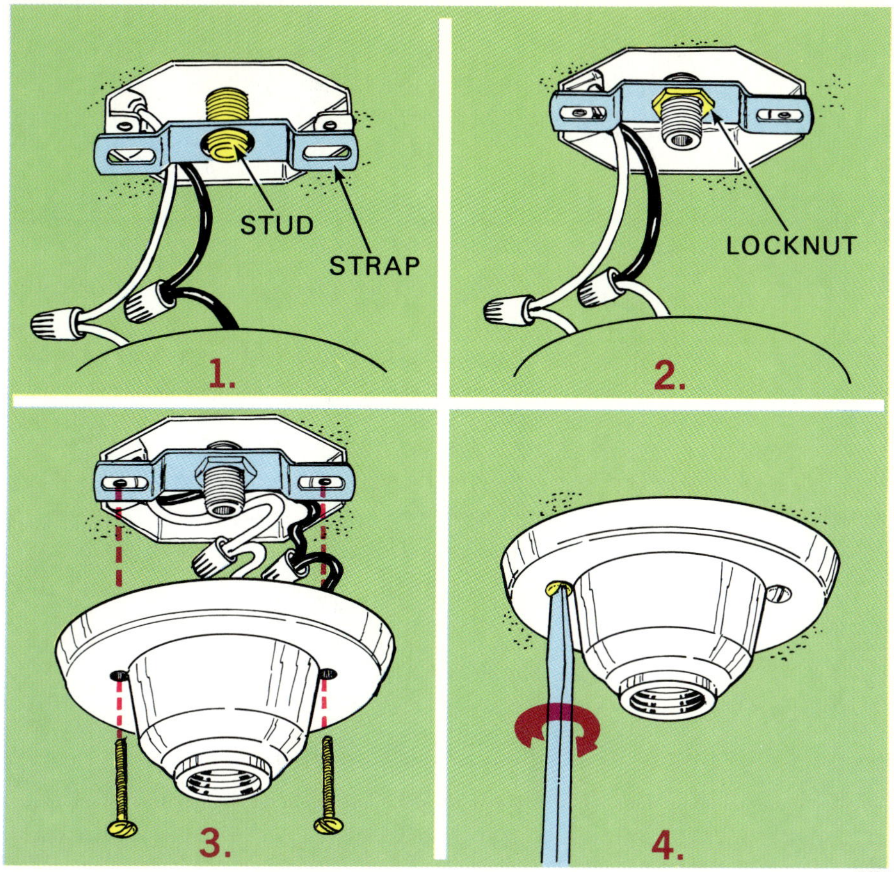

Figure 125. STRAP-TO-STUD MOUNT. (1) Place center hole of strap on stud. (2) Tighten strap on stud with locknut. (3) Position fixture so that holes are in line with holes in strap. (4) Insert bolts and tighten.

5. *Attach fixture to outlet box.*

   (a) If your fixture is made for a **direct-to-box** mount, follow procedures in Figure 123.

   (b) If your fixture is a **strap-to-box** mount, follow procedures in Figure 124.

   (c) If your fixture is a **strap-to-stud** mount, follow procedures in Figure 125.

   (d) If your fixture is a **fixture-to-stud** mount, follow procedures in Figure 126.

6. *Screw light bulb(s) into socket(s).*

7. *Mount lamp globe on fixture.*

8. *Re-establish service on circuit.*

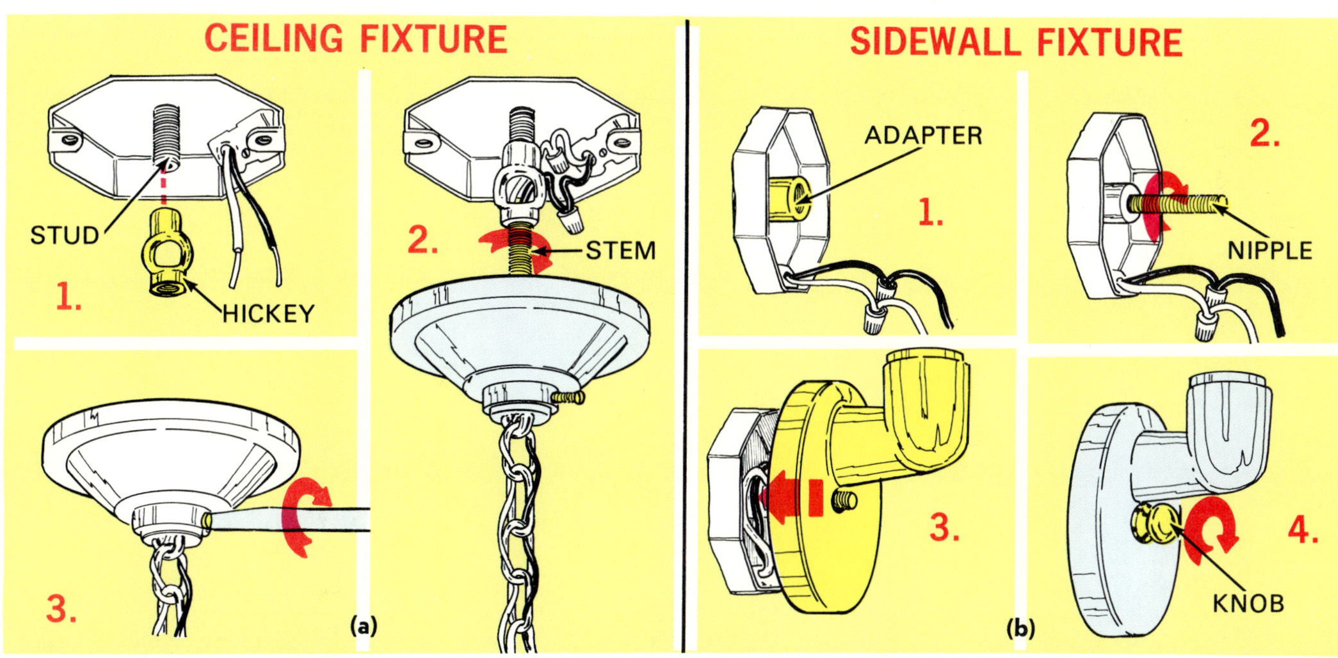

Figure 126. FIXTURE-TO-STUD MOUNT. (a) (1) Attach hickey to stud on outlet box. (2) Thread wires through hickey and attach fixture stem to hickey (connect wires as described under step 3b). (3) Push canopy against ceiling and tighten holding screw. (b) (1) Attach adapter to stud. (2) Screw nipple into adapter. (3) "Accordion" wires and push fixture into position on nipple. (4) Place knob on nipple and tighten until fixture is held securely.

G. Replacing and Maintaining Fluorescent Lighting Fixtures

Fluorescent lighting is quite popular for home and business use. Its popularity is due to several factors, among them are: (1) higher efficiency—more light per watt as compared to incandescent lamps. This becomes increasingly important as energy costs rise, (2) less heat is developed, (3) it is well suited to built-in (recessed) lighting, (4) the tubes are available in a variety of colors, and (5) the tubes average a longer life than incandescent bulbs. Some fluorescent tubes last as long as five years under home operating conditions. However, there are more parts that can go wrong with fluorescent units than with incandescent lamps.

The three principal types of fluorescent lighting fixtures are the rapid-start, the starter-type and the instant-start.

The **rapid-start** is the most popular for home use (Figure 127). It does not require a starter and lights up fairly quickly. Grounding connections are especially important for this type so be sure the grounding screws and connectors are tight.

The **starter-type** (Figure 128) is seen most often on low-wattage fixtures and on older types. This type starts more slowly than the other two. The starter must be replaced occasionally but this is easily done.

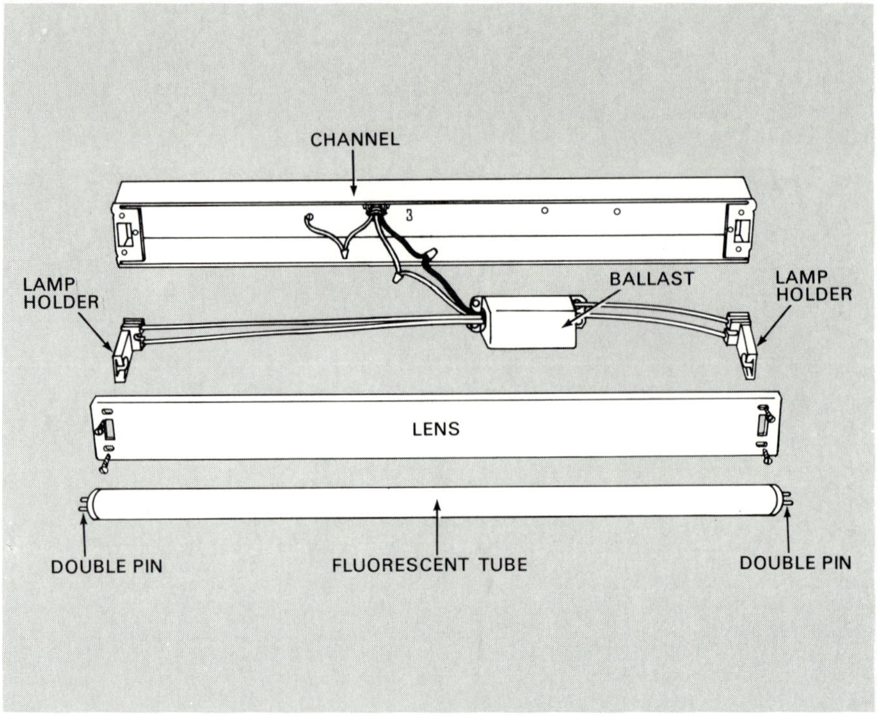

Figure 127. Rapid-start fluorescent lamp requires no starter and lights up within a few seconds. It is the most popular for home use.

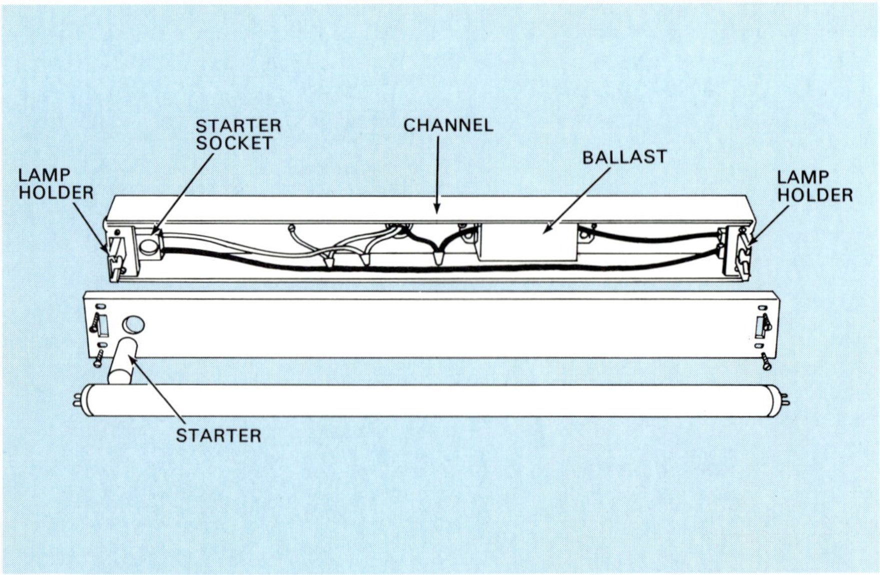

Figure 128. The starter type fluorescent has a small cannister-type starter that requires occasional replacement.

The **instant-start** has few maintenance problems other than tube replacement (Figure 129). The tube has a single pin at each end and one holder has a spring-loaded switch that cuts off current if the pin is not making good contact. The tube should be replaced when it begins to flicker or flash to prevent damage to the ballast.

Both straight-tube and circular-tube fixtures are available in the three types of fluorescent fixtures described above.

Upon completion of this section, you will be able to **replace and maintain straight-tube and circular-tube fluorescent fixtures.**

Procedures for replacing and maintaining fluorescent lighting units are discussed under the following headings:

1. Tools and Materials Needed.
2. Replacing Fluorescent Lighting Fixtures.
3. Installing Fluorescent Tubes in Fixtures.
4. Finding and Correcting Trouble.
5. Replacing a Starter.
6. Replacing a Ballast.

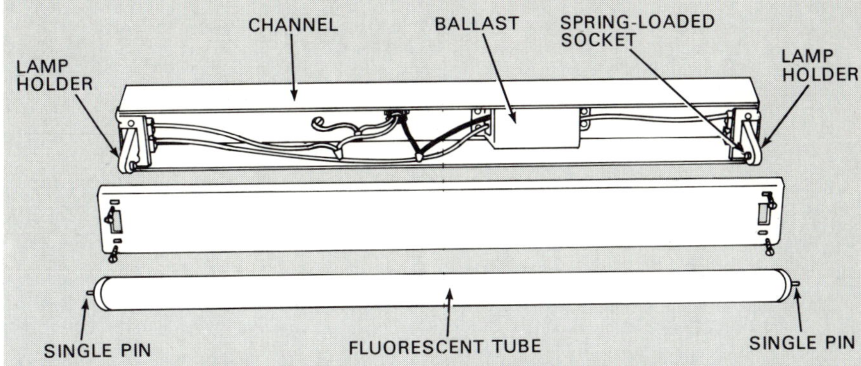

Figure 129. Instant-start fluorescent lamp tubes have only one pin at each end.

TOOLS AND MATERIALS

Tools:

Screwdriver
Pliers
Wire cutters
Knife
Drill

Materials:

Replacement fixture
Fixture wire
Fixture bar and stud
Large metal washer
Molly screws (if 40-watt fixture)
Starter
Ballast

REPLACING FLUORESCENT LIGHTING FIXTURES

Replacing fluorescent lighting fixtures is discussed under the following headings:

a. Replacing Straight-Tube Fluorescent Fixtures.

b. Replacing Circular-Tube Fluorescent Fixtures.

a. Replacing Straight-Tube Fluorescent Lighting Fixtures

Proceed as follows to install or replace a straight-tube fluorescent fixture:

1. *Turn off current at service entrance panel and remove the old fixture from the box.*

2. *Prepare the wire.*

Cut two pieces of stranded fixture wire, each about eight inches long, one black, one white. Cut two eight inch pieces of bare copper or green insulated wire for grounding. Remove half inch of insulation on each end of insulated wires.

3. *Connect fixture wires to house wiring.*

Using wire nuts, attach black jumper wire to black wire, and white jumper wire to white wire. With a wire nut, connect the two lengths of bare copper wire to the bare grounding wire from the cable. Connect the other end of one wire to the grounding screw in the outlet box (Figure 130).

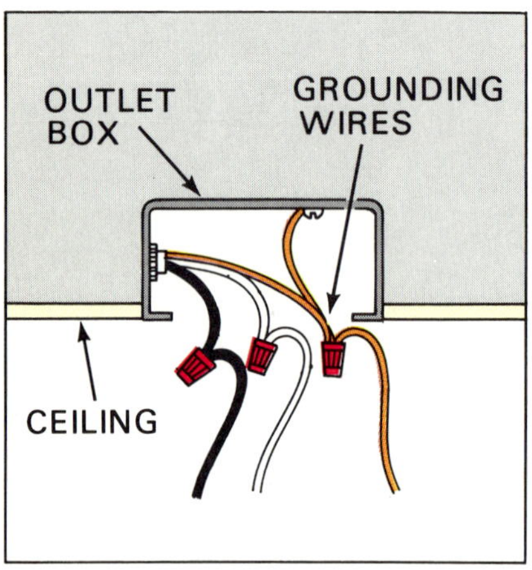

Figure 130. House wiring attached to jumper wires from the fixture.

4. *Attach fixture bar.*

Attach the fixture bar with stud to the outlet box with screws at each end. Pass wires through the hole in the stud (Figure 131).

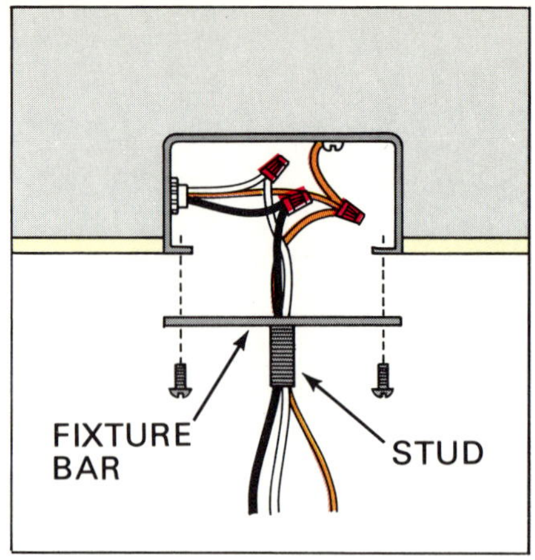

Figure 131. Fixture bar with stud attached.

5. *Place fixture on outlet box (Figure 132).*

Remove the lens from the fluorescent fixture. Remove knockout from the base of the fixture. Raise the assembled fixture to the ceiling so wires and stud pass through the knockout opening.

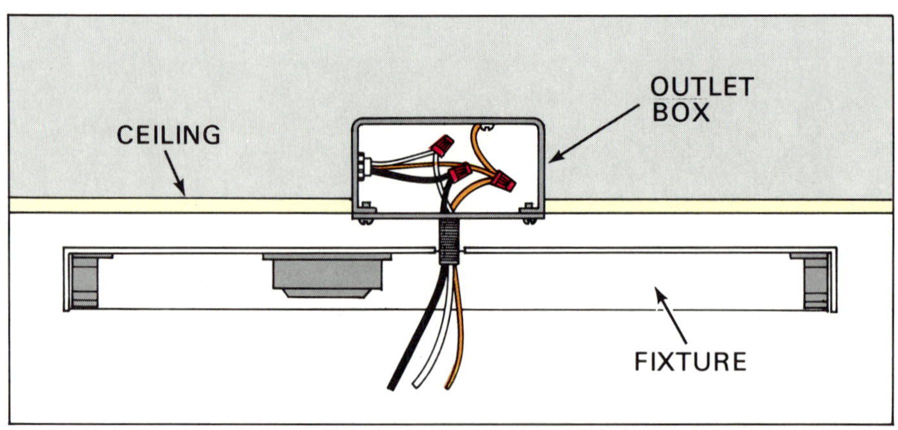

Figure 132. Fixture knockout opening fitted over the stud and wires.

6. *Fasten fixture in place (Figure 133).*

Place a large metal washer over the stud to help brace the fixture. Attach the locknut and tighten.

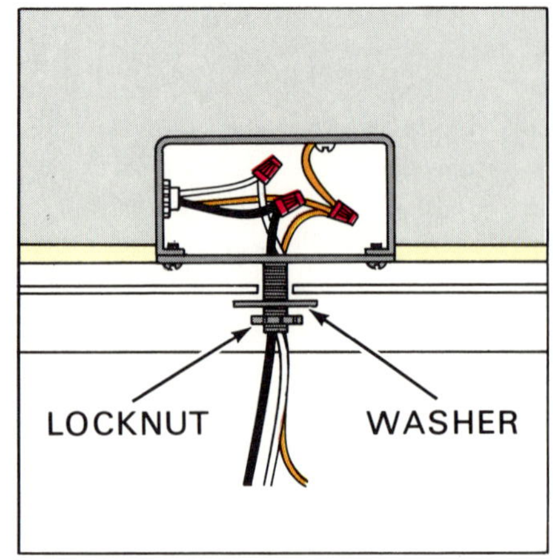

Figure 133. Washer and locknut attached to fixture.

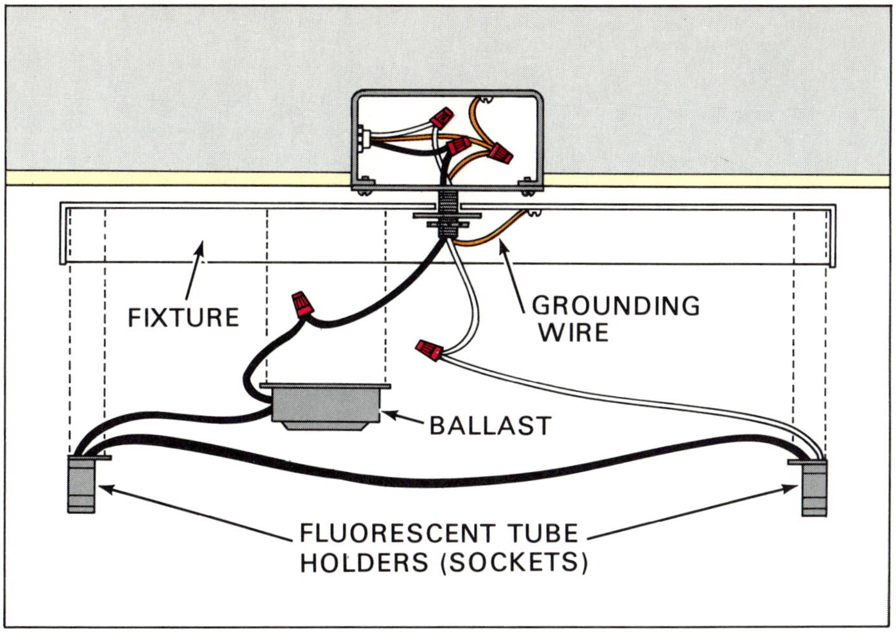

Figure 134. Completed fixture wiring including the grounding wire connection to the grounding screw.

7. *Attach fixture wires (Figure 134).*

   With wire nuts, connect black wire to black wire and white wire to white wire. Connect grounding wire to grounding screw in fixture base.

8. *Provide more support for fixture if required.*

   Forty-watt fixtures require more support than smaller types. Drill a hole at each end of fixture base and insert molly screws to secure the longer base of a 40-watt fixture (Figure 135).

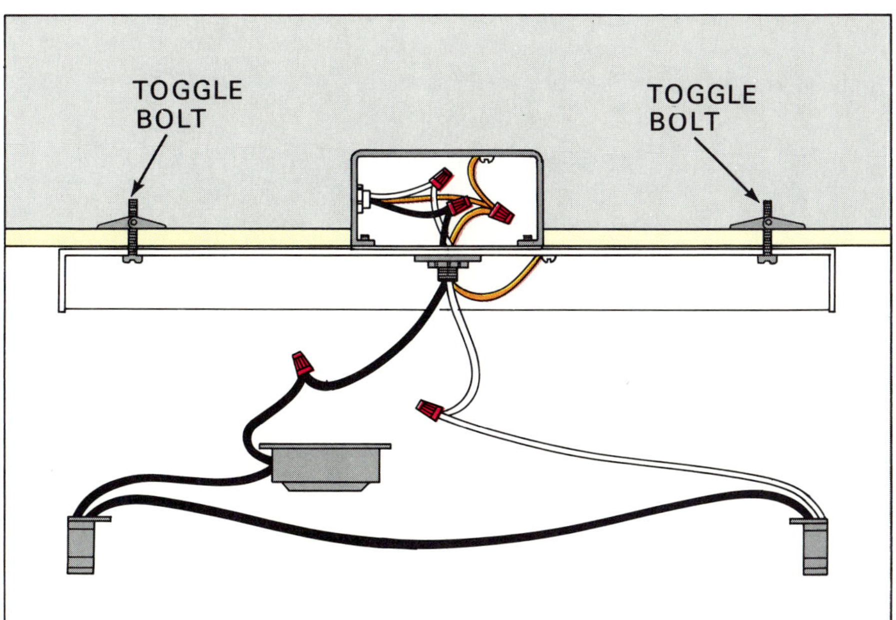

Figure 135. Molly screws provide additional support for 40-watt fixture.

a. Replacing Circular Fluorescent Lighting Fixtures

Wiring connections are similar for circular-tube fluorescent fixture. Proceed as follows (Figure 136):

1. *Connect fixture wires to outlet box wiring.*

   With wire nuts, connect black wire to black wire, white wire to white wire, and grounding wires as for straight tube.

2. *Fasten fixture in place.*

   Place fixture opening over the outlet box stud. Place cap nut on stud and tighten.

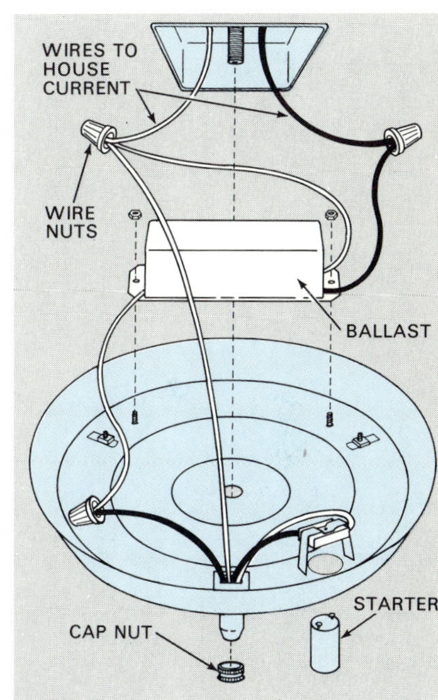

Figure 136. Wiring connections for circular fluorescent fixture.

INSTALLING FLUORESCENT TUBES IN FIXTURES

Installing fluorescent tubes in fixtures is discussed under the following headings:

a. Installing Straight Tubes in Fixtures.

b. Installing Circular Tubes in Fixtures.

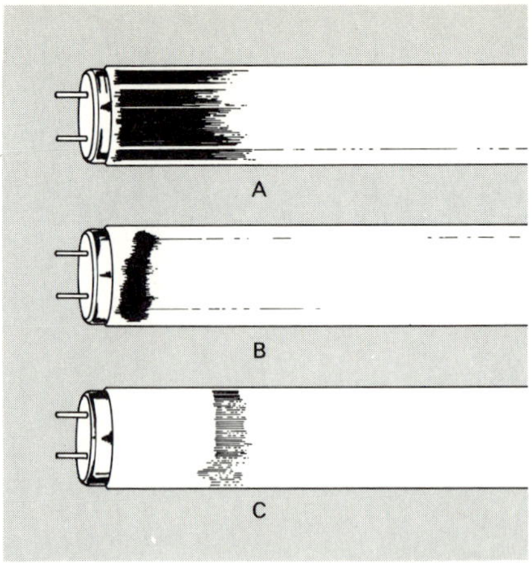

Figure 137. Very dark areas near ends of lamps indicate lamp is near failure (a) and (b). Gray bands do not affect lamp performance (c).

a. Installing Straight Tubes in Fixtures

The need to replace a fluorescent tube is indicated by blackened areas near the ends of the tube (Figure 137).

Most straight fluorescent tubes have two pins at each end. The instant-start type has one pin on each end and one socket is spring-loaded. Be sure the tube is the size required for the fixture (Figure 138). Proceed as follows (Figure 139):

1. *Place pins in socket at each end of fixture.*

2. *Rotate tube about one-quarter turn for correct contact.*

3. *Test for operation.*

If tube does not light, rotate slightly in socket for better contact. Lighting in new tubes may swirl and twist when first installed but will usually correct itself after a few starts.

Figure 138. Markings on the tube indicate size and type of a fluorescent lamp.

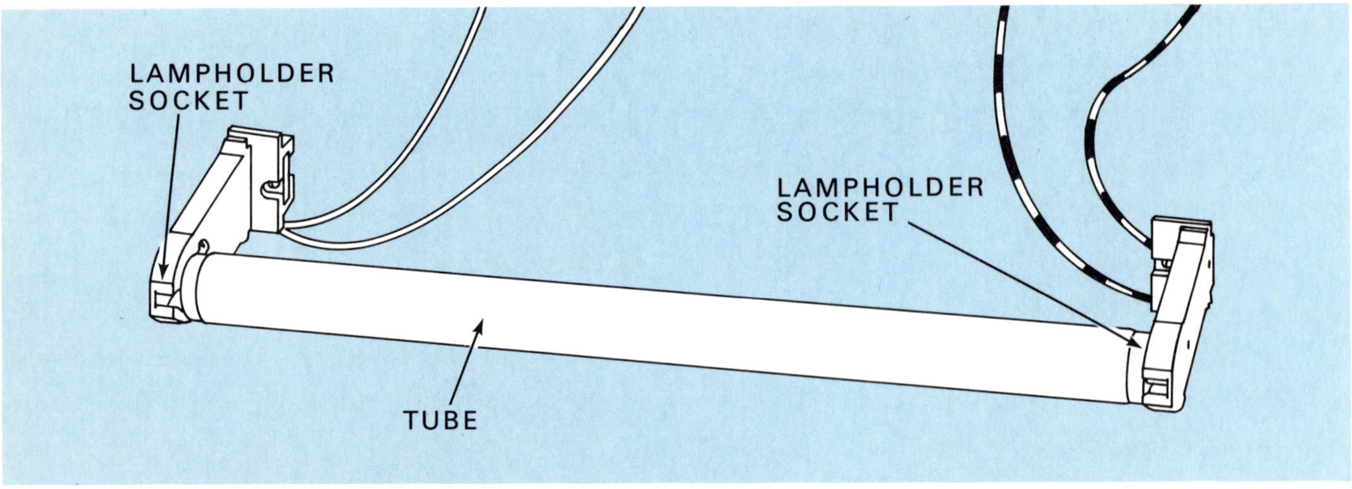

Figure 139. Pins on ends of the lamp fit into the lampholder socket.

b. Installing Circular Fluorescent Tubes in Fixtures

Circular fluorescent tubes have four pins grouped closely to plug directly into the outlet (Figure 140). Be sure the tube is correct size for the fixture.

Proceed as follows:

1. *Plug the lamp into the socket.*
2. *Place lamp in position for mechanical supports to be fastened.*

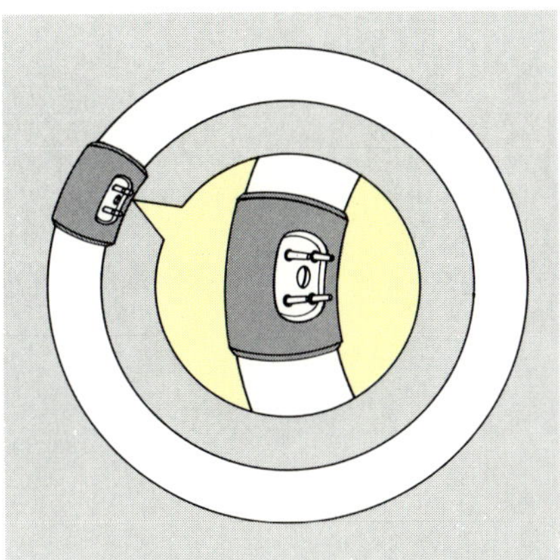

Figure 140. Circular fluorescent lamp has four contacts for plug connections.

3. *Fasten mechanical supports (Figure 141).*

   These are usually spring-loaded clamps that snap into position.

4. *Test for operation.*

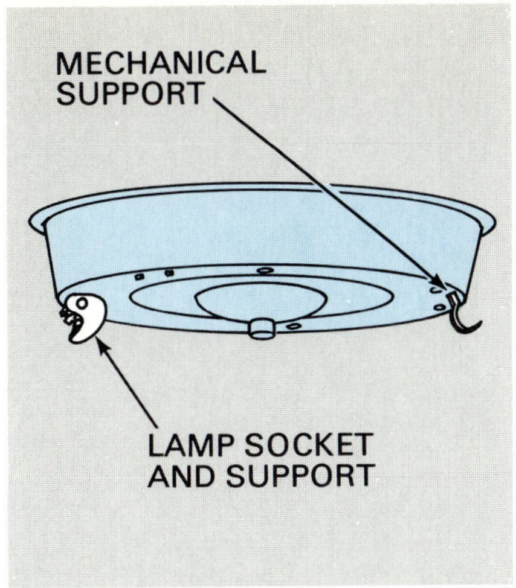

Figure 141. Spring-loaded clamps hold circular lamp in place.

FINDING AND CORRECTING TROUBLE

Table II lists the more common troubles encountered with fluorescent lighting. Note that tube blackening, shown in the first part of the table, applies to all fluorescent types. The remainder of the table shows troubles common to the different types. Consequently, it is important for you to know which type of fluorescent lighting you are checking.

Some of the simple procedures for correcting are included in the table. The remainder are under the headings following the table.

REPLACING A STARTER

Proceed as follows:

1. *Locate position of starter.*

   The starter is usually mounted on the metal plate back of the lamp. It may be immediately under the tube. If so, you will need to remove the tube to have access to the starter.

   There is one starter for each lamp.

2. *Press in on starter body and turn it to left.*

   This should release the starter from the base. Note in Figure 142 that the contacts on the starter fit into keyhole slots in the base.

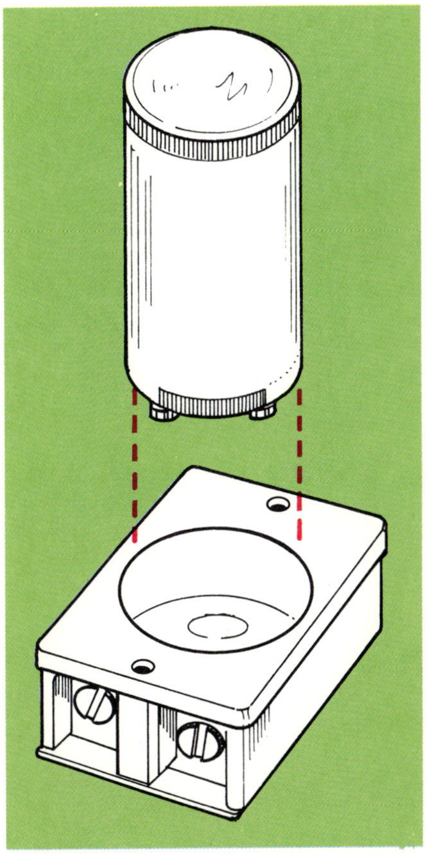

Figure 142. Fluorescent lamp starter and base.

TABLE II. FLUORESCENT LIGHTING MAINTENANCE

Common Causes of Trouble — Correction

Symptoms	Most Probable Cause	How to Correct
ALL FLUORESCENT TYPES		
Dense blackening extends 2 or 3 inches from the base at either or both ends (Figure 137a).	Lamp is worn out.	Replace with new lamp.
Dense black spot, about 1 inch from base, about ½ inch wide, extending half way around tube (Figure 137b).	Lamp nearing end of life.	Replace with new lamp.
Gray bands about 2 inches from either base (Figure 137c).	Normal on some lamps.	Normal condition.
STARTER TYPES		
Lamp blinks "on" and "off."	1. Lamp is worn out.	Replace with new lamp.
	2. Starter may be defective.	Replace with new starter.
	3. Low air temperature may be affecting it.	May require fixture with enclosure to maintain warm air around lamp. This helps lighting efficiency.
Ends of lamp remain lighted.	Defective starter.	Replace with new starter.
Lamp does not start or starts slowly.	1. Pins are not making proper contact in lampholders.	Rotate lamp slightly in holders for better contact.
	2. Defective starter.	Replace with new starter.
	3. Defective ballast.	Replace with new ballast.
	4. Dirt accumulation on tube under damp conditions.	Remove tube and clean.
Light spirals, swirls or snakes.	1. Normal with a new lamp.	Condition will probably correct itself after a few starts.
	2. If lamp is old, it may indicate a defective starter.	Replace with new starter.
RAPID-START TYPES		
Lamp does not start or starts slowly—takes 5 seconds or longer.	1. Pins not making proper contact in lampholders.	Rotate lamp slightly in holders for better contact.
	2. Dirt accumulation on tube under damp conditions.	Remove and clean tube.
	3. Lamps used frequently but for short periods of time.	Operate continuously for several hours
	4. Broken cathode.	Replace with new lamp.
	5. Defective ballast.	Replace with new ballast.
	6. Low circuit voltage.	Have power supplier check voltage.
INSTANT-START TYPES		
Light spirals violently, sometimes orange-colored flashes occur.	Broken cathode.	Replace with new lamp.

3. *Examine new starter to make certain it has the same markings as the one being removed.*
4. *Insert new starter. Press in on starter body and turn clockwise.*
5. *Try out lamp to see if starter works satisfactorily.*

REPLACING A BALLAST

The ballast is usually in the trough underneath the fluorescent lamp (Figure 143). Proceed as follows:

1. *Disconnect circuit at main switch, circuit breaker or fuse panel.*
2. *Remove fluorescent tube from fixture.*
3. *Remove metal cover over the wiring trough.*
4. *Check your new ballast to see if it has the same specifications and the same number of wires as the old ballast.*

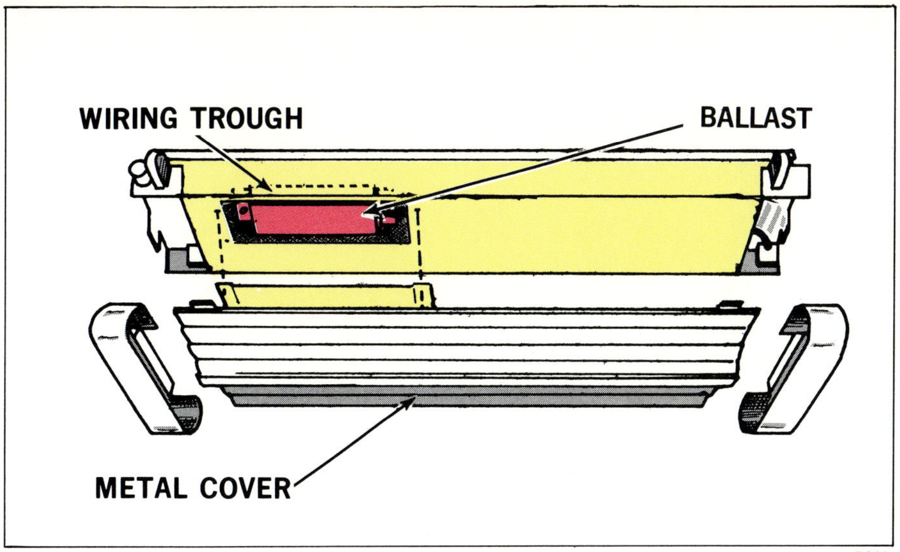

Figure 143. Ballast in wiring trough of fluorescent fixture. Metal cover plate has been removed.

5. *If possible, disconnect the old ballast one wire at a time and reconnect the equivalent wire on the new ballast in its place.*

   This is to avoid making errors on your wiring connections. If you can not do this, study the connections on the new ballast and make sure you know where they connect when the old ballast is removed.

6. *Unfasten and remove old ballast from trough.*
7. *Install and fasten new ballast in place.*
8. *Reassemble trough cover and reinstall fluorescent lamp.*
9. *Re-establish service on circuit.*

NOTES

H. Replacing Ground Fault Circuit Interrupters

Ground fault interrupters are required by the National Electrical Code (1) on all electrical wiring servicing permanent and storable swimming pools (Article 680-5[b]), (2) on residential outdoor receptacles (Article 210-8[a] and 8[b]) on 15- and 20-amp receptacle outlets in use on construction sites.

Equipment is available to plug into an existing receptacle (Figures 144a and 144b), be installed in receptacle outlets (Figure 144c), and replace existing circuit breakers in some installations (Figure 144d).

Each GFCI is equipped with a test button or switch. About every month the test button or switch should be pushed and the GFCI should "trip" off. If the device does not "trip," it is defective and should be replaced.

Upon completion of this section, you will be able to **replace the various types of ground fault circuit interrupters.**

Procedures used in replacing ground fault circuit interrupters are given under the following headings:

1. Tools and Materials Needed.
2. Removing Old Circuit Breaker.
3. Installing New Combination Ground Fault Circuit Interrupter.
4. Installing GFCI in a Branch Circuit.

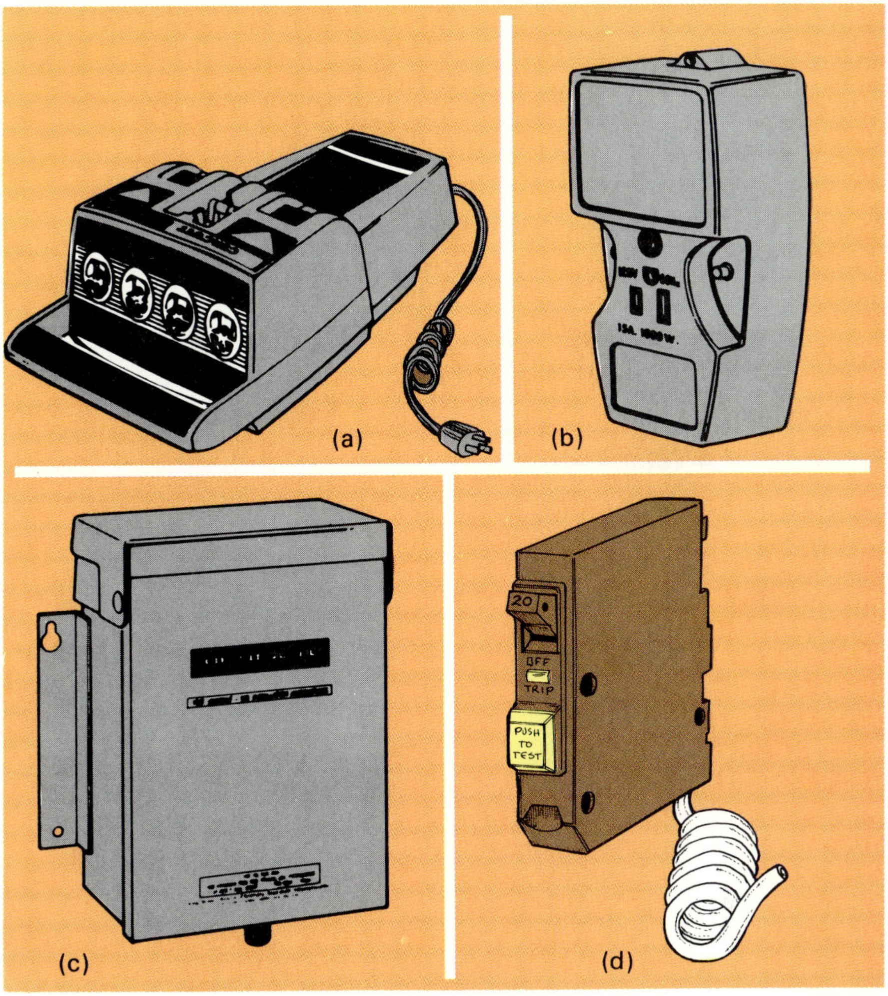

Figure 144. Ground fault circuit interrupters.

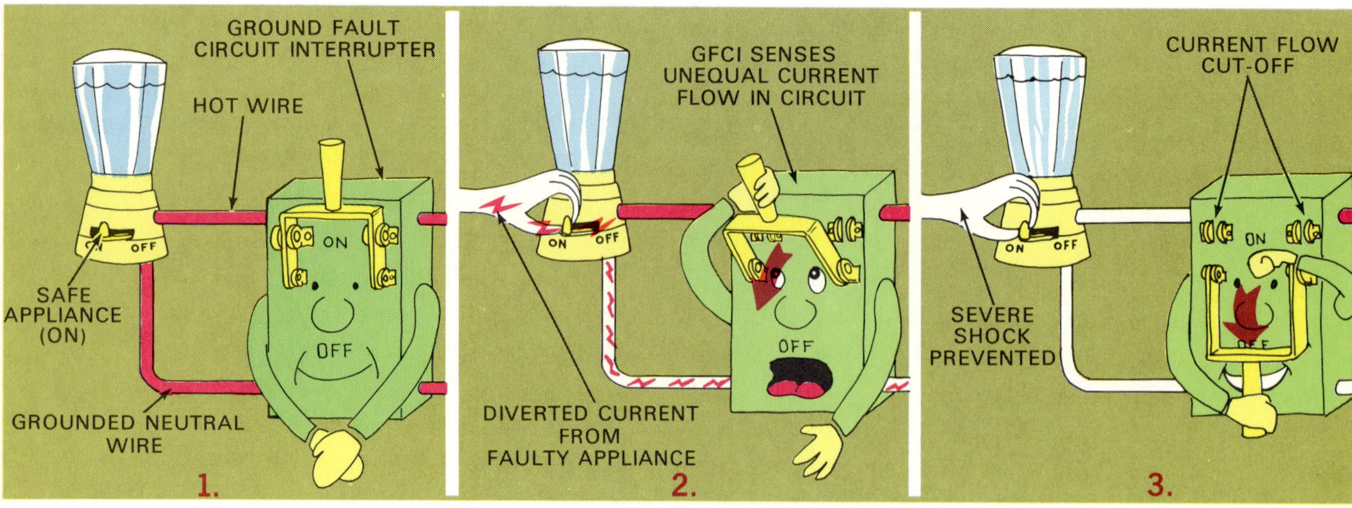

Figure 145. How a ground fault circuit interrupter cuts off power.

TOOLS AND MATERIALS NEEDED

Tools:

Screwdriver
Pliers (preferably thin-nose pliers)

Materials:

Combination ground fault circuit interrupter and circuit breaker.

REMOVING OLD CIRCUIT BREAKER

Follow procedures outlined in the section, "Removing Old Circuit Breaker", page 216.

INSTALLING NEW COMBINATION GROUND FAULT CIRCUIT INTERRUPTER

To install interrupter and circuit breaker, proceed as follows:

1. *Disconnect power from branch circuit.*

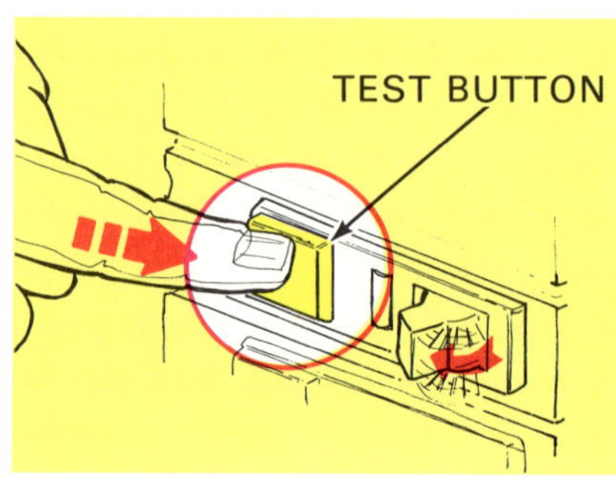

Figure 146. Checking test button on ground fault circuit interrupter.

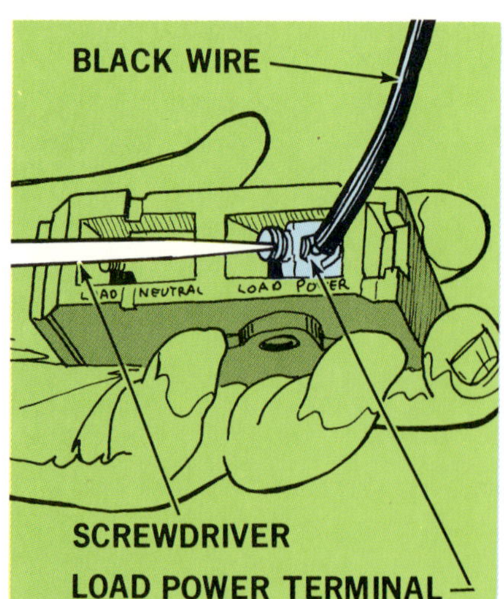

Figure 147. Installing wire on ground fault circuit interrupter.

2. *Insert metal projection on back of device into the slot under the screw of the electrical panel (Figure 69).*

3. *Press electrical contacts onto prong of electrical bus bar (Figure 68).*

4. *Loosen the two terminal screws on the GFCI.*

5. *Insert wire removed for original circuit breaker under one terminal screw marked "load power." (Figure 147). Tighten the screw.*

6. *Remove the original neutral (white) circuit wire from the neutral bar and place it under the other terminal screw marked "load neutral" (Figure 148). Tighten the terminal screw.*

   If you have a green wire (ground), connect it to the grounding bar as USUAL.

7. *Place the end of the white "pigtail" extending from the GFCI under a neutral bar terminal screw and tighten the screw (Figure 149).*

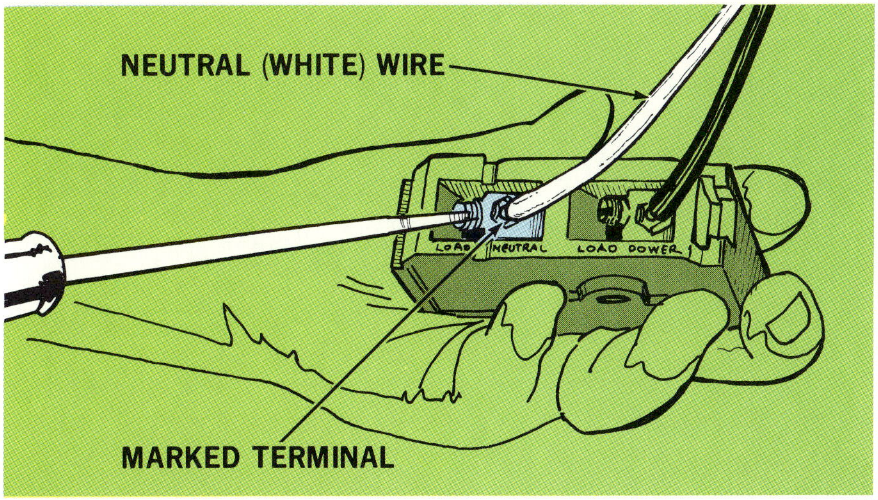

Figure 148. Installing ground wire.

8. *Replace the electrical panel cover plate.*
9. *Turn main switch to "on."*
10. *Turn circuit breaker to "on."*
11. *Push "test" button to see if breaker "trips" (Figure 150).*

    If it does not, the unit is faulty. If it trips and you can not get it to stay on, you have either a faulty GFCI unit, or faulty wiring on that circuit.

Figure 149. Installing "pigtail" wire.

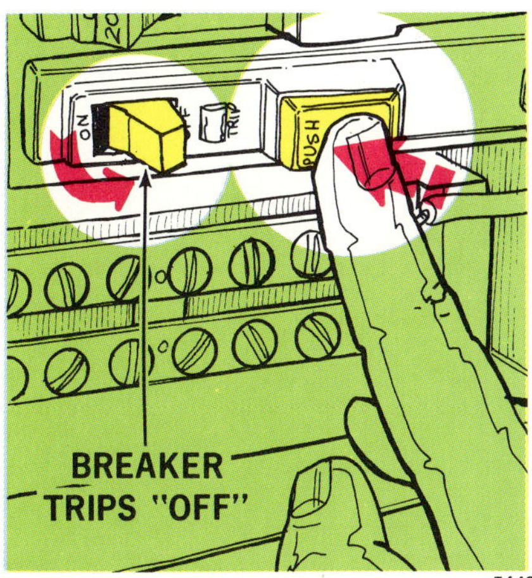

Figure 150. Testing new ground fault interrupter installation.

INSTALLING GFCI IN BRANCH CIRCUITS

To install a GFCI in a branch circuit such as for a swimming pool, proceed as follows:

1. *Disconnect power from branch circuit.*
2. *Break external branch circuit and install GFCI (Figure 151).*

    Place GFCI at least 4 feet from inside wall of swimming pool and at a convenient height for testing.

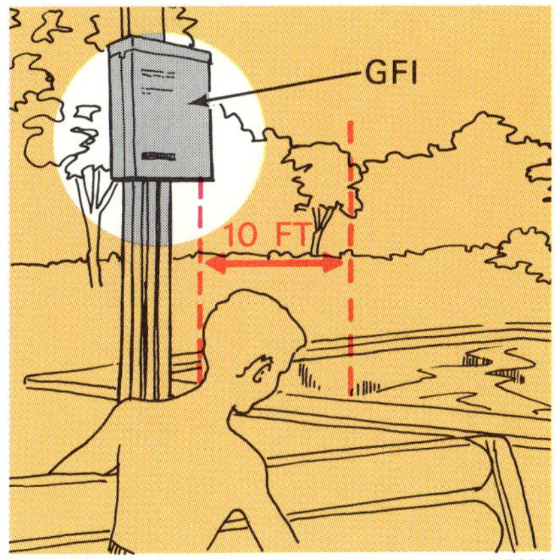

Figure 151. Install GFCI at least 4 feet away from inside wall of pool.

Figure 152. Connect green wires to grounding terminals.

Figure 153. Connections for GFCI.

3. *Insert power supply wires to the left of the GFCI and wires leading to swimming pool to the right (Figure 153).*

4. *Connect green wires to terminals marked "load" (Figure 150).*

5. *Connect white wires together and black wires together (Figure 151).*

6. *Return power to line.*

   Replace fuse or turn circuit breaker on.

7. *Test GFCI (Figure 154).*

   Be sure actuator is on. Push test switch to left and release it. Actuator should snap to off automatically.

   If not, this indicates a ground in the system.

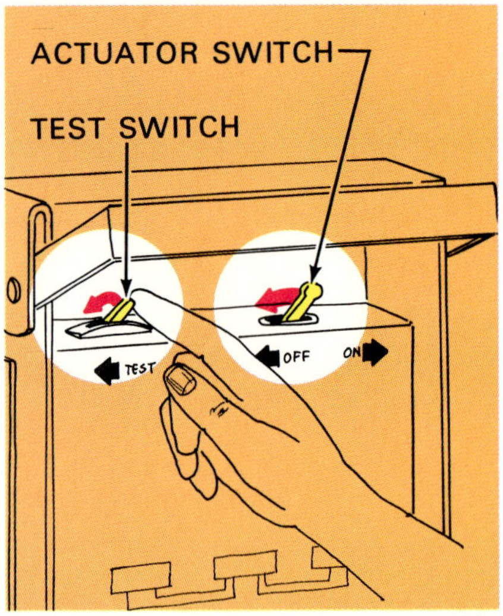

Figure 154. To test the GFCI, push the test switch to the left. The actuator switch should snap to off.

NOTES

NOTES

NOTES

NOTES

NOTES

NOTES

NOTES

NOTES

NOTES

NOTES

Index

Index

Other Bestsellers From TAB

☐ **BASIC ROOF FRAMING—Barnow**
Leading off with an in-depth look at various different types of roof styles, Barnow carefully identifies and describes the framing members of each one. You'll find an invaluable introduction to the tools used in roof framing, including a complete explanation of the steel framing square—the most important tool of the framing carpenter! And you'll also get a look at professional techniques for choosing and buying needed materials with plenty of practical advice on proper material estimating. 192 pp., 250 illus. Large Format (7″ × 10″).
Paper $11.95 **Hard $19.95**
Book No. 2677

☐ **111 YARD AND GARDEN PROJECTS—from boxes and bins to tables and tools—Blandford**
Expert woodworker and metalcrafter Percy Blandford gives you step-by-step building guidance that's based on years of experience as a teacher and author of bestselling project guides. Plus, he provides a wealth of basic know-how on how to choose and work with wood and metal. He gives invaluable tips on constructing tool handles. And he shows how to custom design exactly the outdoor tools, accessories, and gardening aids that are right for your own particular needs!
Paper $16.95 **Hard $25.95**
Book No. 2644

☐ **HOW TO CAST SMALL METAL AND RUBBER PARTS—2nd Edition—Cannon**
Using this excellent sourcebook as a guide, you can easily make high quality, defect-free castings for almost any purpose and at an amazingly low cost! Just some of the countless uses you'll find for this potentially profitable skill . . . making obsolete or vintage car parts, hood ornaments, garden and fireplace tools, kitchen utensils, automotive parts, replacing broken antique parts, reproducing sculpture, plaques and other art . . . all kinds of decorative and useful objects for your own use or to sell! 176 pp., 143 illus.
Paper $9.95 **Hard $15.95**
Book No. 2614

☐ **TILE FLOORS—INSTALLING, MAINTAINING AND REPAIRING—Ramsey**
Now you can easily install resilient or traditional hard tiles on both walls and floors. Find out how to buy quality resilient floor products at reasonable cost . . . and discover the types and sizes of hard tiles available. Get step-by-step instructions for laying out the floor, selecting needed tools and adhesives, cutting tiles, applying adhesives, and more. 192 pp., 200 illus. 4 pages in full color. 7″ × 10″.
Paper $12.95 **Book No. 1998**

☐ **BUILDING OUTDOOR PLAYTHINGS FOR KIDS, with Project Plans—Barnes**
Imagine the delight of your youngsters—children or grandchildren—when you build them their own special backyard play area complete with swings, climbing bars, sandboxes, even an A-frame playhouse their own size or a treehouse where they can indulge in their own imaginary adventures. Best of all, discover how you can make exciting, custom-designed play equipment at a fraction of the cost of ordinary, ready-made swing sets or sandbox units! It's all here in this practical, step-by-step guide to planning and building safe, sturdy outdoor play equipment. 240 pp., 213 illus. 7″ × 10″.
Paper $12.95 **Hard $21.95**
Book No. 1971

☐ **THE COMPUTER FURNITURE PLAN AND PROJECT BOOK—Wiley**
Now, with the help of this first-of-its-kind handbook, even a novice can build good looking, functional, and low-cost computer furniture that's custom-designed for your own special needs—tables, stands, desks, modular or built-in units, even a posture supporting kneeling chair! Computer hobbyist and craftsman Jack Wiley provides all the step-by-step guidance, detailed project plans, show-how illustrations, and practical customizing advice . . . even basic information on tools, materials, and construction techniques. 288 pp., 385 illus. 7″ × 10″.
Paper $15.95 **Hard $23.95**
Book No. 1949

☐ **ROOFING THE RIGHT WAY—A Step-by-Step Guide for the Homeowner—Bolt**
If you're faced with having to replace your roof because of hidden leaks, torn or missing shingles, or simply worn roofing that makes your whole house look shabby and run down . . . don't assume that you'll have to take out another mortgage to pay for the project. The fact is, *almost anyone can install a new or replacement roof easily and at amazingly low cost compared with professional contractor prices*! All the professional techniques and step-by-step guidance you'll need is here in this complete new roofing manual written by an experienced roofing contractor. 192 pp., 217 illus. Large Format (7″ × 10″).
Paper $11.95 **Hard $19.95**
Book No. 2667

☐ **66 FAMILY HANDYMAN® WOOD PROJECTS**
Here are 66 practical, imaginative, and decorative projects . . . literally something for every home and every woodworking skill level from novice to advanced cabinetmaker: room dividers, a free-standing corner bench, china/book cabinet, coffee table, desk and storage units, a built-in sewing center, even your own Shaker furniture reproductions! 210 pp., 306 illus. 7″ × 10″.
Paper $14.95 **Hard $21.95**
Book No. 2632

☐ **UPHOLSTERY TECHNIQUES ILLUSTRATED—Gheen**
Here's an easy-to-follow, step-by-step guide to modern upholstery techniques that covers everything from stripping off old covers and padding to restoring and installing new foundations, stuffing, cushions, and covers. All the most up-to-date pro techniques are included along with lots of time- and money-saving "tricks-of-the-trade" not usually shared by professional upholsterers. 352 pp., 549 illus. 7″ × 10″.
Paper $16.95 **Hard $27.95**
Book No. 2602

☐ **CABINETS AND VANITIES—A BUILDER'S HANDBOOK—Godley**
Here is easy-to-follow, step-by-step detail in everything you need to know to design, build, and install your own customized kitchen cabinets and bathroom vanities and cabinets for a fraction of the price charged by professional cabinetmakers or kitchen remodelers . . . and for less than a third of what you'd spend for the most cheaply made ready-made cabinets and vanities! 142 pp., 126 illus. 7″ × 10″.
Paper $12.95 **Hard $19.95**
Book No. 1982

☐ **HOW TO TROUBLESHOOT AND REPAIR ANY SMALL GAS ENGINE—Dempsey**
Here's a time-, money, and aggravation-saving sourcebook that covers the full range of two and four-cycle gas engines from just about every major American manufacturer—from Briggs & Stratton, to West Bend, and others! With the expert advice and step-by-step instructions provided by master mechanic Dempsey, you'll be amazed at how easily you can solve almost any engine problem. 272 pp., 228 illus.
Paper $10.95 **Hard $18.95**
Book No. 1967

☐ **HARDWOOD FLOORS—INSTALLING, MAINTAINING AND REPAIRING—Ramsey**
Do-it-yourself expert Dan Ramsey gives you all the guidance you need to install, restore, maintain, or repair all types of hardwood flooring at costs far below those charged by professional builders and maintenance services. From details on how to select the type of wood floors best suited to your home, to time- and money-saving ways to keep your floors in tip-top condition . . . nothing has been left out. 160 pp., 230 illus. 4 pages in full color. 7″ × 10″.
Paper $10.95 **Hard $18.95**
Book No. 1928

Other Bestsellers From TAB

☐ **DO-IT-YOURSELF DESIGNER WINDOWS**
If the cost of custom-made draperies puts you in a state of shock . . . if you've had trouble finding window coverings of any kind for cathedral or other problem windows . . . or if you're unsure of what type of window decor would look right in your home . . . here's all the advice and information you've been searching for. It's a complete, hands-on guide to selecting, measuring, making, and installing just about any type of window treatment imaginable. You'll even get an expert's insight into selection and installation of decorative storm windows and thermal windows, stained glass windows, woven or wooden blinds, and workable treatments for problem areas. 272 pp., 414 illus. Large Format (7″ × 10″).
Paper $14.95 **Hard $21.95**
Book No. 1922

☐ **MOPED MAINTENANCE AND REPAIR**
Guaranteed to save you time, money, and inconvenience, it is packed with practical troubleshooting and repair information, preventative maintenance tips, safety advice, even important facts on state moped regulations and operating guidelines. Generously illustrated, it provides solutions to these routine problems that plague every moped rider—flat tires, fouled spark plugs, broken or worn cables and chains, and more. 256 pp., 229 illus.
Paper $14.95 **Book No. 1847**

☐ **DO YOUR OWN DRYWALL—AN ILLUSTRATED GUIDE—Kozloski**
Professional expertise for the amateur builder and remodeler! Proper installation of interior plasterboard or drywall is a must-have skill for successful home building or remodeling. Now, there's a new time- and money-saving alternative: this excellent step-by-step guide to achieving professional-quality drywalling results, the first time and every time! Even joint finishing, the drywalling step most dreaded by do-it-yourselfers, is a snap when you know what you're doing. And this is a guide that leaves absolutely nothing to chance and that leaves no question unanswered. 160 pp., 161 illus.
Paper $10.95 **Hard $17.95**
Book No. 1838

☐ **PRACTICAL LANDSCAPING AND LAWN CARE—Webb**
Make your lawn the envy of the entire neighborhood . . . *without* spending a fortune or putting in never-ending hours of maintenance time! Here's absolutely everything you need to successfully plan, plant, and maintain lawn grasses and groundcovers, vines, and flowering ornamentals . . . annual, biennial, and perinnial flowers . . . shade trees, lawn trees . . . even decorative (and delicious) fruits and berries. It doesn't matter whether your climate is cold and damp or hot and dry . . . whether your soil is sandy, rocky, or gummy clay . . . *everything* you need is here! 240 pp., 84 illus., 7″ × 10″.
Paper $13.95 **Hard $21.95**
Book No. 1818

☐ **HOUSE CRAFTSMANSHIP: A GUIDE TO RESTYLING AND REFURBISHING**
An information-packed sourcebook for every homeowner and do-it-yourselfer! Here are all the practical tips and step-by-step procedures you need to accomplish your restyling and refurbishing projects quickly and easily. Like having a skilled craftsman at your side. This illustrated guide shows you how to do just about every interior and exterior job imaginable, easily and inexpensively. 288 pp., 291 illus. 7″ × 10″.
Paper $14.95 **Hard $22.95**
Book No. 1809

☐ **PANELING WITH SOLID LUMBER, including projects—Ramsey**
Home remodeling expert Dan Ramsey shows you how to use solid wood paneling to give almost any room in your home a new look that's comfortable, convenient, economical, and practically maintenance-free plus gives you an exciting selection of projects. Included are step-by-step directions for building a cedar closet . . . storm doors and shutters . . . a bathroom partition . . . a sauna decor for your bath . . . storage walls . . . a home bar . . . an attic study . . . book cabinets . . . a fireplace mantel . . . a kitchen corner cabinet . . . and more! 192 pp., 288 illus. 7″ × 10″.
Paper $12.95 **Hard $18.95**
Book No. 1868

☐ **58 HOME SHELVING AND STORAGE PROJECTS—Blandford**
From a two-shelf book rack or table-top organizer to a paneled chest, basic room divider, or hall locker . . . from shelves or a spoon rack to a period reproduction of a Shaker cabinet or a Welsh dresser, you'll be amazed at the variety of projects included. And, each one includes easy-to-follow, step-by-step directions, plenty of show-how drawings, and complete materials list. 288 pp., 227 illus. 7″ × 10″.
Paper $14.95 **Book No. 1844**

☐ **BUILD YOUR OWN FITNESS CENTER**
This sourcebook gives you expert guidance on everything from planning your facility to constructing it, step-by-step. It shows you how to choose and install all the needed equipment, and in many cases gives you direction for building your own equipment and accessories for a fraction of the commercial cost! Plus, there's practical advice on the proper use of your equipment once it's installed. 224 pp., 301 illus. 7″ × 10″.
Paper $12.95 **Hard $18.95**
Book No. 1828

☐ **CASTING BRASS—Ammen**
A highly respected metalcasting professional shares his secrets for casting almost any type of object—from art works to mechanical fittings in copper, brass, bronze, or other copper alloy—in this easy-to-follow sourcebook! Whether you're a first-time metalcaster, an experienced amateur, or a metalworking professional, you'll find this book packed with practical information and expert advice that's simply not available in any other metalcasting guide! Plus, the author has included an invaluable section on how to set up and run a successful small brass foundry business! No matter what your interest in brass casting, this is an invaluable sourcebook! 252 pp., 200 illus.
Paper $11.95 **Hard $18.95**
Book No. 1810

☐ **TROUBLE-FREE SWIMMING POOLS**
Here is the ideal sourcebook for anyone thinking of installing a swimming pool—inground or above ground from wading pool size to large indoor public pool. It shows how to plan, excavate, construct, and safely maintain all types and sizes of pools. You'll find out how to have your own pool for as little as $1,000 . . . or how to get more pool for the money no matter how much you're able to spend! 176 pp., 306 illus. 7″ × 10″.
Paper $11.95 **Hard $18.95**
Book No. 1808

*Prices subject to change without notice.

Look for these and other TAB books at your local bookstore.

TAB BOOKS Inc.
P.O. Box 40
Blue Ridge Summit, PA 17214

Send for FREE TAB catalog describing over 1200 current titles in print.